Lecture Notes in Computer Science　　11682

More information about this series at http://www.springer.com/series/7407

Robert Mercaş · Daniel Reidenbach (Eds.)

Combinatorics on Words

12th International Conference, WORDS 2019
Loughborough, UK, September 9–13, 2019
Proceedings

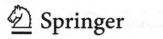 Springer

Editors
Robert Mercaş 🔘
Loughborough University
Loughborough, UK

Daniel Reidenbach 🔘
Loughborough University
Loughborough, UK

ISSN 0302-9743   ISSN 1611-3349   (electronic)
Lecture Notes in Computer Science
ISBN 978-3-030-28795-5   ISBN 978-3-030-28796-2   (eBook)
https://doi.org/10.1007/978-3-030-28796-2

LNCS Sublibrary: SL1 – Theoretical Computer Science and General Issues

This Springer imprint is published by the registered company Springer Nature Switzerland AG
The registered company address is: Gewerbestrasse 11, 6330 Cham, Switzerland

Preface

This volume of *Lecture Notes in Computer Science* contains the proceedings of the 12th International Conference WORDS 2019, which was organized by the Department of Computer Science at Loughborough University and held during September 9–13, 2019, in Loughborough, UK.

WORDS is the main conference series devoted to the mathematical theory of words. In particular, the combinatorial, algebraic and algorithmic aspects of words are emphasized. Motivations may also come from other domains such as theoretical computer science, bioinformatics, digital geometry, symbolic dynamics, numeration systems, text processing, number theory, etc.

The conference WORDS takes place every two years. The first conference of the series was held in Rouen, France in 1997. Since then, the locations of WORDS conferences have been: Rouen, France (1999), Palermo, Italy (2001), Turku, Finland (2003 and 2013), Montréal, Canada (2005 and 2017), Marseille, France (2007), Salerno, Italy (2009), Prague, Czech Republic (2011), and Kiel, Germany (2015).

For the fourth time in the history of WORDS, a refereed proceedings volume was published in the *Lecture Notes in Computer Science* series of Springer. There were 34 submissions, from 17 countries, and each of them was reviewed by three or four referees. The selection process was undertaken by the Program Committee with the help of generous reviewers. From these submissions, 21 papers were selected to be published and presented at WORDS.

In addition to the contributed presentations, WORDS featured a session *In Memoriam: Aldo de Luca*, as well as six invited talks:

- Florin Manea (Kiel University, Germany): "Matching Patterns with Variables"
- Svetlana Puzynina (Saint Petersburg State University, Russia): "Abelian Properties of Words"
- Antonio Restivo (University of Palermo, Italy): "On Sets of Words of Rank Two"
- Gwenaël Richomme (Université Paul-Valéry Montpellier 3, France): "*S*-adicity and Property Preserving Morphisms"
- Aleksi Saarela (University of Turku, Finland): "Independent Systems of Word Equations: From Ehrenfeucht to Eighteen"
- Kristina Vuškovic (University of Leeds, UK): "Structure of Graph Classes and Algorithms"

The present volume also includes the papers of five of these invited talks.

We take this opportunity to warmly thank all the invited speakers and all the authors for their contributions. We are also grateful to all Program Committee members and the additional reviewers for their hard work that led to the selection of papers published in this volume. The reviewing process was facilitated by the EasyChair conference system, created by Andrej Voronkov. Special thanks are due to Alfred Hofmann and Anna Kramer and the Lecture Notes in Computer Science team at Springer for having granted us the opportunity to publish this special issue devoted to WORDS 2019 and for their help during the process. We are also grateful for the support received from the School of Science, and especially the administrative team at Loughborough University, the Institute of Advanced Studies at Loughborough University, the European Association for Theoretical Computer Science, and the journal *Algorithms*. We gratefully acknowledge the contribution by Alex Hewitson and Luke Bellamy, who created the conference website, and Ovidiu Toader (from Brainik), who designed the conference poster. Finally, we are much obliged to a number of people who contributed to the success of the conference: our colleagues Joel Day and Manfred Kufleitner, and our students Laura Hutchinson and Alex Smith. Our warmest thanks for their assistance in the organization of the event!

July 2019

Robert Mercaş
Daniel Reidenbach

Organization

Program Committee Chairs

Robert Mercaş	Loughborough University, UK
Daniel Reidenbach	Loughborough University, UK

Steering Committee

Valérie Berthé	IRIF, CNRS Paris 7, France
Srečko Brlek	Université du Québec à Montréal, Canada
Julien Cassaigne	Institut de Mathématiques de Luminy, France
Maxime Crochemore	King's College London, UK
Aldo de Luca	University of Naples, Italy
Anna Frid	Aix-Marseille Université, France
Juhani Karhumäki	University of Turku, Finland
Jean Néraud	University of Rouen, France
Dirk Nowotka	Kiel University, Germany
Edita Pelantová	Czech Technical University in Prague, Czech Republic
Dominique Perrin	Université Paris-Est Marne-la-Vallée, France
Antonio Restivo	University of Palermo, Italy
Christophe Reutenauer	Université du Québec à Montréal, Canada
Jeffrey Shallit	University of Waterloo, Canada
Mikhail Volkov	Ural Federal University, Russia

Program Committee

Marie-Pierre Beal	Université Paris-Est Marne-la-Vallée, France
Srecko Brlek	Université du Québec à Montréal, Canada
Émilie Charlier	University of Liege, Belgium
Volker Diekert	University of Stuttgart, Germany
Gabriele Fici	University of Palermo, Italy
Anna Frid	Aix-Marseille Université, France
Amy Glen	Murdoch University, Australia
Stepan Holub	Charles University, Czech Republic
Shunsuke Inenaga	Kyushu University, Japan
Robert Mercaş	Loughborough University, UK
Dirk Nowotka	Kiel University, Germany
Edita Pelantova	Czech Technical University in Prague, Czech Republic
Jarkko Peltomäki	University of Turku, Finland
Narad Rampersad	University of Winnipeg, Canada

Daniel Reidenbach   Loughborough University, UK
Jeffrey Shallit   University of Waterloo, Canada
Arseny Shur   Ural Federal University, Russia

Additional Reviewers

Bannai, Hideo   Maslennikova, Marina
Berthe, Valerie   Massuir, Adeline
Cassaigne, Julien   Mol, Lucas
Castiglione, Giuseppa   Mühle, Henri
Catalano, Costanza   Nicaud, Cyril
Clément, Julien   Picantin, Matthieu
Day, Joel   Pin, Jean-Eric
De Luca, Alessandro   Plyushchenko, Andrei
Dolce, Francesco   Pribavkina, Elena
Ferrari, Luca   Prihoda, Pavel
Fleischer, Lukas   Rao, Michael
Fleischmann, Pamela   Restivo, Antonio
Giammarresi, Dora   Reutenauer, Christophe
Hejda, Tomas   Rindone, Giuseppina
Hertrampf, Ulrich   Saarela, Aleksi
Kazda, Alexandr   Salo, Ville
Kitaev, Sergey   Seki, Shinnosuke
Klouda, Karel   Stipulanti, Manon
Kopra, Johan   Sugimoto, Shiho
Kufleitner, Manfred   Vavra, Tomas
Köppl, Dominik   Volkov, Mikhail
Labbé, Sébastien   Weiss, Armin
Leroy, Julien   Whiteland, Markus
Liptak, Zsuzsanna   Žemlička, Jan

On Families of Limit *S*-adic Words (Invited Talk)

Gwenaël Richomme ⓘ

LIRMM, Université Paul-Valéry Montpellier 3,
Université de Montpellier, CNRS, Montpellier, France
gwenael.richomme@lirmm.fr
http://www.lirmm.fr/~richomme/

Abstract. Given a set S of morphisms, an infinite word is limit S-adic if it can be recursively desubstituted using morphisms in S. Substitutive-adicity arises naturally in various studies especially in studies on infinite words with factor complexity bounded by an affine function. In the literature, when a family F of infinite words defined by a combinatorial property P appears to be S-adic for some set S of morphisms, it is very rare that the whole set of limit S-adic words coincides with F. The aim of the talk is to survey such situations in which necessarily morphisms of S preserve the property P of infinite words.

Keywords: S-adicity · Property preserving morphisms

We assume that readers are familiar with combinatorics on words; for omitted definitions see, *e.g.*, [5, 11, 12]. All the infinite words considered in this abstract are right infinite words.

As explained with more details in [3], the terminology S-adic was introduced by S. Ferenczi [8]. Without context, letter S refers to term "substitution" (and we will sometimes use the terminology *substitutive-adicity* instead) and in more precise definitions, it refers to a set S of (nonerasing) morphisms. An infinite word $\mathbf{w}$ is said *S-adic* if there exist a sequence $(f_n)_{n \geq 1}$ of morphisms in S and a sequence of letters $(a_n)_{n \geq 1}$ such that $lim_{n \to +\infty} |f_1 f_2 \cdots f_n(a_{n+1})| = +\infty$ and $\mathbf{w} = lim_{n \to +\infty} f_1 f_2 \cdots f_n(a_{n+1}^\omega)$. The sequence $(f_n)_{n \geq 1}$ is called a *directive word* of $\mathbf{w}$. Assume limits $\mathbf{w}_k = lim_{n \to +\infty} f_k f_{k+1} \cdots f_n(a_{n+1}^\omega)$ exist for all k. Observe that $\mathbf{w}_1 = \mathbf{w}$ and $\mathbf{w}_n = f_n(\mathbf{w}_{n+1})$ for all $n \geq 1$, that is, $\mathbf{w}$ can be infinitely desubstituted using morphisms in S. Following [1] (where is used terminology *limit point* of a sequence of substitutions), we say that $\mathbf{w}$ is a *limit S-adic* word when such sequences $(f_n)_{n \geq 1}$ and $(\mathbf{w})_{n \geq 1}$ exist. Observe that the set of limit S-adic words is the minimal set of infinite words X such that $X = \bigcup_{f \in S} f(X)$ that we could denote by abuse of notation $X = S(X)$ to emphasize the fact that limit S-adic words are generalizations of fixed points of morphisms (and even of morphic sequences).

Limit substitutive-adicity arises naturally in various studies as, for instance, this of Sturmian words (see, *e.g.* [12, Chap. 2] and [13, Chap. 5]). It is well-known (and easy to prove) that any Sturmian word is a limit S_{sturm}-adic word with $S_{sturm} = S_a \cup S_b$, $S_a = \{L_a, R_a\}$, $S_b = \{L_b, R_b\}$, $L_a(a) = a = R_a(a)$, $L_a(b) = ab$, $R_a(b) = ba$,

$L_b(a) = ba$, $R_b(a) = ab$, $L_b(b) = b = R_b(b)$. But not all limit S_{sturm}-adic words are Sturmian. Only the limit S_{sturm}-adic words whose directive word contains infinitely many elements of S_a and infinitely many elements of S_b are Sturmian. This condition can be described using infinite paths with prohibited segments in an automaton (or a graph), here with two states (or vertices), one for each set S_a and S_b. This kind of condition with an automaton to characterize allowed directive words is also used, for instance, in the characterization of words for which the first difference of factor complexity is bounded by 2 [10] or in the characterization of sequences arising from the study of multidimensional continued fraction algorithm (see for instance [4, 6]).

In [14], answering a question of G. Fici, the author characterizes in term of limit S-adicity the family of so-called LSP infinite words, that is the words having all their left special factors as prefixes. For this he determines a suitable set S_{bLSP} of morphisms and an automaton recognizing allowed infinite desubstitutions. As the obtained characterization is quite evolved, a second part of [14] considers the question of finding a simpler S-adic characterization. It is proved that, unfortunately, it does not exist any set of morphisms S such that the family of LSP infinite words is (exactly) the family of limit S-adic words except in the binary case.

The aim of the talk is to consider the question: which are the known families of infinite words defined by a combinatorial property P that correspond to a family of limit S-adic words for some set of S of morphisms? In [14], it was observed that when such a situation arises necessarily morphisms of S preserve the property P of infinite words. In what follows we will say that a morphism preserves words of a family F if it maps any word on this family on another word of this family.

In the case of Sturmian words, morphisms in S_{sturm} indeed preserve Sturmian words. That all S_{sturm}-adic infinite words are not Sturmian comes from the fact that some limit S_{sturm}-adic words are periodic. At this stage, readers should remember that Sturmian words are the aperiodic binary balanced words (where balanced means that the numbers of a occurring in any factors u and v of equal length may differ by at most one). Morphisms in S_{sturm} preserve binary balanced words and consequently limit S_{sturm}-adic words are the infinite binary balanced words.

Actually it is easy to observe that the family of Sturmian words corresponds to a family of limit substitutive-adic words, more precisely to the family of limit S'_{sturm}-adic words where $S'_{sturm} = \bigcup_{n \geq 1} S_a^n S_b \cup S_b^n S_a$. Morphisms of S'_{sturm} are obtained composing morphisms to enforce infinite occurrences of morphisms of each of the sets S_a and S_b in directive words of Sturmian words. One can note that in this case the set of morphisms S'_{sturm} is infinite.

A similar situation holds for the Arnoux-Rauzy words [2]. They are limit-S_E-adic words for a finite set of morphisms that generalize Sturmian words and they correspond to a family of limit substitutive-adic words for an infinite set of morphisms (obtained concatenating morphisms in S_E). The morphisms in S_E^* are morphisms that preserve episturmian words [7, 9] and the set of episturmian words is exactly the set of S_E-adic words.

Although limit S-adicity reveals itself to be a useful tool to study some combinatorial properties, although the notion of property preserving morphisms is often considered to generate words with interesting properties, the family of right infinite balanced words and the family of episturmian words seems to be the unique known families of words that correspond exactly to a family of limit *S*-adic words with *S* a *finite* set of morphisms.

Acknowledgment. Many thanks to J. Leroy and P. Séébold for their remarks on an earlier version of this summary.

References

1. Arnoux, P., Mizutani, M., Sellami, T.: Random product of substitutions with the same incidence matrix. Theor. Comput. Sci. **543**, 68–78 (2014)
2. Arnoux, P., Rauzy, G.: Représentation géométrique de suites de complexité $2n + 1$. Bull. Soc. Math. France **119**(2), 199–215 (1991)
3. Berthé, V., Delecroix, V.: Beyond substitutive dynamical systems: S-adic expansions. In: Akiyama, S. (ed.) Numeration and Substitution 2012. RIMS Kôkyûroku Bessatsu, vol. B46, pp. 81–123 (2014)
4. Berthé, V., Labbé, S.: Factor complexity of S-adic words generated by the Arnoux–Rauzy–Poincaré algorithm. Adv. App. Math. **63**, 90–130 (2015)
5. Berthé, V., Rigo, M. (eds.): Combinatorics, Automata and Number Theory, Encyclopedia of Mathematics and Its Applications, vol. 135. Cambridge University Press (2010)
6. Cassaigne, J., Labbé, S., Leroy, J.: A set of sequences of complexity $2n + 1$. In: Brlek, S., Dolce, F., Reutenauer, C., Vandomme, É. (eds.) WORDS 2017. LNCS, vol. 10432, pp. 144–156. Springer, Cham (2017). https://doi.org/10.1007/978-3-319-66396-8_14
7. Droubay, X., Justin, J., Pirillo, G.: Episturmian words and some constructions of de Luca and Rauzy. Theoret. Comput. Sci. **255**, 539–553 (2001)
8. Ferenczi, S.: Rank and symbolic complexity. Ergodic Theory Dynam. Syst. **16**, 663–682 (1996)
9. Justin, J., Pirillo, G.: Episturmian words and episturmian morphisms. Theoret. Comput. Sci. **276**(1–2), 281–313 (2002)
10. Leroy, J.: An *S*-adic characterization of minimal subshifts with first difference of complexity $p(n + 1) - p(n) \leq 2$. Discrete Math. Theor. Comput. Sci. **16**(1), 233–286 (2014)
11. Lothaire, M.: Combinatorics on Words, Encyclopedia of Mathematics and its Applications, vol. 17. Addison-Wesley (1983). Reprinted in the Cambridge Mathematical Library, Cambridge University Press, UK (1997)
12. Lothaire, M.: Algebraic Combinatorics on Words, Encyclopedia of Mathematics and Its Applications, vol. 90. Cambridge University Press (2002)
13. Pytheas Fogg, N.: In: Berthé, V., Ferenczi, S., Mauduit, C., Siegel, A. (eds.) Substitutions in dynamics, arithmetics and combinatorics. Lecture Notes in Mathematics, vol. 1794. Springer, Heidelberg (2002). https://doi.org/10.1007/b13861
14. Richomme, G.: Characterization of infinite LSP words and endomorphisms preserving the LSP property. Internat. J. Found. Comput. Sci. **30**(1), 171–196 (2019)

Contents

xiv   Contents

Matching Patterns with Variables

Florin Manea[1]([✉])[iD] and Markus L. Schmid[2][iD]

[1] Kiel University, Kiel, Germany
`flmanea@gmail.com`
[2] Trier University, Trier, Germany
`MLSchmid@MLSchmid.de`

Abstract. A pattern α (i.e., a string of variables and terminals) matches a word w, if w can be obtained by uniformly replacing the variables of α by terminal words. The respective matching problem, i.e., deciding whether or not a given pattern matches a given word, is generally NP-complete, but can be solved in polynomial-time for classes of patterns with restricted structure. In this paper we overview a series of recent results related to efficient matching for patterns with variables, as well as a series of extensions of this problem.

Keywords: Combinatorial pattern matching · Patterns with variables · String structural parameters · Efficient algorithms · NP-hardness

1 Introduction

A *pattern with variables*, called simply pattern in the context of this work, is a string that consists of *terminal symbols* (e.g., a, b, c) and *variables* (e.g., x_1, x_2, x_3). The terminal symbols are treated as constants, while the variables are to be uniformly replaced by strings over the set of terminals (i.e., different occurrences of the same variable are replaced by the same string); thus, a pattern is mapped to a terminal word. For example, $x_1 ab x_1 x_2 c x_2 x_1$ can be mapped to `acabaccaaccaaac` and `babbacab` by the replacements $(x_1 \rightarrow ac, x_2 \rightarrow caa)$ and $(x_1 \rightarrow b, x_2 \rightarrow a)$, respectively.

Patterns with variables appear in various areas of theoretical computer science, such as language theory (pattern languages [2]), learning theory (inductive inference [2,24,71,77], PAC-learning [55]), combinatorics on words (word equations [53,69], unavoidable patterns [63]), pattern matching (generalised function matching [1,72]), database theory (extended conjunctive regular path queries [5]), and we can also find them in practice in the form of extended regular expressions with backreferences [12,35,39], used in programming languages like Perl, Java, Python, etc.

Generally, in all these contexts, patterns with variables are used to model various combinatorial pattern matching questions. For instance, searching for a word w in a text t can be expressed as testing whether the pattern xwy can

© Springer Nature Switzerland AG 2019
R. Mercaş and D. Reidenbach (Eds.): WORDS 2019, LNCS 11682, pp. 1–27, 2019.
https://doi.org/10.1007/978-3-030-28796-2_1

be mapped to t and testing whether a word w contains a cube is equivalent to testing whether the pattern xy^3z can be mapped to w, such that y is not mapped to an empty word. Not only problems of testing whether a given word contains a regularity or a motif of a certain form can be expressed by patterns, but also problems asking whether a word can be factorised in a specifically restricted manner can be modelled in this way. For instance, asking whether $x_1^2 x_2^2 \ldots x_k^2$ can be mapped to w, such that none of the variables x_i are mapped to an empty word, is equivalent to asking whether the word w can be factorised into k non-empty squares.

Unfortunately, deciding whether a given arbitrary pattern can be mapped to a given word, the *matching problem*, is NP-complete [2], whether we ask that the variables are mapped to non-empty words or not. This intractability result severely limits the practical application of patterns. Indeed, in many tasks related to applications of patterns, the matching problem is a necessary step, so the tasks become intractable as well. For instance, this is the case for the task of computing so-called descriptive patterns for finite sets of words (see [2,37,38] for more information on descriptive patterns): one cannot solve this problem without solving a series of (general) pattern matching tasks [27]. A more detailed analysis of the complexity of the hardness of the matching problem will be presented in Sect. 3.

On the other hand, some strong restrictions on the structure of patterns yield subclasses for which the matching problem is tractable (i.e., can be solved in polynomial time). This is clearly the case of patterns where the number of different variables in the patterns is bounded by a constant, but more sophisticated and general such subclasses can be defined. We will discuss a series of results related to this topic in Sects. 4.2, 5.1 and 5.2. In our analysis, the most general class of patterns which allow for a polynomial-time pattern matching problem is defined by establishing a deep connection between strings/patterns and graphs, and considering only patterns which correspond to graphs with bounded structural parameters. As such, the subclass of *patterns with bounded treewidth*. The question of finding classes of patterns which can be matched in polynomial time but do not have bounded treewidth seemed interesting to us. We show a natural construction of such patterns in Sect. 6.

We continue this survey with a result showing that considering some of the structural parameters, that lead to efficient pattern matching algorithms, as general structural parameters of strings, may lead to remarkable results in other apparently unrelated domains. We show in Sect. 7 how our results for strings can be used to obtain a state-of-the-art approximation algorithm for computing the *cutwidth of graphs*.

We conclude the survey with a series of extensions. We discuss the problem of *injective pattern matching* as well as the satisfiability problem for word equations with restricted form.

2   Basic Definitions

For detailed definitions regarding combinatorics on words we refer to [62].

We denote our *alphabet* by Σ, the *empty word* by ε, the set of all non-empty words over Σ by Σ^+, the set of all words over Σ by Σ^*, and the *length* of a word w by $|w|$. $(\Sigma^*, \cdot, \varepsilon)$ is the free monoid over Σ with *concatenation* as its binary operation, written $\cdot$. For $w \in \Sigma^*$ and every integers i, j with $1 \leq i \leq j \leq |w|$, let $w[i..j] = w[i] \cdots w[j]$, where $w[k]$ represents the *letter on position* k and $1 \leq k \leq |w|$. A *period* of w is any positive integer p for which $w[i] = w[i+p]$, for all defined positions. Moreover, in this case, w is said to be p-periodic. Its *minimal period* is denoted by $per(w)$ and represents the smallest period of w. For example, $w = $ abacabacabacabacab has periods 8 and 4; in particular, $per(w) = 4$. A word w is called periodic if $per(w) \leq \frac{|w|}{2}$.

The *concatenation* of k words $w_1, w_2, \ldots, w_k$ is written $\Pi_{i=1,k} w_i$. If $w = w_i$ for all integers i with $1 \leq i \leq k$, this represents the kth *power* of w, denoted by w^k; here, w is a *root* of w^k. We can further extend the notion of a power of a word by saying that $w = w[1..per(w)]^{\frac{|w|}{per(w)}}$. We say that w is *primitive* if it cannot be expressed as a power of exponent ℓ of any root, where ℓ is an integer with $\ell > 1$. Conversely, if $w = v^\ell$ for some integer $\ell > 1$, then w is also called a *repetition*. The infinite repetition $vvv \cdots$ of some word v is denoted v^ω.

For any word $w \in \Sigma^+$ with $w = xyz$, we say that y is a *factor* of w. If x is empty, then y is also a *prefix* of w, while when z is empty, then y is also a *suffix*. Whenever we have a factor both as a prefix and as a suffix, the factor is said to be a *border* of the word. Furthermore, every word $u = yzx \in \Sigma^+$ is a *conjugate* of w. Note that, if w is primitive, so is every conjugate of it. If $w = vu$, then $v^{-1}w = u$.

Let $X = \{x_1, x_2, x_3, \ldots\}$ and call every $x \in X$ a *variable*. For a finite alphabet Σ of *terminals* with $\Sigma \cap X = \emptyset$, we define $\mathsf{Pat}_\Sigma = (X \cup \Sigma)^+$ and $\mathsf{Pat} = \bigcup_\Sigma \mathsf{Pat}_\Sigma$. Every $\alpha \in \mathsf{Pat}$ is a *pattern* and every $w \in \Sigma^*$ is a (*terminal*) *word*. Given a word or a pattern v, for the smallest sets $B \subseteq \Sigma$ and $Y \subseteq X$ with $v \in (B \cup Y)^*$, we denote $\mathsf{alph}(v) = B$ and $\mathsf{var}(v) = Y$. For any $x \in \Sigma \cup X$ and $\alpha \in \mathsf{Pat}_\Sigma$, $|\alpha|_x$ denotes the number of occurrences of x in α; for the sake of convenience, we set $|\alpha|_x = 0$ for every symbol x not in $\Sigma \cup X$. For a pattern α, we say that $w = \alpha[i..i + |w|]$ is a maximal terminal factor of α if $\alpha[i-1]$ and $\alpha[i + |w| + 1]$ are either not defined, or are variables.

A *substitution* (*for* α) is a mapping $h : \mathsf{var}(\alpha) \to \Sigma^*$. For every $x \in \mathsf{var}(\alpha)$, we say that x *is substituted by* $h(x)$ and $h(\alpha)$ denotes the word obtained by substituting every occurrence of a variable x in α by $h(x)$ and leaving the terminals unchanged. We say that the pattern α *matches* $w \in \Sigma^+$ if $h(\alpha) = w$ for some substitution $h : \mathsf{var}(\alpha) \to \Sigma^*$. Substitutions of the form $h : \mathsf{var}(\alpha) \to \Sigma^+$, i.e., the empty word is excluded from the range of the substitution, are also called *non-erasing*; in order to emphasize that the substitution by the empty word is allowed, we also use the term *erasing* substitution.

Example 1. Let $\beta = x_1 a x_2 b x_2 x_1 x_2$ be a pattern and let $u = $ bacbabbbbacbb and $v = $ abaabbababab be terminal words. The pattern β matches both u and v, witnessed by the substitutions h with $h(x_1) = $ bacb, $h(x_2) = $ b and g with $g(x_1) = g(x_2) = $ ab, respectively. Moreover, β also matches the word $w = $

acbbcbcb by the *erasing* substitution h with $h(x_1) = \varepsilon$, $h(x_2) = $ cb; it can be easily verified that there is no non-erasing substitution that maps β to w.

The *matching problem*, denoted by MATCH, is to decide for a given pattern α and word w, whether there exists a substitution h with $h(\alpha) = w$. The variant where we are only concerned with non-erasing substitutions is called the *non-erasing case* of the matching problem; we also use the term *erasing-case* in order to emphasize that substitution by the empty word is allowed. Another special variant is the *terminal-free case* of the matching problem, where the input patterns are terminal-free, i.e., they do not contain any occurrences of terminal symbol. We shall briefly discuss some particularities of these different special cases of the matching problem in Sect. 3. Note that in the sections on efficient algorithms, namely Sects. 5.1, 5.2, and 6, we only consider the non-erasing case (with terminal symbols) of the matching problem. The presented results can easily be generalised to the general setting, but we prefer the respective framework for the ease of the presentation.

For any $P \subseteq $ Pat, the *matching problem for P* (or MATCH *for P*, for short) is the matching problem, where the input patterns are from P. In the sections of this paper we will introduce and discuss several interesting families of patterns.

As we discuss efficient algorithms, it is important to describe the computational model we use in this work. This is the standard unit-cost RAM with logarithmic word size. Also, all logarithms appearing in our time complexity evaluations are in base 2. For the sake of generality, we assume that whenever we are given as input a word $w \in \Sigma^*$ of length n, the symbols of w are in fact integers from $\{1, 2, \ldots, n\}$ (i.e., $\Sigma = $ alph$(w) \subseteq \{1, 2, \ldots, n\}$), and w is seen as a sequence of integers. This is a common assumption in the area of algorithmics on words (see, e.g., the discussion in [54]). Clearly, our algorithmic results hold canonically for constant alphabets, as well.

3   The Hardness of the Matching Problem

First, we recall that there are several different variants of the matching problem: the most general case (substitution by the empty word and occurrences of terminals in the patterns are possible), the non-erasing case (with terminal symbols), the terminal-free (erasing) case, and finally the terminal free non-erasing case. As we shall see, these differences do not matter too much if we are only concerned with the matching problem of patterns. However, in other contexts of patterns with variables (e. g., other decision problems, learning theory), these differences are most crucial and we therefore briefly provide some background.

For the class of the so-called *pattern languages*, i. e., the sets of all words that match a pattern, the difference between the erasing and the non-erasing case is important, since these classes of formal languages differ quite substantially with respect to basic decision problems. For example, in the non-erasing case, two patterns describe the same language if and only if the patterns are identical (up to a renaming of variables), while it is open whether the equivalence problem is even decidable in the erasing-case (see, e. g., Sect. 6 in [70], or [76]). Moreover, the inclusion problem, which is undecidable for both the erasing and the non-erasing case (see [36,52]), can be decided for terminal-free patterns in the erasing case, while for terminal-free non-erasing patterns the decidability status is open (intuitively speaking, this has to do with the fact that avoidability questions of the form "does pattern β necessarily occur in long enough words over a k-letter alphabet?" can be expressed as inclusion for two languages given by terminal-free non-erasing patterns). Finally, also whether patterns (or descriptive patterns) can be inferred from positive data strongly depends on whether the erasing or non-erasing case is considered, or whether or not terminal symbols in the patterns are allowed (see [37,38,75,77]).

For the matching problem (note that this corresponds to the membership problem for pattern languages), whether we consider erasing or non-erasing substitution, or whether or not we disallow terminal symbols in the patterns, has little impact on its computational hardness. In fact, that the matching problem for patterns with variables is NP-complete has been independently discovered in different communities and for slightly different problem variants (see, e. g., the introductions of [28,30] for some remarks on the history of the investigation of the matching problem).

If we consider the most general case, i. e., erasing substitutions and terminals in the patterns, then a hardness-reduction is rather simple. For example, the Boolean formula

$$((v_1, v_2, v_3), (v_2, v_4, v_5), (v_3, v_1, v_3), (v_4, v_1, v_2))$$

in 3-CNF (without negated variables) is 1-in-3 satisfiable (i. e., satisfiable with exactly one literal per clause set to *true*) if and only if the following word w is matched by the pattern α:

$$
\begin{aligned}
w ={} & \text{a} \quad \text{b} \quad \text{a} \quad \text{b} \quad \text{a} \quad \text{b} \quad \text{a} \\
\alpha ={} & x_1 x_2 x_3 \ \text{b} \ x_2 x_4 x_5 \ \text{b} \ x_3 x_1 x_3 \ \text{b} \ x_4 x_1 x_2
\end{aligned}
$$

We can further observe that this simple reduction also shows that the matching problem is hard even for binary terminal alphabets and under the restriction that variables are substituted by single symbols (or the empty word) only. This directly raises the questions under which restrictions the matching problem remains hard. For example, a problem instance has a large number of natural parameters (length of the pattern, length of the word, number of variables, number of occurrences per variable, alphabet size, length of words substituted for variables) and in addition to that, it comes in four natural variants resulting from whether we consider the erasing or non-erasing case, and whether or not we allow terminals in the pattern. In the above reduction, the number of variables, the number of occurrences per each variable and the word length are unbounded.

All these numerous restricted problem variants have been thoroughly investigated in [29] and it turns out that the matching problem remains NP-hard under rather strong restrictions. We cite the following result as an example and refer to [29] for further details.

Theorem 1 ([29]). *The erasing case of the matching problem for patterns with variables is* NP-*complete, even if* $\Sigma = \{a, b\}$, *every variable has at most 2 occurrences and every variable can only be substituted by a single symbol or the empty word.*

This result also holds as stated for terminal-free patterns. In the non-erasing case, however, it holds when the bound on the substitution words is 3 instead of 1, and in the non-erasing and terminal-free case the result holds when additionally the bounds on the occurrences per variable and alphabet size are 3 and 4, respectively.

The only polynomial-time solvable cases of the matching problem obtained by restricting the numerical parameters mentioned above are trivial ones. More precisely, the matching problem can be easily solved for unary alphabets (in this case, we only have to solve an equation in the integers and with integer coefficients, which are given in unary encoding), or if every variable has only one occurrence (the patterns are then *regular*, see Sect. 4), or if the number of variables or the length of the input word is bounded by a constant (the former is obvious, while the latter, in the erasing case, requires a slightly more careful argument [45]).

In particular, this also points out that Theorem 1 describes some kind of dichotomy, i.e., if we would further restrict the alphabet size, or the maximum number of occurrences per variable to 1, then we would obtain a polynomial-time solvable variant (even if all other parameters are unrestricted); similarly, if we allow variables to be substituted by single symbols only, but not the empty word, then the matching problem becomes efficiently solvable as well (regardless of the alphabet size).

Generally, by brute-force algorithms, the matching problem can be solved in time $|\alpha|^{O(|w|)}$ or $|w|^{O(|\alpha|)}$, making it polynomial-time solvable provided that there is a constant upper bound on $|w|$ or $|\alpha|$ (in fact, a bound on $|\text{var}(\alpha)|$ is sufficient). However, this constant upper bound occurs in the exponent, which means that even for rather low such bounds, say 7, the corresponding polynomial-time algorithms are most likely impractical for larger problem instances. This leads to the question whether exponential-time algorithms are possible whose running-times are such that the exponential part *exclusively* depends on, say $|\text{var}(\alpha)|$, but not on $|w|$, i.e., running-times of the form $f(|\text{var}(\alpha)|) \times g(|\alpha|, |w|)$, where g is a polynomial and f is some *computable* function (exponential, or even double-exponential etc.). Such a running-time is polynomial for upper bounded $|\text{var}(\alpha)|$, but the degree of the polynomial is always the same independent from the actual upper bound. If a problem has an algorithm with such a running-time, then it is called *fixed-parameter tractable* (with respect to the bounded parameter); see the textbooks [23, 32] for more information on parameterised complexity. Whether the matching problem for patterns with variables allows fixed-parameter tractability for some parameters has been thoroughly investigated in [31]. Although there are some more or less trivial cases of fixed-parameter tractability, the main insight provided by [31] is of a negative nature and can be summarised in the following way.

Theorem 2 ([31]). *All variants of the matching problem parameterised by $|\alpha|$ are W[1]-hard. The erasing case of the matching problem parameterised by $|w|$ is W[1]-hard.*[1]

Note that since $|\alpha|$ and $|w|$ are rather general parameters, this result covers other parameters as well, e.g., $|\Sigma|$ or $|\mathsf{var}(\alpha)|$. In the non-erasing case, $|w|$ is an upper bound for $|\mathsf{var}(\alpha)|$; thus, treating $|w|$ as a parameter means that $|\mathsf{var}(\alpha)|$ is also a parameter and therefore the matching problem is fixed-parameter tractable by the obvious brute-force algorithm. We refer to [31] for further such simple fixed-parameter tractable case.

Consequently, even strong restrictions of the obvious numerical parameters of instances of the matching problem, i.e., number of variables, alphabet size, occurrences per variable etc., does not yield interesting efficiently matchable subclasses of patterns with variables. However, as discussed in the next section, looking deeper into the structure of patterns will help.

4 Structural Restrictions for Patterns

From an intuitive point of view it is clear that not only the mere length of a pattern or the number of its variables should have an impact on the matching complexity, but also the actual order of the variables. For example, it has been observed rather early in [81] that if the variable occurrences in the patterns are sorted, e.g., as in $x_1 a x_1 x_2 x_2 \mathsf{ab} x_2 x_2 a x_3 x_4 c x_4$, then they can be matched efficiently "from-left-to-right" (more precisely, it is observed in [81] that matching such patterns can be done in logarithmic space).

A systematic investigation of such structural restrictions has been done in the last decade and numerous efficiently matchable subclasses of patterns have been found. In the following, we first present a unifying approach based on graph morphisms and the concept of treewidth. Then, we define and summarise several structural parameters for patterns and respective subclasses of patterns.

4.1 Pattern Matching by Graph Morphisms

The following general framework for matching patterns with variables has been developed in [78]. For a pattern $\alpha \in (X \cup \Sigma^*)$, the *standard graph representation* of α is the undirected graph $\mathcal{G}_\alpha^{\mathsf{pat}} = (V_\alpha, E_\alpha)$, where $V_\alpha = \{1, 2, \ldots, |\alpha|\}$ and $E_\alpha = E_\alpha^{\mathsf{equ}} \cup E_\alpha^{\mathsf{nei}}$ with $E_\alpha^{\mathsf{equ}} = \{\{i, i+1\} \mid 1 \leq i \leq |\alpha| - 1\}$ being the set of *neighbour edges* and $E_\alpha^{\mathsf{equ}} = \{\{i, j\} \mid \alpha[i..j] = x\beta x, x \in X, |\beta|_x = 0\}$ being the set of *equality edges* (see Fig. 1 for an illustration).

In a similar way, we can also encode words $w \in \Sigma^*$ as graph structures $\mathcal{G}_w^{\mathsf{wo}}$, where every factor $w[i..j]$ of w is represented by a vertex (i, j), equality edges are drawn between (i, j) and (i', j') if $w[i..j] = w[i'..j']$, and neighbour edges if $j+1 = i'$. It has been shown in [78] that α matches w if and only if there is a graph

---

[1] Problems that are hard for the parameterised complexity class W[1] are strongly believed to be not fixed-parameter tractable.

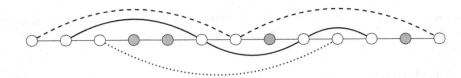

Fig. 1. The standard graph representation G_α^{pat} for $\alpha = x_1x_2x_3\mathsf{bb}x_2x_1\mathsf{a}x_2x_3x_2\mathsf{c}x_1$; the dashed, straight and dotted equality edges correspond to occurrences of x_1, x_2 and x_3, respectively; the grey vertices correspond to occurrences of terminal symbols.

morphism from $\mathcal{G}_\alpha^{\mathsf{pat}}$ to $\mathcal{G}_w^{\mathsf{wo}}$. Moreover, the concept of the treewidth for graphs now also applies to patterns (i. e., the treewidth of a pattern is the treewidth of its standard graph representation), which is of relevance since the graph morphism problem can be solved in polynomial-time provided that the source graphs have bounded treewidth.[2] Consequently, we can conclude the following algorithmic meta-theorem.

Theorem 3 ([78]). *If a class P of patterns has bounded treewidth, then the matching problem for P can be solved in polynomial-time.*

Due to the generality of the statement of Theorem 3, the polynomial-time matching algorithm that it implies is of little practical value, even for rather simple classes of patterns. On the other hand, its theoretical relevance is demonstrated by the fact that it covers almost all known classes of patterns with a polynomial-time matching problem.[3] After an additional remark regarding [78], we shall briefly define and compare those efficiently matchable classes of patterns in the next subsection.

Remark 1. Technically, the matching problem reduces to the morphism problem for (simple) relational structures instead of undirected graphs. However, since we are here only interested in the treewidth of these structures, we can as well only talk about the underlying undirected graphs.

Moreover, the actual meta-theorem of [78] is stronger in the sense that there the treewidth of patterns is not defined with respect to the standard graph representation, but with respect to a slightly more general graph representations (i. e., we allow any way of drawing the equality edges as long as all vertices corresponding to the same variable form a connected component).

4.2   Efficiently Matchable Classes of Patterns

The most obvious way to restrict patterns is to limit their number of (repeated) variables or the number of occurrences per variable. In this regard, let var_k and repv_k be the class of patterns with at most k variables and with at most k

---

[2] See [23, 32] for a formal definition of the treewidth.

[3] See Sect. 6 for the respective exceptions.

repeated variables, respectively. Due to Theorem 1, we already know that bounding the number of occurrences per variable does not in general yield polynomial-time matchable classes. The only exception are patterns with at most one occurrence per variable, which are called *regular* patterns and are denoted by reg, e. g., $x_1 a x_2 b a c x_3 a$ is a regular pattern. Regular patterns have been first considered in [81] and their name is motivated by the fact that the corresponding pattern languages are regular languages.

Next, we define the so-called scope coincidence degree (see [78]). For every $y \in \text{var}(\alpha)$, the *scope of* y *in* α is defined by $\text{sc}_\alpha(y) = \{i, i+1, \ldots, j\}$, where i is the leftmost and j the rightmost occurrence of y in α. The scopes of some variables $y_1, y_2, \ldots, y_k \in \text{var}(\alpha)$ *coincide in* α if $\bigcap_{1 \leq i \leq k} \text{sc}_\alpha(y_i) \neq \emptyset$. By $\text{scd}(\alpha)$, we denote the *scope coincidence degree* of α, which is the maximum number of variables in α such that their scopes coincide, and by scd_k, we denote the class of patterns with scope coincidence degree of at most k. See Fig. 2 for an example of the scope coincidence degree. An important special class is scd_1, which has been first introduced in [81] as the class of *non-cross* patterns (denoted by nc). Intuitively speaking, the variables in non-cross patterns are sorted, e. g., $x_1 a x_1 x_2 b a x_2 c x_3 a x_3 x_3$.

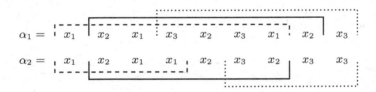

Fig. 2. Two pattern α_1 and α_2 with $\text{scd}(\alpha_1) = 3$ and $\text{scd}(\alpha_2) = 2$. The scopes of variable x_1 (dashed line), x_2 (straight line) and x_3 (dotted line) are highlighted.

Next, we define the locality number, which is a general string-parameter, and which has been first introduced in [15]. A word is k-local if there exists an order of its symbols such that, if we *mark* the symbols in the respective order (which is called a *marking sequence*), at each stage there are at most k contiguous blocks of marked symbols in the word. This k is called the *marking number* of that marking sequence. The *locality number* of a word is the smallest k for which that word is k-local, or, in other words, the minimum marking number over all marking sequences. For example, the marking sequence $\sigma = (a, g, c)$ marks $w = \text{agagcac}$ as follows (marked blocks are illustrated by overlines): $\overline{a}g\overline{a}gcac$, $\overline{a}\overline{g}\overline{a}\overline{g}c\overline{a}c$, $\overline{a}\overline{g}\overline{a}\overline{g}c\overline{a}c$, $\overline{agagcac}$; thus, the marking number of σ is 3. In fact, all marking sequences for w have a marking number of 3, except (g, a, c), for which it is 2: $a\overline{g}a\overline{g}cac$, $\overline{agag}cac$, $\overline{agagcac}$. Thus, the locality number of w, denoted by $\text{loc}(w)$, is 2. When we measure the locality number for patterns, we simply ignore all terminal symbols, e. g., $\text{loc}(abx_1 x_2 a x_1 x_2 c x_3 x_1 a x_3) = \text{loc}(x_1 x_2 x_1 x_2 x_3 x_1 x_3) = 2$. The class of patterns with locality number at most k is denoted by loc_k.

The next classes have been first considered in [78] and are based on possible nesting structures of variables. For a pattern α, we call two variables $x, y \in \text{var}(\alpha)$

entwined if α contains $xyxy$ or $yxyx$ as a subsequence. A pattern α is *nested*, if no two variables in α are entwined; the class of nested patterns is denoted by nest. A proper subclass of nest, considered in [15], are the so-called *strongly nested* patterns (denoted by snest), which are inductively defined as follows: any pattern $\alpha \in \mathsf{var}_1$ is strongly nested; if α_1 and α_2 are strongly nested and variable-disjoint patterns, x is a variable not in $\mathsf{var}(\alpha_1) \cup \mathsf{var}(\alpha_2)$ and $\beta_1, \beta_2 \in (\{x\} \cup \Sigma)^*$, then $\alpha_1\alpha_2$ and $\beta_1\alpha_1\beta_2$ are strongly nested patterns. For example, the pattern $\alpha = x_1x_2\mathtt{a}x_2x_1\mathtt{b}x_3x_4\mathtt{a}x_3$ is strongly nested, whereas αx_1 is nested, but not strongly nested anymore.

If, for every $x, y \in \mathsf{var}(\alpha)$, $\alpha = \beta x \gamma_1 y \gamma_2 x \gamma_3 y \delta$ implies $\gamma_2 = \varepsilon$, then α is called *closely entwined*, and a pattern α is *mildly entwined* if it is closely entwined and, for every $x \in \mathsf{var}(\alpha)$, if $\alpha = \beta x \gamma x \delta$ with $|\gamma|_x = 0$, then γ is nested. We denote the class of mildly entwined patterns by ment. The main motivation for the somewhat peculiar class of mildly entwined patterns is that mildly entwined patterns are exactly those patterns that have a standard graph representation that is outer-planar (see [78]).[4] It is known that outer-planar graphs have a rather low treewidth of at most 2. Since the concept of outer-planarity generalises to k-outer-planarity and k-outer-planar graphs have a treewidth of at most $3k-1$, we can also define the classes outp_k of k-outer-planar patterns (i. e., their standard graph representation is k-outer-planar). In this regard note that $\mathsf{outp}_1 = \mathsf{ment}$. See Fig. 3 for an example of a mildly-entwined pattern.

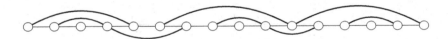

Fig. 3. The standard graph representation G_α^{pat} for $\alpha = x_1x_3x_4x_3x_1x_2x_3x_5\mathtt{b}x_5x_2x_5x_6\mathtt{a}x_6x_2$. By definition, α is mildly entwined. Furthermore, since no vertex is completely "surrounded" by edges, the shown embedding is outer-planar.

It can be easily verified that all of the pattern classes defined above have bounded treewidth; thus, by application of Theorem 3, they can be matched efficiently. For some of them this upper bound on the treewidth is rather low (e. g., reg, nc, ment), while for those classes obtained by bounding a structural parameter, e. g., repv_k, scd_k, loc_k, the bound on the treewidth also grows with this parameter. Figure 4 shows how these pattern classes relate to each other and how they form infinite hierarchies within the class of all patterns (denoted by Pat).

In a sense, Fig. 4 is a "tractability map" for the matching problem of patterns with variables. For the classes that have low treewidth, we can expect matching algorithms that are rather efficient. On the other hand, these classes are quite

---

[4] A graph is outer-planar if it has a planar embedding with all vertices lying on the outer face.

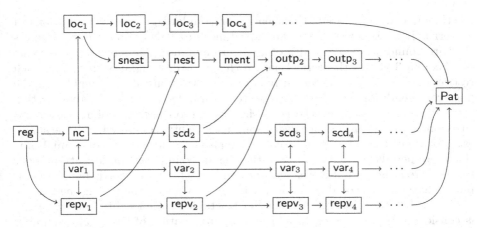

Fig. 4. An overview of efficiently matchable classes of patterns. By $A \rightarrow B$, we denote $A \subset B$; pairs without arrow are incomparable. Note that $\mathsf{nc} = \mathsf{scd}_1$ and $\mathsf{ment} = \mathsf{outp}_1$.

restricted (compared to the full class of patterns) and are most likely only applicable for very special pattern matching tasks. An obvious approach to matching general patterns would be to first perform a preprocessing that identifies a "low class" of the tractability map that contains the input pattern and then uses the most efficient algorithm for matching it. In this regard, it is even an asset that most of the different efficiently matchable classes and hierarchies of classes are incomparable: it is possible that an input pattern has a very large locality number of 100, but can nevertheless be matched efficiently, because its standard graph representation is 2-outerplanar; on the other hand, a pattern could have a large scope coincidence degree and a large number of variables, but at the same time a very low locality number. It might even be a worthwhile research task to experimentally analyse a large corpus of (random) patterns with respect to the classes of the tractability map in which they are contained.

Remark 2. Bounding the structural parameters defined above yields polynomial-time matchable classes of patterns; thus, the question arises whether the matching problem is also fixed-parameter tractable with respect to those parameters. However, Theorem 2 already states that this is most likely not the case for parameter $|\mathsf{var}(\alpha)|$, and since $|\mathsf{var}(\alpha)|$ is an upper bound for the number of repeated variables, the scope coincidence degree, the outer-planarity and the locality number of α, it is also highly unlikely that we can achieve fixed-parameter tractability with respect to those parameters.

4.3   Computing Structural Parameters for Patterns

Since the structural restrictions of patterns surveyed above are all meant to be exploited algorithmically, the task of checking them (or computing the respective parameters) is an important issue. In this regard, note that in general computing the treewidth of a graph is an NP-hard problem and it is also not known whether it can be computed efficiently for standard graph representations of patterns. This also emphasises the importance of *easily computable* parameters that are bounding the treewidth of a pattern and also points out why the value of Theorem 3 is of a theoretical nature that provides guidance in finding such restrictions with higher practical relevance. Restrictions like the regularity, the non-cross condition, number of (repeated) variables and the different nesting properties can be easily checked for. Moreover, also the scopes of a pattern and therefore its scope coincidence degree can be efficiently computed, and the smallest k for which a graph is k-outerplanar can also be computed in polynomial time (for more details see [79]). On the other hand, computing the locality number seems more difficult and it was left open in [15] whether or not is hard to compute. This gap was closed in [13] where it was shown that computing the locality number is NP-hard, but fixed-parameter tractable (if the locality number or $|\Sigma|$ is considered a parameter); in addition, approximation of the locality number has also been investigated in [13] (note that these result will be discussed in more detail in Sect. 7).

5   Faster Pattern Matching

In this section we will overview some efficient matching algorithms developed for various classes of patterns, some defined already in the previous sections, and some defined via some other natural structural restrictions. Most of the result of this paper were shown in [15,16,26].

5.1   Patterns with Low Scope Coincidence Degree

We start with several definitions. The *one-variable blocks* in a pattern are maximal contiguous blocks of occurrences of the same variable. A pattern α with m one-variable blocks can be written as $\alpha = w_0 \Pi_{i=1,m}(z_i^{k_i} w_i)$ with $z_i \in \mathsf{var}(\alpha)$ for $i \in \{1, 2, \ldots, m\}$ and $z_i \neq z_{i+1}$, whenever $w_i = \varepsilon$ for $i \in \{1, 2, \ldots, m-1\}$. The number of one-variable blocks is a natural complexity measure that we will consider.

Example 2. The pattern $\alpha = x_1 x_2 x_2 \mathsf{a} x_2 x_2 x_2 x_3 \mathsf{a} x_3 x_2 x_2 x_3 x_3$ has 7 one-variable blocks: $x_1, x_2 x_2, x_2 x_2 x_2, x_3, x_3, x_2 x_2, x_3 x_3$.

As discussed in the previous sections, prominent subclasses of patterns for which MATCH can be solved in polynomial time are the classes of patterns with a bounded number of (repeated) variables (var_k and repv_k), of regular patterns (reg), of non-cross patterns (nc), and of patterns with a bounded scope coincidence degree (scd_k). However, the known respective algorithms are rather poor considering their running times. For example, for var_k, the matching problem can be solved in $O(\frac{mn^{k-1}}{(k-1)!})$, where m and n are the lengths of the pattern and the word (see [47]). For patterns with a scope coincidence degree of at most k, an $O(mn^{2(k+3)}(k+2)^2)$ time algorithm can be derived using the general matching technique described by Theorem 3, where m and n are the lengths of the pattern and the word, respectively, and the proof that the matching problem for non-cross patterns is in P (see [81]) leads to an $O(n^4)$-time algorithm. Hence, for all these classes, we consider the following refinement of the problem of showing that the matching problem for a class of patterns is in P.

Problem 1. Let K be a class of patterns for which the matching problem can be solved in polynomial time. Find an efficient algorithm that solves the matching problem for K.

The main class of patters considered in the following is that of patterns with bounded scope coincidence degree, and its subclasses.

If the scope coincidence degree is bounded by 1, i.e., non-cross patterns, we can decide whether a pattern α having m one-variable blocks matches a word w of length n in $O(mn \log n)$ time. This result can be achieved via a general dynamic programming approach, which tries to match prefixes of the pattern α to the prefixes of the word w. This general approach is rather standard but the big gain is that it can be implemented efficiently by a detailed combinatorial analysis of the possible matches between the one-variable blocks occurring in α and factors of w. For instance, if the shortest factor of α containing all occurrences of a variable x starts with a one-variable block containing at least two occurrences the variable x, we can efficiently find the matches of this factor by exploiting a major result from [14], which states that the primitively rooted squares contained in a word of length n can be listed optimally in $O(n \log n)$. As each match for a factor starting with two occurrences of a variable starts with a primitively rooted square, the respective matches can be found efficiently. The result regarding primitively rooted squares can be extended to show that, given a word w of length n and a word v with length shorter than n, the word w contains $O(n \log n)$ factors of the form uvu with uv primitive, and all these factors can be found optimally in $O(n \log n)$ time. This allows us to find efficiently the matches for one-variable that the shortest factor of α which contains all occurrences of x and starts with xvx, for all choices of a variable x such that v is a non-empty terminal string.

Theorem 4 ([26]). *The matching problem for* nc *is solvable in* $O(mn \log n)$ *time, where* w *is the input word of length* n *and* m *is the number of one-variable blocks occurring in the pattern.*

Two particular subclasses of non-cross patterns are of interest: the regular patterns reg and the one-variable patterns var₁ (see also Fig. 4). It is not hard to show that regular patterns can be matched in linear time $O(|\alpha| + |w|)$, by iteratively using the Knuth-Morris-Pratt algorithm to identify greedily the terminal factors occurring in the pattern, in their orders of occurrences. All factors of a word w that match a given regular pattern α can be detected in linear time too.

More interesting is the case of one-variable patterns. The simplest example of one-variable patterns are the repetitions, i.e., patterns of the form x^k. Checking whether a word is a match for a pattern x^k can be done in linear time. Moreover, a compact representation of all periodic factors of a word w can be also obtained in linear time by identifying the (at most $|w|$) so-called runs inside w [4]. With this, a compact representation of occurrences of x^k in w can also be obtained in linear time. More complex one-variable patterns are the pseudo-repetitions (see [41,43,44] and the references therein). These are patterns from $\{x, x^R\}^*$, where x^R is a variable that is always substituted by the reverse image of the string substituting x. Checking whether a string matches a given pseudo-repetition can be done in linear time [44]. The following general result can be shown for one-variable patterns, see [59]. Given a pattern $\alpha = v_1 x v_2 x \cdots v_{r-1} x v_r$ such that x is a variable and $v_1, v_2, \ldots, v_r$ are terminal strings, a compact representation of all P instances of α in the input string w of length n can be computed in $O(rn)$ time, so that one can report those occurrences in $O(P)$ time. The same result holds also for the case when some of the occurrences of x in such a pattern are replaced by x^R. It is worth noting that using this algorithm to find the factors of a given word that match the shortest factor of α containing all occurrences of a variable x inside a non-cross pattern in our approach for matching nc does not lead to a faster matching algorithm in that case.

When considering general patterns with bounded scope coincidence degree, one can show, using a similar dynamic programming approach as in the case of non-cross patterns, that the matching problem for scd_k is solvable in $O(\frac{mn^{2k}}{((k-1)!)^2})$ time, where n is the length of the input word and m is, again, the number of one-variable blocks occurring in the pattern. One should note that in this case it seems hard to use the combinatorial insights used for non-cross patterns (thus, the $\log n$ factor is replaced by an n factor in the evaluation of the time complexity), but, still, this algorithm is significantly faster than the previously known solution.

Theorem 5 ([26]). *The matching problem for scd_k is solvable in* $O\left(\frac{mn^{2k}}{((k-1)!)^2}\right)$ *time, where w is the input word of length n and m is the number of one-variable blocks occurring in the pattern.*

Next we consider the classes repv_k. For the basic case of $k = 1$, the matching problem can be solved in $O(n^2)$ time, where n is the length of the input word. The idea of this algorithm is to guess the length ℓ of the repeated variable x, and then to partition the suffix array of the input word into clusters, such that all suffixes in a cluster start with the same factor of length ℓ. Essentially, in a match between the pattern and the word, where x is mapped to a factor of length ℓ, the positions where the factors matching x occur in the input word belong to the same cluster. Using this idea, the desired complexity is then reached, again via dynamic programming.

Theorem 6 ([26]). *The matching problem for repv_k is solvable in quadratic time.*

Further, one can use this result to show that the matching problem for the general class of patterns repv_k is solvable in $O(\frac{n^{2k}}{((k-1)!)^2})$ time. This algorithm is better than the one that could have been obtained by using the fact that patterns with at most k repeated variables have the scope coincidence degree bounded by $k+1$, and then directly applying our previous algorithm solving the matching problem for scd_{k+1}.

Theorem 7 ([26]). *The matching problem for repv_k is solvable in* $\mathrm{O}\left(\frac{n^{2k}}{((k-1)!)^2}\right)$ *time, where n is the length of the input word.*

Note that the classes of non-cross patterns and of patterns with a bounded scope coincidence degree or with a bounded number of repeated variables are of special interest, since for them we can compute so-called descriptive patterns (see [2,81]) in polynomial time. A pattern α is *descriptive* (with respect to, say, non-cross patterns) for a finite set S of words if it can generate all words in S and there exists no other non-cross pattern that describes the elements of S in a better way. Computing a descriptive pattern, which is NP-complete in general, means to infer a pattern common to a finite set of words, with applications for inductive inference of pattern languages (see [71]). For example, our algorithm for computing non-cross patterns can be used in order to obtain an algorithm that computes a descriptive non-cross pattern in time $\mathrm{O}(\sum_{w\in S}(m^2|w|\log|w|))$, where m is the length of a shortest word of S (see [27] for details).

The algorithms, except the ones for the basic cases of regular and non-cross patterns and patterns with only one repeated variable, still have an exponential dependency on the number of repeated variables or the scope coincidence degree. Therefore, only for very low constant bounds on these parameters can these algorithms be considered efficient. Naturally, finding a polynomial time algorithm for which the degree of the polynomial does not depend on the number of repeated variables or on the scope coincidence degree would be desirable. However, by Remark 2 such algorithms are very unlikely.

Finally we recall a result regarding *gapped repeats and palindromes*. A gapped repeat (palindrome) is an instance of a terminal-free pattern xyx (respectively, xyx^R). For $\alpha \geq 1$, an α-gapped repeat in a word w is a factor uvu of w such that $|uv| \leq \alpha|u|$; the two factors u in such a repeat are called arms, while the factor v is called gap. Such a repeat is called maximal if its arms cannot be extended simultaneously with the same symbol to the right or, respectively, to the left. In a sense, α-gapped repeats are instances of the pattern xyx where length constraints are imposed on the strings that substitute x and y. In [42] it was shown that the number of maximal α-gapped repeats that may occur in a word is upper bounded by $18\alpha n$. Using this, an algorithm finding all the maximal α-gapped repeats of a word in $\mathrm{O}(\alpha n)$ was defined; this result is optimal, in the worst case, as there are words that have $\Theta(\alpha n)$ maximal α-gapped repeats. Comparable results were developed for the case of α-gapped palindromes, i.e., factors uvu^R with $|uv| \leq \alpha|u|$. On the one hand, these results were relevant as they provided optimal algorithms for the identification of α-gapped repeats and palindromes, and closed an open problem from [57,58] (see also [42] and the references therein for more on gapped repeats and palindromes). On the other hand, they point towards the study of MATCH for patterns with (linear) length constraints on the images of the variables.

5.2 Patterns with Low Locality Number

Intuitively, the notion of k-locality (already introduced in Sect. 4.2) involves marking the variables in the pattern in some arbitrary order until all the variables are marked. The pattern is k-local if this marking can be done while never creating more than k marked blocks. Variables which only occur adjacent to those which are already marked can be marked "for free" – without creating any new blocks, and thus a valid marking sequence allows a sort-of parsing of the pattern whilst maintaining a degree of closeness (locality) to the parts already parsed. The notion of k-locality was introduced and further analysed in [15]. With respect to pattern matching, the main result proven in that paper is the following:

Theorem 8 ([15]). MATCH *for* loc_k *can be decided in* $O(mkn^{\max(3k-1,2k+1)})$ *time, where m is the length of the input pattern and n is the length of the input word.*

To solve the matching problem for loc_k we use the following idea. Using a simple dynamic programming approach we can show that, given a pattern $\beta \in (X \cup \Sigma)^*$ of length m, we can decide in $O(m^{2k}k)$ time whether $\beta \in \mathsf{loc}_k$, and if the answer is positive, we can produce in the same time a marking sequence witnessing that β is k-local. As such we can keep track of the marked factors in the pattern, while executing the marking according to the computed marking sequence. We also need now to keep track to which factors of the input word the marked factors correspond. Then we try to assign every new variable so that it fits nicely around the already matched factors. This is done efficiently using a data structure from [59], mentioned also above: given a word w and a one-variable pattern γ (so, $|\mathsf{var}(\gamma)| = 1$), one can produce a compact representation of all the g factors of w matching γ in $O(|\gamma||w|)$ time; moreover, we can obtain all the g factors of w matching γ in $O(|g|)$ time. This allows us to test efficiently which factors of w match any of the one-variable blocks of β, and, ultimately, to assign a value to each variable. In comparison to the algorithm from [78] for patterns of bounded treewidth, which firstly constructs relational structures from α and w, and solves the homomorphism problem on these relational structures (see Sect. 4.1), the above algorithm exploits directly the locality structure present in the patterns. The advantage of this more focussed approach is that it allows for a considerable improvement in the required time, reducing the exponent of n from $4k + 4$ to $3k - 1$.

6 Efficient Pattern Matching Beyond Bounded Treewidth

In [16] the authors tried to identify classes of patterns that do not have bounded treewidth but can still be matched in polynomial time. The idea behind defining such classes was relatively simple: consider generalised repetitions of patterns.

One simple observation is that, if we can match patterns from a class $\mathcal{C}$ in polynomial time, then we can also match repetitions of these patterns in polynomial time: if we wish to check whether α^k matches a word w, where α is chosen from the class $\mathcal{C}$ for which we can solve MATCH efficiently, then we can firstly check whether $w = v^k$ for some word v, and then check whether α matches v, so we can also match α^k efficiently. Moreover, it can be observed that most parameters that lead to efficiently matchable classes, e. g., the scope coincidence degree or locality, are defined independently from the terminal symbols, i. e., via the word obtained after removing all terminals, which shall be called *skeleton* in the following (e. g., the skeleton of $x_1ax_2bax_1x_2b$ is $x_1x_2x_1x_2$). As a result, it is possible that a pattern, that is *not* a repetition of any $\alpha \in \mathcal{C}$, has nevertheless a skeleton that is a repetition of a skeleton from $\mathcal{C}$. For example, $ax_1(x_2)^3x_3bx_3x_1(x_2)^2bx_2a(x_3)^2$ is not a repetition of a non-cross pattern, but its skeleton $(x_1(x_2)^3(x_3)^2)^2$ is. In [16] it is shown that, for some important classes $\mathcal{C}$ of patterns, including loc_k and scd_k, for constant k, the polynomial time solvability of MATCH does not only extend from $\mathcal{C}$ to exact repetitions, but also to such skeleton-repetitions, called $\mathcal{C}$-*repetitions*.

Theorem 9 ([16]). *For $\mathcal{C} \in \{nc, reg, loc_k, scd_k\}$, solving the matching problem for the class of $\mathcal{C}$-repetitions can be done in polynomial time.*

It is interesting to note that the general treewidth-based framework of polynomial time matching of patterns does not seem to cover a very simple and natural aspect: repetitions of the same pattern. More precisely, if $\mathcal{C}$ is one of the known efficiently matchable classes of patterns, then a repetition α^k for some $\alpha \in \mathcal{C}$ is usually not in $\mathcal{C}$ anymore. In fact, it can be shown that even for patterns α with bounded and very low treewidth, the treewidth of repetitions α^k can be unbounded.

Theorem 10 ([16]). *Let $\mathcal{C}$ be a class of patterns that contains reg. Then the class of $\mathcal{C}$-repetitions contains patterns with arbitrarily large treewidth.*

In particular, the previous theorem holds for the class reg of regular patterns, arguably the simplest class allowing an unbounded number of variables (note that patterns with a constant number of variables can trivially be matched in polynomial-time). In the same paper it is shown that if the notion of repetition is relaxed further, by considering a setting where the order in which the variables appear is no longer constrained at all (i.e., considering *abelian repetitions* instead of repetitions), then the matching problem is NP-complete. This holds even in the minimal case when the number of repetitions is restricted to two, and that the pattern which is repeated is regular.

7 From Locality to Graph Parameters

Following the ideas of Sect. 3 we explore further the connection between string and graph parameters. The main idea behind such a connection is to reach it by "flattening" a graph into a sequential form, or by "inflating" a string into a graph, so that algorithmic techniques available for each one of these become applicable for the other one as well. In this section, following [13], we are concerned with certain structural parameters (and the problems of computing them) for graphs and strings: the *cutwidth* $\mathsf{cw}(G)$ of a graph G (i.e., the maximum number of "stacked" edges if the vertices of a graph are drawn on a straight line), the *path-width* $\mathsf{pw}(G)$ of a graph G (i.e., the minimum width of a tree decomposition the tree structure of which is a path), and the *locality number* $\mathsf{loc}(\alpha)$ of a string α (explained in more detail in Sect. 4.2). By CUTWIDTH, PATHWIDTH and LOC, we denote the corresponding natural decision problems (i.e., decide whether a given graph has a pathwidth/cutwidth, or a given string has a locality number of at most k, for given k) and with the prefix MIN, we refer to the minimisation variants. The two former graph-parameters are very classical. Pathwidth is a simple (yet still hard to compute) subvariant of treewidth, which measures how much a graph resembles a path. The problems PATHWIDTH and MINPATHWIDTH are intensively studied (in terms of exact, parameterised and approximation algorithms) and have numerous applications (see the surveys and textbook [7,9,56]). CUTWIDTH is the best known example of a whole class of so-called *graph layout problems* (see the survey [20,73] for detailed information), which are studied since the 1970s and were originally motivated by questions of circuit layouts.

In comparison, the locality number seems a rather simple parameter directly defined on strings, but, however, it bounds the treewidth of the string (in the sense defined in Sect. 4.1), and the corresponding marking sequences can be seen as instructions for a dynamic programming algorithm for matching the pattern. In this way, it resembles a bit to the way the pathwidth and treewidth of graphs are used in algorithmic settings. Moreover, compared to other "tractability-parameters" of strings, it seems to cover best the treewidth of a string, but it also cannot be efficiently computed compared to the other simpler parameters.

Going more into detail, for LOC, exact exponential-time algorithms are not hard to be devised [15] but whether it can be solved in polynomial-time, or whether it is at least fixed-parameter tractable was left open in the paper where this measure was introduced. On the other hand, PATHWIDTH and CUTWIDTH are known NP-complete problems, fixed-parameter tractable with respect to parameter $\mathsf{pw}(G)$ or $\mathsf{cw}(G)$, respectively (even with "linear" fpt-algorithms with running-time $g(k)\,\mathrm{O}(n)$ [8,10,82]). With respect to approximation, their minimisation variants have received a lot of attention, mainly because they yield (like many other graph parameters) general algorithmic approaches for numerous graph problems, i.e., a good linear arrangement or path-decomposition can often be used to design a dynamic programming (or even divide and conquer) algorithm for other problems. The best known approximation algorithms for the problems MINPATHWIDTH and MINCUTWIDTH (with approximations ratios of $\mathrm{O}(\sqrt{\log(\mathsf{opt})}\log(n))$ and $\mathrm{O}(\log^2(n))$, respectively) follow from approximations of vertex separators (see [25]) and edge separators (see [60]), respectively.

There are two natural approaches to represent a word α over alphabet Σ as a graph $G_\alpha = (V_\alpha, E_\alpha)$: (1) $V_\alpha = \{1, 2, \ldots, |\alpha|\}$ and the edges are somehow used to represent the actual symbols (note that this is the case for the standard graph representation of patterns defined in Sect. 4.1), or (2) $V_\alpha = \Sigma$ and the edges are somehow used to represent the positions of α. A reduction of type (2) can be defined such that $|E_\alpha| = O(|\alpha|)$ and $\mathsf{cw}(G_\alpha) = 2\,\mathsf{loc}(\alpha)$, and a reduction of type (1) can be defined such that $|E_\alpha| = O(|\alpha|^2)$ and $\mathsf{loc}(\alpha) \leq \mathsf{pw}(G_\alpha) \leq 2\,\mathsf{loc}(\alpha)$. Since these reductions are parameterised reductions and also allow to transfer approximation results, one may conclude that LOC is fixed-parameter tractable if parameterised by $|\Sigma|$ (note that for parameter $|\Sigma|$ a simple, but less efficient fpt-algorithm is trivially obtained by simply enumerating all marking sequences) or by the locality number, and also that there is a polynomial-time $O(\sqrt{\log(\mathsf{opt})}\log(n))$-approximation algorithm for MINLOC.

In addition, one can represent an arbitrary multi-graph $G = (V, E)$ by a word α_G over alphabet V with $|\alpha_G| = |E|$ and $\mathsf{cw}(G) = \mathsf{loc}(\alpha)$. This describes a Turing-reduction from CUTWIDTH to LOC which also allows to transfer approximation results between the minimisation variants. As a result, LOC is NP-complete. Finally, by plugging together the reductions from MINCUTWIDTH to MINLOC and from MINLOC to MINPATHWIDTH, one obtains a reduction which transfers approximation results from MINPATHWIDTH to MINCUTWIDTH, which yields an $O(\sqrt{\log(\mathsf{opt})}\log(n))$-approximation algorithm for MINCUTWIDTH. This result from [13] improved, for the first time since 1999, the best approximation for CUTWIDTH from [60]. Interestingly, this improvement appeared as a side-product of relating string-parameters with graph-parameters.

Theorem 11 ([13])**.** *There is an* $O(\sqrt{\log(\mathsf{opt})}\log(h))$*-approximation algorithm (running in polynomial time) for* MINCUTWIDTH *on multigraphs with* h *edges. In particular, this yields an* $O(\sqrt{\log(\mathsf{opt})}\log(n))$*-approximation algorithm for* MINCUTWIDTH *for graphs.*

Moreover, this approach allows also for establishing a direct connection between cutwidth and pathwidth, which preserves the good algorithmic properties, and has not yet been reported in the literature so far. This is rather surprising, since CUTWIDTH and PATHWIDTH have been jointly investigated in the context of exact and approximation algorithms, especially in terms of balanced vertex and edge separators. We think that a reason for overlooking this connection might be that it is less obvious on the graph level and becomes more apparent if linked via the string parameter of locality, emphasising, as such, the value of such mixed approaches.

8 Extensions

8.1 Injectivity

In our setting, the substitutions that map variables to words are *not* required to be injective, i.e., different variables can be mapped to the same word. However, the requirement of injectivity is natural in some contexts. For example, in the pattern matching community, the first mentioning of pattern matching with variables concerns the case where variables have to be substituted by single symbols and in an injective way. More precisely, this *parameterised pattern matching* was introduced in [3] to formalise the problem of detecting code clones (i.e., we want to find code segments that are created by copying some code blocks and renaming program variables (this renaming will be injective, since otherwise the semantic of the code might change)). More generally speaking, the injectivity condition is appropriate whenever we know a priori that different variables should always refer to different words (e.g., when matching the pattern

$$x_1 \text{ name: } y \; ; \; \text{address: } z \; ; \; x_2 \text{ name: } y \; ; \; \text{address: } z \; x_3$$

in order to check whether there is a repetition of some name-address data tuple, then it is likely that we can assume injectivity).

Depending on the actual variant, the injectivity condition can make the matching problem harder or easier. In [26], it is shown that it is NP-hard to decide for a given word w and an integer k whether w can be factorised into at least k *pairwise different* factors. This immediately implies that the injectivity condition makes the matching problem NP-hard even for the "trivial" pattern class $\{x_1 x_2 \ldots x_n \mid n \geq 1\}$ (note that this is even a subset of the class reg of regular patterns). On the other hand, if we have an upper bound on $|\Sigma|$ and $\max\{|h(x)| \mid x \in X\}$ (recall that this case is still NP-hard even for bounds 2 and 1, respectively; see Theorem 1) then also the total number of possible substitution words is bounded; thus, the injectivity condition bounds the total number of variables and therefore the matching problem becomes tractable (see [29]). A similar observation can be made with respect to fixed-parameter tractability if we parameterise by $|\Sigma|$ and $\max\{|h(x)| \mid x \in X\}$ (see [31]).

8.2   Word Equations

A *word equation* is an equality $\alpha = \beta$, where α and β are patterns with variables, e.g., $\alpha = x_1 \mathsf{ab} x_2$ and $\beta = \mathsf{a} x_1 x_2 \mathsf{b}$ define the equation $x_1 \mathsf{ab} x_2 = \mathsf{a} x_1 x_2 \mathsf{b}$. A *solution* to an equation $\alpha = \beta$ is a substitution $h : (\mathsf{var}(\alpha) \cup \mathsf{var}(\beta)) \to \Sigma^*$ (in the sense defined in Sect. 2) that satisfies $h(\alpha) = h(\beta)$. For the example equation from above, the solutions are the substitutions h with $h(x_1) = \mathsf{a}^k$, for $k \geq 0$, and $h(x_2) = \mathsf{b}^\ell$, for $\ell \geq 0$.

The study of word equations (or the existential theory of equations over free monoids) is an important topic found at the intersection of algebra and computer science, with significant connections to, e.g., combinatorial group or monoid theory [21,65,66], unification [48,49,80]), and, more recently, data base theory [33,34].

The central computational problem for word equations is the satisfiability problem, i.e., the problem of deciding whether a given word equation $\alpha = \beta$ has a solution or not. In this regard, the matching problem for patterns with variables describes just the special case of the satisfiability problem for word equations where one side of the equation is a terminal word, e.g., $x_1 \mathsf{ab} x_1 x_2 \mathsf{c} x_2 x_1 = \mathsf{babbacab}$ is an instance of the matching problem already mentioned in the introduction, phrased as a word equation. Consequently, the satisfiability problem is intractable, even for very strongly restricted cases (see

Theorems 1 and 2). Also note that it has been shown in [22] that the solvability problem remains NP-hard if every variable has at most two occurrences in $\alpha\beta$ (called *quadratic* equations), but the proof of [22] actually talks about the matching problem for patterns with at most two occurrences per variable.

While the matching problem for patterns with variables is trivially decidable, it is not at all obvious how to solve the satisfiability problem for word equations. In fact, the question whether it is decidable was initially approached with the expectation that it will be answered in the negative. It was, however, shown to be decidable by Makanin [67] (see Chap. 12 of [64] for a survey). Later it was shown that the satisfiability problem is in PSPACE by Plandowski [74]; a new proof of this result was obtained in [51], based on a new simple technique called recompression. There are also cases when the satisfiability problem is tractable. For instance, word equations with only one variable can be solved in linear time in the size of the equation, see [50]; equations with two variables can be solved in time $O(|\alpha\beta|^5)$, see [19].

Given the fact that there are many structural restrictions of patterns that yield tractability (with respect to the matching problem, see Sect. 4), the question naturally arises how the complexity of the satisfiability problem for word equations (which are essentially equations of patterns) behaves if these restrictions are applied to word equations. More precisely, while each class of patterns with NP-hard matching problem yields a class of word equations with NP-hard satisfiability problem, the hardness of the satisfiability problem for equations with sides in some efficiently matchable class of patterns is no longer immediate. An investigation of that question was initiated in [68], where the following results were obtained. Firstly, the satisfiability problem for non-cross word equations (i.e., word equations for which both sides are non-cross) remains NP-hard. In particular, solving non-cross equations $\alpha = \beta$ where each variable occurs at most three times, at most twice in α and exactly once in β, is NP-hard (note that this constitutes the first NP-hardness result for word equations that is not a direct conclusion from a hardness result for the matching problem). Secondly, the satisfiability of one-repeated variable equations (i.e., at most one variable occurs more than once in $\alpha\beta$, but arbitrarily many other variables occur only once) having at least one non-repeated variable on each side, was shown to be trivially in P.

In [18], it is shown that it is (still) NP-hard to solve regular ordered word equations. More precisely, these are word equations where each side is a regular pattern and the order of the variables in both sides is the same (it is, however, possible that some variables only occur on one side of the equation), e.g., $x_1 a x_2 b a x_3 x_4 = b x_1 x_3 a a x_4$ is a regular ordered word equation. They are particular cases of both quadratic equations and non-cross equations, so the reductions showing the hardness of solving these more general equations do not carry over. In particular, note that the class of regular patterns is arguably the most simple class of patterns in terms of their matching complexity (see Sect. 5.1).

The respective hardness reduction relied on some deep word-combinatorics ideas. As a first step, a reachability problem for a certain type of (regulated) string rewriting systems was introduced, and showed it is NP-complete. This was achieved via a reduction from the problem 3-PARTITION [40], which is strongly NP-complete. Then it was shown that this reachability problem can be reduced to the satisfiability of regular-ordered word equations; in this reduction the applications of the rewriting rules of the system were encoded into the periods of the words assigned to the variables in a solution to the equation. The main technicality was to make sure to only use one occurrence of each variable per side, and moreover to even have the variables in the same order in both sides. This result exhibits the arguably structurally-simplest class of word equations for which the satisfiability problem is NP-hard.

The main open problem in the area of word equations remains, even for simple subclasses such as regular equations or quadratic equations, to show that the satisfiability problem of word equations of the respective types is in NP (note that this was already explicitly posed as an open question for the class of quadratic word equations in [22]).

9   Conclusions

In this work we tried to survey several results related to the problem of matching patterns with variables, that seem important to us. While this work is clearly not exhaustive, it is aimed to offer a basic understanding of the problems and state of the art in this area.

From an algorithmic point of view, the results we covered provide a wide variety of classes of patterns with variables, for which MATCH can be efficiently solved. Moreover, as explained in Sect. 4.3, it is usually easy to check whether a pattern belongs to one of these classes. So, putting it all together, one could use the following approach when trying to match a pattern, rather than just using an exponential time algorithm (based, e.g., on general SMT-solvers, or on the theory of string solving [6,83]). First, check if the pattern belongs to one of the classes for which efficient matching algorithms are known and, then, use this algorithm; only use a general algorithm when no customised one can be applied. Identifying more natural pattern classes for which MATCH can be solved efficiently appears, as such, as a rather useful task. Following the practically motivated challenges that arise from the area of string solving, one could also try to find efficient matching algorithms for various classes of patterns, enhanced with various constraints: regular constraints, length constraints, etc.

As an important part of this survey deals with polynomial time algorithms, it is natural to also ask whether they are optimal or not. This kind of questions are the focus of the area of fine-grained complexity (see, e.g., the survey [11] and the citations therein). It would be interesting to see, using tools from this area, whether one can show lower bounds for the MATCH problems for different classes of patterns.

In the light of the results from [13], it seems that exploring the connections between string parameters and parameters for other classes of objects could lead to some interesting results in both worlds. So, it also seems like an interesting challenge to explore what the structural parameters of strings that we explored here (and maybe some other new ones) mean when various other types of data are represented as strings, and what consequences can be derived from such a representation.

Finally, the area of word equations abounds with open problems. As mentioned, it is not even clear whether the satisfiability of regular or quadratic equations is in NP. So even if we restrict to equations with structurally simple left and right hand sides, the complexity of solving equations is not known. Such problems become even more involved when we consider equations with various types of constraints (e.g., length or regular). For instance, the decidability of general word equations with length constraints is a long standing open problem, but it is already an interesting open question for simpler cases (once again: regular or quadratic equations); see, e.g., [17,46,61], and the references therein. It seems interesting to us whether some of the ideas used in matching patterns can be transferred to solving (simplified) word equations, with or without constraints.

References

1. Amir, A., Nor, I.: Generalized function matching. J. Discrete Algorithms **5**, 514–523 (2007)
2. Angluin, D.: Finding patterns common to a set of strings. J. Comput. Syst. Sci. **21**, 46–62 (1980)
3. Baker, B.S.: Parameterized pattern matching: algorithms and applications. J. Comput. Syst. Sci. **52**, 28–42 (1996)
4. Bannai, H., I, T., Inenaga, S., Nakashima, Y., Takeda, M., Tsuruta, K.: The "runs" theorem. SIAM J. Comput. **46**(5), 1501–1514 (2017)
5. Barceló, P., Libkin, L., Lin, A.W., Wood, P.T.: Expressive languages for path queries over graph-structured data. ACM Trans. Database Syst. **37**, 31 (2012)
6. Barrett, C., et al.: CVC4. In: Gopalakrishnan, G., Qadeer, S. (eds.) CAV 2011. LNCS, vol. 6806, pp. 171–177. Springer, Heidelberg (2011). https://doi.org/10.1007/978-3-642-22110-1_14
7. Bodlaender, H.L.: A tourist guide through treewidth. Acta Cybern. **11**(1–2), 1–21 (1993)
8. Bodlaender, H.L.: A linear-time algorithm for finding tree-decompositions of small treewidth. SIAM J. Comput. **25**(5), 1305–1317 (1996). https://doi.org/10.1137/s0097539793251219
9. Bodlaender, H.L.: A partial k-arboretum of graphs with bounded treewidth. Theor. Comput. Sci. **209**(1–2), 1–45 (1998). https://doi.org/10.1016/S0304-3975(97)00228-4
10. Bodlaender, H.L.: Fixed-parameter tractability of treewidth and pathwidth. In: Bodlaender, H.L., Downey, R., Fomin, F.V., Marx, D. (eds.) The Multivariate Algorithmic Revolution and Beyond. LNCS, vol. 7370, pp. 196–227. Springer, Heidelberg (2012). https://doi.org/10.1007/978-3-642-30891-8_12
11. Bringmann, K.: Fine-grained complexity theory (tutorial). In: Niedermeier, R., Paul, C. (eds.) 36th International Symposium on Theoretical Aspects of Computer Science (STACS 2019). Leibniz International Proceedings in Informatics (LIPIcs), vol. 126, pp. 4:1–4:7. Schloss Dagstuhl-Leibniz-Zentrum fuer Informatik, Dagstuhl (2019). https://doi.org/10.4230/LIPIcs.STACS.2019.4. http://drops.dagstuhl.de/opus/volltexte/2019/10243

12. Câmpeanu, C., Salomaa, K., Yu, S.: A formal study of practical regular expressions. Int. J. Found. Comput. Sci. **14**, 1007–1018 (2003)
13. Casel, K., Day, J.D., Fleischmann, P., Kociumaka, T., Manea, F., Schmid, M.L.: Graph and string parameters: connections between pathwidth, cutwidth and the locality number. CoRR, to appear in Proceedings of the ICALP 2019, abs/1902.10983 (2019). http://arxiv.org/abs/1902.10983
14. Crochemore, M.: An optimal algorithm for computing the repetitions in a word. Inf. Process. Lett. **12**(5), 244–250 (1981)
15. Day, J.D., Fleischmann, P., Manea, F., Nowotka, D.: Local patterns. In: 37th IARCS Annual Conference on Foundations of Software Technology and Theoretical Computer Science, FSTTCS 2017, pp. 24:1–24:14 (2017)
16. Day, J.D., Fleischmann, P., Manea, F., Nowotka, D., Schmid, M.L.: On matching generalised repetitive patterns. In: Hoshi, M., Seki, S. (eds.) DLT 2018. LNCS, vol. 11088, pp. 269–281. Springer, Cham (2018). https://doi.org/10.1007/978-3-319-98654-8_22
17. Day, J.D., Ganesh, V., He, P., Manea, F., Nowotka, D.: The satisfiability of word equations: decidable and undecidable theories. In: Potapov, I., Reynier, P.-A. (eds.) RP 2018. LNCS, vol. 11123, pp. 15–29. Springer, Cham (2018). https://doi.org/10.1007/978-3-030-00250-3_2
18. Day, J.D., Manea, F., Nowotka, D.: The hardness of solving simple word equations. In: Proceedings of the MFCS 2017. LIPIcs, vol. 83, pp. 18:1–18:14 (2017)
19. Dąbrowski, R., Plandowski, W.: Solving two-variable word equations. In: Díaz, J., Karhumäki, J., Lepistö, A., Sannella, D. (eds.) ICALP 2004. LNCS, vol. 3142, pp. 408–419. Springer, Heidelberg (2004). https://doi.org/10.1007/978-3-540-27836-8_36
20. Díaz, J., Petit, J., Serna, M.: A survey of graph layout problems. ACM Comput. Surv. **34**(3), 313–356 (2002). https://doi.org/10.1145/568522.568523
21. Diekert, V., Jez, A., Kufleitner, M.: Solutions of word equations over partially commutative structures. In: Proceedings of the 43rd International Colloquium on Automata, Languages, and Programming, ICALP 2016. Leibniz International Proceedings in Informatics (LIPIcs), vol. 55, pp. 127:1–127:14 (2016)
22. Robson, J.M., Diekert, V.: On quadratic word equations. In: Meinel, C., Tison, S. (eds.) STACS 1999. LNCS, vol. 1563, pp. 217–226. Springer, Heidelberg (1999). https://doi.org/10.1007/3-540-49116-3_20
23. Downey, R.G., Fellows, M.R.: Fundamentals of Parameterized Complexity. TCS. Springer, London (2013). https://doi.org/10.1007/978-1-4471-5559-1
24. Erlebach, T., Rossmanith, P., Stadtherr, H., Steger, A., Zeugmann, T.: Learning one-variable pattern languages very efficiently on average, in parallel, and by asking queries. Theoret. Comput. Sci. **261**, 119–156 (2001)
25. Feige, U., HajiAghayi, M., Lee, J.R.: Improved approximation algorithms for minimum weight vertex separators. SIAM J. Comput. **38**(2), 629–657 (2008). https://doi.org/10.1137/05064299x
26. Fernau, H., Manea, F., Mercas, R., Schmid, M.L.: Pattern matching with variables: fast algorithms and new hardness results. In: 32nd International Symposium on Theoretical Aspects of Computer Science, STACS 2015, pp. 302–315 (2015)
27. Fernau, H., Manea, F., Mercas, R., Schmid, M.L.: Revisiting Shinohara's algorithm for computing descriptive patterns. Theoret. Comput. Sci. **733**, 44–54 (2018)
28. Fernau, H., Schmid, M.L.: Pattern matching with variables: a multivariate complexity analysis. In: Fischer, J., Sanders, P. (eds.) CPM 2013. LNCS, vol. 7922, pp. 83–94. Springer, Heidelberg (2013). https://doi.org/10.1007/978-3-642-38905-4_10

29. Fernau, H., Schmid, M.L.: Pattern matching with variables: a multivariate complexity analysis. Inf. Comput. **242**, 287–305 (2015)
30. Fernau, H., Schmid, M.L., Villanger, Y.: On the parameterised complexity of string morphism problems. In: Proceedings of the 33rd IARCS Annual Conference on Foundations of Software Technology and Theoretical Computer Science, FSTTCS. Leibniz International Proceedings in Informatics (LIPIcs), vol. 24, pp. 55–66 (2013)
31. Fernau, H., Schmid, M.L., Villanger, Y.: On the parameterised complexity of string morphism problems. Theory Comput. Syst. **59**(1), 24–51 (2016)
32. Flum, J., Grohe, M.: Parameterized Complexity Theory. TTCSAES. Springer, Heidelberg (2006). https://doi.org/10.1007/3-540-29953-X
33. Freydenberger, D.D.: A logic for document spanners. In: Proceedings of the 20th International Conference on Database Theory, ICDT 2017. Leibniz International Proceedings in Informatics (LIPIcs)
34. Freydenberger, D.D., Holldack, M.: Document spanners: from expressive power to decision problems. Theory Comput. Syst. **62**(4), 854–898 (2018)
35. Freydenberger, D.D.: Extended regular expressions: succinctness and decidability. Theory Comput. Syst. **53**, 159–193 (2013)
36. Freydenberger, D.D., Reidenbach, D.: Bad news on decision problems for patterns. Inf. Comput. **208**(1), 83–96 (2010)
37. Freydenberger, D.D., Reidenbach, D.: Existence and nonexistence of descriptive patterns. Theor. Comput. Sci. **411**(34–36), 3274–3286 (2010)
38. Freydenberger, D.D., Reidenbach, D.: Inferring descriptive generalisations of formal languages. J. Comput. Syst. Sci. **79**(5), 622–639 (2013)
39. Friedl, J.E.F.: Mastering Regular Expressions, 3rd edn. O'Reilly, Sebastopol (2006)
40. Garey, M.R., Johnson, D.S.: Computers and Intractability: A Guide to the Theory of NP-Completeness. W. H. Freeman & Co., New York (1979)
41. Gawrychowski, P., Manea, F., Nowotka, D.: Testing generalised freeness of words. In: STACS 2014. LIPIcs, vol. 25, pp. 337–349. Schloss Dagstuhl-Leibniz-Zentrum fuer Informatik (2014)
42. Gawrychowski, P., I, T., Inenaga, S., Köppl, D., Manea, F.: Tighter bounds and optimal algorithms for all maximal α-gapped repeats and palindromes - finding all maximal α-gapped repeats and palindromes in optimal worst case time on integer alphabets. Theory Comput. Syst. **62**(1), 162–191 (2018)
43. Gawrychowski, P., Manea, F., Mercas, R., Nowotka, D.: Hide and seek with repetitions. J. Comput. Syst. Sci. **101**, 42–67 (2019). https://doi.org/10.1016/j.jcss.2018.10.004
44. Gawrychowski, P., Manea, F., Mercas, R., Nowotka, D., Tiseanu, C.: Finding pseudo-repetitions. In: 30th International Symposium on Theoretical Aspects of Computer Science, STACS 2013, Kiel, Germany, 27 February-2 March 2013. LIPIcs, vol. 20, pp. 257–268 (2013)
45. Geilke, M., Zilles, S.: Learning relational patterns. In: Kivinen, J., Szepesvári, C., Ukkonen, E., Zeugmann, T. (eds.) ALT 2011. LNCS (LNAI), vol. 6925, pp. 84–98. Springer, Heidelberg (2011). https://doi.org/10.1007/978-3-642-24412-4_10
46. Halfon, S., Schnoebelen, P., Zetzsche, G.: Decidability, complexity, and expressiveness of first-order logic over the subword ordering. In: Proceedings of the 32nd Annual ACM/IEEE Symposium on Logic in Computer Science, LICS 2017, pp. 1–12. IEEE Computer Society (2017)
47. Ibarra, O.H., Pong, T.C., Sohn, S.M.: A note on parsing pattern languages. Pattern Recogn. Lett. **16**, 179–182 (1995)
48. Jaffar, J.: Minimal and complete word unification. J. ACM **37**(1), 47–85 (1990)

49. Jeż, A.: Context unification is in PSPACE. In: Esparza, J., Fraigniaud, P., Husfeldt, T., Koutsoupias, E. (eds.) ICALP 2014. LNCS, vol. 8573, pp. 244–255. Springer, Heidelberg (2014). https://doi.org/10.1007/978-3-662-43951-7_21
50. Jeż, A.: One-variable word equations in linear time. Algorithmica **74**, 1–48 (2016)
51. Jeż, A.: Recompression: a simple and powerful technique for word equations. J. ACM **63**, 4 (2016)
52. Jiang, T., Salomaa, A., Salomaa, K., Yu, S.: Decision problems for patterns. J. Comput. Syst. Sci. **50**(1), 53–63 (1995)
53. Karhumäki, J., Plandowski, W., Mignosi, F.: The expressibility of languages and relations by word equations. J. ACM **47**, 483–505 (2000)
54. Kärkkäinen, J., Sanders, P., Burkhardt, S.: Linear work suffix array construction. J. ACM **53**, 918–936 (2006)
55. Kearns, M.J., Pitt, L.: A polynomial-time algorithm for learning k-variable pattern languages from examples. In: Proceedings of the Second Annual Workshop on Computational Learning Theory, COLT 1989, Santa Cruz, CA, USA, 31 July–2 August 1989, pp. 57–71 (1989)
56. Kloks, T. (ed.): Treewidth, Computations and Approximations. LNCS, vol. 842. Springer, Heidelberg (1994). https://doi.org/10.1007/BFb0045375
57. Kolpakov, R., Kucherov, G.: Searching for gapped palindromes. Theor. Comput. Sci. **410**(51), 5365–5373 (2009)
58. Kolpakov, R., Podolskiy, M., Posypkin, M., Khrapov, N.: Searching of gapped repeats and subrepetitions in a word. In: Kulikov, A.S., Kuznetsov, S.O., Pevzner, P. (eds.) CPM 2014. LNCS, vol. 8486, pp. 212–221. Springer, Cham (2014). https://doi.org/10.1007/978-3-319-07566-2_22
59. Kosolobov, D., Manea, F., Nowotka, D.: Detecting one-variable patterns. In: Proceedings of the 24th International Symposium on String Processing and Information Retrieval , SPIRE 2017, Palermo, Italy, 26–29 September 2017, pp. 254–270 (2017)
60. Leighton, T., Rao, S.: Multicommodity max-flow min-cut theorems and their use in designing approximation algorithms. J. ACM **46**(6), 787–832 (1999). https://doi.org/10.1145/331524.331526
61. Lin, A.W., Majumdar, R.: Quadratic word equations with length constraints, counter systems, and presburger arithmetic with divisibility. In: Lahiri, S.K., Wang, C. (eds.) ATVA 2018. LNCS, vol. 11138, pp. 352–369. Springer, Cham (2018). https://doi.org/10.1007/978-3-030-01090-4_21
62. Lothaire, M.: Combinatorics on Words. Cambridge University Press, Cambridge (1997)
63. Lothaire, M.: Algebraic Combinatorics on Words, chap. 3. Cambridge University Press, Cambridge, New York (2002)
64. Lothaire, M.: Algebraic Combinatorics on Words. Cambridge University Press, Cambridge, New York (2002)
65. Lyndon, R.C.: Equations in free groups. Trans. Am. Math. Soc. **96**, 445–457 (1960)
66. Lyndon, R.C., Schupp, P.E.: Combinatorial Group Theory. Springer, Heidelberg (1977)
67. Makanin, G.S.: The problem of solvability of equations in a free semigroup. Matematicheskii Sbornik **103**, 147–236 (1977)
68. Manea, F., Nowotka, D., Schmid, M.L.: On the solvability problem for restricted classes of word equations. In: Brlek, S., Reutenauer, C. (eds.) DLT 2016. LNCS, vol. 9840, pp. 306–318. Springer, Heidelberg (2016). https://doi.org/10.1007/978-3-662-53132-7_25

69. Mateescu, A., Salomaa, A.: Finite degrees of ambiguity in pattern languages. RAIRO Inf. Théor. Appl. **28**, 233–253 (1994)

70. Mateescu, A., Salomaa, A.: Aspects of classical language theory. In: Rozenberg, G., Salomaa, A. (eds.) Handbook of Formal Languages, pp. 175–251. Springer, Heidelberg (1997). https://doi.org/10.1007/978-3-642-59136-5_4

71. Ng, Y.K., Shinohara, T.: Developments from enquiries into the learnability of the pattern languages from positive data. Theoret. Comput. Sci. **397**, 150–165 (2008)

72. Ordyniak, S., Popa, A.: A parameterized study of maximum generalized pattern matching problems. Algorithmica **75**, 1–26 (2016)

73. Petit, J.: Addenda to the survey of layout problems. Bull. EATCS **105**, 177–201 (2011). http://eatcs.org/beatcs/index.php/beatcs/article/view/98

74. Plandowski, W.: An efficient algorithm for solving word equations. In: Proceedings of the 38th Annual ACM Symposium on Theory of Computing, STOC 2006, pp. 467–476 (2006)

75. Reidenbach, D.: A non-learnable class of e-pattern languages. Theor. Comput. Sci. **350**(1), 91–102 (2006)

76. Reidenbach, D.: An examination of ohlebusch and ukkonen's conjecture on the equivalence problem for e-pattern languages. J. Automata Lang. Comb. **12**(3), 407–426 (2007)

77. Reidenbach, D.: Discontinuities in pattern inference. Theor. Comput. Sci. **397**(1–3), 166–193 (2008)

78. Reidenbach, D., Schmid, M.L.: Patterns with bounded treewidth. Inf. Comput. **239**, 87–99 (2014)

79. Schmid, M.L.: A note on the complexity of matching patterns with variables. Inf. Process. Lett. **113**(19–21), 729–733 (2013)

80. Schulz, K.U.: Word unification and transformation of generalized equations. J. Autom. Reason. **11**, 149–184 (1995)

81. Shinohara, T.: Polynomial time inference of pattern languages and its application. In: Proceedings of 7th IBM Symposium on Mathematical Foundations of Computer Science, MFCS, pp. 191–209 (1982)

82. Thilikos, D.M., Serna, M.J., Bodlaender, H.L.: Cutwidth I: a linear time fixed parameter algorithm. J. Algorithms **56**(1), 1–24 (2005). https://doi.org/10.1016/j.jalgor.2004.12.001

83. Zheng, Y., Ganesh, V., Subramanian, S., Tripp, O., Berzish, M., Dolby, J., Zhang, X.: Z3str2: an efficient solver for strings, regular expressions, and length constraints. Formal Methods Syst. Des. **50**(2–3), 249–288 (2017)

Abelian Properties of Words

Svetlana Puzynina$^{(\boxtimes)}$

Saint Petersburg State University,
7–9 Universitetskaya emb., 199034 Saint Petersburg, Russia
s.puzynina@gmail.com

Abstract. Abelian properties of words is a widely studied field in combinatorics on words. Two finite words are abelian equivalent if for each letter they contain the same numbers of occurrences of this letter. In this paper, we give a short overview of some directions of research on abelian properties of words, and discuss in more detail two new problems: small abelian complexity of two-dimensional words, and abelian subshifts.

Keywords: Abelian properties of infinite words · Words complexity · Subshifts

1 Introduction

Abelian approach is a new and very effective tool in the theory of words. Two finite words are called *abelian equivalent* if they can be obtained from each other by permutations of letters. For example, the words 001012 and 100021 are abelian equivalent. The study of abelian properties of words dates back to Erdös's question whether there is an infinite word avoiding abelian squares [24]. Abelian powers and their avoidability in infinite words is a natural generalization of analogous questions for ordinary powers. The answer to Erdős's question has been given by Evdokimov, who provided a construction of an abelian square-free word [25], and later by Keränen over the minimal alphabet [37]. Nowadays in combinatorics on words, the study of abelian properties of words is an extremely popular subject. Various abelian properties of words, including abelian complexity, avoidance, powers, periods, have been widely studied recently, see, e.g. [4,11,15,37,41,47,49–55]. In this paper, we give a short overview of some directions of research on abelian properties of words, and discuss in more detail two new problems: small abelian complexity of two-dimensional words, and abelian subshifts.

Nivat's conjecture, introduced at ICALP 1997, is a two-dimensional analog of a classical one-dimensional theorem of Morse and Hedlund. In their seminal paper *Symbolic Dynamics* from 1938 [42], they introduced the notion of complexity of an infinite word as a function $p(n)$ which counts, for each integer n, the number of its distinct factors (i.e., blocks of consecutive letters) of length n. The

Partially supported by Russian Foundation of Basic Research (grant 18-31-00118).

R. Mercaş and D. Reidenbach (Eds.): WORDS 2019, LNCS 11682, pp. 28–45, 2019.
https://doi.org/10.1007/978-3-030-28796-2_2

factor complexity provides a useful measure of randomness of an infinite word and more generally of the subshift it generates. They further proved that an infinite word is periodic if and only if its complexity function satisfies $p(n) \leq n$ for some n. The latter result is referred to as Morse-Hedlund theorem.

In two dimensions, the complexity is a function $p_w(m, n)$ counting for each $m, n \in \mathbb{N}$ the number of distinct rectangular $m \times n$ blocks. Nivat's conjecture states that if the complexity p_w of a two-dimensional word w satisfies $p_w(n, m) \leq nm$, then w has a periodicity vector [44]. Weak forms of the conjecture for $p_w(n, m) \leq \frac{nm}{c}$ for different constants $c > 0$ were proved by Epifanio, Koskas, Mignosi [23], by Quas, Zamboni [48], by Cyr, Kra [16], and in asymptotic form by Kari, Szabados [36]; see also [21]. Minimal complexity of aperiodic two-dimensional words was explored by Berthé and Vuillon [7]. Remarkably, the conjecture does not hold in higher dimensions. In this paper, we are interested in abelian version of Nivat's conjecture in the case of recurrent two-dimensional words, i.e., words where each fragment occurs infinitely many times, and in the structure of recurrent two-dimensional words of small abelian complexity.

Another topic we discuss in the paper is abelian subshifts. A *subshift* Ω_x generated by an infinite word x can be defined as the set of infinite words y such that each factor of y is a factor of x. We consider an abelian version of the notion of subshift: We define the *abelian subshift* $\mathcal{A}_x$ of an infinite word x as the set of infinite words y such that, for each factor u of y, there exists a factor v of x with $u \sim_{ab} v$. Clearly, $\Omega_x \subseteq \mathcal{A}_x$ for any word x. We are interested in when the equality holds, and we study the general structure of abelian subshifts. As an example, it is not hard to see that for Sturmian words $\Omega_x = \mathcal{A}_x$ (so, the abelian subshift is small compared to Ω_x). Moreover, the property $\Omega_x = \mathcal{A}_x$ characterizes Sturmian words among uniformly recurrent binary words (see Theorem 11). On the other hand, it is easy to see that the abelian subshift of the Thue-Morse word TM is $\{\varepsilon, 0, 1\} \cdot \{01, 10\}^{\mathbb{N}}$ (see Example 4). So, contrary to Sturmian words, the abelian subshift of the Thue-Morse is huge compared to Ω_{TM}: basically, it is a morphic image of the full binary shift. In general, the abelian subshift of an infinite word might have a pretty complicated structure. T. Hejda, W. Steiner, and L.Q. Zamboni studied the abelian subshift of the Tribonacci word TR. They announced that $\mathcal{A}_{TR} \setminus \Omega_{TR} \neq \emptyset$ but that Ω_{TR} is the only minimal subshift contained in $\mathcal{A}_{TR}$ [31,59].

The paper is organized as follows. In Sect. 2, we give some overview of abelian properties of words which are well studied. In particular, we discuss abelian complexity and avoidance. In Sect. 3, we introduce new results concerning small abelian complexity of two-dimensional words. In particular, we show that, contrary to the one-dimensional case, there exist aperiodic words with abelian complexity 1 for some block sizes. Further we show that for recurrent words abelian complexity cannot be bounded by 2 and moreover the abelian complexity at least 3 must be achieved for infinitely many block sizes. In Sect. 4, we discuss a recent notion of abelian subshift and study its properties. For the abelian subshifts of binary words, we prove that if an aperiodic binary uniformly recurrent word is not Sturmian, then its abelian subshift contains infinitely many minimal subshifts. We further study the abelian subshifts of some generalisations of Sturmian words: We show that abelian subshift of an aperiodic recurrent balanced word is a finite union of minimal subshifts. We also characterize abelian subshifts of aperiodic words having factor complexity $n + C$ for each n. Depending on the word, their abelian subshifts contain either finitely many or uncountably many distinct minimal subshifts.

2   Overview

In this section we first give some definitions and notation, and then give an overview of some results on abelian properties of words; in particular, we discuss abelian complexity and abelian avoidance.

2.1   Definitions and Notation

Let Σ be a finite non-empty set called an alphabet. A *finite word* over the alphabet Σ is any finite sequence of its symbols. By an *infinite word* w we mean an element $w = w_0 w_1 w_2 \cdots \in \Sigma^{\mathbb{N}}$. Although in the one-dimensional case we usually consider one-way infinite words, most of the notions and results we discuss extend to biinfintie words, i.e., elements from $\Sigma^{\mathbb{Z}}$.

The set of finite words over an alphabet Σ is denoted by Σ^* and the set of non-empty words is denoted by Σ^+. We let $|w|$ denote the length of a word $w \in \Sigma^*$. The empty word is denoted by ε and by convention we set $|\varepsilon| = 0$. The set of words of length n over the alphabet Σ is denoted by Σ^n. A *factor* of an finite or infinite word is any finite sequence of its consecutive letters; we let $\mathcal{L}(x)$ denote the set of factors of x.

An infinite word w is called *recurrent* if each its factor occurs in it infinitely many times. An infinite word w is called *uniformly recurrent* if for each integer n there exists an integer N such that each factor of w of length N contains all factors of w of length n. In other words, a word is uniformly recurrent if each its factor occurs in it with bounded gap. An infinite word w is called *eventually periodic* if there exist integers N and T such that $w_{n+T} = w_n$ for each $n \geq N$. An infinite word w is called *purely periodic* if $w_{n+T} = w_n$ for each $n \geq 0$. A word is *aperiodic* if it is not ultimately periodic.

Given a finite word u over an alphabet Σ and a letter $a \in \Sigma$, we let $|u|_a$ denote the number of occurrences of a in u. Two finite words u and v are *abelian equivalent*, denoted by $u \sim_{ab} v$, if $|u|_a = |v|_a$ for each letter a. In other words, u and v are permutations of one another. It is straightforward that abelian equivalence is indeed an equivalence relation on the set of finite words.

Parikh vector of a finite word v over an alphabet $\Sigma = \{a_1, \ldots, a_k\}$ is defined as $PV(v) = (|v|_{a_1}, \ldots, |v|_{a_k})$. Clearly, two words are abelian equivalent if their Parikh vectors coincide. The set of Parikh vectors of factors of length n of an infinite word is then denoted by $PV(n)$. A *frequency* of a in a finite word v is $\text{freq}_v(a) = \frac{|v|_a}{|v|}$. A (bi-)infinite word w has *uniform frequency* of a letter a if the ratio $\frac{|w_k \cdots w_{k+n-1}|}{n}$ has a limit $\text{freq}_w(a)$ when $n \to \infty$, uniformly in k. A (finite or infinite) word w is *C-balanced* for an integer $C > 0$, if for every letter a and any two factors u, v of w of the same length, one has $||u|_a - |v|_a| \leq C$. For $C = 2$, the constant is usually omitted and the word is called *balanced*.

2.2   Abelian Complexity

For each infinite word $w = w_0 w_1 w_2 \cdots \in \Sigma^{\mathbb{N}}$, the complexity or *factor complexity* $p_w(n)$ is a function counting the number of distinct blocks $w_i w_{i+1} \cdots w_{i+n-1} \in \Sigma^n$ of length n occurring in w. A celebrated theorem of Morse and Hedlund gives a link between periodicity and complexity:

Theorem 1 (Morse and Hedlund [42]). *Let w be a one-dimensional word. If there exists n such that $p_w(n) \leq n$, then w is eventually periodic.*

Clearly, periodic words have bounded complexity. Words satisfying $p_x(n) = n + 1$ for each $n \geq 0$ are called Sturmian words, and hence are regarded as the simplest aperiodic words. Sturmian words admit various types of characterizations of geometric and combinatorial nature, e.g., they can be defined via balance, morphisms, rotations, etc. (see [43] and Chap. 2 in [40]).

In a similar way the *abelian complexity* $a_w(n)$ of an infinite word w is defined as a function which counts the number of distinct abelian classes of factors of length n occurring in w. It easy to see that, similarly to factor complexity, the abelian complexity also gives a characterization of periodicity:

Lemma 1 [14]. *A infinite word w is purely periodic if and only if there exists n such that $a_w(n) = 1$.*

Clearly, if an infinite word w is ultimately periodic, then its abelian complexity is bounded. On the other hand, there exist aperiodic words of bounded abelian complexity. For example, Sturmian words are aperiodic and have abelian complexity 2 for each n, and moreover this is a characterization:

Theorem 2 [14]. *Let x be an aperiodic binary infinite word. Then x is a Sturmian word if and only if $a_x(n) = 2$ for every $n \geq 1$.*

Therefore, Sturmian words are aperiodic words of minimal abelian complexity as well. The maximal abelian complexity is realized, for example, by words with maximal factor complexity. We have:

Proposition 1. *For all infinite words x over Σ, $|\Sigma| = k$ and for all $n \geq 0$,*

$$1 \leq a_x(n) \leq \binom{n+k-1}{k-1}.$$

The following proposition relating C-balance to bounded abelian complexity is straightforward:

Proposition 2. *Let x be an infinite word. Then the abelian complexity of x is bounded if and only if x is C-balanced for some $C > 0$.*

Example 1. The Thue-Morse word $\boldsymbol{TM} = 0110100110010110\ldots$ can be defined as the fixed point of the substitution $\mu : 0 \mapsto 01, 1 \mapsto 10$. It is easy to see that its abelian complexity satisfies:

$$a_{TM}(n) = \begin{cases} 2, & \text{if } n \text{ is odd}; \\ 3, & \text{if } n \text{ is even}. \end{cases}$$

Example 2. Another example of an aperiodic word of bounded complexity is the Tribonacci word $\boldsymbol{TR}$, which can be defined as the fixed point of the morphism $0 \mapsto 01, 1 \mapsto 02, 2 \mapsto 0$. This is also the simplest example of Arnoux-Rauzy words. For every $n \geq 1$, the abelian complexity of the Tribonacci word satisfies $a_{TR}(n) \in \{3, 4, 5, 6, 7\}$. Moreover, each of these five values is assumed [53].

In [52] it has been proved that there are recurrent words of abelian complexity 3, but there are no recurrent words of abelian complexity 4 [15]. However, there are recurrent words with ultimately constant complexity c for every c [55]. We will now discuss the (abelian) complexity of morphic words. A mapping $\varphi : \Sigma^* \mapsto \Delta^*$ is called a *morphism* if $\varphi(uv) = \varphi(u)\varphi(v)$ for all $u, v \in \Sigma^*$. The notion of a morphism extends naturally to infinite words. An infinite word is a fixed point of a morphism $\varphi : \Sigma^* \mapsto \Sigma^*$ if $\varphi(w) = w$. A well known classification of complexities of fixed points of morphisms finished by Pansiot [45] says that there are 5 classes of possible complexity growth: $\Theta(1)$, $\Theta(n)$, $\Theta(n \log n)$, $\Theta(n \log \log n)$ and $\Theta(n^2)$.

The abelian complexity of purely morphic words is more complicated and is completely classified only for fixed points of binary morphisms (more precisely, complexity limsup has been classified). Note first that the balance function of primitive morphisms has been characterized by Adamczewski [1]. As an immediate corollary of this characterization, we get a classification of complexities of fixed points of primitive binary morphisms. If we write $f(x) = \Omega'(g(x))$ if $\limsup_{x \to \infty} f(x)/g(x) > 0$, then the complexity of a pure morphic word is either $\Theta(1)$, or $O \cap \Omega'(\log n)$, or $O \cap \Omega'(n \log_{\theta_1} \theta_2)$, where θ_1 and θ_2 are the first and second most significant eigenvalues of φ. A classification of abelian complexities of non-primitive binary morphisms is due to Blanchet-Sadri, Fox and Rampersad [8] completed by Whiteland [58]: it can be either $\Theta(1)$, or $\Theta(n)$, or $\Theta(n/\log n)$, or $\Theta(n^{\log_k l})$ with $1 < k < l$, or it can fluctuate between $\Theta(1)$ and $\Theta(\log(n))$.

2.3   Abelian Avoidance

Avoidability of powers and patterns is a well studied area in combinatorics on words. In this subsection, we provide some results on avoidability of abelian powers. An *abelian square* is a nonempty word of the form uv, where u and v are abelian equivalent. More generally, an *abelian cube* (resp. an *abelian kth power*) is a word of the form $u_1 u_2 u_3$ (resp. $u_1 u_2 \cdots u_k$), where $u_i \sim_{ab} u_j$ for all $i, j \in \{1, \ldots, k\}$. The study of abelian avoidance started with a question of Erdős who asked whether it is possible to construct an infinite word containing no abelian squares as factors [24].

A k-power is a particular case of an abelian k-power. So, unavoidability of k-powers implies unavoidability of abelian k-powers, but not vise versa. This implies, for example, that abelian squares are unavoidable for binary alphabet. The following theorem gives the minimal sizes of the alphabet for avoiding abelian powers:

Theorem 3 [18,37].

1. *There exists an infinite word over an alphabet of size 4 with no abelian squares.*
2. *There exists an infinite ternary word over with no abelian cube factor.*
3. *There exists an infinite binary word with no abelian 4th power.*
   The sizes of the alphabets are optimal.

The first part of the theorem has been proved by Keränen in 1992 [37], the other two parts by Dekking in 1979 [18].

The summary of results on avoidability of (abelian) k-powers is provided in Table 1.

Table 1. Minimal sizes of the alphabets over which the corresponding powers are avoidable.

	Usual	Abelian
Squares	3	4
Cubes	2	3
4-powers	2	2

Abelian squares are unavoidable over a binary alphabet; however, one can wonder whether it is possible to construct an infinite binary word containing only a finite number of abelian squares (as it is the case for ordinary squares, where there exists an infinite binary word containing only 00, 11 and 0101 as squares). The answer to this question is known, and it is negative. However, in the ternary case, it is possible to construct infinite words containing only a finite number of abelian squares:

Theorem 4 [22,51]. *The following holds true:*

1. *Every infinite binary word contains arbitrarily long abelian squares.*
2. *There exist infinite ternary words with no abelian square of period larger or equal to 6.*

We refer to [22] for part 1 of the theorem, and to [51] for part 2 of the theorem.

The following conjecture is believed to be true, but is still unproved:

Conjecture 1 (Mäkela, [38]). There exists an infinite ternary word whose abelian squares are only 00, 11, 22.

Another conjecture of Mäkela stated that there exists an infinite binary word containing only 000 and 111 as abelian cubes. The conjecture has been shown to be false in [50], but the following modification of Mäkela's question is still open:

Problem 1. Is it possible to construct an infinite binary word containing only a finite number of abelian cubes?

2.4   Other Abelian Properties

In this subsection we discuss some other results on abelian properties of infinite words, such as abelian periods, returns, borders and abelian richness, without going into detail.

Constantinescu and Ilie introduced in 2006 the following generalization to the abelian case of the notion of a classical period. We write $PV(u) \subset PV(v)$ if $PV(u)$ is component-wise smaller than or equal to $PV(v)$. A word w has an *abelian period* p with preperiod h if $w = u_0 u_1 \cdots u_{m-1} u_m$ such that $PV(u_0) \subset PV(u_1) = \cdots = PV(u_{m-1}) \supset PV(u_m)$ and $|u_0| = h$, $|u_1| = p$. The words u_0 and u_l are called *head* and *tail*, respectively.

Recall that the classical theorem of Fine & Wilf states that if a word w has two (classical) periods p and q and length $|w| \geq p + q - \gcd(p, q)$, then w has also period $\gcd(p, q)$ [29]. Moreover, this value is optimal, in the sense that for any p and q it is possible to construct a word with periods p and q and length $|w| = p + q - \gcd(p, q) - 1$ such that $\gcd(p, q)$ is not a period of w. In particular, if a word w has two coprime periods p and q and length $|w| \geq p + q - 1$, then w is a power of a single letter. The following gives an abelian analog of Fine and Wilf's theorem:

Theorem 5 [13]. *If a word w has coprime abelian periods p and q and length $|w| \geq 2pq - 1$, then w is a power of a single letter.*

The latter result has been recently generalized to the case when the abelian periods p and q are not coprime:

Theorem 6 [56]. *If a word w has abelian periods $p = p'd$ and $q = q'd$ and length $|w| \geq 2p'q'd - 1$ for integers d', p', q', then the number of letters occurring in w is at most d.*

Moreover, if the difference $||v_0| - |u_0||$ of the lengths of the heads of the two periods p and q is not a multiple of d, then the previous bound can be reduced to $2p'q'd - 2$.

The following theorem gives a relation between bounded abelian complexity and avoiding abelian powers:

Theorem 7 [52]. *An infinite word with bounded abelian complexity contains abelian k-powers for any k.*

In the case of Sturmian words something stronger holds: For every Sturmian word w and positive integer k, each sufficiently long factor of w begins in an Abelian k-power [52]. However, it is possible to construct a uniformly recurrent binary word with bounded abelian complexity such that none of its prefixes is an abelian square [10].

Another concept which gives a productive abelian version is return words. Return words constitute a powerful tool for studying various problems in combinatorics on words, symbolic dynamical systems and number theory. Given a factor v of an infinite word w, a return word to v (in w) is a factor u of w such that uv is a factor of w beginning and ending in v and having no other (internal) occurrence of v. In other words, the set of all return words to v is the set of all distinct words beginning with an occurrence of v and ending just before the next occurrence of v. The notion of return words can be regarded as a discrete analogue of the first return map in dynamical systems. Return words are used, for example, to characterize Sturmian words: A binary recurrent infinite word w is Sturmian if and only if each factor u of w has two returns in w [57].

The notion of a return word is naturally generalized to the abelian setting: Given a factor u of an infinite word x, let $n_1 < n_2 < n_3 < \ldots$ be all the integers n_i such that $w_{n_i} \cdots w_{n_{i+|u|-1}}$ is abelian equivalent u. Then we call each $w_{n_i} \cdots w_{n_{i+|u|-1}}$ a *semi-abelian return* to u. By an *abelian return* to u we mean an abelian class of $w_{n_i} \cdots w_{n_{i+1}-1}$. We note that in both cases these definitions depend only on the abelian class of u. Each of these notions of abelian returns gives rise to a characterization of Sturmian words. Moreover, the characterisations are the same in terms of abelian and semi-abelian returns:

Theorem 8 [47]. *A binary recurrent infinite word x is Sturmian if and only if each factor u of x has two or three (semi-)abelian returns in x.*

For further results on abelian returns we refer to [41,54].

We finish this subsection by a problem of counting abelian factors in a word. Counting a maximal number of factors of different type (palindrome, square, run) is a rather popular subject in combinatorics on words. Clearly, a maximal number of factors a word of length n can contain is $\Theta(n^2)$. It is easy to see that a word $a^{n/2}b^{n/2}$ contains $\Theta(n^2)$ distinct abelian factors. The question is, can we construct an infinite word such that each its factor contains $\Theta(n^2)$ distinct abelian squares? The answer is negative for binary alphabets [3] and is open for nonbinary alphabets:

Problem 2. Does there exist a constant $C > 0$ and an infinite word w, such that for each n an each factor of w has at least Cn^2 distinct abelian factors?

A related problem concerning finite words which are rich in squares has been considered in [27].

There are many other abelian aspects of words which are not covered in this survey. For example, abelian borders [11,12], abelian critical factorization theorem [4], abelian properties of Sturmian words [28] to name just a few.

3   Small Abelian Complexity of Two-Dimensional Words

In Subset. 2.2 we discussed connections between periodicity and complexity in the one-dimensional case. Due to a celebrated result of Morse and Hedlund, Theorem 1, if a complexity of an infinite word w satisfies $p_w(n) \leq n$ for some n, then the word is ultimately periodic. In two dimensions, a similar assertion is known as Nivat's conjecture:

Conjecture 2 (Nivat's conjecture [44]**).** Let w be a two-dimensional word, n, m two numbers such that the complexity p_w satisfies $p_w(n, m) \leq nm$. Then w has a periodicity vector.

The problem considered in this section concerns abelian complexity of multidimentional words and in particular an abelian analog of Nivat's conjecture under recurrence condition. A closely related paper studies balance in two-dimensional words [6]. It is worth noting that, contrary to one-dimensional case, in two dimensions balance is not equivalent to bounded abelian complexity.

Without the condition on recurrence, the problem in the abelian setting is trivial. Indeed, there exist aperiodic two-dimensional words of abelian complexity 2, for example a binary word that has only one occurrence of 1. Clearly, the word from the example is not recurrent. The main result of this section is Theorem 10, stating that for a two-dimensional aperiodic recurrent word w there exist integers m and n such that $a_w(m, n) \geq 3$; moreover there are infinitely many such pairs (m, n). The results of this sections are based on [46].

3.1   Two-Dimensional Words

A two-dimensional word w is called *periodic* if there exist integers (m, n) such that $w(x, y) = w(x + m, y + n)$ for each pair (x, y) of integers. A two-dimensional word w is called *fully periodic* if there exist two noncollinear integer vectors (m_1, n_1), (m_2, n_2) such that $w(x, y) = w(x + m_i, y + n_i)$ for each pair (x, y) of integers and $i = 1, 2$. By an $m \times n$ factor of a two-dimensional word we mean a rectangular block of the form

$$
\begin{array}{ccc}
w_{x,y+n-1} & \cdots & w_{x+m-1,y+n-1} \\
\vdots & & \vdots \\
w_{x,y} & \cdots & w_{x+m-1,y}
\end{array}
$$

for some integers x, y.

Similarly to the one-dimensional case, the *abelian complexity* $a_w(m, n)$ of a two-dimensional word w is defined as the number of abelian classes of $m \times n$ blocks. A two-dimensional word is *balanced* if for each pair (m, n) of integers, each letter a and any two $m \times n$-factors u and v of the word it holds $||u|_a - |v|_a| \leq 1$. Sometimes C-balance is defined if the equality $||u|_a - |v|_a| \leq C$ holds. We remark that although in the one-dimensional case bounded abelian complexity is equivalent to C-balance for some constant C, in two dimensions it is not true:

Consider, for example, a word with alternating horizontal lines of 1's and of 0's. This word has bounded abelian complexity but is not balanced.

A two-dimensional word w is called *recurrent* if each its factor occurs in it infinitely many times. An infinite word w is called *uniformly recurrent* if for each integer n there exists an integer N such that each square $N \times N$ factor of w contains all square factors of w of size $n \times n$.

We will need a few technical definitions. We call an (m, n)-*lattice nested in* $(x, y) \in \mathbb{Z}^2$ the set $\{(x, y) + (mi, nj) | i, j \in \mathbb{Z}\}$. We remark that we can assume that $0 \leq x < m, 0 \leq y < n$, and that this way $\mathbb{Z}^2$ is split into mn many (m, n)-lattices.

3.2 Small Abelian Complexity of Recurrent Two-Dimensional Words

In this section we study small abelian complexity of two-dimensional words. First we consider the case of abelian complexity equal to 1 for some blocks. In particular, we study the structure of such words and show that, contrary to the one-dimensional case, there exist aperiodic words with abelian complexity 1 for some block sizes. Then we show that for recurrent two-dimensional words abelian complexity cannot be bounded by 2, moreover, the value at least 3 must be achieved for an infinite number of block sizes. We also show that there exist aperiodic recurrent two-dimensional words with abelian complexity bounded by 3.

The following theorem describes the structure of two-dimensional words having abelian complexity 1 for some block size:

Theorem 9. *Let w be a two-dimensional word, and let $a_w(m, n) = 1$ for some integers m and n. Then in each (m, n)-lattice w is either $(0, n)$-periodic or $(m, 0)$-periodic, i.e., we have either $w(x, y) = w(x + m, y)$ for each point (x, y) from the lattice, or $w(x, y) = w(x, y + n)$ for each point (x, y) from the lattice.*

The following proposition shows that, contrary to one-dimensional case (Lemma 1), there exist aperiodic two-dimensional words that have abelian complexity 1 for some values m and n:

Proposition 3. *There exists an aperiodic two-dimensional word w with abelian complexity $a_w(m, n) = 1$ for infinitely many pairs (m, n).*

We construct such a word of blocks 2×2 each of them being either $\begin{smallmatrix} a & b \\ b & a \end{smallmatrix}$ or $\begin{smallmatrix} b & a \\ a & b \end{smallmatrix}$, in a way that in $(4, 2)$-lattices nested in $(x, y) \in \{0, 1\}^2$ the word is $(4, 0)$-periodic, and in $(4, 2)$-lattices nested in $(x, y) \in \{2, 3\} \times \{0, 1\}$ it is $(0, 2)$-periodic. Now we fill in the blocks $\{0, 1\} \times \{2i, 2i+1\}, i \in \mathbb{Z}$, choosing one of the two blocks for any i in arbitrary way, and continue by $(4, 0)$-periodicity to fill into the blocks $\{4j, 4j + 1\} \times \{2i, 2i + 1\}, i, j \in \mathbb{Z}$. In the similar way we fill in the remaining blocks: we can fill in the blocks $\{2j+2, 2j+3\} \times \{0, 1\}, j \in \mathbb{Z}$ choosing one of the two blocks for any j in arbitrary way, and continue by $(0, 2)$-periodicity to fill into the blocks $\{4j+2, 4j+3\} \times \{2i, 2i+1\}, i, j \in \mathbb{Z}$. Here is an fragment of such a word, with letters from the $(4, 2)$-lattices nested in $(x, y) \in \{0, 1\}^2$ marked by bold:

a b	a b	a b	b a	a b	b a
b a	b a	**b a**	a b	**b a**	a b
b a	a b	**b a**	b a	**b a**	b a
a b	b a	a b	a b	a b	a b
a b	a b	a b	b a	a b	b a
b a	b a	**b a**	a b	**b a**	a b
a b	a b	a b	b a	a b	b a
b a	b a	**b a**	a b	**b a**	a b

This word is aperiodic and has abelian complexity 1 in 4×2 rectangles, and moreover for each $n', m' \in \mathbb{N}$ we have $a_w(4n', 2m') = 1$; the proof is straightforward.

Although the words with abelian complexity one have some periodicity structure (in lattices), it is possible to construct words of complexity 2 or 3 without periodicity in lattices:

Proposition 4. *There exist recurrent aperiodic two-dimensional words that have abelian complexity 2 or 3 and are aperiodic in any lattice.*

Surprisingly, the structure of such words is not related to Sturmian words. To build such an example, we start with any binary word $x \in \{0, 1\}^{\mathbb{Z}}$. We build a two-dimensional word w by 2×2 blocks in the following way:

$$
\begin{matrix} w_{2i,2j+1} & w_{2i+1,2j+1} \\ w_{2i,2j} & w_{2i+1,2j} \end{matrix} = \begin{cases} \begin{matrix} 1 & 0 \\ 0 & 1 \end{matrix}, & \text{if } x_i = x_j, \\[2mm] \begin{matrix} 0 & 1 \\ 1 & 0 \end{matrix}, & \text{if } x_i \neq x_j. \end{cases}
$$

The complexity can be checked straightforwardly by considering several cases. Actually, this is the smallest possible complexity:

Theorem 10. *Let w be a two-dimensional aperiodic recurrent word. Then there exist infinitely many pairs (m, n) for which $a_w(m, n) \geq 3$.*

3.3 Small Complexity of Uniformly Recurrent Two-Dimensional Words

An interesting open question concerns the (non-abelian) Nivat's conjecture under the condition of recurrence. It has been recently proved that if we assume that an infinite word is uniformly recurrent and for some (m, n) its complexity is bounded by mn, then it has a periodicity vector [35]. On the other hand, Julien Cassaigne characterized infinite words of complexity $mn + 1$ [9]. In particular, he showed that none of such sequences is uniformly recurrent. So the question is, what is the minimal complexity of aperiodic two-dimensional uniformly recurrent words? We remark that one well-known family of two-dimensional words of low complexity include Sturmian words, i.e., words obtained by a rotation on a thorus [7]: Let α, β be real numbers, with 1, α, β rationally independent, and $0 < \alpha + \beta < 1$. The two-dimensional Sturmian word s over the three-letter alphabet $\{1, 2, 3\}$ (with parameters α, β, ρ) is defined as

$$s_{m,n} = i \Leftrightarrow (m\alpha + n\beta + \rho \text{ modulo } 1) \in I_i,$$

where $I_3 = [0, \alpha), I_2 = [\alpha, \alpha + \beta), I_1 = [\alpha + \beta, 1)$ (the intervals may also be all half-open on the left). These words have no periodicity vector and they have complexity $p(m, n) = mn + m + n$. A *projection* of a two-dimensional Sturmian word defined by $v(m, n) = 1$ if and only if $s(m, n) = 1$ or 3 and $v(m, n) = 2$ if and only if $s(m, n) = 2$ gives an example of a two-dimensional uniformly recurrent word of complexity y $p(m, n) = mn + n$ for sufficiently large m and n. A challenging open problem, which is a modification of Nivat's conjecture, is to find uniformly recurrent aperiodic two-dimensional words of the lowest complexity.

4 Abelian Subshifts

Part of the results of this section were introduced at DLT 2018 [34], the others are based on a recent joint work with Markus Whiteland. In the section, we undertake a general study of the notion of abelian subshift. First, we consider the abelian subshifts of binary words. We prove that if an aperiodic binary uniformly recurrent word is not Sturmian, then its abelian subshift contains infinitely many minimal subshifts. Secondly, we characterize the abelian subshifts of aperiodic recurrent balanced words; they are a finite union of minimal subshifts. Finally, we characterize abelian subshifts of aperiodic words having factor complexity $n + C$ for each n. Depending on the word, its abelian subshift contains either finitely many or uncountably many distinct minimal subshifts.

4.1 Preliminaries and Notation

A *subshift* $X \subseteq \Sigma^{\mathbb{N}}$, $X \neq \emptyset$, is a closed set (with respect to the product topology of $\Sigma^{\mathbb{N}}$) which is invariant under the shift operator σ (defined by $\sigma(a_0 a_1 a_2 \cdots) = a_1 a_2 \cdots$), that is, $\sigma(X) \subseteq X$. We call $\Sigma^{\mathbb{N}}$ the *full shift* over Σ. A subshift $X \subseteq \Sigma^{\mathbb{N}}$ is called *minimal* if X does not contain any proper subshifts. For a subshift $X \subseteq \Sigma^{\mathbb{N}}$ we let $\mathcal{L}(X) = \cup_{y \in X} \mathcal{L}(y)$. Let $x \in \Sigma^{\mathbb{N}}$. We let Ω_x denote the *shift orbit closure* of x, that is, the set $\{y \in \Sigma^{\mathbb{N}} : \mathcal{L}(y) \subseteq \mathcal{L}(x)\}$. Thus $\mathcal{L}(\Omega_x) = \mathcal{L}(x)$ for an infinite word $x \in \Sigma^{\mathbb{N}}$. It is known that Ω_x is minimal if and only if x is uniformly recurrent. See [39] for more on the topic.

We turn to the main notion of this section. For a subshift $X \subseteq \Sigma^{\mathbb{N}}$ we define the *abelian subshift* of X as $\mathcal{A}_X = \{y \in \Sigma^{\mathbb{N}} : \forall u \in \mathcal{L}(y) \exists v \in \mathcal{L}(X) : u \sim_{\text{ab}} v\}$. Observe that for any $x \in \Sigma^{\mathbb{N}}$ the abelian subshift $\mathcal{A}_x$ is indeed a subshift. We make preliminary observations on abelian subshifts of periodic infinite words.

Proposition 5. *For any periodic word x, the abelian subshift $\mathcal{A}_x$ is finite.*

In general, the abelian subshift of an ultimately periodic word can be huge.

Example 3. Let $x = 0011(001101)^\omega$. It is readily verified that $\boldsymbol{TM} \in \mathcal{A}_x$ so that $\mathcal{A}_x = \mathcal{A}_{TM} = \{\varepsilon, 0, 1\}\{01, 10\}^{\mathbb{N}}$.

4.2   On Abelian Subshifts of Binary Words

In this subsection we show that for a uniformly recurrent word x, its abelian subshift $\mathcal{A}_x$ contains exactly one minimal subshift if and only if x is a Sturmian word, aperiodic or periodic. We remark that purely periodic balanced words are sometimes also called Sturmian. We show that if x is an aperiodic binary uniformly recurrent word which is not Sturmian, then $\mathcal{A}_x$ contains infinitely many minimal subshifts.

The following theorem gives a characterization of Sturmian words in terms of abelian subshifts.

Theorem 11. *Let $x \in \{0, 1\}^{\mathbb{N}}$ be a uniformly recurrent aperiodic word. Then $\mathcal{A}_x$ contains exactly one minimal subshift if and only if x is Sturmian.*

We remark that the characterization does not extend to non-binary alphabets: Let $f = 010010100\ldots$ be the Fibonacci word and let $\varphi : 0 \mapsto 02, 1 \mapsto 12$. Then for $w = \varphi(f)$ one has $\mathcal{A}_w = \Omega_w$ (see Theorem 14).

Theorem 12. *Let x be a binary uniformly recurrent word which is not aperiodic or periodic Sturmian. Then $\mathcal{A}_x$ contains infinitely many minimal subshifts.*

The proof consists of three parts treated in different ways: if x has no frequencies of letters, the proof is almost immediate. If it has rational frequencies, then using so-called standard factors we can show that its abelian subshift contains uncountably many infinite subshifts. The hardest case is the case of irrational frequency, the proof is geometric and different for balanced and unbalanced case. We omit the details.

Example 4 (Thue–Morse word). Consider the abelian subshift of the Thue-Morse word $\boldsymbol{TM}$. For odd lengths $\boldsymbol{TM}$ has two abelian factors, and for even lengths three. Further, the number of occurrences of 1 in each factor differs by at most 1 from half of its length. It is easy to see that any factor of any word in $\{\varepsilon, 0, 1\} \cdot \{01, 10\}^{\mathbb{N}}$ has the same property, i.e., $\{\varepsilon, 0, 1\} \cdot \{01, 10\}^{\mathbb{N}} \subseteq \mathcal{A}_{TM}$. In fact, equality holds: $\mathcal{A}_{TM} = \{\varepsilon, 0, 1\} \cdot \{01, 10\}^{\mathbb{N}}$. Indeed, let $\boldsymbol{x} \in \mathcal{A}_{TM}$. Then $\boldsymbol{x}$ has blocks of each letter of length at most 2 (since there are no factors 000 and 111). Moreover, between two consecutive occurrences of 00 there must occur 11, and vice versa (otherwise we have a factor $001010\cdots0100$, where the number of occurrences of 1 differs by more than 1 from half of its length). Clearly, such word is in $\{\varepsilon, 0, 1\} \cdot \{01, 10\}^{\mathbb{N}}$. So, for the Thue-Morse word, its subshift is huge compared to Ω_{TM}: basically, it is a morphic image of the full binary shift.

4.3   On Abelian Subshifts of Minimal Complexity Words and Related Words

In this subsection we are interested in extending the characterization of $\Omega_x = \mathcal{A}_x$ from Theorem 11 to nonbinary alphabet. Natural idea is to check natural generalizations of Sturmian words to nonbinary alphabet. Following the definition of Sturmian words as words of minimal complexity, we study abelian subshifts of nonbinary words of minimal complexity. As Sturmian words can be defined as balanced words, it is reasonable to check nonbinary balanced words. Finally, using the definition via palindromic closures, we discuss Arnoux-Rauzy words.

First we study the abelian subshifts of aperiodic nonbinary words of minimal complexity. Over an alphabet Σ, the minimal complexity is $n + |\Sigma| - 1$. The structure of words of complexity $n + C$ is related to the structure of Sturmian words and is well understood [19,26,33]. We start with infinite words for which $p(n) = n + 2$ for all $n \geq 1$. Observe that this implies that we deal with ternary words.

Theorem 13 ([26] **as formulated in** [33]). *A word $u \in \{0,1,2\}$ has factor complexity $p_u(n) = n + 2$ for all $n \geq 1$ if and only if u is of the form (up to permuting the letters)*

1. *$u = 2s$ for some Sturmian word $s \in \{0,1\}^{\mathbb{N}}$, or $u \in \Omega_{\varphi(s)}$, where s is a Sturmian word and φ is defined by*
2. *$0 \mapsto 02$, $1 \mapsto 12$;*
3. *$0 \mapsto 0$, $1 \mapsto 12$.*

Using this description, we obtain a complete characterization of abelian subshifts of ternary words of minimal complexity. In fact, their abelian subshifts either contain one, or uncountably many minimal subshifts:

Theorem 14. *Let u be a word of factor complexity $n + 2$ for all $n \geq 1$. If u is as in Theorem 13, item 1 or item 2, then $\mathcal{A}_u = \Omega_u$. If u is as in item 3, then $\mathcal{A}_u$ contains uncountably many minimal subshifts.*

Surprisingly, for alphabet of size greater than 3 there are always only finitely many subshifts:

Theorem 15. *Let u be a recurrent word of factor complexity $n + C$ for all $n \geq 1$, where $C > 2$. Then $\mathcal{A}_u$ contains exactly two minimal subshifts.*

The proof is based on the characterization of words of factor complexity $n + C$ for all $n \geq 1$ from [26].

Now we turn to aperiodic uniformly recurrent balanced words and their abelian subshifts. We can prove that abelian subshift of such word is a finite union of minimal subshifts. Our results rely heavily on the characterization of aperiodic recurrent balanced words by Hubert [32].

Theorem 16. *Let u be aperiodic recurrent and balanced. Then $\mathcal{A}_u$ is the union of finitely many minimal subshifts.*

The techniques used in the proof of the above theorem give us the following proposition. We remark that the words in question are not necessarily balanced.

Proposition 6. *For each $k \geq 1$ there exists an aperiodic word x_k such that $\mathcal{A}_{x_k}$ equals the union of k distinct minimal subshifts.*

We remark that although the structure of aperiodic balanced words is clear, the structure of periodic balanced words is a mystery. The following conjecture by Fraenkel, 1973, remains open despite efforts of different scientists: The unique (up to a permutation of letters) balanced word on $k \geq 3$ letters with all distinct frequencies of letters is $(F_k)^\omega$, where $F_k = F_{k-1}kF_{k-1}$ and $F_2 = 121$ [30]. The conjecture has been verified for $k \leq 7$ (see [5] and references therein).

In the end of this subsection we discuss Arnoux-Rauzy words, which are another generalization of Sturmian words to larger alphabet. One of the ways to define Arnoux-Rauzy words is via palindromic closures. The following basics on Arnoux-Rauzy words are well-known and mostly taken from [2,20]. In fact, this is a generalization of the facts about Sturmian words given for binary words in [17].

A finite word $v = v_0 \cdots v_{n-1}$ is a *palindrome* if it is equal to its reversal, i.e., $v = v_{n-1} \cdots v_0$. The *right palindromic closure* of a finite word u, denoted by $u^{(+)}$, is the shortest palindrome that has u as a prefix. The *iterated (right) palindromic closure operator* ψ is defined recursively by the following rules:

$$\psi(\varepsilon) = \varepsilon, \quad \psi(va) = (\psi(v)a)^{(+)}$$

for all $v \in \Sigma^*$ and $a \in \Sigma$. We let $\mathrm{Pref}_n w$ denote the prefix of w of length n, i.e., $\mathrm{Pref}_n w = w_0 \cdots w_{n-1}$. The definition of ψ may be extended to infinite words u over Σ as $\psi(u) = \lim_n \psi(\mathrm{Pref}_n u)$, i.e., $\psi(u)$ is the infinite word having $\psi(\mathrm{Pref}_n u)$ as its prefix for every $n \in \mathbb{N}$.

Let Δ be an infinite word on the alphabet Σ such that every letter occurs infinitely often in Δ. The word $c = \psi(\Delta)$ is then called a *characteristic (or standard) Arnoux-Rauzy word* and Δ is called the *directive sequence* of c. An infinite word u is called an Arnoux-Rauzy word if it has the same set of factors as a (unique) characteristic Arnoux-Rauzy word, which is called the characteristic word of u. The directive sequence of an Arnoux-Rauzy word is the directive sequence of its characteristic word. It is not hard to see that the Tribonacci word (see Example 2) is an Arnoux-Rauzy word with the directive sequence $(123)^\omega$.

Apparently, the structure of abelian subshifts of Arnoux-Rauzy words is rather complicated. For example, it is not hard to see that for any Arnoux-Rauzy word with a characteristic word c its abelian subshift contains $20c$ (here we assume that 0 is the first letter of Δ and 2 is the third letter occurring in Δ for the first time, i.e., Δ has a prefix of the form $0\{0,1\}^*1\{0,1\}^*2$). On the other hand, $20c \notin \Omega_c$.

T. Hejda, W. Steiner, and L.Q. Zamboni studied the abelian subshift of the Tribonacci word $\boldsymbol{TR}$. They announced that $\mathcal{A}_{TR} \setminus \Omega_{TR} \neq \emptyset$ but that Ω_{TR} is the only minimal subshift contained in $\mathcal{A}_{TR}$ [31,59].

An interesting open question is to understand the general structure of abelian subshifts of Arnoux-Rauzy words:

Problem 3. Characterize abelian subshifts of Arnoux-Rauzy words.

References

1. Adamczewski, B.: Balances for fixed points of primitive substitutions. Theor. Comput. Sci. **307**(1), 47–75 (2003)
2. Arnoux, P., Rauzy, G.: Représentation géométrique de suites de complexité $2n+1$. Bull. Soc. Math. France **119**, 199–215 (1991)
3. Avgustinovich, S., Cassaigne, J., Karhumäki, J., Puzynina, S., Saarela, A.: On abelian saturated infinite words. Theoret. Comput. Sci. (to appear). https://doi.org/10.1016/j.tcs.2018.05.013
4. Avgustinovich, S., Karhumäki, J., Puzynina, S.: On abelian versions of critical factorization theorem. RAIRO - Theor. Inf. Applic. **46**, 3–15 (2012)
5. Bark, J., Varju, P.: Partitioning the positive integers to seven beatty sequences. Indag. Math. **14**(2), 149–161 (2003)
6. Berthé, V., Tijdeman, R.: Balance properties of multi-dimensional words. Theor. Comput. Sci. **273**, 197–224 (2002)
7. Berthé, V., Vuillon, L.: Tilings and rotations: a two-dimensional generalization of Sturmian sequences. Discrete Math. **223**, 27–53 (2000)
8. Blanchet-Sadri, F., Fox, N., Rampersad, N.: On the asymptotic Abelian complexity of morphic words. Adv. Appl. Math. **61**, 46–84 (2014)
9. Cassaigne, J.: Double sequences with complexity $mn+1$. J. Autom. Lang. Comb. **4**(3), 153–170 (1999)
10. Cassaigne, J., Richomme, G., Saari, K., Zamboni, L.Q.: Avoiding abelian powers in binary words with bounded abelian complexity. Int. J. Found. Comput. Sci. **22**(4), 905–920 (2011)
11. Charlier, E., Harju, T., Puzynina, S., Zamboni, L.Q.: Abelian bordered factors and periodicity. Eur. J. Comb. **51**, 407–418 (2016)
12. Christodoulakis, M., Christou, M., Crochemore, M., Iliopoulos, C.S.: Abelian borders in binary words. Discrete Appl. Math. **171**, 141–146 (2014)
13. Constantinescu, S., Ilie, L.: Fine and Wilf's theorem for abelian periods. Bull. Eur. Assoc. Theor. Comput. Sci. **89**, 167–170 (2006)
14. Coven, E.M., Hedlund, G.A.: Sequences with minimal block growth. Math. Syst. Theory **7**, 138–153 (1973)
15. Currie, J., Rampersad, N.: Recurrent words with constant Abelian complexity. Adv. Appl. Math. **47**, 116–124 (2011)
16. Cyr, V., Kra, B.: Nonexpansive $\mathbb{Z}^2$-subdynamics and Nivat's conjecture. Trans. Am. Math. Soc. **367**(9), 6487–6537 (2015)
17. de Luca, A.: Sturmian words: structure, combinatorics, and their arithmetics. Theor. Comput. Sci. **183**, 45–82 (1997)
18. Dekking, F.M.: Strongly non-repetitive sequences and progression-free sets. J. Comb. Theory Ser. A **27**(2), 181–185 (1979)
19. Didier, G.: Caractérisation des N-écritures et application à l'étude des suites de complexité ultimement $n + c^{ste}$. Theoret. Comp. Sci. **215**(1–2), 31–49 (1999)
20. Droubay, X., Justin, J., Pirillo, G.: Episturmian words and some constructions by de Luca and Rauzy. Theor. Comput. Sci. **255**, 539–553 (2001)

21. Durand, F., Rigo, M.: Multidimensional extension of the Morse-Hedlund theorem. Eur. J. Comb. **34**(2), 391–409 (2013)
22. Entringer, R.C., Jackson, D.E., Schatz, J.A.: On nonrepetitive sequences. J. Comb. Theory (A) **16**, 159–164 (1974)
23. Epifanio, C., Koskas, M., Mignosi, F.: On a conjecture on bi-dimensional words. Theor. Comput. Sci. **299**, 123–150 (2003)
24. Erdös, P.: Some unsolved problems. Magyar Tud. Akad. Mat. Kutató Int. Közl. **6**, 221–254 (1961)
25. Evdokimov, A.A.: Strongly asymmetric sequences generated by a finite number of symbols. Dokl. Akad. Nauk SSSR **179**, 1268–1271 (1968)
26. Ferenczi, S., Mauduit, C.: Transcendence of numbers with a low complexity expansion. J. Number Theory **67**, 146–161 (1997)
27. Fici, G., Mignosi, F., Shallit, J.: Abelian-square-rich words. Theor. Comput. Sci. **684**, 29–42 (2017)
28. Fici, G., et al.: Abelian powers and repetitions in Sturmian words. Theor. Comput. Sci. **635**, 16–34 (2016)
29. Fine, N.J., Wilf, H.S.: Uniqueness theorems for periodic functions. Proc. Am. Math. Soc. **16**, 109–114 (1965)
30. Fraenkel, A.S.: Complementing and exactly covering sequences. J. Comb. Theory Ser. A **14**(1), 8–20 (1973)
31. Hejda, T., Steiner, W., Zamboni, L.Q.: What is the abelianization of the tribonacci shift? In: Workshop on Automatic Sequences, Liège, May 2015
32. Hubert, P.: Suites équilibrées. Theor. Comput. Sci. **242**(1–2), 91–108 (2000)
33. Kaboré, I., Tapsoba, T.: Combinatoire de mots récurrents de complexité $n + 2$. ITA **41**(4), 425–446 (2007)
34. Karhumäki, J., Puzynina, S., Whiteland, M.: On abelian subshifts. In: DLT 2018, pp. 453–464 (2018)
35. Kari, J. Moutot, E.: Decidability and Periodicity of Low Complexity Tilings. https://arxiv.org/abs/1904.01267
36. Kari, J., Szabados, M.: An algebraic geometric approach to nivat's conjecture. In: Halldórsson, M.M., Iwama, K., Kobayashi, N., Speckmann, B. (eds.) ICALP 2015, Part II. LNCS, vol. 9135, pp. 273–285. Springer, Heidelberg (2015). https://doi.org/10.1007/978-3-662-47666-6_22
37. Keränen, V.: Abelian squares are avoidable on 4 letters. In: Kuich, W. (ed.) ICALP 1992. LNCS, vol. 623, pp. 41–52. Springer, Heidelberg (1992). https://doi.org/10.1007/3-540-55719-9_62
38. Keränen, V.: New abelian square-free DT0L-languages over 4 letters (2003, manuscript)
39. Lind, D., Marcus, B.: An Introduction to Symbolic Dynamics and Coding. Cambridge University Press, New York (1995)
40. Lothaire, M.: Algebraic Combinatorics on Words. Cambridge University Press, Cambridge (2002)
41. Masáková, Z., Pelantová, E.: Enumerating abelian returns to prefixes of sturmian words. In: Karhumäki, J., Lepistö, A., Zamboni, L. (eds.) WORDS 2013. LNCS, vol. 8079, pp. 193–204. Springer, Heidelberg (2013). https://doi.org/10.1007/978-3-642-40579-2_21
42. Morse, M., Hedlund, G.: Symbolic dynamics. Am. J. Math. **60**, 815–866 (1938)
43. Morse, M., Hedlund, G.: Symbolic dynamics II: sturmian sequences. Am. J. Math. **62**, 1–42 (1940)
44. Nivat M.: Invited talk at ICALP'97 (1997)

45. Pansiot, J.-J.: Complexité des facteurs des mots infinis engendrés par morphismes itérés. In: Paredaens, J. (ed.) ICALP 1984. LNCS, vol. 172, pp. 380–389. Springer, Heidelberg (1984). https://doi.org/10.1007/3-540-13345-3_34

46. Puzynina, S.: Small abelian complexity of two-dimensional infinite words (submitted)

47. Puzynina, S., Zamboni, L.Q.: Abelian returns in sturmian words. J. Comb. Theory Ser. A **120**(2), 390–408 (2013)

48. Quas, A., Zamboni, L.Q.: Periodicity and local complexity. Theor. Comput. Sci. **319**(1–3), 229–240 (2004)

49. Rampersad, N., Rigo, M., Salimov, P.: A note on abelian returns in rotation words. Theor. Comput. Sci. **528**, 101–107 (2014)

50. Rao, M., Rosenfeld, M.: Avoidability of long k-abelian repetitions. Math. Comput. **85**(302), 3051–3060 (2016)

51. Rao, M., Rosenfeld, M.: Avoiding two consecutive blocks of same size and same sum over $\mathbb{Z}^2$. SIAM J. Discrete Math. **32**(4), 2381–2397 (2018)

52. Richomme, G., Saari, K., Zamboni, L.Q.: Abelian complexity in minimal subshifts. J. London Math. Soc. **83**, 79–95 (2011)

53. Richomme, G., Saari, K., Zamboni, L.Q.: Balance and abelian complexity of the tribonacci word. Adv. Appl. Math. **45**, 212–231 (2010)

54. Rigo, M., Salimov, P., Vandomme, E.: Some properties of abelian return words. J. Integer Seq. **16** (2013). Article number 13.2.5

55. Saarela, A.: Ultimately constant abelian complexity of infinite words. J. Autom. Lang. Comb. **14**(3–4), 255–258 (2009)

56. Simpson, J.: An abelian periodicity lemma. Theor. Comput. Sci. **656**, 249–255 (2016)

57. Vuillon, L.: A characterization of Sturmian words by return words. Eur. J. Comb. **22**, 263–275 (2001)

58. Whiteland, M.A.: Asymptotic abelian complexities of certain morphic binary words. J. Autom. Lang. Comb. **24**(1), 89–114 (2019)

59. Zamboni, L.Q.: Personal communication (2018)

On Sets of Words of Rank Two

Giuseppa Castiglione, Gabriele Fici, and Antonio Restivo[✉]

Dipartimento di Matematica e Informatica, Università di Palermo,
Via Archirafi 34, Palermo, Italy
{giuseppa.castiglione,gabriele.fici,antonio.restivo}@unipa.it

Abstract. Given a (finite or infinite) subset X of the free monoid A^* over a finite alphabet A, the rank of X is the minimal cardinality of a set F such that $X \subseteq F^*$. A submonoid M generated by k elements of A^* is k-maximal if there does not exist another submonoid generated by at most k words containing M. We call a set $X \subseteq A^*$ primitive if it is the basis of a $|X|$-maximal submonoid. This extends the notion of primitive word: indeed, $\{w\}$ is a primitive set if and only if w is a primitive word. By definition, for any set X, there exists a primitive set Y such that $X \subseteq Y^*$. The set Y is therefore called a primitive root of X. As a main result, we prove that if a set has rank 2, then it has a unique primitive root. This result cannot be extended to sets of rank larger than 2.

For a single word w, we say that the set $\{x, y\}$ is a *binary root* of w if w can be written as a concatenation of copies of x and y and $\{x, y\}$ is a primitive set. We prove that every primitive word w has at most one binary root $\{x, y\}$ such that $|x| + |y| < \sqrt{|w|}$. That is, the binary root of a word is unique provided the length of the word is sufficiently large with respect to the size of the root.

Our results are also compared to previous approaches that investigate pseudo-repetitions, where a morphic involutive function θ is defined on A^*. In this setting, the notions of θ-power, θ-primitive and θ-root are defined, and it is shown that any word has a unique θ-primitive root. This result can be obtained with our approach by showing that a word w is θ-primitive if and only if $\{w, \theta(w)\}$ is a primitive set.

Keywords: Repetition · Pseudo-repetition · Hidden repetition · Primitive set · Binary root · k-maximal monoid

1 Introduction

The notion of *rank* plays an important role in combinatorics on words. Given a subset X of the free monoid A^* over a finite alphabet A, the rank of X, in symbols $r(X)$, is defined as the smallest number of words needed to express all words of X, i.e., as the minimal cardinality of a set F such that $X \subseteq F^*$. Notice that this minimal set F may not be unique. For instance, the set $X = \{aabca, aa, bcaaa\}$ has rank 2 and there exist two distinct sets $F_1 = \{aa, bca\}$ and $F_2 = \{a, bc\}$ such that $X \subseteq F_1^*$ and $X \subseteq F_2^*$. It is worth noticing that since $r(X) \leq min\{|X|, |A|\}$, $r(X)$ is always finite even if X is an infinite set.

© Springer Nature Switzerland AG 2019
R. Mercaş and D. Reidenbach (Eds.): WORDS 2019, LNCS 11682, pp. 46–59, 2019.
https://doi.org/10.1007/978-3-030-28796-2_3

A set X is said to be *elementary* if $r(X) = |X|$. The notion of rank – and the related notion of elementary set – have been investigated in several papers (cf. [16–18]). In particular, in [16] it is shown that the problem to decide whether a finite set is elementary is co-NP-complete.

In this paper, we introduce the notion of primitiveness for a *set* of words, which is closely related to that of rank. We first define the notion of k-*maximal* submomoid. A submonoid M of A^*, generated by k elements, is k-maximal if there does not exist another submonoid generated by at most k words containing M. We then call a set $X \subseteq A^*$ *primitive* if it is the basis of a $|X|$-maximal submonoid. Notice that if X is primitive, then $r(X) = |X|$, i.e., X is elementary. The converse is not in general true: there exist elementary sets that are not primitive. For instance, the set $F_1 = \{aa, bca\}$ is elementary, but it is not primitive since $F_1^* \subseteq F_2^* = \{a, bc\}^*$. The set F_2, instead, is primitive.

The notion of primitive set can be seen as an extension of the classical notion of primitive word. Indeed, given a word $w \in A^*$, the set $\{w\}$ is primitive if and only if the word w is primitive. For instance, the set $\{abab, abababab\}$ is not elementary, the set $\{abab\}$ is elementary but not primitive, and the set $\{ab\}$ is primitive.

We have from that definition that for every set X, there exists a primitive set Y such that $X \subseteq Y^*$. The set Y is therefore called a *primitive root* of X. However, the primitive root of a set is not, in general, unique. Consider for instance the set $X = \{abcbab, abcdcbab, abcdcdcbab\}$. It has rank 3, hence it is elementary, yet it is not primitive. Indeed, $X \subseteq \{ab, cb, cd\}^*$. The set $\{ab, cb, cd\}$ is primitive, and it is a primitive root of X. However, it is not the only primitive root of X: the set $\{abc, dc, bab\}$ is primitive and $X \subseteq \{abc, dc, bab\}^*$, hence $\{abc, dc, bab\}$ is another primitive root of X. In the special case of sets of rank 1, clearly these always have a unique primitive root. For instance, the primitive root of the set $\{abab, abababab\}$ is the set $\{ab\}$.

As a main result, we prove that if a set has rank 2, then it has a unique primitive root. This is equivalent to say that for every pair of nonempty words $\{x, y\}$ such that $xy \neq yx$ there exists a unique primitive set $\{u, v\}$ such that x and y can be written as concatenations of copies of u and v. The proof is based on the algebraic properties of k-maximal submonoids of a free monoid.

In this investigation, we also take into account another notion of rank, that of *free rank* (in the literature, in order to avoid ambiguity, the notion of rank we gave above is often referred to as the *combinatorial rank*). The free rank of a set X is the cardinality of the basis of the minimal *free* submonoid containing X. Closely related to the notion of free rank is the *defect theorem*, which states that if X is not a *code* (i.e., X^* is not a free submonoid), then the free rank of X is strictly smaller than its cardinality. We are specially interested in the case $k = 2$ (that is, the case of 2-maximal submonoids) and we use the fact that, in this special case, the notions of free rank and (combinatorial) rank coincide. A fundamental step in our argument is Theorem 14, which states that the intersection of two 2-maximal submonoids is either the empty word or a submonoid generated by one primitive word. As a consequence, for every submonoid M generated by two words that do not commute, there exists a unique 2-maximal submonoid containing M. This is equivalent to the fact that every set of rank 2 has a unique primitive root. One of the examples we gave above shows that this result is no longer true for sets of rank 3 or larger—this highlights the very special role of sets of rank 1 or 2.

From these results we derive some consequences on the combinatorics of a *single* word. Given a word w, we say that $\{x, y\}$ is a binary root of w if w can be written as a concatenation of copies of x and y and $\{x, y\}$ is a primitive set. We prove that every primitive word w has at most one binary root $\{x, y\}$ such that $|x| + |y| < \sqrt{|w|}$. That is, the binary root of a word is unique provided the length of the word is sufficiently large with respect to the size of the root. The notion of binary root of a single word may be seen as a way to capture a hidden "repetitive structure", which encompasses the classical notion of integer repetition (non-primitive word). Indeed, the existence in a word w of a "short" (with respect to $|w|$) binary root reveals some hidden repetition in the word.

As described in the last section, our results can also be compared to previous approaches that investigate *pseudo-repetitions*, where an involutive morphism (or antimorphism) θ is defined on the set of words A^*. This idea stems from the seminal paper of Czeizler, Kari and Seki [4], where originally θ was the Watson-Crick complementarity function and the motivation was the discovery of hidden repetitive structures in biological sequences. A word w is called a θ-power if there exists a word v such that w can be factored using copies of v and $\theta(v)$—otherwise the word w is called θ-primitive. If v is a θ-primitive word, then it is called the θ-primitive root of w. Of course, since the same applies to the word $\theta(w)$, these definitions can be given in terms of the pair $\{w, \theta(w)\}$ and considering as the root the pair $\{v, \theta(v)\}$. With our results, we generalize this setting by considering as a root any pair of words $\{x, y\}$, i.e., dropping the relation between the components of the pair.

2   Preliminaries

Given a finite nonempty set A, called the *alphabet*, with A^* (resp. $A^+ = A^* \backslash \{\varepsilon\}$) we denote the *free monoid* (resp. *free semigroup*) generated by A, i.e., the set of all finite words (resp. all finite nonempty words) over A.

The length $|w|$ of a word $w \in A^*$ is the number of its symbols. The length of the empty word ε is 0. For a word $w = uvz$, with $u, v, z \in A^*$, we say that v is a *factor* of w. Such a factor is called *internal* if $u, z \neq \varepsilon$, a *prefix* if $u = \varepsilon$, or a *suffix* if $z = \varepsilon$. A word w is *primitive* if $w = v^n$ implies $n = 1$, otherwise it is called *a power*. Equivalently, w is primitive if and only if it is not an internal factor of w^2.

It is well known in combinatorics on words (see, e.g., [14]) that given two words x and y we have $xy = yx$ if and only if x and y are powers of the same word.

Given a subset X of A^*, we let X^* denote the submonoid of A^* generated by X (under concatenation). Conversely, given a submonoid M of A^*, there exists a unique set X that generates M and is minimal for set inclusion. In fact, X is the set

$$X = (M \setminus \{\varepsilon\}) \setminus (M \setminus \{\varepsilon\})^2, \tag{1}$$

i.e., X is the set of nonempty words of M that cannot be written as a product of two nonempty words of M. The set X will be referred to as the *minimal generating set* of M, or the set of *generators* of M.

Let M be a submonoid of A^* and X its minimal generating set. M is said to be *free* if any word of M can be *uniquely* expressed as a product of elements of X. The minimal generating set of a free submonoid M of A^* is called a *code*; it is referred to as *the basis* of M. It is easy to see that a set X is a code if and only if, for every $x, y \in X$, $x \neq y$, one has $xX^* \cap yX^* = \emptyset$. We say that X is a *prefix code* (resp. a *suffix code*) if for all $x, y \in X$, one has $x \cap yA^* = \emptyset$ (resp. $x \cap A^*y = \emptyset$). A code is a *bifix code* if it is both a prefix and a suffix code. It follows from elementary automata theory that if X is a prefix code, then there exists a DFA $\mathcal{A}_X$ recognizing X^* whose set of states Q_X verifies (cf. [2]):

$$|Q_X| \leq \sum_{x \in X} |x| - |X| + 1.$$

A submonoid M of A^* is called *pure* (cf. [19]) if for all $w \in A^*$ and $n \geq 1$,

$$w^n \in M \Rightarrow w \in M.$$

By a result of Tilson [20], any nonempty intersection of free submonoids of A^* is free. As a consequence, for any subset $X \subseteq A^*$, there exists the smallest free submonoid containing X.

Here we mention the well-known Defect Theorem (cf. [1], [14, Chap. 1], [15, Chap. 6]), a fundamental result in the theory of codes that provides a relation between a given subset X of A^* and the basis of the minimal free submonoid containing X (called the *free hull* of X).

Theorem 1 (Defect Theorem). *Let X be a finite nonempty subset of A^*. Let Y be the basis of the free hull of X. Then either X is a code, and $Y = X$, or*

$$|Y| \leq |X| - 1.$$

As in [9], given a set $X \subseteq A^*$, we let $r_f(X)$ denote the cardinality of the basis of the free hull of X, called the *free rank* of X. Notice that for any subset $X \subseteq A^*$, X and X^* have the same free rank. Furthermore, by $r(X)$ we denote the *combinatorial rank* (or simply rank) of X, defined by:

$$r(X) = \min\{|Y| \mid Y \subseteq A^*, X \subseteq Y^*\}.$$

With this notation, the Defect Theorem can be stated as follows.

Theorem 2. *Let X be a finite nonempty subset of A^*. Then $r_f(X) \leq |X|$, and the equality holds if and only if X is a code.*

Note that, for any $X \subseteq A^+$, one has

$$r(X) \leq r_f(X) \leq |X|.$$

Example 3. Let $X = \{aa, ba, baa\}$. One can prove that X is a code, hence we have $r_f(X) = 3$, while $r(X) = 2$ since $X \subset \{a, b\}^*$. For $X = \{aa, aaa\}$, we have $r(X) = r_f(X) = 1$.

Remark 4. If $|X| = 2$ then $r_f(X) = r(X)$. So for sets of cardinality 2 we will not specify if we refer to the free rank or to the (combinatorial) rank.

Moreover, from the complexity point of view, Néraud proved that deciding if a set has rank 2 can be done in polynomial time [17], whereas for general rank k it is an NP-hard problem [16].

The *dependency graph* (cf. [9]) of a finite set $X \subset A^+$ is the graph $G_X = (X, E_X)$ where $E_X = \{(u, v) \in X \times X \mid uX^* \cap vX^* \neq \emptyset\}$. Notice that if X is a code, then G_X has no edge. Furthermore, if (u, v) is an edge, then u is a prefix of v or vice versa. In [8] and [9], the following useful lemma is proved.

Lemma 5 (Graph Lemma). *Let $X \subseteq A^+$ be a finite set that is not a code. Then*

$$r_f(X) \leq c(X) < |X|,$$

where $c(X)$ is the number of connected components of G_X.

Example 6. Let $X = \{a, ab, abc, bca, acb, cba\}$. We have $acba = a \cdot cba = acb \cdot a$ and $abca = a \cdot bca = abc \cdot a$. The basis of the free hull of X is $Y = \{a, ab, bc, cb\}$, hence $r_f(X) = 4$. Furthermore, $r(X) = 3$ and $c(X) = 4$, as shown in Fig. 1.

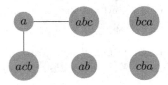

Fig. 1. The dependency graph of $X = \{a, ab, abc, bca, acb, cba\}$.

3 k-Maximal Monoids

With $\mathcal{M}_k$ we denote the family of submonoids of A^* having at most k generators in A^+. The following definition is fundamental for the theory developed in this paper.

Definition 7. *A submonoid $M \in \mathcal{M}_k$ is k-maximal if for every $M' \in \mathcal{M}_k$, $M \subseteq M'$ implies $M = M'$.*

In other words, M is k-maximal if it is not possible to find another submonoid generated by at most k words containing M.

Example 8. Let $A = \{a, b, c\}$. The submonoid $M = \{a, abca\}^*$ is not 2-maximal since $abca$ can be factored with a and bc, hence M is contained in $\{a, bc\}^*$. On the contrary, $\{a, bc\}^*$ is 2-maximal since, obviously, a and bc cannot be factored using two common factors.

Example 9. Let $A = \{a, b, c, d\}$. The submonoid $\{a, cbd, dbd\}^*$ is 3-maximal, whereas $\{a, cbd, dcbd\}^*$ is not 3-maximal since it is contained in $\{a, d, cb\}^*$.

Proposition 10. *Let M be a k-maximal submonoid and X its minimal generating set. Then, X is a bifix code.*

Proof. By contradiction, if X is not prefix (resp. not suffix) then there exist $u, v \in X$ and $t \in A^+$ such that $v = ut$ (resp $v = tu$). It follows that $X^* \subseteq (X \setminus \{v\} \cup \{t\})^*$, whence $X^* = M$ is not k-maximal. $\qquad\square$

Remark 11. By Proposition 10, it follows that if X^* is k-maximal, then $r(X) = r_f(X) = k$. The inverse implication does not hold in general. For example, the submonoid $X^* = \{a, cbd, dcbd\}^*$ of Example 9 has both rank and free rank equal to 3 and is bifix, but it is not 3-maximal.

Proposition 12. *Let M be a k-maximal submonoid. Then M is a pure submonoid.*

Proof. We have to show that, for every $z \in A^*$, if $z^n \in M$, for some $n \geq 1$, then $z \in M$. Let X be the minimal generating set of M. If $z^n \in M$, for some $n > 1$, then $z \in X$ or the set $X \cup \{z\}$ is not a code. By the Defect Theorem (Theorem 1), there exist $u_1, u_2, \ldots, u_k \in A^+$ such that $(X \cup \{z\})^* \subseteq \{u_1, u_2, ..., u_k\}^*$. Since $X^* \subseteq \{u_1, u_2, \ldots, u_k\}^*$ and X^* is k-maximal, we have that $X = \{u_1, u_2, \ldots, u_k\}$. Therefore, $X \cup \{z\} \subseteq X^*$, hence $z \in X^*$.

As a direct consequence of Proposition 12, we have that a k-maximal submonoid is generated by primitive words. However, not any set of k primitive words generates a k-maximal monoid (e.g., $X = \{ab, ba\}^*$ is not 2-maximal since it is contained in $\{a, b\}^*$).

Submonoids generated by two words, i.e., the elements of $\mathcal{M}_2$, are of special interest for our purposes. They have been extensively studied in literature (cf. [10, 12, 13, 17]) and play an important role in some fundamental aspects of combinatorics on words.

The reader may observe that, as a consequence of some well-known results in combinatorics on words, the submonoids in $\mathcal{M}_1$ have the following important property: If x^* and u^* are 1-maximal submonoids (i.e., x and u are primitive words) then $x^* \cap u^* = \{\varepsilon\}$. Next Theorem 14, which represents the main result of this section, can be seen as a generalization of this result to the case of 2-maximal submonoids.

It is known (see [10]) that if X and U both have rank 2, then the intersection $X^* \cap U^*$ is a free monoid generated either by at most two words or by an infinite set of words.

Example 13. Let $X_1 = \{abca, bc\}$ and $U_1 = \{a, bcabc\}$. One can verify that $X_1^* \cap U_1^* = \{abcabc, bcabca\}^*$. Let $X_2 = \{aab, aba\}$ and $U_2 = \{a, baaba\}$. Then $X_2^* \cap U_2^* = (a(abaaba)^* baaba)^*$.

In the previous example, we have two submonoids that are not 2-maximal. Indeed, $X_1^*, U_1^* \subseteq \{a, bc\}^*$ and $X_2^*, U_2^* \subseteq \{a, b\}^*$. We now address the question of finding the generators of the intersection of two 2-maximal submonoids.

Theorem 14. *Let $X^* = \{x, y\}^*$ and $U^* = \{u, v\}^*$ be two 2-maximal submonoids. If $X^* \cap U^* \neq \{\varepsilon\}$, then there exists a word $z \in A^+$ such that $X^* \cap U^* = z^*$. Moreover, z is primitive, that is, $X^* \cap U^*$ is 1-maximal.*

Proof. If $X \cap U = \{z\}$ then $X^* \cap U^* = z^*$. Indeed, if $y = v = z$ and $X^* \cap U^* \neq z^*$ we have the following graph G_Z for $Z = \{x, u, z\}$:

since $\{x, z\}$ and $\{u, z\}$ are bifix sets. Hence, by the Graph Lemma, $r_f(Z) \leq c(Z) = 2$, contradicting the 2-maximality of X and U.

If $X \cap U = \emptyset$, let us consider the set $Z = X \cup U$. We have that $r_f(Z) > 2$ since X^* and U^* are 2-maximal, and, by the Defect Theorem (Theorem 1), $r_f(Z) < 4$ since Z^* is not free (as $X^* \cap U^*$ contains a nonempty word). Hence, the free rank of Z is equal to 3.

Let z be a generator of $X^* \cap U^*$. So, $z = x_1 x_2 \cdots x_m = u_1 u_2 \cdots u_n$, with $m, n \geq 1$, $x_i \in X$ and $u_j \in U$. Clearly, since z is a generator, for every $p < m$ and $q < n$ one has $x_1 x_2 \cdots x_p \neq u_1 u_2 \cdots u_q$. Moreover, we can suppose, without loss of generality, that $x_1 = x$ and $u_1 = u$. We want to prove that z is the unique generator of $X^* \cap U^*$. By contradiction, suppose that there exists another generator $z' \neq z$ of $X^* \cap U^*$, and let $z' = x_1' x_2' \cdots x_r' = u_1' u_2' \cdots u_s'$. If $x_1' \neq x_1 = x$, then $x_1' = y$ and we have $xZ^* \cap uZ^* \neq \emptyset$ and $yZ^* \cap u_1' Z^* \neq \emptyset$. In both cases ($u_1' = u$ or $u_1' = v$), we have that the graph G_Z has two edges, i.e., $c(Z) = 2$, which is impossible by the Graph Lemma. So $x_1 = x_1' = x$. In the same way we prove that $u_1 = u_1' = u$, and therefore in the graph G_Z there is only one edge, namely the one joining x and u.

Let $h = \max\{i \mid x_j = x_j' \ \forall j \leq i\}$ and $k = \max\{i \mid u_j = u_j' \ \forall j \leq i\}$. The hypothesis that $z \neq z'$ implies that $h < m$ and $k < n$. We show that this leads to a contradiction, and then we conclude that $z = z'$ is the unique generator of $X^* \cap U^*$.

Without loss of generality, we can suppose that $x_1 x_2 \cdots x_h$ is a prefix of $u_1 u_2 \cdots u_k$. Hence, there exists a nonempty word t such that $x_1 x_2 \cdots x_h t = u_1 u_2 \cdots u_k$. By definition of h, $x_{h+1} \neq x'_{h+1}$, and we can suppose that $x_{h+1} = x$ and $x'_{h+1} = y$. Then,

$$tu_{k+1} \cdots u_n = x_{h+1} \cdots x_m = x \cdots x_m$$
$$tu'_{k+1} \cdots u'_s = x'_{h+1} \cdots x'_r = y \cdots x'_r.$$

Set $Z_t = X \cup U \cup \{t\}$. We have

$$tZ_t^* \cap xZ_t^* \neq \emptyset$$
$$tZ_t^* \cap yZ_t^* \neq \emptyset.$$

Thus, the graph G_{Z_t} contains the edges depicted in figure:

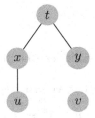

By the Graph Lemma, then, the free rank of Z_t is at most 2, and this contradicts the 2-maximality of X^* and U^*.

Finally, let us prove that z is primitive. Since X^* and Y^* are 2-maximal, by Proposition 12 they are both pure, hence also their intersection z^* is pure. But it is immediate that z^* is pure if and only if z is primitive. $\square$

Example 15. Consider the two 2-maximal monoids $\{abcab, cb\}^*$ and $\{abc, bcb\}^*$. Their intersection is $\{abcabcbcb\}^*$. The intersection of $\{a, bc\}^*$ and $\{a, cb\}^*$ is a^*.

We have shown that the intersection of two 2-maximal submonoids is generated by at most one element. Moreover, we know that the intersection of two 1-maximal submonoids is the empty word, i.e., it is generated by zero elements. Thus, it is natural to ask if in general, for every $k \geq 1$, the intersection of two k-maximal submonoids is generated by at most $k - 1$ elements. The following examples, suggested to us by Štěpán Holub, provide a negative answer to this question.

Example 16. The intersection of the two 3-maximal monoids $\{abc, dc, bab\}^*$ and $\{ab, cb, cd\}^*$ is infinitely generated by $abc(dc)^*bab$. The intersection of the two 4-maximal monoids $\{a, b, cd, ce\}^*$ and $\{ac, bc, da, ea\}^*$ is $\{acea, bcea, acda, bcda\}^*$.

Thus, our Theorem 14 is specific for rank 2 and cannot be generalized to larger k.

For an upper bound on the length of the word that generates the intersection of two 2-maximal submonoids, we have the following proposition.

Proposition 17. *With the hypotheses of Theorem 14,*

$$|z| < (|x| + |y|)(|u| + |v|).$$

Proof. Let $\mathcal{A}_X$ (resp. $\mathcal{A}_U$) be the minimal DFA recognizing X^* (resp. U^*) and Q_X (resp. Q_U) its set of states. Since X and U are bifix codes, we have $|Q_X| < |x| + |y|$ and $|Q_U| < |u| + |v|$. Then the automaton $\mathcal{A}$ recognizing $X^* \cap U^*$ has a set of states Q such that $|Q| < (|x| + |y|)(|u| + |v|)$. By Theorem 14, $\mathcal{A}$ is composed by only one cycle, labeled by z. Thus, $|z| < (|x| + |y|)(|u| + |v|)$.   □

For all our examples, the bound is much smaller than the previous one, hence we pose the following

Problem 18. Find a tight bound on the length of z in terms of the lengths of x and y and u and v.

4   Primitive Sets

We now show how the previous results can be interpreted in the terminology of combinatorics on words.

Let us start with the remark that a word $x \in A^+$ is primitive if and only if

$$x \in u^*, u \in A^+ \Rightarrow x = u.$$

With our definition of maximality, we have that a word $x \in A^+$ is primitive if and only if the monoid x^* is 1-maximal. Inspired by this observation, we give the following definition.

Definition 19. *A finite set $X \subseteq A^*$ is primitive if it is the basis of a $|X|$-maximal submonoid.*

Remark 20. The definition of primitive set does not coincide with that of elementary set. A set X is said to be elementary if $r(X) = |X|$. If X is primitive, then $r(X) = |X|$, i.e., it is elementary. But there exist elementary sets that are not primitive. For instance, the set $\{aa, bca\}$ is elementary, but it is not primitive since $\{aa, bca\}^* \subseteq \{a, bc\}^*$. The set $\{a, bc\}$, instead, is primitive.

From the definition of primitive set, we have that for every set X there exists a primitive set Y such that $X \subseteq Y^*$. The set Y is therefore called a *primitive root* of X. However, the primitive root of a set is not, in general, unique. Consider for instance the set $X = \{abcbab, abcdcbab, abcdcdcbab\}$. It has rank 3, hence it is elementary, yet it is not primitive. Indeed, $X \subseteq \{ab, cb, cd\}^*$. The set $\{ab, cb, cd\}$ is primitive, and it is a primitive root of X. However, it is not the only primitive root of X: the set $\{abc, dc, bab\}$ is primitive and $X \subseteq \{abc, dc, bab\}^*$, hence $\{abc, dc, bab\}$ is another primitive root of X. In the special case of sets of rank 1, clearly these always have a unique primitive root. For instance, the primitive root of the set $\{abab, abababab\}$ is the set $\{ab\}$.

However, as a consequence of Theorem 14 we have the following result.

Theorem 21. *A set X of rank 2 has a unique primitive root.*

Proof. If $\{u_1, u_2\}$ and $\{v_1, v_2\}$ are two primitive roots of X then $X^* \subseteq \{u_1, u_2\}^* \cap \{v_1, v_2\}^*$. Hence, by Theorem 14, $X \subseteq \{z\}^*$, for some primitive word z, i.e. $r(X) = 1$, a contradiction. □

In what follows, we find convenient call a primitive set of cardinality 2 a *primitive pair*.

Example 22. The words $abca$ and bc are primitive words, yet the pair $\{abca, bc\}$ is not a primitive pair, since $\{abca, bc\}^* \subseteq \{a, bc\}^*$, hence $\{abca, bc\}^*$ is not 2-maximal. The pair $\{abcabc, bcabca\}$ can be written as concatenations of copies of both $\{abca, bc\}$ and $\{a, bcabc\}$. However, there is a unique way to decompose each word of the pair $\{abcabc, bcabca\}$ as a concatenation of words of a primitive pair, and this pair is $\{a, bc\}$. Indeed, the primitive root of $\{abcabc, bcabca\}$ is $\{a, bc\}$.

As it is well known, a primitive word x does not have internal occurrences in xx. The next result, whose proof is omitted for brevity, provides a similar property in the case of a primitive set of two words.

Theorem 23. *Let $\{x, y\}$ be a primitive pair. Then neither xy nor yx occurs internally in a word of $\{x, y\}^3$.*

Example 24. Let $x = abcabca$, $y = bcaabcabc$. Then xy has an internal occurrence in yxx, yet $\{x, y\} \subset \{a, bc\}^*$. This example shows that the hypothesis that $\{x, y\}$ is primitive cannot be replaced by simply requiring that x and y are primitive words.

Differently to the case of a single primitive word, the converse of Theorem 23 does not hold. For example, $\{abcaa, bc\}$ is not primitive ($\{abcaa, bc\}^* \subseteq \{a, bc\}^*$), yet neither $abcaabc$ nor $bcabcaa$ occurs internally in a word of $\{x, y\}^3$.

5   Binary Root of a Single Primitive Word

In this section, we derive some consequences on the combinatorics of a *single* word. In particular, we introduce the notion of binary root of a primitive word, and we show how this notion may be useful to reveal some hidden repetitive structure in the word.

Let w be a nonempty word. If w is not primitive, then it can be written in a unique way as a concatenation of copies of a primitive word r, called the *root* of w. However, if w is primitive, one can ask whether it can be written as a concatenation of copies of two words x and y. If we further require that $\{x, y\}$ is a primitive set, then we call $\{x, y\}$ a *binary root* of the word w. Note that the binary root of a single word is not, in general, unique. For instance, for $w = abcbac$ we have $w = ab \cdot cbac = abcb \cdot ac$ and $\{ab, cbac\}$ and $\{abcb, ac\}$ are both primitive pairs, i.e., they are both binary roots of w. However, if we additionally require that the size $|x| + |y|$ of the binary root $\{x, y\}$ is "short" with respect to the length of w, then we obtain again the uniqueness. This is shown in the next theorem.

Theorem 25. *Let w be a primitive word. Then w has at most one binary root $\{x, y\}$ such that $|x| + |y| < \sqrt{|w|}$.*

Proof. Suppose by contradiction there exists another binary root $\{u, v\}$ of w with $|u| + |v| < \sqrt{|w|}$. Take $X = \{x, y\}$ and $U = \{u, v\}$. By Theorem 14, there exists a primitive word z and an integer n such that $w = z^n$. As w is primitive, $w = z$ and $n = 1$. By Proposition 17, we have that $|w| < (|x| + |y|)(|u| + |v|) < \sqrt{|w|} \cdot \sqrt{|w|} = |w|$, a contradiction.

$\square$

The following example shows a word w that has binary roots of different sizes, but only one of size less than $\sqrt{|w|}$.

Example 26. Consider the primitive word $w = abcaabcabc$ of length 10. The pair $\{a, bc\}$ is the only binary root of w of size smaller than $\sqrt{|w|}$.

Asking for a tight bound in the statement of Theorem 25 is of course a problem intimately related to Open Problem 18.

We observe that both the classical notion of root and that of binary root are related to some repetitive structure inside the word. If w is not primitive, the length of its root reveals its repetitive structure in the sense that, if such a length is much smaller than the length of w, then the word w can be considered highly repetitive. If w is primitive, the size of its binary root plays an analogous role. This could be illustrated by the following (negative) example. Consider a word w over the alphabet A such that all the letters of w are distinct, so that $|w| = |A|$. This word is not repetitive at all, and it has $|w| - 1$ different binary roots $\{x, y\}$, all of size $|w|$, corresponding to the trivial factorizations $w = xy$. Thus, the absence of repetitions in a word is related to the large size of its binary roots. On the contrary, the existence in a word w of a "short" (with respect to $|w|$) binary root corresponds to the existence of some hidden repetitive structure in the word. This approach generalizes some already-considered notions of hidden repetitions (cf. [5–7]).

We think that the notion of a binary root can be further explored and may have applications, e.g., in the area of string algorithms.

Notice that the minimal length of a binary root (intended as the sum of the lengths of the two components of the pair) is affected by the combinatorial properties of the word. For example, if w is a square-free word, then w cannot have a binary root $\{x, y\}$ such that $|x| + |y| < |w|/4$, since otherwise w would contain a square (xx, yy, $xyxy$ or $yxyx$). The previous remark suggests a possible link between the notion of a binary root and the classical notion of *binary pattern*, which has been deeply investigated in combinatorics on words and fully classified by Cassaigne [3] (see also [15, Chap. 3] for a survey).

6   Connections with Pseudo-Primitive Words

We now show how the notion of a primitive pair can be seen as a generalization of the notion of a pseudo-primitive word, with respect to an involutive (anti-)morphism θ, as introduced in [4].

A map $\theta : A^* \to A^*$ is a *morphism* (resp. *antimorphism*) if for each $u, v \in A^*$, $\theta(uv) = \theta(u)\theta(v)$ (resp. $\theta(uv) = \theta(v)\theta(u)$)—$\theta$ is an *involution* if $\theta(\theta(a)) = a$ for every $a \in A$.

Let θ be an involutive morphism or antimorphism other than the identity function. We say that a word $w \in A^*$ is a θ-*power* of t if $w \in t\{t, \theta(t)\}^*$. A word w is θ-*primitive* if there exists no nonempty word t such that w is a θ-power of t and $|w| > |t|$.

Theorem 27 ([4]). *Given a word $w \in A^*$ and an involutive (anti-)morphism θ, there exists a unique θ-primitive word $u \in A^*$ such hat w is a θ-power of u. The word u is called the θ-root of w.*

Example 28. Let $\theta : \{a, b, c\}^* \to \{a, b, c\}^*$ the involutive morphism defined by $\theta(a) = b$, $\theta(b) = a$ and $\theta(c) = c$. The θ-*root* of the word $abcabcbac$ is abc.

If θ is an involutive morphism, we show that Theorem 27 can be obtained as a consequence of Theorem 21. If θ is an involutive antimorphism, we obtain a slightly different formulation, from which we derive a new property of θ-primitive words.

Given a morphism θ and a set $X \subseteq A^*$, $\theta(X)$ denotes the set $\{\theta(u) \mid u \in X\}$. We say that X is θ-*invariant* if $\theta(X) \subseteq X$.

We have the following propositions.

Proposition 29. *Let θ be involutive. If $\{x, y\}$ is θ-invariant, then so is its root.*

Example 30. Let θ be as in Example 28. The pair $\{abcabcbac, abcbacabc\}$ is θ-invariant. However, it is not a primitive pair. Its binary root is the pair $\{abc, bac\}$, which is θ-invariant since $\theta(abc) = bac$.

Remark 31. Let θ be an involutive morphism. Then $\{x, y\}$ is θ-invariant if and only if $y = \theta(x)$. If θ is an involutive antimorphism, then $\{x, y\}$ is θ-invariant if and only if either $y = \theta(x)$ or $x = \theta(x)$ and $y = \theta(y)$. In the last case, x and y are called θ-*palindromes*.

Example 32. Let $\theta : \{a, b, c\}^* \mapsto \{a, b, c\}^*$ be the involutive antimorphism defined by $\theta(a) = a$, $\theta(b) = b$, $\theta(c) = c$. The pair $\{abcbbcba, abcba\}$ is θ-invariant. Its binary root is $\{a, bcb\}$, which is θ-invariant since composed by θ-palindromes. With the same θ, the pair $\{abbbbabba, abbabbbba\}$ is θ-invariant and its binary root is $\{abb, bba\}$, which is θ-invariant since $\theta(abb) = bba$.

Proposition 33. *Let $w \in A^*$ and θ be an involutive morphism of A^*. Then, w is θ-primitive if and only if the pair $\{w, \theta(w)\}$ is a primitive pair.*

Proof. Let us suppose, by contradiction, that $\{w, \theta(w)\}$ is a primitive pair and w is not θ-primitive. Then there exists t such that $w \in \{t, \theta(t)\}^*$. Hence, $\theta(w) \in \{t, \theta(t)\}^*$, so the pair $\{w, \theta(w)\}$ is not primitive. Conversely, let us suppose that w is θ-primitive and $\{w, \theta(w)\}$ is not a primitive pair. Denote by $\{u, v\}$ its binary root. Since $\{w, \theta(w)\}$ is θ-invariant, then $\{u, v\}$ is θ-invariant, i.e., $v = \theta(u)$. Hence, $w \in \{u, \theta(u)\}^*$, i.e., w is not θ-primitive. □

From Theorem 21 and Proposition 33 we derive Theorem 27 when θ is an involutive morphism.

Now, let us consider the case of antimorphisms. Reasoning analogously as we did in the proof of Proposition 33, we can prove the following result.

Proposition 34. *Let $w \in A^*$ and θ an involutive antimorphism of A^*. If the pair $\{w, \theta(w)\}$ is a primitive pair, then w is θ-primitive.*

The converse does not hold in general, as the following example shows.

Example 35. Let θ be the antimorphic involution of Example 32. The word $w = abbaabbacbc$ is θ-primitive, whereas the pair $\{w, \theta(w)\} = \{abbaabbacbc, cbcabbaabba\}$ is not a primitive pair, since its binary root is the pair $\{abba, cbc\}$.

Finally, we can state the following proposition, which provides a factorization property of θ-primitive words.

Proposition 36. *Let $w \in A^*$ and θ an involutive antimorphism. If w is θ-primitive and $\{w, \theta(w)\}$ is not a primitive pair, then there exist two θ-palindromes p and q such that $w \in \{p, q\}^*$.*

Proof. Suppose that $\{w, \theta(w)\}$ is not a primitive pair and denote by $\{u, v\}$ its binary root. Since $\{w, \theta(w)\}$ is θ-invariant, then so is $\{u, v\}$ by Proposition 29, and $v \neq \theta(u)$ since w is θ-primitive. Then, $u = \theta(u)$ and $v = \theta(v)$ are θ-palindromes. $\qquad\square$

Finally, we point out that our Theorem 23 can be viewed as a generalization of the following result of Kari, Masson and Seki [11]:

Theorem 37 (Theorem 12 of [11]). *Let x be a nonempty θ-primitive word. Then neither $x\theta(x)$ nor $\theta(x)x$ occurs internally in a word of $\{x, \theta(x)\}^3$.*

Acknowledgments. We thank Štěpán Holub for useful discussions and in particular for suggesting us the important Example 16.

References

1. Berstel, J., Perrin, D., Perrot, J.F., Restivo, A.: Sur le théorème du défaut. J. Algebra **60**(1), 169–180 (1979)
2. Berstel, J., Perrin, D., Reutenauer, C.: Codes and Automata (Encyclopedia of Mathematics and Its Applications), 1st edn. Cambridge University Press, New York (2009)
3. Cassaigne, J.: Motifs évitables et régularités dans les mots. Ph.D. thesis, Université Paris VI (1994)
4. Czeizler, E., Kari, L., Seki, S.: On a special class of primitive words. Theoret. Comput. Sci. **411**(3), 617–630 (2010)
5. Gawrychowski, P., Manea, F., Mercas, R., Nowotka, D.: Hide and seek with repetitions. J. Comput. Syst. Sci. **101**, 42–67 (2019)

6. Gawrychowski, P., Manea, F., Mercas, R., Nowotka, D., Tiseanu, C.: Finding pseudo-repetitions. In: Portier, N., Wilke, T. (eds.) STACS 2013, Proceedings. LIPIcs, vol. 20, pp. 257–268. Schloss Dagstuhl - Leibniz-Zentrum fuer Informatik (2013)
7. Gawrychowski, P., Manea, F., Nowotka, D.: Discovering hidden repetitions in words. In: Bonizzoni, P., Brattka, V., Löwe, B. (eds.) CiE 2013. LNCS, vol. 7921, pp. 210–219. Springer, Heidelberg (2013). https://doi.org/10.1007/978-3-642-39053-1_24
8. Harju, T., Karhumäki, J.: On the defect theorem and simplifiability. Semigroup Forum **33**(1), 199–217 (1986)
9. Harju, T., Karhumäki, J.: Many aspects of defect theorems. Theoret. Comput. Sci. **324**(1), 35–54 (2004)
10. Karhumäki, J.: A note on intersections of free submonoids of a free monoid. Semigroup Forum **29**(1), 183–205 (1984)
11. Kari, L., Masson, B., Seki, S.: Properties of pseudo-primitive words and their applications. Int. J. Found. Comput. Sci. **22**(2), 447–471 (2011)
12. Le Rest, E.B., Le Rest, M.: Sur la combinatoire des codes à deux mots. Theoret. Comput. Sci. **41**(C), 61–80 (1985)
13. Lentin, A., Schützenberger, M.: A combinatorial problem in the theory of free monoids. In: Proceedings of the University of North Carolina, pp. 128–144 (1967)
14. Lothaire, M.: Combinatorics on Words. Addison-Wesley, Boston (1983)
15. Lothaire, M.: Algebraic Combinatorics on Words. Cambridge University Press, Cambridge (2002)
16. Néraud, J.: Elementariness of a finite set of words is co-NP-complete. ITA **24**, 459–470 (1990)
17. Néraud, J.: Deciding whether a finite set of words has rank at most two. Theoret. Comput. Sci. **112**(2), 311–337 (1993)
18. Néraud, J.: On the rank of the subsets of a free monoid. Theoret. Comput. Sci. **99**(2), 231–241 (1992)
19. Restivo, A.: On a question of McNaughton and Papert. Inf. Control **25**(1), 93–101 (1974)
20. Tilson, B.: The intersection of free submonoids of a free monoid is free. Semigroup Forum **4**(1), 345–350 (1972)

Independent Systems of Word Equations: From Ehrenfeucht to Eighteen

Aleksi Saarela$^{(\boxtimes)}$ (iD)

Department of Mathematics and Statistics, University of Turku,
20014 Turku, Finland
amsaar@utu.fi

Abstract. A system of equations is called independent if it is not equivalent to any of its proper subsystems. We consider the following decades-
old question: If we fix the number of variables, then what is the maximal
size of an independent system of constant-free word equations? This can
be easily answered in the trivial cases of one and two variables, but all
other cases remain open, even the three-variable case, where the conjectured answer is as small as three. We survey some historical as well
as more recent results related to this question, starting with the one
known as Ehrenfeucht's compactness property: Every infinite system is
equivalent to a finite subsystem, and consequently an independent system cannot be infinite. We also discuss several variations and related
questions on word equations. Finally, we pay special attention to the following result from 2018: The maximal size of an independent system of
three-variable equations is at most 18. This is the first such finite upper
bound, but hopefully it will not be the last.

Keywords: Combinatorics on words · Word equation ·
Independent system

1 Introduction

The following two questions are among the most important open problems in
combinatorics on words: Is the satisfiability problem of word equations (that
is, the problem of deciding whether a given word equation has a solution) NP-
complete? What is the maximal size of an independent system of constant-free
n-variable word equations? The satisfiability problem is known to be NP-hard, it
was proved to be decidable by Makanin [20] and in PSPACE by Plandowski [25],
and a simpler PSPACE algorithm was given by Jeż [16]. The question about the
maximal sizes of independent systems, on the other hand, is the topic of this
article.

We start in Sect. 2 by giving formal definitions and simple examples of word
equations and independent systems. We continue in Sect. 3 by describing Ehrenfeucht's conjecture and the idea of its proof. In Sects. 4 and 5, we introduce the
main open questions and survey some results related to them. Rather than just

© Springer Nature Switzerland AG 2019
R. Mercaş and D. Reidenbach (Eds.): WORDS 2019, LNCS 11682, pp. 60–67, 2019.
https://doi.org/10.1007/978-3-030-28796-2_4

presenting the current state-of-the-art, we try to give a rough picture of how the research has progressed from 1983 to 2018. We conclude in Sect. 6 by describing some variations of the main questions.

2 Preliminaries

Let Ξ be an alphabet of n variables and Σ an alphabet of constants. An n-variable word equation is a pair $(u, v) \in (\Xi \cup \Sigma)^* \times (\Xi \cup \Sigma)^*$. The equation (u, v) is constant-free if $u, v \in \Xi^*$, and it is trivial if $u = v$. The length of (u, v) is $|uv|$ and it is denoted by $|(u, v)|$.

A constant-preserving morphism $h : (\Xi \cup \Sigma)^* \to \Sigma^*$ is a solution of an equation (u, v) if $h(u) = h(v)$. The set of all solutions of (u, v) is denoted by $\mathrm{Sol}((u, v))$. The morphism h is periodic if $h(\Xi) \subseteq w^*$ for some $w \in \Sigma^*$.

Example 1. First, let $\Xi = \{x\}$ and $\Sigma = \{a, b\}$. The equation $(xaxbab, abaxbx)$ has two solutions f and g defined by $f(x) = \varepsilon$ and $g(x) = ab$:

$$f(xaxbab) = \varepsilon \cdot a \cdot \varepsilon \cdot bab = abab = aba \cdot \varepsilon \cdot b \cdot \varepsilon = f(abaxbx),$$
$$g(xaxbab) = ab \cdot a \cdot ab \cdot bab = abaabbab = aba \cdot ab \cdot b \cdot ab = g(abaxbx).$$

Then, let $\Xi = \{x, y, z\}$ and $\Sigma = \{a, b\}$. The constant-free equation (xyz, zyx) has infinitely many solutions. For example, for all $p, q \in \Sigma^*$ and $i, j, k \geq 0$, the morphism h defined by $h(x) = (pq)^i p$, $h(y) = (qp)^j q$, $h(z) = (pq)^k p$ is a solution of this equation because

$$h(xyz) = (pq)^i p \cdot (qp)^j q \cdot (pq)^k p = (pq)^k p \cdot (qp)^j q \cdot (pq)^i p = h(zyx).$$

A system of equations is a set of equations. A morphism is a solution of a system if it is a solution of every equation in the system. The set of solutions of a system S is denoted by $\mathrm{Sol}(S)$, so $\mathrm{Sol}(S) = \bigcap_{E \in S} \mathrm{Sol}(E)$. Two equations or systems are equivalent if they have the same set of solutions. A subset (proper subset) of a system S is called a subsystem (proper subsystem, respectively) of S. A system is independent if it is not equivalent to any of its proper subsystems. Clearly, a system S is independent if and only if for every $E \in S$, there exists a morphism h such that $h \notin \mathrm{Sol}(E)$ but $h \in \mathrm{Sol}(E')$ for all $E' \in S \setminus \{E\}$.

Example 2. Let $\Xi = \{x, y, z\}$ and $\Sigma = \{a, b\}$. The system of equations $S = \{(xyz, zyx), (xyyz, zyyx)\}$ is independent and has a nonperiodic solution h defined by $h(x) = a$, $h(y) = b$, $h(z) = a$. To see independence, note that S is not equivalent to (xyz, zyx) because the morphism h defined by $h(x) = a$, $h(y) = b$, $h(z) = aba$ is a solution of (xyz, zyx) but not of S, and S is not equivalent to $(xyyz, zyyx)$ because the morphism h defined by $h(x) = a$, $h(y) = b$, $h(z) = abba$ is a solution of $(xyyz, zyyx)$ but not of S.

3   Ehrenfeucht's Conjecture

A subset K of a language L is a *test set* of L if there does not exist morphisms f, g such that $f(w) = g(w)$ for all $w \in K$ but $f(w) \neq g(w)$ for some $w \in L$. Ehrenfeucht conjectured at the beginning of the 1970s that every language has a finite test set. In 1983, Culik and Karhumäki [3] proved that this conjecture can be equivalently formulated as follows: Every system of word equations is equivalent to a finite subsystem. The conjecture was proved in 1985 by Albert and Lawrence [1] (an independent proof was given by Guba [9]), giving the following theorem, still known as *Ehrenfeucht's conjecture* or *Ehrenfeucht's compactness property*.

Theorem 1. *Every system of word equations is equivalent to a finite subsystem.*

The idea of the proof can be described as follows: Words can be turned into numbers by interpreting them as representations of integers in a suitable k-ary number system. A word equation can then be turned into a multivariate polynomial that has roots corresponding to all solutions of the word equation. A system of word equations then corresponds to a polynomial ideal. If an infinite system of word equations is not equivalent to any of its finite subsystems, then we get an infinite chain $I_1 \subsetneq I_2 \subsetneq I_3 \subsetneq \cdots$ of polynomial ideals, which contradicts a result from commutative algebra called *Hilbert's basis theorem*.

4   Size of Independent Systems

In 1983, Culik and Karhumäki [3] asked the following question, which is still open.

Question 1. Is it true that every independent system of three constant-free three-variable equations has only periodic solutions?

The following even more difficult question is also well-known.

Question 2. Let $\mathrm{IS}(n)$ be the maximal size of an independent system of constant-free n-variable word equations (or $\mathrm{IS}(n) = \infty$ if there is no maximum). How large is $\mathrm{IS}(n)$?

It is easy to prove that $\mathrm{IS}(1) = 1$, $\mathrm{IS}(2) = 2$, and $\mathrm{IS}(n) \geq n$ for all n. A positive answer to Question 1 would imply $\mathrm{IS}(3) = 3$. It follows from Theorem 1 that an independent system of word equations cannot be infinite. However, this does not prove even $\mathrm{IS}(n) < \infty$, because in principle it might be possible that there are arbitrarily large finite independent systems.

Some lower bounds better than the trivial $\mathrm{IS}(n) \geq n$ are known. In 1994, Karhumäki and Plandowski [17] gave examples of independent systems showing that $\mathrm{IS}(n) = \Omega(n^4)$. The hidden constant was improved by Karhumäki and Saarela [18], but no examples larger than $\Theta(n^4)$ have been found. The constructions can be said to be based on the fact that $(ababa)^k = (ab)^k a(ba)^k$ for all

$k \leq 2$ but not for any $k \geq 3$. Plandowski [24] pointed out that if we could find words $u_0, \ldots, u_m$ such that $u_0^k = u_1^k \ldots u_m^k$ for all $k \leq 3$ but not for any $k \geq 4$, then it would follow that $\text{IS}(n) = \Omega(n^5)$. However, these kinds of equalities had been studied before, for example in [10] and [13], and it was suspected that such words $u_0, \ldots, u_m$ do not exist. This was proved later, as will be discussed in Sect. 5.

In the three-variable case, some results restricting the form of equations in an independent system have been proved. For example, an equation (u, v) is called *balanced* if every variable occurs in u as many times as in v, and Harju and Nowotka [12] proved that if a system of two constant-free three-variable equations is independent and has a nonperiodic solution, then both equations are balanced. For more results, see [5] and [6].

We can try to find upper bounds that depend on the length of the equations in the system. If $E_1, \ldots, E_k$ is an independent system, then trivially k is at most exponential with respect to $\max\{|E_1|, \ldots, |E_k|\}$, simply because the number of equations of a certain length is exponential. The first nontrivial bound of a similar type was proved by Saarela [26]: In the case of constant-free three-variable equations, k is at most quadratic with respect to $\min\{|E_1|, \ldots, |E_k|\}$. This bound was improved to a linear one by Holub and Žemlička [14]. These two results were proved with the help of polynomials and linear algebra, so there are some similarities to the proof of Theorem 1, but the way in which polynomials were used is very different. The linear bound has been improved since then, as we shall see in Sect. 5, but the articles [26] and [14] contain also some results about n-variable equations that are still the best ones known.

5   Recent Results

In 2016, Nowotka and Saarela [21] found a new way to apply an old characterization of three-generator subsemigroups of a free semigroup by Budkina and Markov [2] (or alternatively a similar result by Spehner [29,30]) to analyze independent systems. Specifically, they found that an upper bound for the number of solutions a one-variable word equation with only finitely many solutions can have implies a (worse) upper bound for $\text{IS}(3)$.

The question about the maximal finite number of solutions of one-variable equations had been considered before [7,8,19], but the above connnection made it even more important. It had been proved that the number of solutions is at most logarithmic with respect to the number of occurrences of the variable in the equation [19]. From this it now followed that the size of an independent system of constant-free three-variable equations is at most logarithmic with respect to the length of the shortest equation in the system, thus improving the previous linear bound.

It had been conjectured that a one-variable equation with only finitely many solutions has at most two solutions. In view of [21], this conjecture would have implied that $\text{IS}(3) \leq 18$. In 2016, however, the attempts to prove the conjecture led to the discovery of the counterexample equation (first published in [22])

$$(xaxbxaabbabaxbabaabbab, abaabbabaxbabaabbxaxbx)$$

with exactly three solutions $h(x) = \varepsilon$, $h(x) = ab$ and $h(x) = abaabbab$, and to a weaker version of the conjecture: A one-variable equation with only finitely many solutions has at most three solutions. This weaker conjecture would have implied that $IS(3) \leq 23$. In the journal version [22] of the conference article [21], the conditional result was improved so that even the weaker conjecture would imply $IS(3) \leq 18$.

In 2017, Saarela [27] proved that there does not exist words $u_0, \ldots, u_m$ such that $u_0^k = u_1^k \ldots u_m^k$ for all $k \leq 3$ but not for any $k \geq 4$ (a stronger result was proved later in the journal version [28]). This meant that one particular approach for improving the lower bound $IS(n) = \Omega(n^4)$ was impossible, as was mentioned in Sect. 4. However, the more significant consequence of [27] was that Nowotka and Saarela found a way to apply a method similar to the one used in [27] to study one-variable equations. In 2018, the weaker conjecture about one-variable equations was finally proved, thus proving that $IS(3) \leq 18$ [23].

Theorem 2. *A one-variable word equation has either infinitely many solutions or at most three.*

Theorem 3. $3 \leq IS(3) \leq 18$.

The rough idea behind the proof of Theorem 2 is that given an equation (u, v) on the variable x, we can assume that the equation is in a certain kind of normal form, and then we find a morphism $\sigma : \Sigma^* \to \mathbb{Z}$ that satisfies certain properties, for example $\sigma(h(x)) = 0$ for all $h \in \text{Sol}((u, v))$. Then we study the images of prefixes of $h(u)$ and $h(v)$ under σ, how they match and how they change when the solution h changes. This eventually leads to a proof of Theorem 2.

Theorem 3 follows from Theorem 2 and the result in [22]. The idea behind the proof of the result in [22] is that if $E_1, \ldots, E_k$ is an independent system of constant-free three-variable equations and $h_1, \ldots, h_k$ are morphisms such that $h_i \in \text{Sol}(E_j)$ for all $i \neq j$ but not for $i = j$, then the morphisms h_i can be classified into a small number of families by the results in [2], and, given Theorem 2, not too many of them can be in the same family. Improving the result in [22] seems like a potential path towards a smaller upper bound than 18.

6   Variations

If we consider equations in a free semigroup instead of a free monoid, that is, we require that a solution h cannot map a variable to the empty word, then how large can an independent system of constant-free n-variable equations be? The largest known examples have size $\Theta(n^3)$ [17]. Independent systems can be studied also in other semigroups, see the survey of Harju, Karhumäki and Plandowski [11].

A finite sequence of nontrivial equations $E_1, \ldots, E_n$ is a *decreasing chain* if

$$\text{Sol}(E_1) \supsetneq \text{Sol}(E_1, E_2) \supsetneq \cdots \supsetneq \text{Sol}(E_1, \ldots, E_n).$$

and an *increasing chain* if

$$\mathrm{Sol}(E_1, \ldots, E_n) \subsetneq \mathrm{Sol}(E_2, \ldots, E_n) \subsetneq \cdots \subsetneq \mathrm{Sol}(E_n).$$

Similar definition can be given for infinite chains. Chains have been studied by Honkala [15] and Czeizler [4], for example. If the number of variables is fixed, then how long can chains be? Clearly, $E_1, \ldots, E_n$ is a decreasing chain if and only if $E_n, \ldots, E_1$ is an increasing chain, so if there exists a finite maximal length, then it is the same for both types of chains. However, if there are arbitrarily long chains, then the situation is more complicated: Ehrenfeucht's conjecture is equivalent to the fact that decreasing chains cannot be infinite, but no such result is known for increasing chains. In the case of constant-free three-variable equations, there are chains of length 7 [18], and the results in [22] and [23] imply an upper bound of 24.

Finally, independent systems can be considered also in the case of equations with constants. Independent systems can be larger in this case, but Ehrenfeucht's conjecture still holds. Moreover, an independent n-variable system of equations with constants cannot be larger than an independent $(n + 2)$-variable system of constant-free equations. This is because we can assume that the alphabet of constants is binary, and then replace the constant letters with new variables, and this preserves independence. Thus the question of the maximal size of independent systems does not become fundamentally different if we allow constants.

References

1. Albert, M.H., Lawrence, J.: A proof of Ehrenfeucht's conjecture. Theoret. Comput. Sci. **41**(1), 121–123 (1985). https://doi.org/10.1016/0304-3975(85)90066-0
2. Budkina, L.G., Markov, A.A.: F-semigroups with three generators. Mat. Zametki **14**, 267–277 (1973)
3. Culik II, K., Karhumäki, J.: Systems of equations over a free monoid and Ehrenfeucht's conjecture. Discrete Math. **43**(2–3), 139–153 (1983). https://doi.org/10.1016/0012-365X(83)90152-8
4. Czeizler, E.: Multiple constraints on three and four words. Theoret. Comput. Sci. **391**(1–2), 14–19 (2008). https://doi.org/10.1016/j.tcs.2007.10.026
5. Czeizler, E., Karhumäki, J.: On non-periodic solutions of independent systems of word equations over three unknowns. Int. J. Found. Comput. Sci. **18**(4), 873–897 (2007). https://doi.org/10.1142/S0129054107005030
6. Czeizler, E., Plandowski, W.: On systems of word equations over three unknowns with at most six occurrences of one of the unknowns. Theoret. Comput. Sci. **410**(30–32), 2889–2909 (2009). https://doi.org/10.1016/j.tcs.2009.01.023
7. Dąbrowski, R., Plandowski, W.: On word equations in one variable. Algorithmica **60**(4), 819–828 (2011). https://doi.org/10.1007/s00453-009-9375-3
8. Obono, S.E., Goralcik, P., Maksimenko, M.: Efficient solving of the word equations in one variable. In: Prívara, I., Rovan, B., Ruzička, P. (eds.) MFCS 1994. LNCS, vol. 841, pp. 336–341. Springer, Heidelberg (1994). https://doi.org/10.1007/3-540-58338-6_80
9. Guba, V.S.: Equivalence of infinite systems of equations in free groups and semigroups to finite subsystems. Mat. Zametki **40**(3), 321–324 (1986). https://doi.org/10.1007/BF01142470

10. Hakala, I., Kortelainen, J.: On the system of word equations $x_1^i x_2^i \cdots x_m^i = y_1^i y_2^i \cdots y_n^i$ $(i = 1, 2, \cdots)$ in a free monoid. Acta Inform. **34**(3), 217–230 (1997). https://doi.org/10.1007/s002360050081
11. Harju, T., Karhumäki, J., Plandowski, W.: Independent systems of equations. In: Lothaire, M. (ed.) Algebraic Combinatorics on Words, pp. 443–472. Cambridge University Press, Cambridge (2002)
12. Harju, T., Nowotka, D.: On the independence of equations in three variables. Theoret. Comput. Sci. **307**(1), 139–172 (2003). https://doi.org/10.1016/S0304-3975(03)00098-7
13. Holub, Š.: Local and global cyclicity in free semigroups. Theoret. Comput. Sci. **262**(1–2), 25–36 (2001). https://doi.org/10.1016/S0304-3975(00)00156-0
14. Holub, Š., Žemlička, J.: Algebraic properties of word equations. J. Algebra **434**, 283–301 (2015). https://doi.org/10.1016/j.jalgebra.2015.03.021
15. Honkala, J.: On chains of word equations and test sets. Bull. EATCS **68**, 157–160 (1999)
16. Jeż, A.: Recompression: a simple and powerful technique for word equations. J. ACM **63**(1), Art. 4, 51 (2016). https://doi.org/10.1145/2743014
17. Karhumäki, J., Plandowski, W.: On the defect effect of many identities in free semigroups. In: Paun, G. (ed.) Mathematical Aspects of Natural and Formal Languages, pp. 225–232. World Scientific (1994). https://doi.org/10.1142/9789814447133_0012
18. Karhumäki, J., Saarela, A.: On maximal chains of systems of word equations. Proc. Steklov Inst. Math. **274**, 116–123 (2011). https://doi.org/10.1134/S0081543811060083
19. Laine, M., Plandowski, W.: Word equations with one unknown. Int. J. Found. Comput. Sci. **22**(2), 345–375 (2011). https://doi.org/10.1142/S0129054111008088
20. Makanin, G.S.: The problem of the solvability of equations in a free semigroup. Mat. Sb. (N.S.) **103**(2), 147–236 (1977). English translation in Math. USSR Sb. **32**, 129–198 (1977)
21. Nowotka, D., Saarela, A.: One-unknown word equations and three-unknown constant-free word equations. In: Brlek, S., Reutenauer, C. (eds.) DLT 2016. LNCS, vol. 9840, pp. 332–343. Springer, Heidelberg (2016). https://doi.org/10.1007/978-3-662-53132-7_27
22. Nowotka, D., Saarela, A.: One-variable word equations and three-variable constant-free word equations. Int. J. Found. Comput. Sci. **29**(5), 935–950 (2018). https://doi.org/10.1142/S0129054118420121
23. Nowotka, D., Saarela, A.: An optimal bound on the solution sets of one-variable word equations and its consequences. In: Proceedings of the 45th ICALP. LIPIcs, vol. 107, pp. 136:1–136:13. Schloss Dagstuhl-Leibniz-Zentrum fuer Informatik (2018). https://doi.org/10.4230/LIPIcs.ICALP.2018.136
24. Plandowski, W.: Test sets for large families of languages. In: Ésik, Z., Fülöp, Z. (eds.) DLT 2003. LNCS, vol. 2710, pp. 75–94. Springer, Heidelberg (2003). https://doi.org/10.1007/3-540-45007-6_6
25. Plandowski, W.: Satisfiability of word equations with constants is in PSPACE. J. ACM **51**(3), 483–496 (2004)
26. Saarela, A.: Systems of word equations, polynomials and linear algebra: a new approach. Eur. J. Comb. **47**, 1–14 (2015). https://doi.org/10.1016/j.ejc.2015.01.005
27. Saarela, A.: Word equations where a power equals a product of powers. In: Proceedings of the 34th STACS. LIPIcs, vol. 66, pp. 55:1–55:9. Schloss Dagstuhl-Leibniz-Zentrum fuer Informatik (2017). https://doi.org/10.4230/LIPIcs.STACS.2017.55

28. Saarela, A.: Word equations with kth powers of variables. J. Comb. Theory Ser. A **165**, 15–31 (2019). https://doi.org/10.1016/j.jcta.2019.01.004
29. Spehner, J.C.: Quelques problémes d'extension, de conjugaison et de présentation des sous-monoïdes d'un monoïde libre. Ph.D. thesis, Univ. Paris (1976)
30. Spehner, J.C.: Les systemes entiers d'équations sur un alphabet de 3 variables. In: Semigroups, pp. 342–357 (1986)

Parikh Determinants

Adrian Atanasiu[1], Ghajendran Poovanandran[2(✉)], and Wen Chean Teh[3]

[1] Faculty of Mathematics and Computer Science, Bucharest University,
Str. Academiei 14, 010014 Bucharest, Romania
aadrian@gmail.com
[2] School of Mathematics, Actuarial and Quantitative Studies,
Asia Pacific University of Technology & Innovation,
Technology Park Malaysia, 57000 Bukit Jalil, Kuala Lumpur, Malaysia
ghajendran@staffemail.apu.edu.my
[3] School of Mathematical Sciences, Universiti Sains Malaysia,
11800 USM, Malaysia
dasmenteh@usm.my

Abstract. Parikh matrices, introduced by Mateescu et al. in 2001, are generalization of the classical Parikh vectors. These special matrices are often utilized in the combinatorial study of words as an elegant tool to compute the number of occurrences of certain subwords in a word. In this paper, we study the determinant of a certain submatrix of a Parikh matrix, where the submatrix preserves the information contained in the original matrix. We present a formula to compute such a determinant, which we term as the Parikh determinant, for any given word. By using a classical result on Parikh matrices, we establish Parikh determinants as a natural combinatorial characteristic of words. Consequently, a new general identity involving the number of occurrences of certain subwords of a word is obtained. Finally, we address some related observations and possible future directions of this study.

Keywords: Parikh matrices · Subwords · Core words · Determinants

1 Introduction

In combinatorics of words, it is of interest to characterize a word by using numerical quantities. One such numerical quantity that is often investigated in the literature is the number of occurrences of a word as a subword of another word. In dealing with studies of this nature, the Parikh matrix mapping, which was introduced by Mateescu et al. in [11], is a useful tool.

Parikh matrices are a generalization of the classical Parikh vectors [13]. The Parikh matrix of a word is an upper triangular matrix which contains information on the number of occurrences of certain subwords of that word. In general, a Parikh matrix does not uniquely determine a word. Nevertheless, Parikh matrices and their variants [1,6–8,16] have led to various new investigations in the combinatorial study of words (for example, see [2–5,9,10,12,14,15,17–22]).

© Springer Nature Switzerland AG 2019
R. Mercaş and D. Reidenbach (Eds.): WORDS 2019, LNCS 11682, pp. 68–79, 2019.
https://doi.org/10.1007/978-3-030-28796-2_5

In this work, we study a natural algebraic property of Parikh matrices. Particularly, we scrutinize the determinant of a certain submatrix of a Parikh matrix—we term such determinants as Parikh determinants. The submatrix considered preserves the essential information from the original Parikh matrix. Our main contributions would be as follows:

1. A general formula to compute the Parikh determinant of any given word;
2. A new general identity involving the number of occurrences of certain subwords of a word.

The remainder of this paper is structured as follows. Section 2 provides the basic terminology and preliminaries. Section 3, which is the core section of this paper, introduces and studies Parikh determinants. By using the Leibniz formula for determinants, a general formula to compute the Parikh determinant of a word is developed. This formula is then generalized to compute the Parikh determinant of a word raised to an arbitrary power. Also, the Parikh determinant of a word is shown to be equivalent to the number of occurrences of a certain subword of the word—thus realizing a new general identity involving subword occurrences in a word. Finally, Sect. 4 contains some final remarks and possible future directions of this study.

2   Preliminaries

The set of all positive integers up to k is denoted by $[k]$. We denote a sequence of integers $i_1, i_2, \ldots, i_n$ by $\langle i_1, i_2, \ldots, i_n \rangle$. The length of a sequence α is denoted by $|\alpha|$. Given a set X, we denote by $X^{<\omega}$ the set of all finite sequences over X.

Suppose A is a $k \times k$ square matrix. We denote the (i, j)-entry of A by $A_{i,j}$. The determinant of A is denoted by $\|A\|$. For all integers $1 \leq i, j \leq k$, we denote by $A(i, j)$ the submatrix obtained by deleting the i-th row and j-th column of A.

Suppose Σ is a finite nonempty alphabet. The set of all words over Σ is denoted by Σ^*. The unique empty word is denoted by λ. Given a word $w = a_1 a_2 \cdots a_k$ ($a_i \in \Sigma$ for all $1 \leq i \leq k$), we denote by $\mathrm{mi}(w)$ the *mirror image* of w, that is $\mathrm{mi}(w) = a_k a_{k-1} \cdots a_1$. If $v, w \in \Sigma^*$, the concatenation of v and w is denoted by vw. An *ordered alphabet* is an alphabet $\Sigma = \{a_1, a_2, \ldots, a_k\}$ with a total ordering on it. For example, if $a_1 < a_2 < \cdots < a_k$, then we may write $\Sigma = \{a_1 < a_2 < \cdots < a_k\}$. For all integers $1 \leq i \leq j \leq k$, let $a_{i,j}$ denote the word $a_i a_{i+1} \cdots a_j$. Frequently, we will abuse notation and use Σ to stand for both the ordered alphabet and its underlying alphabet. Suppose $\Gamma \subseteq \Sigma$. The projective morphism $\pi_\Gamma : \Sigma^* \to \Gamma^*$ is defined by

$$\pi_\Gamma(a) = \begin{cases} a, & \text{if } a \in \Gamma \\ \lambda, & \text{otherwise.} \end{cases}$$

A word v is a *scattered subword* (or simply *subword*) of $w \in \Sigma^*$ if and only if there exist $x_1, x_2, \ldots, x_n, y_0, y_1, \ldots, y_n \in \Sigma^*$ (possibly empty) such that $v =$

$x_1 x_2 \cdots x_n$ and $w = y_0 x_1 y_1 \cdots y_{n-1} x_n y_n$. If the letters in v occur contiguously in w (that is $y_1 = y_2 = \ldots = y_{n-1} = \lambda$), then v is a *factor* of w. The number of occurrences of a word v as a subword of w is denoted by $|w|_v$. Two occurrences of v are considered different if and only if they differ by at least one position of some letter. For example, $|bcbcc|_{bc} = 5$ and $|aabcbc|_{abc} = 6$. By convention, $|w|_\lambda = 1$ for all $w \in \Sigma^*$.

For any integer $k \geq 2$, let $\mathcal{M}_k$ denote the multiplicative monoid of $k \times k$ upper triangular matrices with nonnegative integral entries and unit diagonal.

Definition 1. *Suppose* $\Sigma = \{a_1 < a_2 < \cdots < a_k\}$ *is an ordered alphabet, where* $k \geq 1$. *The Parikh matrix mapping with respect to* Σ, *denoted by* Ψ_Σ, *is the morphism*

$$\Psi_\Sigma : \Sigma^* \to \mathcal{M}_{k+1}$$

defined as follows: $\Psi_\Sigma(\lambda) = I_{k+1}$; *if* $\Psi_\Sigma(a_q) = M$, *then*

- $M_{i,i} = 1$ *for each* $1 \leq i \leq k+1$;
- $M_{q,q+1} = 1$; *and*
- *all other entries of the matrix* $\Psi_\Sigma(a_q)$ *are zero.*

Matrices of the form $\Psi_\Sigma(w)$ *for* $w \in \Sigma^*$ *are called Parikh matrices.*

Remark 1. The Parikh vector $\Psi(w) = (|w|_{a_1}, |w|_{a_2}, \ldots, |w|_{a_k})$ of a word $w \in \Sigma^*$ is embedded in the second diagonal of the Parikh matrix $\Psi_\Sigma(w)$.

Theorem 1 [11]. *Suppose* $\Sigma = \{a_1 < a_2 < \cdots < a_k\}$ *is an ordered alphabet and* $w \in \Sigma^*$. *The matrix* $\Psi_\Sigma(w) = M$ *has the following properties:*

- $M_{i,i} = 1$ *for each* $1 \leq i \leq k+1$;
- $M_{i,j} = 0$ *for each* $1 \leq j < i \leq k+1$;
- $M_{i,j+1} = |w|_{a_{i,j}}$ *for each* $1 \leq i \leq j \leq k$.

Example 1. Suppose $\Sigma = \{a < b < c < d\}$ and $w = dcabac$. Then

$$\Psi_\Sigma(w) = \Psi_\Sigma(d)\Psi_\Sigma(c)\Psi_\Sigma(a)\Psi_\Sigma(b)\Psi_\Sigma(a)\Psi_\Sigma(c)$$

$$= \begin{pmatrix} 1 & 0 & 0 & 0 & 0 \\ 0 & 1 & 0 & 0 & 0 \\ 0 & 0 & 1 & 0 & 0 \\ 0 & 0 & 0 & 1 & 1 \\ 0 & 0 & 0 & 0 & 1 \end{pmatrix} \begin{pmatrix} 1 & 0 & 0 & 0 & 0 \\ 0 & 1 & 0 & 0 & 0 \\ 0 & 0 & 1 & 1 & 0 \\ 0 & 0 & 0 & 1 & 0 \\ 0 & 0 & 0 & 0 & 1 \end{pmatrix} \cdots \begin{pmatrix} 1 & 0 & 0 & 0 & 0 \\ 0 & 1 & 0 & 0 & 0 \\ 0 & 0 & 1 & 1 & 0 \\ 0 & 0 & 0 & 1 & 0 \\ 0 & 0 & 0 & 0 & 1 \end{pmatrix}$$

$$= \begin{pmatrix} 1 & 2 & 1 & 1 & 0 \\ 0 & 1 & 1 & 1 & 0 \\ 0 & 0 & 1 & 2 & 0 \\ 0 & 0 & 0 & 1 & 1 \\ 0 & 0 & 0 & 0 & 1 \end{pmatrix} = \begin{pmatrix} 1 & |w|_a & |w|_{ab} & |w|_{abc} & |w|_{abcd} \\ 0 & 1 & |w|_b & |w|_{bc} & |w|_{bcd} \\ 0 & 0 & 1 & |w|_c & |w|_{cd} \\ 0 & 0 & 0 & 1 & |w|_d \\ 0 & 0 & 0 & 0 & 1 \end{pmatrix}.$$

3   Parikh Determinant of a Word

Since Parikh matrices are upper triangular matrices with ones in the main diagonal, the determinant of a Parikh matrix is always one—therefore it is not interesting to further study its determinant. Clearly, if the last row and the first column of a Parikh matrix is removed, the information (on the word) contained in the original Parikh matrix is still preserved by the resulting submatrix. Furthermore, the determinant of the resulting submatrix is no longer trivially one. This motivates the following notion, which will be the central object of our study.

Definition 2. *Suppose Σ is an ordered alphabet and $w \in \Sigma^*$. Let $M = \Psi_\Sigma(w)$. The Parikh determinant of the word w (with respect to Σ), denoted by $\Delta_\Sigma(w)$, is the determinant of the matrix $M(k+1,1)$.*

Remark 2. The idea of studying the determinant of the submatrix obtained by deleting the last row and the first column of a Parikh matrix appears also in [12] but with a single goal to prove the nonnegativity of *canonical subword histories*.

We now develop a formula (Theorem 2) to compute the Parikh determinant of a word for an arbitrary ordered alphabet. However, we first need the following lemma, which follows as a special case of the Leibniz formula for the determinant.

Lemma 1. *Suppose A is a $k \times k$ matrix such that $A_{i,j} = 0$ whenever $j \leq i - 2$ and $A_{i,i-1} = 1$ for all integers $2 \leq i \leq k$. Let*

$$S = \{\langle i_1, i_2, \ldots, i_n \rangle \in [k]^{<\omega} \mid i_1 < i_2 < \cdots < i_n = k\}.$$

Then,

$$\|A\| = \sum_{\langle i_1, i_2, \ldots, i_n \rangle \in S} (-1)^{n+k} A_{1,i_1} A_{i_1+1,i_2} \cdots A_{i_{n-1}+1,i_n}.$$

Proof. By the Leibniz formula of determinant, we have

$$\|A\| = \sum_{\sigma \in \mathrm{Sym}([k])} \left(\mathrm{sgn}(\sigma) \prod_{i=1}^{k} A_{i,\sigma(i)} \right) \tag{1}$$

where $\mathrm{Sym}([k])$ denotes the set of permutations on $[k]$ and

$$\mathrm{sgn}(\sigma) = \begin{cases} 1 & \text{if } \sigma \text{ is even,} \\ -1 & \text{if } \sigma \text{ is odd.} \end{cases}$$

Since $A_{i,j} = 0$ whenever $j \leq i - 2$, it follows that for every $\sigma \in \mathrm{Sym}([k])$ such that $\sigma(i) \leq i - 2$ for some $1 \leq i \leq k$, the term $\mathrm{sgn}(\sigma) \prod_{i=1}^{k} A_{i,\sigma(i)} = 0$. Thus, define

$$R = \{\sigma \in \mathrm{Sym}([k]) \mid \sigma(i) \geq i - 1 \text{ for all } 1 \leq i \leq k\},$$

then Eq. (1) can be written as

$$\|A\| = \sum_{\sigma \in R} \left(\text{sgn}(\sigma) \prod_{i=1}^{k} A_{i,\sigma(i)} \right) \tag{2}$$

For every $\sigma \in R$, define $r(\sigma) = \langle \sigma(i_1), \sigma(i_2), \ldots, \sigma(i_n) \rangle$ where $1 \leq i_j \leq k$ such that $i_j < i_{j+1}$ for every integer $1 \leq j \leq n-1$ and $\sigma(i_j) \geq i_j$ for every integer $1 \leq j \leq n$.

Observation 1. Let $\sigma \in R$ be arbitrary. For every integer $1 \leq i \leq k-1$,

- if $\sigma(i) = i$, then $\sigma(i+1) \geq i+1$;
- if $\sigma(i) > i$, let $i' = \sigma(i)$, then $\sigma(j) = j-1$ for every integer $i+1 \leq j \leq i'$ and $\sigma(i'+1) \geq i'+1$.

Suppose $\sigma \in R$ and let $\langle \sigma(i_1), \sigma(i_2), \ldots, \sigma(i_n) \rangle = r(\sigma)$. Then by the above observation, it follows that $i_{j+1} = \sigma(i_j) + 1$ for every integer $1 \leq j \leq n-1$. Thus $\sigma(i_j) < \sigma(i_{j+1})$ for every integer $1 \leq j \leq n-1$. Notice that $\sigma(i) \geq 1$ for all integers $1 \leq i \leq k$, thus $i_1 = 1$. Also, we have $i, \sigma(i) \leq k$ for all integers $1 \leq i \leq k$, thus it must be the case that $\sigma(i_n) = k$. By the above reasons, it holds that $\{ r(\sigma) \mid \sigma \in R \} = S$.

The following claim holds by the definition $\text{sgn}(\sigma) = (-1)^{N(\sigma)}$ where $N(\sigma)$ is the number of inversions in σ, a standard observation that $N(\sigma) = k - |r(\sigma)|$, and the fact that $k - |r(\sigma)|$ and $k + |r(\sigma)|$ have equal parity.

Claim 1. For every $\sigma \in R$, we have $\text{sgn}(\sigma) = (-1)^{|r(\sigma)|+k}$.

By the hypothesis, we have $A_{i,i-1} = 1$ for all integers $2 \leq i \leq k$. Thus by (2), our observation and the claim, it holds that

$$\|A\| = \sum_{\substack{i_1, i_2, \ldots, i_n \\ \langle \sigma(i_1), \sigma(i_2), \ldots, \sigma(i_n) \rangle = r(\sigma) \\ \sigma \in R}} (-1)^{n+k} A_{i_1, \sigma(i_1)} A_{i_2, \sigma(i_2)} \cdots A_{i_n, \sigma(i_n)}$$

$$= \sum_{\substack{i_1, i_2, \ldots, i_n \\ \langle \sigma(i_1), \sigma(i_2), \ldots, \sigma(i_n) \rangle = r(\sigma) \\ \sigma \in R}} (-1)^{n+k} A_{i_1, \sigma(i_1)} A_{\sigma(i_1)+1, \sigma(i_2)} \cdots A_{\sigma(i_{n-1})+1, \sigma(i_n)}$$

$$= \sum_{\langle i_1, i_2, \ldots, i_n \rangle \in S} (-1)^{n+k} A_{1, i_1} A_{i_1+1, i_2} \cdots A_{i_{n-1}+1, i_n}.$$

Thus the conclusion holds.

The following notation will be frequently used in the subsequent part of this section.

Definition 3. *Suppose* $\Sigma = \{ a_1 < a_2 < \cdots < a_k \}$ *and* $w \in \Sigma^*$. *For every* $\langle i_1, i_2, \ldots, i_n \rangle \in [k]^{<\omega}$ *and integer* m *such that* $m \leq i_1 < i_2 < \cdots < i_n$, *we define*

$$|w|_{\langle m, i_1, i_2, \ldots, i_n \rangle} = |w|_{a_{m}, i_1} |w|_{a_{i_1}+1, i_2} \cdots |w|_{a_{i_{n-1}}+1, i_n}.$$

Theorem 2. *Suppose $\Sigma = \{a_1 < a_2 < \cdots < a_k\}$ and $w \in \Sigma^*$. Let*

$$S = \{\langle i_1, i_2, \ldots, i_n \rangle \in [k]^{<\omega} \mid i_1 < i_2 < \cdots < i_n = k\}.$$

Then,

$$\Delta_\Sigma(w) = \sum_{\langle i_1, i_2, \ldots, i_n \rangle \in S} (-1)^{n+k} |w|_{\langle 1, i_1, i_2, \ldots, i_n \rangle}.$$

Proof. Let $M = \Psi_\Sigma(w)$ and $A = M(k+1, 1)$. Then the result directly holds by Lemma 1 since $\Delta_\Sigma(w) = \|A\|$ and

$$A_{i,j} = \begin{cases} |w|_{a_{i,j}} & \text{if } j \geq i, \\ 1 & \text{if } j = i - 1, \\ 0 & \text{if } j \leq i - 2. \end{cases}$$

Example 2. Suppose $\Sigma = \{a < b < c\} = \{a_1 < a_2 < a_3\}$ and $w \in \Sigma^*$. Then

$$\{\langle i_1, i_2, \ldots, i_n \rangle \in \{1, 2, 3\}^{<\omega} \mid i_1 < i_2 < \cdots < i_n = 3\}$$
$$= \{\langle 1, 2, 3 \rangle, \langle 1, 3 \rangle, \langle 2, 3 \rangle, \langle 3 \rangle\}.$$

Therefore,

$$\Delta_\Sigma(w) = |w|_{a_{1,1}} |w|_{a_{2,2}} |w|_{a_{3,3}} - |w|_{a_{1,1}} |w|_{a_{2,3}} - |w|_{a_{1,2}} |w|_{a_{3,3}} + |w|_{a_{1,3}}$$
$$= |w|_a |w|_b |w|_c - |w|_a |w|_{bc} - |w|_{ab} |w|_c + |w|_{abc}.$$

We now use the formula of the Parikh determinant of a word to develop an identity (Theorem 6) involving the number of occurrences of certain subwords of a word. The following notion and known result are essential for our purpose.

Definition 4. *Suppose $\Sigma = \{a_1 < a_2 < \cdots < a_k\}$ and $w \in \Sigma^*$. Let $M = \Psi_\Sigma(w)$. The alternate Parikh matrix of w, denoted by $\overline{\Psi}_\Sigma(w)$, is the matrix A such that $A_{i,j} = (-1)^{i+j} M_{i,j}$ for all $1 \leq i, j \leq k + 1$.*

Theorem 3 [11]. *Suppose $\Sigma = \{a_1 < a_2 < \cdots < a_k\}$ and $w \in \Sigma^*$. Then*

$$[\Psi_\Sigma(w)]^{-1} = \overline{\Psi}_\Sigma(\text{mi}(w)).$$

For a given ordered alphabet, the following result establishes a connection between the corresponding Parikh determinant of a word and the number of occurrences of a certain subword in that word.

Theorem 4. *Suppose $\Sigma = \{a_1 < a_2 < \cdots < a_k\}$ and $w \in \Sigma^*$. Then*

$$\Delta_\Sigma(w) = |w|_{a_k a_{k-1} \cdots a_1}.$$

Proof. Let $A = \Psi_\Sigma(w)$. Then

$$A^{-1} = \frac{1}{\|A\|} \operatorname{adj}(A).$$

Since $\|\Psi_\Sigma(x)\| = 1$ for any $x \in \Sigma^*$, it follows that $A^{-1} = \operatorname{adj}(A)$. By definition, the (i,j)-entry of $\operatorname{adj}(A)$ is $(-1)^{i+j}\|A(j,i)\|$.

Notice that the $(1, k+1)$-entry of $\operatorname{adj}(A)$ is $(-1)^{k+2}\|A(k+1,1)\|$. Therefore, by the definition of Parikh determinant, we have

$$A^{-1}_{1,k+1} = (-1)^{k+2}\|A(k+1,1)\| = (-1)^{k+2}\Delta_\Sigma(w).$$

Let $B = \Psi_\Sigma(\operatorname{mi}(w))$ and $B' = \overline{\Psi}_\Sigma(\operatorname{mi}(w))$. By Theorem 3, it holds that

$$(-1)^{k+2}\Delta_\Sigma(w) = A^{-1}_{1,k+1} = B'_{1,k+1} = (-1)^{k+2}B_{1,k+1}.$$

It remains to see that $\Delta_\Sigma(w) = B_{1,k+1} = |\operatorname{mi}(w)|_{a_1 a_2 \cdots a_k} = |w|_{a_k a_{k-1} \cdots a_1}$, thus the conclusion holds.

At this point, we highlight the following observation, which is essential in the proof of Theorem 4.

Remark 3. Suppose Σ is an ordered alphabet and $w \in \Sigma^*$. Let $A = \Psi_\Sigma(w)$ and $B = \Psi_\Sigma(\operatorname{mi}(w))$. For all integers $1 \le i, j \le k+1$, we have $B_{i,j} = \|A(j,i)\|$. That is to say, every minor of $\Psi_\Sigma(w)$ uniquely corresponds to an entry in $\Psi_\Sigma(\operatorname{mi}(w))$.

The following are immediate consequences of Theorem 4.

Remark 4. Suppose $\Sigma = \{a_1 < a_2 < \cdots < a_k\}$.

1. For every positive integer n, there exists $w \in \Sigma^*$ such that $\Delta_\Sigma(w) = n$ (take $w = a_k a_{k-1} \cdots a_2 a_1^n$).
2. For any $w \in \Sigma^*$, if $|w|_{a_i} = 0$ for some $1 \le i \le k$, then $\Delta_\Sigma(w) = 0$.
3. For any $u, v \in \Sigma^*$,
   a. if $|u|_{a_k} = 0$, then $\Delta_\Sigma(uv) = \Delta_\Sigma(v)$;
   b. if $|v|_{a_1} = 0$, then $\Delta_\Sigma(uv) = \Delta_\Sigma(u)$.

Proposition 1. *Suppose $\Sigma = \{a_1 < a_2 < \cdots < a_k\}$ and $u, v \in \Sigma^*$. For all integers $1 \le i \le k-1$, define $\Sigma_i = \{a_1 < a_2 < \cdots < a_i\}$ and $\Sigma\backslash\Sigma_i = \{a_{i+1} < a_{i+2} < \cdots < a_k\}$. Then*

$$\Delta_\Sigma(uv) = \Delta_\Sigma(u) + \Delta_\Sigma(v) + \sum_{i=1}^{k-1}\Delta_{\Sigma_i}(\pi_{\Sigma_i}(v)) \cdot \Delta_{\Sigma\backslash\Sigma_i}(\pi_{\Sigma\backslash\Sigma_i}(u)).$$

Proof. Let $y = a_k a_{k-1} \ldots a_1$ and for all integers $1 \le i \le k-1$, let $y_i = a_k a_{k-1} \ldots a_{i+1}$ and $\overline{y}_i = a_i a_{i-1} \ldots a_1$. Then the result is straightforward by Theorem 4 and the fact that

$$|uv|_y = |u|_y + |v|_y + \sum_{i=1}^{k-1}|u|_{y_i}|v|_{\overline{y}_i}.$$

The following formula computes, for any positive integer p, the Parikh determinant of a word to the power of p.

Theorem 5. *Suppose $\Sigma = \{a_1 < a_2 < \cdots < a_k\}$ and $w \in \Sigma^*$. Let*

$$S = \{\langle i_1, i_2, \ldots, i_n \rangle \in [k]^{<\omega} \mid i_1 < i_2 < \cdots < i_n = k\}.$$

Then for any positive integer p, we have

$$\Delta_\Sigma(w^p) = \sum_{\langle i_1, i_2, \ldots, i_n \rangle \in S} (-1)^{n+k} \binom{p+n-1}{p-1} |w|_{\langle 1, i_1, i_2, \ldots, i_n \rangle}.$$

Proof. We argue by induction on the power of w. The base step holds by Theorem 2. For the induction step, assume the hypothesis is true up to power $p-1$. We show that it is true for power p.

For every integer $1 \le i \le k-1$, define $\Sigma_i = \{a_1 < a_2 < \cdots < a_i\}$ and $\Sigma \backslash \Sigma_i = \{a_{i+1} < a_{i+2} < \cdots < a_k\}$. By Proposition 1, it holds that

$$\Delta_\Sigma(w^p) = \Delta_\Sigma(w w^{p-1})$$

$$= \Delta_\Sigma(w) + \Delta_\Sigma(w^{p-1}) + \sum_{i=1}^{k-1} \Delta_{\Sigma_i}(\pi_{\Sigma_i}(w^{p-1})) \cdot \Delta_{\Sigma \backslash \Sigma_i}(\pi_{\Sigma \backslash \Sigma_i}(w)). \tag{3}$$

For every integer $1 \le i \le k-1$, define

$$R_i = \{\langle r_1, r_2, \ldots, r_{n_i} \rangle \in [i]^{<\omega} \mid r_1 < r_2 < \cdots < r_{n_i} = i\}$$

and

$$T_i = \{\langle t_1, t_2, \ldots, t_{m_i} \rangle \in [k]^{<\omega} \mid i+1 \le t_1 < t_2 < \cdots < t_{m_i} = k\}.$$

Then for every integer $1 \le i \le k-1$, by the induction hypothesis, it holds that

$$\Delta_{\Sigma_i}(\pi_{\Sigma_i}(w^{p-1})) = \sum_{\langle r_1, r_2, \ldots, r_{n_i} \rangle \in R_i} (-1)^{n_i+i} \binom{p+n_i-2}{p-2} |\pi_{\Sigma_i}(w)|_{\langle 1, r_1, r_2, \ldots, r_{n_i} \rangle}$$

and

$$\Delta_{\Sigma \backslash \Sigma_i}(\pi_{\Sigma \backslash \Sigma_i}(w)) = \sum_{\langle t_1, t_2, \ldots, t_{m_i} \rangle \in T_i} (-1)^{m_i+k-i} |\pi_{\Sigma \backslash \Sigma_i}(w)|_{\langle i+1, t_1, t_2, \ldots, t_{m_i} \rangle}.$$

For every integer $1 \le i \le k$, if $\langle r_1, r_2, \ldots, r_{n_i} \rangle \in R_i$ and $\langle t_1, t_2, \ldots, t_{m_i} \rangle \in T_i$, then

$$|\pi_{\Sigma_i}(w)|_{\langle 1, r_1, r_2, \ldots, r_{n_i} \rangle} |\pi_{\Sigma \backslash \Sigma_i}(w)|_{\langle i+1, t_1, t_2, \ldots, t_{m_i} \rangle}$$

$$= |w|_{\langle 1, r_1, r_2, \ldots, r_{n_i} \rangle} |w|_{\langle i+1, t_1, t_2, \ldots, t_{m_i} \rangle}$$

$$= |w|_{\langle 1, r_1, r_2, \ldots, r_{n_i}, t_1, t_2, \ldots, t_{m_i} \rangle} \text{ (since } r_{n_i} = i < i+1 \le t_1).$$

Therefore, for every integer $1 \le i \le k$, we have

$$\Delta_{\Sigma_i}(\pi_{\Sigma_i}(w^{p-1})) \cdot \Delta_{\Sigma \backslash \Sigma_i}(\pi_{\Sigma \backslash \Sigma_i}(w))$$

$$= \sum_{\langle r_1, \ldots, r_{n_i} \rangle \in R_i} \sum_{\langle t_1, \ldots, t_{m_i} \rangle \in T_i} (-1)^{n_i+m_i+k} \binom{p+n_i-2}{p-2} |w|_{\langle 1, r_1, \ldots, r_{n_i}, t_1, \ldots, t_{m_i} \rangle}. \tag{4}$$

Observe that

$$\bigcup_{i=1}^{k-1} \{\langle r_1, \ldots, r_{n_i}, t_1, \ldots, t_{m_i}\rangle \mid \langle r_1, \ldots, r_{n_i}\rangle \in R_i \text{ and } \langle t_1, \ldots, t_{m_i}\rangle \in T_i\} = S.$$

Furthermore, for every $\langle i_1, i_2, \ldots, i_n\rangle \in S$, if $1 \le h \le n-1$, then $\langle i_1, i_2, \ldots, i_h\rangle \in R_{i_h}$ and $\langle i_{h+1}, i_{h+2}, \ldots, i_n\rangle \in T_{i_h}$. Therefore, we obtain from (4) the equality

$$\sum_{i=1}^{k-1} \Delta_{\Sigma_i}(\pi_{\Sigma_i}(w^{p-1})) \cdot \Delta_{\Sigma \setminus \Sigma_i}(\pi_{\Sigma \setminus \Sigma_i}(w))$$

$$= \sum_{\langle i_1, i_2, \ldots, i_n\rangle \in S} (-1)^{n+k} \left[\sum_{h=1}^{n-1} \binom{p+h-2}{p-2}\right] |w|_{\langle 1, i_1, i_2, \ldots, i_n\rangle}. \tag{5}$$

By (3), (5), and the induction hypothesis on $\Delta_\Sigma(w^{p-1})$ and $\Delta_\Sigma(w)$, we have $\Delta_\Sigma(w^p)$

$$= \sum_{\langle i_1, \ldots, i_n\rangle \in S} (-1)^{n+k} |w|_{\langle 1, i_1, \ldots, i_n\rangle} + \sum_{\langle i_1, \ldots, i_n\rangle \in S} (-1)^{n+k} \binom{p+n-2}{p-2} |w|_{\langle 1, i_1, \ldots, i_n\rangle}$$

$$+ \sum_{\langle i_1, \ldots, i_n\rangle \in S} (-1)^{n+k} \left[\sum_{h=1}^{n-1} \binom{p+h-2}{p-2}\right] |w|_{\langle 1, i_1, \ldots, i_n\rangle}$$

$$= \sum_{\langle i_1, l \ldots, i_n\rangle \in S} (-1)^{n+k} \left[\underbrace{\binom{p-2}{p-2}}_{=1} + \binom{p+n-2}{p-2} + \sum_{h=1}^{n-1} \binom{p+h-2}{p-2}\right] |w|_{\langle 1, i_1, \ldots, i_n\rangle}$$

$$= \sum_{\langle i_1, \ldots, i_n\rangle \in S} (-1)^{n+k} \left[\sum_{h=0}^{n} \binom{p+h-2}{p-2}\right] |w|_{\langle 1, i_1, \ldots, i_n\rangle}.$$

Finally, by induction on n and Pascal's rule, it can be shown that

$$\sum_{h=0}^{n} \binom{p+h-2}{p-2} = \binom{p+n-1}{p-1}.$$

Thus the conclusion holds.

Finally, by Theorems 4 and 5, the following identity holds.

Theorem 6. *Suppose Σ is an alphabet and $a_1, a_2, \ldots, a_k$ are distinct letters in Σ. Let*

$$S = \{\langle i_1, i_2, \ldots, i_n\rangle \in [k]^{<\omega} \mid i_1 < i_2 < \cdots < i_n = k\}.$$

For any positive integer p, we have

$$|w^p|_{a_k a_{k-1} \cdots a_1} = \sum_{\langle i_1, i_2, \ldots, i_n\rangle \in S} (-1)^{n+k} \binom{p+n-1}{p-1} |w|_{\langle i_1, i_2, \ldots, i_n\rangle}.$$

In particular,

$$|w|_{a_k a_{k-1} \cdots a_1} = \sum_{\langle i_1, i_2, \ldots, i_n \rangle \in S} (-1)^{n+k} |w|_{\langle i_1, i_2, \ldots, i_n \rangle}.$$

Notice that for the binary alphabet $\{a < b\}$, Theorem 6 gives rise to the well-known identity $|w|_a |w|_b = |w|_{ab} + |w|_{ba}$.

4 Final Remarks

The Parikh determinant is natural, both as an algebraic aspect of Parikh matrices, as well as a combinatorial characteristic of words. In this short section, we present some observations and possible future directions of this study.

4.1 Parikh Determinant, Parikh Vector and Parikh Matrix

Given a word w, the Parikh determinant, Parikh vector, and Parikh matrix of the word w describe w in terms of the number of occurrences of certain subwords. For the binary alphabet, the combined information of the Parikh vector and Parikh determinant of a word is equivalent to that given by the Parikh matrix of the word. This is due to the fact that $\Delta_{\{a<b\}}(w) = |w|_{ba}$ (by Theorem 4) and the equality $|w|_a |w|_b = |w|_{ab} + |w|_{ba}$. Expectedly, for larger alphabets, this is not true. For example, $\Delta_{\{a<b<c\}}(cbcaa) = 2 = \Delta_{\{a<b<c\}}(ccaba)$ and $\Psi(cbcaa) = (2,1,2) = \Psi(ccaba)$, however

$$\Psi_{\{a<b<c\}}(cbcaa) = \begin{pmatrix} 1 & 2 & 0 & 0 \\ 0 & 1 & 1 & 1 \\ 0 & 0 & 1 & 2 \\ 0 & 0 & 0 & 1 \end{pmatrix} \neq \begin{pmatrix} 1 & 2 & 1 & 0 \\ 0 & 1 & 1 & 0 \\ 0 & 0 & 1 & 2 \\ 0 & 0 & 0 & 1 \end{pmatrix} = \Psi_{\{a<b<c\}}(ccaba).$$

4.2 An Alternative Way to Compute Parikh Determinants

Let $\Sigma = \{a_1 < a_2 < \cdots < a_k\}$ and $w \in \Sigma^*$. Theorem 4 establishes the fact that the Parikh determinant of w is equivalent to the value $|w|_{a_k a_{k-1} \cdots a_1}$. In [19], given any word $v \in \Sigma^*$, the essential subword of w whose letters account for the value $|w|_v$ is identified. Formally, the following notion was introduced.

Definition 5. *Suppose Σ is an alphabet and $v, w \in \Sigma^*$. The v-core of w, denoted by $\mathrm{core}_v(w)$, is the unique shortest-length subword w' of w which satisfies $|w'|_v = |w|_v$. We say that w is a v-core word if and only if $\mathrm{core}_v(w) = w$.*

Example 3. Suppose $\Sigma = \{a, b, c\}$ and consider the word $w = cbabbca$. Then, $\mathrm{core}_a(w) = aa$, $\mathrm{core}_{bc}(w) = bbbc$, $\mathrm{core}_{abc}(w) = abbc$ and $\mathrm{core}_{cba}(w) = cbabbca$. Note that since $\mathrm{core}_{cba}(w) = w$, it follows that w is a cba-core word.

Thus, for $\Sigma = \{a_1 < a_2 < \cdots < a_k\}$, the Parikh determinant of a word over Σ can be computed by systematically analyzing the $a_k a_{k-1} \cdots a_1$-core of the word. We illustrate this by an example for the quarternary alphabet.

Example 4. Suppose $\Sigma = \{a, b, c, d\}$. Consider the word $w = abdcddcbadccbabc$. Then $\Delta_\Sigma(w) = |w|_{dcba} = |\,\mathrm{core}_{dcba}(w)|_{dcba}$. Thus to compute the value $|w|_{dcba}$, we analyze the word $\mathrm{core}_{dcba}(w) = dcddcbadccba$ as follows:

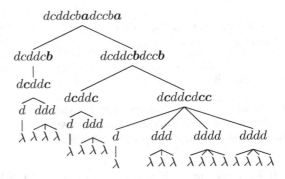

The first level (root) of the tree is the word w. The second level of the tree is obtained by first identifying the position of letters a in w. Then, for each of those letters a, the projection with respect to $\{d, c, b\}$, of the factor right before it, is represented as a node. Continuing like this, the third level corresponds to the projection with respect to $\{d, c\}$, fourth level to $\{d\}$, and finally the fifth level, the empty alphabet.

We can easily see that the value $|w|_{dcba}$ is equal to the number of leaves of the rooted tree, which is 20. Therefore, $\Delta_\Sigma(w) = 20$.

Ultimately, it remains to be investigated whether a more efficient algorithm to compute the Parikh determinant of a word can be obtained.

4.3   Generalization of Parikh Determinants

Given a word w, Parikh matrices can only be used to obtain the value $|w|_u$ for any word u such that u does not contain any repeated letters. As an improvement, the notion of extended Parikh matrices was introduced in [16]. The notion of Parikh determinants can be easily generalized to the context of extended Parikh matrices.

Moreover, a result analogous to Theorem 3 has been proven for extended Parikh matrices (see [16, Theorem 16]). Therefore, a more general identity involving subword occurrences (as in Theorem 6) is attainable.

Acknowledgements. This work was completed during the sabbatical leave of the third author from 15 Nov 2018 to 14 Aug 2019, supported by Universiti Sains Malaysia.

References

1. Atanasiu, A., Atanasiu, R.F.: Enriching Parikh matrix mappings. Int. J. Comput. Math. **90**(3), 511–521 (2013)

2. Atanasiu, A., Martín-Vide, C., Mateescu, A.: On the injectivity of the Parikh matrix mapping. Fund. Inform. **49**(4), 289–299 (2002)
3. Atanasiu, A., Poovanandran, G., Teh, W.C.: Parikh matrices for powers of words. Acta Inform. 1–15 (2018)
4. Atanasiu, A., Teh, W.C.: A new operator over Parikh languages. Int. J. Found. Comput. Sci. **27**(06), 757–769 (2016)
5. Bera, S., Mahalingam, K.: Some algebraic aspects of Parikh q-matrices. Int. J. Found. Comput. Sci. **27**(04), 479–499 (2016)
6. Černý, A.: Generalizations of Parikh mappings. RAIRO Theor. Inform. Appl. **44**(2), 209–228 (2010)
7. Clark, A., Watkins, C.: Some alternatives to Parikh matrices using string kernels. Fund. Inform. **84**(3–4), 291–303 (2008)
8. Egecioglu, O., Ibarra, O.H.: A matrix Q-analogue of the Parikh map. In: Levy, J.-J., Mayr, E.W., Mitchell, J.C. (eds.) TCS 2004. IIFIP, vol. 155, pp. 125–138. Springer, Boston, MA (2004). https://doi.org/10.1007/1-4020-8141-3_12
9. Mahalingam, K., Bera, S., Subramanian, K.G.: Properties of Parikh matrices of binary words obtained by an extension of a restricted shuffle operator. Int. J. Found. Comput. Sci. **29**(3), 403–3413 (2018)
10. Mateescu, A.: Algebraic aspects of Parikh matrices. In: Karhumäki, J., Maurer, H., Păun, G., Rozenberg, G. (eds.) Theory Is Forever. LNCS, vol. 3113, pp. 170–180. Springer, Heidelberg (2004). https://doi.org/10.1007/978-3-540-27812-2_16
11. Mateescu, A., Salomaa, A., Salomaa, K., Yu, S.: A sharpening of the Parikh mapping. Theor. Inform. Appl. **35**(6), 551–564 (2001)
12. Mateescu, A., Salomaa, A., Yu, S.: Subword histories and Parikh matrices. J. Comput. Syst. Sci. **68**(1), 1–21 (2004)
13. Parikh, R.J.: On context-free languages. J. Assoc. Comput. Mach. **13**, 570–581 (1966)
14. Poovanandran, G., Teh, W.C.: Elementary matrix equivalence and core transformation graphs for Parikh matrices. Discrete Appl. Math. **251**, 276–289 (2018)
15. Salomaa, A.: Criteria for the matrix equivalence of words. Theoret. Comput. Sci. **411**(16), 1818–1827 (2010)
16. Şerbănuţă, T.F.: Extending Parikh matrices. Theoret. Comput. Sci. **310**(1–3), 233–246 (2004)
17. Şerbănuţă, V.N.: On Parikh matrices, ambiguity, and prints. Internat. J. Found. Comput. Sci. **20**(1), 151–165 (2009)
18. Şerbănuţă, V.N., Şerbănuţă, T.F.: Injectivity of the Parikh matrix mappings revisited. Fund. Inform. **73**(1), 265–283 (2006)
19. Teh, W.C.: On core words and the Parikh matrix mapping. Int. J. Found. Comput. Sci. **26**(1), 123–142 (2015)
20. Teh, W.C.: Parikh matrices and Parikh rewriting systems. Fund. Inform. **146**, 305–320 (2016)
21. Teh, W.C., Atanasiu, A.: On a conjecture about Parikh matrices. Theoret. Comput. Sci. **628**, 30–39 (2016)
22. Teh, W.C., Atanasiu, A., Poovanandran, G.: On strongly M-unambiguous prints and Şerbănuţă's conjecture for Parikh matrices. Theoret. Comput. Sci. **719**, 86–93 (2018)

Critical Exponent of Infinite Balanced Words via the Pell Number System

Aseem R. Baranwal$^{(\boxtimes)}$ (iD) and Jeffrey Shallit (iD)

School of Computer Science, University of Waterloo, Waterloo, ON N2L 3G1, Canada
{aseem.baranwal,shallit}@uwaterloo.ca

Abstract. In a recent paper of Rampersad et al., the authors conjectured that the smallest possible critical exponent of an infinite balanced word over a 5-letter alphabet is 3/2. We prove this result, using a formulation of first-order logic, the Pell number system, and a machine computation based on finite-state automata.

Keywords: Critical exponent · Balanced word ·
Automatic theorem-proving

1 Introduction

In this paper, we prove a result about the critical exponent of infinite balanced words, using a formulation of first-order logic, the Pell number system, and a machine computation based on finite-state automata. To our knowledge, this is the first result in combinatorics on words to be proved using this approach via the Pell number system.

1.1 Preliminaries

Let w denote a word over the alphabet Σ. If w is finite, then $|w|$ denotes its length, and $|w|_a$ denotes the number of occurrences of the symbol a in w, where $a \in \Sigma$. We let $\mathrm{Fac}(w)$ denote the set of all factors of w.

Definition 1. *A word w over the alphabet Σ is* balanced *if for every symbol $a \in \Sigma$, and every pair of words $u, v \in \mathrm{Fac}(w)$ with $|u| = |v|$, we have $||u|_a - |v|_a| \leq 1$.*

The class of *Sturmian words* and the class of infinite aperiodic balanced words coincide over a binary alphabet. Vuillon [14] provides a survey on some previous work on balanced words, and Berstel et al. [3] provide a survey on Sturmian words.

Definition 2. *Let $w = w_0 w_1 \cdots w_{n-1}$ be a finite word of length n. Then $p \in \mathbb{N}$ is a* period *of w if $w_i = w_{i+p}$ for all i with $0 \leq i < n - p$.*

© Springer Nature Switzerland AG 2019
R. Mercaş and D. Reidenbach (Eds.): WORDS 2019, LNCS 11682, pp. 80–92, 2019.
https://doi.org/10.1007/978-3-030-28796-2_6

We say that a word u has *exponent* e and write $u = z^e$, where $e = |u|/p$ is a positive rational number, and z is the prefix of u of length p; here z is sometimes called a *fractional root* of u. A word may have multiple periods, exponents, and fractional roots. We say u is *primitive* if its only integer exponent is 1. If u is a finite nonempty word, then u^ω denotes the infinite word $uuu\cdots$.

Example 1. The word $w = \mathtt{alfalfa}$ has three periods: $p_1 = 3$, $p_2 = 6$, and $p_3 = 7$. The corresponding exponents are $e_1 = 7/3$, $e_2 = 7/6$, and $e_3 = 1$. In this example, w is a primitive word since its only integer exponent is 1.

Definition 3. *The* critical exponent *of an infinite word* $\mathbf{w}$ *is defined to be the supremum of the set of all rational numbers* r *such that there exists a finite nonempty factor of* $\mathbf{w}$ *with exponent* r. *More formally,*

$$E(\mathbf{w}) = \sup\{r \in \mathbb{Q} : \text{there exist words}\, x, y \in \mathrm{Fac}(\mathbf{w})\, \text{with}\, |y| > 0\, \text{and}\, y = x^r\}.$$

1.2   Previous Work

Rampersad et al. [12] gave a method to construct infinite balanced words from binary Sturmian words, using a characterization of recurrent aperiodic balanced words given by Hubert [7]. Their method is based on the notion of the constant gap property.

Definition 4. *An infinite word* $\mathbf{w}$ *has the* constant gap property *if, for each symbol* a, *there is a positive integer* d *such that the distance between successive occurrences of* a *in* $\mathbf{w}$ *is always* d.

For example, $(0102)^\omega = 010201020102\cdots$ has the constant gap property because the distance between consecutive 0's is always 2, while the distance between consecutive 1's (resp., 2's) is always 4.

Sturmian words $\mathbf{c}_{\alpha,\beta}$ can be defined in terms of two real parameters α, β with $0 \le \alpha, \beta < 1$, and α irrational. Then

$$\mathbf{c}_{\alpha,\beta}[n] := \lfloor \alpha(n+1) + \beta \rfloor - \lfloor \alpha n + \beta \rfloor.$$

A Sturmian word is called *characteristic* if $\beta = 0$, and is written as $\mathbf{c}_\alpha$. In this case, it is well-known that an alternative characterization for these words can be given in terms of the continued fraction expansion of $\alpha = [d_0, d_1, d_2, \ldots]$ where $d_i \in \mathbb{N}$ for $i \ge 0$ and $d_i \ge 1$ for $i \ge 1$. Then $\mathbf{c}_\alpha$ is produced as the limit of the sequence of *standard words* s_n defined as follows:

$$s_0 = 0, \quad s_1 = 0^{d_1 - 1}1, \quad s_n = s_{n-1}^{d_n} s_{n-2} \text{ for } n \ge 2.$$

Theorem 1 [7]. *A recurrent aperiodic infinite word* $\mathbf{x}$ *is balanced if and only if* $\mathbf{x}$ *is obtained from a Sturmian word* $\mathbf{u}$ *over* $\{0,1\}$ *by the following procedure: replace the 0's in* $\mathbf{u}$ *by a periodic sequence* $\mathbf{y}$ *with constant gaps over some alphabet* A *and replace the 1's in* $\mathbf{u}$ *by a periodic sequence* $\mathbf{y}'$ *with constant gaps over some alphabet* B, *disjoint from* A.

The authors of [12] defined certain infinite balanced words $\mathbf{x}_k$ for $3 \leq k \leq 10$ constructed from a Sturmian characteristic word $\mathbf{c}_\alpha$, where α, $\mathbf{y}$ and $\mathbf{y}'$ are carefully chosen. Table 1 shows the choices for $\mathbf{x}_3$, $\mathbf{x}_4$, and $\mathbf{x}_5$. Here $\varphi = (1 + \sqrt{5})/2$ is the golden ratio. The authors also proved that $E(\mathbf{x}_3) = 2 + \frac{\sqrt{2}}{2}$ and $E(\mathbf{x}_4) = 1 + \frac{\varphi}{2}$; furthermore, they showed that $E(\mathbf{x}_3)$ is the least possible critical exponent over an alphabet of 3 symbols. Based on computations, they also suggested that the least possible critical exponents for balanced words over a k-letter alphabet is $(k-2)/(k-3)$ for $k \geq 5$. In this paper we take the first step towards this conjecture by proving the result for $k = 5$. (Very recently a proof that the critical exponent for $\mathbf{x}_4$ is actually minimal was announced by Peltomäki.)

Table 1. α, y, and y' used for the construction of $\mathbf{x}_k$.

k	α	y	y'
3	$\sqrt{2} - 1$	$(01)^\omega$	2^ω
4	$1/\varphi^2$	$(01)^\omega$	$(23)^\omega$
5	$\sqrt{2} - 1$	$(0102)^\omega$	$(34)^\omega$

1.3   Automatic Theorem Proving Using Walnut

The authors in [12] employed a computational approach using the automatic theorem-proving software Walnut [8]. The approach is based on the methods of Du et al. [5,9], using Theorems 2 and 3 below. The nth term of an arbitrary Sturmian characteristic word $\mathbf{c}_\alpha$, and consequently the generated infinite balanced word $\mathbf{x}_k$, can be computed by a finite automaton that takes the Ostrowski α-representation [11] of n as input.

Theorem 2 [1, Theorem 9.1.15]. *Let $N \geq 1$ be an integer with Ostrowski α-representation $b_j b_{j-1} \cdots b_0$. Then $\mathbf{c}_\alpha[N] = 1$ if and only if $b_j b_{j-1} \cdots b_0$ ends with an odd number of 0's.*

Theorem 3 [12, Theorem 12]. *Let α be a quadratic irrational and let $\mathbf{c}_\alpha$ be the Sturmian characteristic word with slope α. Let $\mathbf{x}$ be any word obtained by replacing the 0's in $\mathbf{c}_\alpha$ with a periodic sequence $\mathbf{y}$ and replacing the 1's with a periodic sequence $\mathbf{y}'$. Then $\mathbf{x}$ is Ostrowski α-automatic.*

Using Walnut, we can constructively decide first-order predicates. When a predicate consisting of free variables is provided to Walnut, it also generates an automaton accepting values for the free variables that will satisfy the predicate. For predicates without any free variables, Walnut produces any of the two special automata, the true, and the false automaton, depending on whether the predicate is a tautology, or a contradiction respectively.

2   Building the Automata

We determine the critical exponent of $\mathbf{x}_5$ using the computational approach described above. The Ostrowski α-numeration system for $\mathbf{x}_5$ is defined by the *Pell* numbers, similar to how the numeration system for $\mathbf{x}_4$ is defined by the *Fibonacci* numbers. To enable `Walnut` to work with this new numeration system, we require a deterministic finite automaton that reads its input in the Pell number system and recognizes the addition relation $\{(x, y, z) \in \mathbb{N}^3 : x + y = z\}$. Hieronymi and Terry [6] showed that this is indeed possible when α is a quadratic irrational, and we have $\alpha = \sqrt{2} - 1$ for $\mathbf{x}_5$, which satisfies this condition. Once we have the adder, a second automaton with output that can compute $\mathbf{x}_5$ is required for enabling `Walnut` to understand first-order predicates involving $\mathbf{x}_5$.

2.1   Pell Number System

The Pell numbers are defined by the recurrence relation $P_n = 2P_{n-1} + P_{n-2}$ for $n \geq 2$ with $P_0 = 0$ and $P_1 = 1$. The first few terms of this sequence are

$$0, 1, 2, 5, 12, 29, 70, 169, \ldots$$

and form sequence [A000129](#) in Sloane's *Encyclopedia* [13]. We use this sequence of numbers to define a non-standard positional numeral system in the family of Ostrowski numeration systems [11] with $\alpha = \sqrt{2} - 1$. Given an integer N, we can express it as an integer linear combination of Pell numbers as follows: $N = \sum_{0 \leq i < n} d_i P_{i+1}$. To ensure that this representation is unique, we impose the following conditions on the d_i:

1. The least significant digit $d_0 \in \{0, 1\}$.
2. For all $i > 0$ we have $d_i \in \{0, 1, 2\}$.
3. If $d_i = 2$, then $d_{i-1} = 0$.

In this case, the word $d_{n-1}d_{n-2} \cdots d_0$ is said to be the *canonical Pell representation* of an integer N, and we write it as $(N)_P$. For example, 157 has canonical Pell representation $(201100)_P$. Other representations of 157 include 122100, 201021, and 122021, but they do not conform to the conditions given above, and hence they are not canonical.

2.2   Automaton for the Addition Relation in Pell-Base

To build the automaton that can recognize the addition relation in the Pell number system, we use Theorem 4, a corollary to the Myhill-Nerode theorem [10] based on the idea of Brzozowski derivative [4].

Definition 5. *Given a function* $f : \Sigma^* \rightarrow \mathbb{R}$ *for an alphabet* Σ*, we define its Hankel matrix* $\mathcal{H} \in \mathbb{R}^{\Sigma^* \times \Sigma^*}$ *as follows:*

$$\mathcal{H} = \begin{array}{c} \\ \epsilon \\ a \\ b \\ aa \\ \\ \end{array} \begin{array}{cccccc} \epsilon & a & b & aa & \cdots \\ \left(\begin{array}{ccccc} f(\epsilon) & f(a) & f(b) & f(aa) & \cdots \\ f(a) & f(aa) & f(ab) & f(aaa) & \cdots \\ f(b) & f(ba) & f(bb) & f(baa) & \cdots \\ f(aa) & f(aaa) & f(aab) & f(aaaa) & \cdots \\ \vdots & \vdots & \vdots & \vdots & \vdots & \ddots \end{array} \right) \end{array},$$

where ϵ denotes the empty string. The matrix is indexed by words $u, v \in \Sigma^*$ such that $\mathcal{H}_{uv} = f(uv)$.

Theorem 4 *(Myhill-Nerode [10]). Let $\mathcal{L} = \{w_1, w_2, \ldots\}$ be a language over the finite alphabet Σ. Let $\mathcal{H}$ be a binary Hankel matrix indexed by the words $u, v \in \Sigma^*$ such that*

$$\mathcal{H}_{uv} = \begin{cases} 1, & \text{if } uv \in \mathcal{L}; \\ 0, & \text{otherwise.} \end{cases}$$

Then $\mathcal{L}$ is regular if and only if the number of distinct rows in $\mathcal{H}$ is finite. Furthermore, the number of distinct rows equals the minimal number of states of a deterministic finite automaton recognizing $\mathcal{L}$.

For the indices of $\mathcal{H}$, we use a list of words over the alphabet $\Sigma_3 = \{0, 1, 2\}$ sorted in the *radix* order. The radix order for two words x and y is defined by $x < y$, if $|x| < |y|$, or there exist symbols $a, b \in \Sigma$ such that $|x| = |y|$, $x = uax'$, $y = uby'$, and $a < b$. The adder automaton takes as input 3 integers x, y, z in canonical Pell representation in parallel, and reaches an accepting state if and only if $x + y = z$. To achieve this, we require a generalization of the notion of Pell representation to r-tuples of integers for $r \geq 1$. A representation for $(x_1, x_2, \ldots, x_r)$ consists of a string of symbols z over some alphabet Σ, such that a well-defined projection $\pi_i(z)$ over the ith coordinate gives a canonical Pell representation of x_i. To handle this, first, we pad the canonical Pell representations of smaller integers with leading 0's so that in the r-tuple, strings representing all x_i have the same length. Our goal is to represent a triplet (x, y, z) that will serve as an input to the adder automaton.

Example 2. Let $(x, y, z) = (65, 15, 80)$ be an integer triplet to be input to our adder automaton. After padding with sufficient leading 0's, we have $(65)_{10} = (020110)_P$, $(15)_{10} = (001011)_P$, and $(80)_{10} = (100200)_P$. Next, we project these representations as a series of ternary triplets, i.e., each digit belonging to the alphabet Σ_3. Hence $(65, 15, 80)$ is represented as

$$[0, 0, 1][2, 0, 0][0, 1, 0][1, 0, 2][1, 1, 0][0, 1, 0],$$

where the first digits of the triplets spell out 020110, the canonical Pell representation of 65. Similar claims about the second and third digits hold for 15 and 80, respectively. This projection is necessary to our method because, in order to recognize the addition relation, the automaton must be able to read all three integers x, y, z in parallel.

Since each triplet has digits $\in \Sigma_3 = \{0, 1, 2\}$, it follows that our input alphabet $\mathcal{P}$ for the adder automaton has size $3^3 = 27$. Next, we use a set of radix-ordered strings over $\mathcal{P}$ as indices for our binary Hankel matrix $\mathcal{H}_{\mathcal{P}}$. The value in row u and column v of $\mathcal{H}_{\mathcal{P}}$ $(u, v \in \mathcal{P})$ is 1 if uv denotes a series of triplets over Σ_3 that is a projection for an integer triplet (x, y, z) such that $x + y = z$, and 0 otherwise. Finally, we *learn* the deterministic finite automaton with a combination of membership and equivalence queries using the *Angluin L** algorithm [2]. The adder automaton contains 16 states over an alphabet of size 27, and hence it is infeasible to show here. The full automaton is publicly available on GitHub.[1]

Before we proceed, following an idea suggested to us by Luke Schaeffer [5], we prove the correctness of our adder automaton using `Walnut`. The proof is inductive, using the definition of the successor of an integer. Of course, the successor of an integer x is $x + 1$, but this makes use of the addition relation, which we have not yet proved. Instead, we define the successor in a different way.

Definition 6. *Given two integers x and y, we say y is the successor of x if $x < y$, and $(z \leq x$ or $z \geq y)$ for all $z \in \mathbb{Z}$.*

The canonical Pell representation ensures that `Walnut` can perform comparisons on two integers in Pell representation easily, based on only their radix ordering. Below we present the `Walnut` commands to compute the inductive proof. First, we define the *successor* relation.

```
1   def pell_successor "?msd_pell
2       x < y & (Az (z <= x) | (z >= y))";
```

The above command produces an automaton accepting pairs (x, y) such that y is a successor of x. `Walnut` also stores this definition and allows us to use it in other predicates. Next, we check the base case of the induction. For all $x, z \in \mathbb{Z}$, $x + 0 = z$ if and only if $x = z$. For the addition relation, 0 is the identity element.

```
1   eval base_proof "?msd_pell Ax,z ((x + 0 = z) <=> (x = z))";
```

The predicate `base_proof` produces the `true` automaton signifying that the predicate is true. We now verify our adder using the definition of successor. For all $x, y, z, u, v \in \mathbb{Z}$, if u is the successor of y and v is the successor of z, then we have $x + y = z$ if and only if $x + u = v$.

```
1   eval inductive_proof "?msd_pell Ax,y,z,u,v
2       ($pell_successor(y, u) & $pell_successor(z, v)) =>
3       ((x + y = z) <=> (x + u = v))";
```

The predicate `inductive_proof` also produces the `true` automaton. This completes the proof of correctness for our automaton recognizing the addition relation in Pell-base.

---

[1] Corresponding `Walnut` code is available at https://github.com/aseemrb/walnut.

2.3 Automaton for Computing $\mathbf{x}_5$

Using the adder automaton we created above, we now build a deterministic finite automaton *with output* that can compute $\mathbf{x}_5$. Each state of this automaton is associated with an output symbol from the alphabet $\Sigma_5 = \{0, 1, 2, 3, 4\}$ of $\mathbf{x}_5$. It takes as input an integer N in canonical Pell representation, and halts at the state with output $\mathbf{x}_5[N]$. By Theorem 3 we know that $\mathbf{x}_5$ is an α-automatic sequence for $\alpha = \sqrt{2} - 1$. Hence, using Theorem 2 we first create an automaton for $\mathbf{c}_\alpha$, which is given by the limit of the sequence of finite words s_n defined as follows:

$$s_0 = 0, \quad s_1 = 01, \quad s_n = s_{n-1}^2 s_{n-2} \text{ for } n \geq 2.$$

This definition comes from the fact that the continued fraction expansion of $\alpha = \sqrt{2} - 1$ is $[0, \overline{2}]$. The automaton given in Fig. 1 produces the sequence $\mathbf{c}_\alpha$. The label on each state denotes the output associated with that state. When given a positive integer $N > 0$ as input, this automaton halts at the state with label $\mathbf{c}_\alpha[N]$. Here $\mathbf{c}_\alpha$ is indexed from 1. The first few characters of $\mathbf{c}_\alpha$, and consequently of $\mathbf{x}_5$ constructed from $\mathbf{c}_\alpha$ are given below.

$$\mathbf{c}_\alpha = 010100101001010100101001010 \cdots$$
$$\mathbf{x}_5 = 0314023041032403104230140324 0 \cdots$$

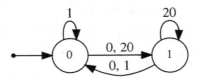

Fig. 1. Pell-base automaton for $\mathbf{c}_\alpha$.

Generating $\mathbf{x}_5$ from $\mathbf{c}_\alpha$ is a simple replacement of 0's and 1's by the constant-gap sequences $\mathbf{y} = (0102)^\omega$ and $\mathbf{y}' = (34)^\omega$ from Table 1. We start indexing $\mathbf{x}_5$ from 0. Let $\mathbf{z}_l$ denote the prefix of $\mathbf{c}_\alpha$ with length l. Then for $(i \geq 0)$, the value of $\mathbf{x}_5[i]$ is a function of $\mathbf{c}_\alpha[i + 1]$, $|\mathbf{z}_{i+1}|_0$, and $|\mathbf{z}_{i+1}|_1$. Figure 2 shows the Pell-base automaton that generates the word $\mathbf{x}_5$. The labels on the states denote the output symbol for that state. Recall that the canonical Pell representation for an integer cannot end in a 2, and cannot have a 2 immediately followed by a 1 or 2 based on the restrictions we have imposed. For this reason, we display certain transitions in the automaton to consist of two symbols. This is done to remove those intermediate states from display that do not produce any output. In practice, the automaton will never halt at such states because the input is always a valid canonical Pell representation.

As an example, consider the canonical Pell representation of $(25)_{10}$, which is $(2001)_P$. When the automaton is given the input string 2001, it halts at a state with label 3, signifying that $\mathbf{x}_5[25] = 3$.

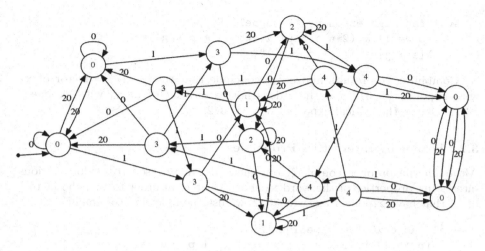

Fig. 2. Pell-base automaton for $\mathbf{x}_5$.

3 Writing the Proof

The authors in [12] proposed a hypothesis about the critical exponent of $\mathbf{x}_5$, but were not able to prove it. We prove their hypothesis using the automata created in Sect. 2. The hypothesis is that the critical exponent of the infinite balanced word $\mathbf{x}_5$ is $E(\mathbf{x}_5) = 3/2$. The predicates used to prove this hypothesis do not contain any free variables. In `Walnut`, such predicates evaluate to either the `true` or the `false` automaton. Please see [8] for further details.

3.1 Proving the Hypothesis

We complete the proof in three steps. First, we test whether there exist integers i, n, p such that a length-n factor of $\mathbf{x}_5$ starting at index i and having period p has exponent at most $3/2$. This predicate produces the `true` automaton.

```
1   eval fac_low_exponent "?msd_pell Ei,p,n
2       (p >= 1) & (2*n <= 3*p) & (Aj (j + p < n) =>
3       X[i + j] = X[i + j + p])";
```

Next, we test whether there exist integers i, n, p such that a length-n factor of $\mathbf{x}_5$ starting at index i and having period p has exponent exactly equal to $3/2$. This predicate also produces the `true` automaton.

```
1   eval fac_ex_exponent "?msd_pell Ei,p,n
2       (p >= 1) & (2*n = 3*p) & (Aj (j + p < n) =>
3       X[i + j] = X[i + j + p])";
```

Finally, we test whether there exist integers i, n, p such that a length-n factor of $\mathbf{x}_5$ starting at index i and having period p has exponent greater than $3/2$. This predicate `fac_high_exponent` produces the `false` automaton.

```
1  eval fac_high_exponent "?msd_pell Ei,p,n
2      (p >= 1) & (2*n > 3*p) & (Aj (j + p < n) =>
3      X[i + j] = X[i + j + p])";
```

Combining the results above, we conclude that there exists a factor of x_5 with exponent $= 3/2$ and there does not exist a factor of x_5 with exponent $>3/2$. Hence, the critical exponent of x_5 is $3/2$.

3.2   Exploring Interesting Properties

Although the proof is complete, our method can be used to explore various interesting properties of the word x_5. For example, in order to find the factors of x_5 that have exponent exactly $=3/2$, we use the following command.

```
1  eval fac_cex5 "?msd_pell En
2      (p >= 1) & (2*n = 3*p) & (Aj (j + p < n) =>
3      X[i + j] = X[i + j + p])";
```

Note that this predicate has two free variables i and p. The corresponding automaton is given in Fig. 3, which accepts pairs of integers (i, p), such that there exists a factor w of x_5 with period p, starting at index i. The automaton suggests that all factors w of x_5 that have exponent $3/2$, have period $p = 4$ and length $|w| = 6$.

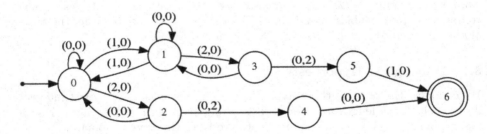

Fig. 3. Pairs (i, p) such that factors of x_5 with starting index i and period p have exponent $= 3/2$.

Another interesting property to explore could be the possible periods p, for which a factor of x_5 is "almost" a $3/2$-power. There are many ways to define this property. To formalize this, let $w = zz'$ be a factor of x_5 with length n, where z' is a prefix of z. The period of w is $p = |z|$. For $p > 10$, we define a factor to be an "almost" $3/2$-power if $n \geq 3p/2 - 2$.

```
1  eval almost_ce_period "?msd_pell Ei
2      (p > 10) &
3      (2*n + 4 >= 3*p) &
4      (Aj (j + p < n) => X[i + j] = X[i + j + p])";
```

Note that this predicate has free variables n and p, which indicate the length and period of w respectively. The automaton produced for this predicate is shown in Fig. 4. We observe that for $p > 10$, all pairs (n, p) have the form:

$$\begin{pmatrix} 1 \\ 1 \end{pmatrix} \begin{pmatrix} 1 \\ 0 \end{pmatrix} \begin{pmatrix} 1 \\ 1 \end{pmatrix} \left\{ \begin{pmatrix} 2 \\ 0 \end{pmatrix} \begin{pmatrix} 0 \\ 0 \end{pmatrix} \right\}^{*} \left\{ \begin{pmatrix} 0 \\ 0 \end{pmatrix}, \begin{pmatrix} 1 \\ 0 \end{pmatrix} \begin{pmatrix} 1 \\ 0 \end{pmatrix} \right\}.$$

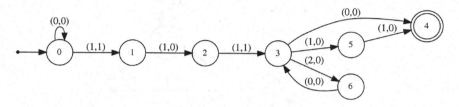

Fig. 4. Pairs (n, p) characterizing factors of $\mathbf{x}_5$ that are "almost" 3/2-powers.

This shows that there exist infinitely many factors of $\mathbf{x}_5$ with this property. We also note that, as p approaches infinity, the exponent of these factors approaches 3/2, which is the critical exponent of $\mathbf{x}_5$.

4   Breadth-First Search

We have proved the existence of a balanced word over $\Sigma_5 = \{0, 1, 2, 3, 4\}$ of critical exponent 3/2. It now remains to show that this exponent 3/2 is optimal for the alphabet Σ_5. To do this, we use a computer program that employs the usual breadth-first search technique. We use the following simple observations to narrow the search space: first, we assume the first letter is 0. Second, we impose the restriction that the first occurrence of the letter i occurs before the first occurrence of j if $i < j$. With these restrictions, the longest balanced word of critical exponent $<3/2$ is of length 44, and there are exactly 5 of them:

```
01203104120130410213014021031401203104120130
01203240210320421023042012302401203240210320
01230240120324021032042102304201230240120324
01231421023124102132412013214201231421023124
01231430132143103213410312341301231430132143
```

5   Future Prospects

5.1   Other Words Characterized by Pell-Base

The authors in [12] determine the value of the critical exponent of the infinite balanced word $\mathbf{x}_3$, $E(\mathbf{x}_3) = 2 + \sqrt{2}/2$ using a manual case-based proof. Given that the value of $\alpha = \sqrt{2} - 1$ for $\mathbf{x}_3$ is the same as that for $\mathbf{x}_5$, we can easily determine the value of $E(\mathbf{x}_3)$ using our method. The constant gap words used for constructing $\mathbf{x}_3$ from $\mathbf{c}_\alpha$ are $\mathbf{y} = (01)^\omega$ and $\mathbf{y}' = 2^\omega$ (see Table 1). Using the same procedure as described in Sect. 2.3, we build an automaton with output that produces the sequence $\mathbf{x}_3$ (see Fig. 5). The claimed value in this case is irrational, unlike $E(\mathbf{x}_5) = 3/2$, which means that it is never actually attained by any factor of $\mathbf{x}_3$.

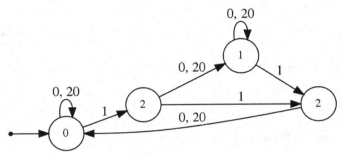

Fig. 5. Pell-base automaton for $\mathbf{x}_3$.

In the subsequent `Walnut` commands, let X denote the automaton in Fig. 5. First, we compute the periods p such that a repetition with exponent $\geq 13/5$ and period p occurs in $\mathbf{x}_3$.

```
1  eval periods_of_high_powers "?msd_pell Ei
2      (p >= 1) & (Aj (5*j <= 8*p) => X[i + j] = X[i + j + p])";
```

The language accepted by the produced automaton is 0*110000*, which is the Pell-base representation of numbers of the form $P_n + P_{n-1}$, for $n \geq 5$. The following command lets us save this as a regular expression.

```
1  reg pows msd_pell "0*110000*";
```

Next, we compute pairs of integers (n, p) such that $\mathbf{x}_3$ has a factor of length $n + p$ with period p, and this factor cannot be extended to a longer factor of length $n + p + 1$ with the same period.

```
1  def maximal_reps "?msd_pell Ei
2      (Aj (j < n) => X[i + j] = X[i + j + p]) &
3      (X[i + n] != X[i + n + p])";
```

Finally, we compute integer pairs (n, p) where p is of the form 0*110000*, and $n + p$ is the maximum possible length of any factor with period p.

```
1  eval highest_powers "?msd_pell
2      (p >= 1) & $pows(p) & $maximal_reps(n, p) &
3      (Am $maximal_reps(m, p) => m <= n)";
```

The predicate `highest_powers` produces an automaton accepting integer pairs (n, p) that have the form

$$\begin{pmatrix} 0 \\ 0 \end{pmatrix}^* \begin{pmatrix} 2 \\ 1 \end{pmatrix} \begin{pmatrix} 0 \\ 1 \end{pmatrix} \begin{pmatrix} 2 \\ 0 \end{pmatrix} \begin{pmatrix} 0 \\ 0 \end{pmatrix} \left\{ \begin{pmatrix} 2 \\ 0 \end{pmatrix} \begin{pmatrix} 0 \\ 0 \end{pmatrix} \right\}^* \left\{ \begin{pmatrix} 0 \\ 0 \end{pmatrix}, \begin{pmatrix} 1 \\ 0 \end{pmatrix} \begin{pmatrix} 1 \\ 0 \end{pmatrix} \right\}.$$

As clear from the pattern of pairs (n, p), for $m \geq 5$, when $p = P_m + P_{m-1}$, then we have $n = P_{m+1} - 2$. Thus, we have the exponent as the ratio of length $n + p$ and period p,

$$e = \frac{P_{m+1} + P_m + P_{m-1} - 2}{P_m + P_{m-1}} = 2 + \frac{P_m - 2}{P_m + P_{m-1}}. \tag{1}$$

Let $a_k/b_k = [d_0, d_1, d_2, \ldots, d_k]$ be the *convergents* of $\alpha = \sqrt{2} - 1$. Then for $k \geq 0$, we have $a_k = P_k$ and $b_k = P_{k+1}$. Hence, the following bound holds:

$$\left| \alpha - \frac{P_k}{P_{k+1}} \right| < \frac{1}{P_{k+1} P_{k+2}} < \frac{1}{P_{k+1}^2}. \tag{2}$$

Substituting $k = m - 1$ in (2), we bound the value of P_m/P_{m-1}. We also know that P_m/P_{m-1} converges to the *silver ratio*, $\sigma = \sqrt{2} + 1$. Thus, we have,

$$e = 2 + \frac{P_m - 2}{P_m + P_{m-1}} < 2 + \frac{\sqrt{2} + 1 + 1/P_{m-1}^2 - 2/P_{m-1}}{\sqrt{2} + 2 - 1/P_{m-1}^2}.$$

For $m \geq 5$, as $m \to \infty$, the value of e is increasing, and tends to $2 + \sqrt{2}/2$. Thus, $e < 2 + \sqrt{2}/2$, which proves the hypothesis.

In theory, one might also hope to determine the values of $E(\mathbf{x}_8)$ and $E(\mathbf{x}_9)$ using our method, since the corresponding continued fraction expansions of α end in a repeating 2. The actual result depends on the practical limitations of run-time and the memory available for the corresponding machine computation.

5.2   Open Problems

The obvious open problem to pursue is the implementation of an automaton that recognizes the addition relation for words in the general Ostrowski α-numeration system (see [6]). Assuming that the machine computation is feasible, we might be able to obtain analogous proofs for balanced words over larger alphabets.

Acknowledgments. We thank Narad Rampersad and Luke Schaeffer for their helpful comments. We are also grateful to the referees who read the paper and offered many useful suggestions.

References

1. Allouche, J.P., Shallit, J.: Automatic Sequences. Cambridge University Press, Cambridge (2003)
2. Angluin, D.: Learning regular sets from queries and counterexamples. Inf. Comput. **75**(2), 87–106 (1987)
3. Berstel, J., Séébold, P.: Sturmian words. In: Lothaire, M. (ed.) Algebraic Combinatorics on Words, Encyclopedia of Mathematics and Its Applications, vol. 30, Chap. 2, pp. 45–110. Cambridge University Press, Cambridge (2002)
4. Brzozowski, J.A.: Derivatives of regular expressions. J. ACM **11**(4), 481–494 (1964)
5. Du, C.F., Mousavi, H., Schaeffer, L., Shallit, J.: Decision algorithms for Fibonacci automatic words, III: enumeration and abelian properties. Int. J. Found. Comput. Sci. **27**(8), 943–963 (2016)
6. Hieronymi, P., Terry, A.: Ostrowski numeration systems, addition, and finite automata. Notre Dame J. Formal Logic **59**(2), 215–232 (2018)
7. Hubert, P.: Suites équilibrées. Theoret. Comput. Sci. **242**(1–2), 91–108 (2000)
8. Mousavi, H.: Automatic theorem proving in Walnut. Preprint: https://arxiv.org/abs/1603.06017 (2016)
9. Mousavi, H., Schaeffer, L., Shallit, J.: Decision algorithms for Fibonacci-automatic words, I: basic results. RAIRO Inform. Théor. App. **50**(1), 39–66 (2016)
10. Nerode, A.: Linear automaton transformations. Proc. Am. Math. Soc. **9**(4), 541–544 (1958)
11. Ostrowski, A.: Bemerkungen zur Theorie der diophantischen Approximationen. Abh. Math. Semin. Univ. Hamburg **1**(1), 77–98 (1922)
12. Rampersad, N., Shallit, J., Vandomme, E.: Critical exponents of infinite balanced words. Theoret. Comput. Sci. (2018). https://doi.org/10.1016/j.tcs.2018.10.017
13. Sloane, N.J.A., et al.: The on-line encyclopedia of integer sequences (2018). https://oeis.org
14. Vuillon, L.: Balanced words. Bull. Belgian Math. Soc. **10**(5), 787–805 (2003)

Repetitions in Infinite Palindrome-Rich Words

Aseem R. Baranwal[✉][iD] and Jeffrey Shallit[iD]

School of Computer Science, University of Waterloo, Waterloo, ON N2L 3G1, Canada
{aseem.baranwal,shallit}@uwaterloo.ca

Abstract. Rich words are those containing the maximum possible number of distinct palindromes. Several characteristic properties of rich words have been studied; yet the analysis of repetitions in rich words still involves some interesting open problems. We consider lower bounds on the repetition threshold of infinite rich words over 2- and 3-letter alphabets, and construct a candidate infinite rich word over the alphabet $\Sigma_2 = \{0, 1\}$ with a small critical exponent of $2 + \sqrt{2}/2$. This represents the first progress on an open problem of Vesti from 2017.

Keywords: Critical exponent · Repetitions · Rich words · Palindrome

1 Introduction

Palindromes—words equal to their reversal—are among the most widely studied repetitions in words. The class of palindrome-rich words, or simply rich words—those words containing, as factors, the maximum possible number of palindromes—was introduced in [4,9,11]. Since then, rich words have received much attention in the combinatorics on words literature; see, for example, [5,13,24].

1.1 Preliminaries

In this section we provide the preliminary definitions and results that we use throughout the paper, along with the motivation behind our work.

Definition 1. *A finite word w is* rich *if it contains $|w|$ distinct nonempty palindromes. An infinite word w is* rich *if all its factors are rich.*

We say that a word $u = z^e$ has *exponent* e and *period* $p = |z|$, where $e = |u|/p$ is a positive rational number that denotes the number of times z is repeated. We say u is *primitive* if its only integer exponent is 1. The word w is an *overlap* if $w = uuu'$ where u' is a nonempty prefix of u.

Example 1. The word $u = 00010001$ is rich, because it has 8 distinct nonempty palindromes as factors, while the word $v = 00101100$ is not rich. The word u has period 4 and exponent 2, since $u = z^e$, where $z = 0001$ and $e = 2$.

R. Mercaş and D. Reidenbach (Eds.): WORDS 2019, LNCS 11682, pp. 93–105, 2019.
https://doi.org/10.1007/978-3-030-28796-2_7

Definition 2 [17]. *For a given alphabet* Σ, *a mapping* φ *on* Σ^* *is an* involutive antimorphism *if* $\varphi(uv) = \varphi(v)\varphi(u)$, *and* $\varphi^2(u) = u$ *for all* $u, v \in \Sigma^*$.

Definition 3. *The* critical exponent *of an infinite word* w *is defined to be the supremum of the set of all rational numbers* e *such that there exists a finite nonempty factor of* w *with exponent* e.

Definition 4. *The* repetition threshold *on an alphabet of size* k *is the infimum of the set of exponents* e *such that there exists an infinite word that avoids greater than* e-*powers.*

The repetition threshold can also be characterized as the smallest possible critical exponent of a word over an alphabet of size k. Dejean gave a famous conjecture about this threshold in [10], which was proven by Currie and Rampersad [8], and independently by Rao [20]. The repetition threshold can also be studied for a limited class of infinite words. For example, Rampersad et al. studied this threshold for infinite balanced words in [19]. In this paper, we study the repetition threshold $RT(k)$ for infinite rich words over an alphabet of size k.

1.2 Previous Work

Let the word w be the fixed point of a given involutive antimorphism Θ. We say w is a Θ-*palindrome* if $w = \Theta(w)$. The set of Θ-palindromic factors of a word w is denoted by $\mathrm{Pal}_\Theta(w)$. In 2013, Pelantová and Starosta introduced the idea of Θ-palindromic defect.

Definition 5. *The* Θ-palindromic defect *of a finite word* w, *denoted by* $D_\Theta(w)$, *is defined as*

$$D_\Theta(w) = |w| + 1 - \gamma_\Theta(w) - |\mathrm{Pal}_\Theta(w)|,$$

where $\gamma_\Theta(w) = |\{\{a, \Theta(a)\} : a \in \Sigma, a \text{ occurs in } w \text{ and } a \neq \Theta(a)\}|$. *For an infinite word* w, *the* Θ-*palindromic defect is the supremum of the set of* $D_\Theta(u)$, *where* u *is a factor of* w.

Further, they proved that all recurrent words with a finite Θ-palindromic defect contain infinitely many overlapping factors [17]. This result leads to the following theorem.

Theorem 1. *All infinite rich words contain a square.*

Theorem 1 provides a lower bound on the repetition threshold for infinite rich words over a k-letter alphabet; namely $RT(k) \geq 2$. In [25], Vesti gives both upper and lower bounds on the length of the longest square-free rich words, and proposes the open problem of determining the repetition threshold for infinite rich words.

2   Results over the Binary Alphabet

We construct an infinite binary rich word and determine the value of its critical exponent. We further conjecture that this value is the repetition threshold for the binary alphabet, based on supporting evidence from computation. We define the word $\mathbf{r}$ as the image of a fixed point, $\mathbf{r} = \tau(\varphi^\omega(0)) = 001001100100110\cdots$, where the morphisms φ and τ are defined as follows:

$$
\begin{aligned}
\varphi: 0 &\to 01 & \tau: 0 &\to 0 \\
1 &\to 02 & 1 &\to 01 \\
2 &\to 022, & 2 &\to 011.
\end{aligned}
$$

2.1   Automatic Theorem-Proving

We utilize the automatic theorem-proving software `Walnut`, written by Hamoon Mousavi, to constructively decide first-order predicates concerning the word $\mathbf{r}$ [15]. To enable `Walnut` to work with the word $\mathbf{r}$, we require an automaton with output that produces $\mathbf{r}$.

The Pell numbers are defined by the recurrence $P_0 = 0$, $P_1 = 1$, and $P_n = 2P_{n-1} + P_{n-2}$. This sequence of numbers is used to define a non-standard positional numeral system in the family of Ostrowski numeration systems [16] with $\alpha = \sqrt{2} - 1$. Given an integer N, we can express it as an integer linear combination of Pell numbers as follows: $N = \sum_{0 \leq i < n} d_i P_{i+1}$. To ensure that this representation is unique, we impose the following conditions on the d_i:

1. The least significant digit $d_0 \in \{0, 1\}$.
2. For all $i > 0$ we have $d_i \in \{0, 1, 2\}$.
3. If $d_i = 2$, then $d_{i-1} = 0$.

In this case, the word $d_{n-1}d_{n-2}\cdots d_0$ is said to be the *canonical Pell representation* of the integer N, and we write it as $(N)_P$. Computing the lengths $L_i = |\tau(\varphi^i(0))|$ for $i \geq 0$, we note that $L_0 = 1$, $L_1 = 3$, and $L_i = 2L_{i-1} + L_{i-2}$ for $i \geq 2$. This suggests that the word $\mathbf{r}$ might be *Pell-automatic* (meaning that there exists an automaton that takes as input an integer N represented in the Pell number system, and outputs the symbol in $\mathbf{r}$ at index N), and indeed, we prove that this is the case. The Pell number system is a non-standard positional number system in the family of Ostrowski numeration systems [16]. In [2], the authors construct a Pell adder automaton to enable writing first-order predicates in this number system. We use this adder to decide predicates about the constructed word. The `Walnut` version equipped with the adder is available on GitHub.[1]

2.2   Constructing the Automaton

We construct an automaton with output for the word $\mathbf{r}$ using the same methods as used by the authors in [2] to construct the Pell adder automaton. The method

---

[1] Repository: https://github.com/aseemrb/Walnut/.

utilizes a combination of membership and equivalence queries as described in the $L*$ algorithm given by Angluin [1]. Figure 2 represents the constructed automaton. Note that this automaton consists of 4 states, and we have not restricted the Pell representations of the input to be canonical, meaning that the input may end with a 2, and a non-zero digit may follow a 2. However, in practice, an input will always be in the canonical Pell representation. The node labels in the figure represent the state and the corresponding output symbol.

In [2], the authors use induction to mechanically prove the correctness of the adder automaton that they construct. Here, we prove that the automaton in Fig. 2 produces the same word as given by $\tau(\varphi^\omega(0))$. To do this, first, we restrict this automaton to only consider canonical Pell representations. Thus, the least significant digit is <2, and a 2 is always followed by a 0. This gives the automaton in Fig. 1. Let f and g be the morphisms associated with this automaton, and let $\mathbf{s} = g(f^\omega(0))$ denote the infinite word produced. The morphisms f and g are given by

$$
\begin{aligned}
f: 0 &\to 012 & g: 0 &\to 0 \\
1 &\to 304 & 1 &\to 0 \\
2 &\to 0 & 2 &\to \epsilon \\
3 &\to 354 & 3 &\to 1 \\
4 &\to 3 & 4 &\to \epsilon \\
5 &\to 032, & 5 &\to 1.
\end{aligned}
$$

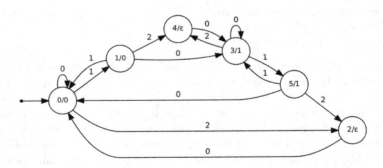

Fig. 1. Automaton for the infinite word $\mathbf{s}$. Here ϵ denotes the empty word.

2.3 Proof of Equivalence of the Morphisms

In this section, we prove that the automaton in Fig. 1 produces the same infinite word as that produced by morphisms φ and τ, that is, $\mathbf{r} = \mathbf{s}$. We need two lemmas to prove this equivalence.

Lemma 1. *For all $n \geq 2$, we have $g(f^n(0)) = g(f^{n-1}(0))g(f^{n-2}(3))g(f^{n-1}(0))$.*

Proof. We prove this by induction on n. For $n = 2$, we have that

$$g(f^2(0)) = g(f^1(0))g(3)g(f^1(0)) = 00100.$$

So the base case holds. Next, we construct the induction hypothesis,

$$H_1 : g(f^k(0)) = g(f^{k-1}(0))g(f^{k-2}(3))g(f^{k-1}(0)), \forall k \leq n.$$

For the inductive step, consider $g(f^{n+1}(0))$. Using the definition of the morphisms f and g, we have that,

$$\begin{aligned}
g(f^{n+1}(0)) &= g(f^n(0))g(f^n(1))g(f^n(2)) \\
&= g(f^n(0))g(f^{n-1}(3))g(f^{n-1}(0))g(f^{n-1}(4))g(f^n(2)) \\
&= g(f^n(0))g(f^{n-1}(3))g(f^{n-1}(0))g(f^{n-2}(3))g(f^{n-1}(0)). \quad (1)
\end{aligned}$$

Using the induction hypothesis H_1 in Eq. (1), we get

$$g(f^{n+1}(0)) = g(f^n(0))g(f^{n-1}(3))g(f^n(0)).$$

This completes the proof.

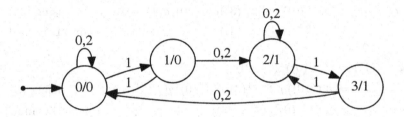

Fig. 2. Guessed automaton for the infinite word **r**.

Lemma 2. *For all $n \geq 2$, we have $g(f^n(3)) = g(f^{n-1}(3))g(f^{n-2}(0))g(f^{n-1}(3))$.*

Proof. The proof is similar to that of Lemma 1, by induction on n. For $n = 2$, we have

$$g(f^2(3)) = g(f^1(3))g(0)g(f^1(3)) = 11011.$$

So the base case holds. We have the induction hypothesis,

$$H_2 : g(f^k(3)) = g(f^{k-1}(3))g(f^{k-2}(0))g(f^{k-1}(3)), \forall k \leq n.$$

For the inductive step, consider $g(f^{n+1}(3))$. Using the definition of the morphisms f and g, we have that

$$\begin{aligned}
g(f^{n+1}(3)) &= g(f^n(3))g(f^n(5))g(f^n(4)) \\
&= g(f^n(3))g(f^{n-1}(0))g(f^{n-1}(3))g(f^{n-1}(2))g(f^n(4)) \\
&= g(f^n(3))g(f^{n-1}(0))g(f^{n-1}(3))g(f^{n-2}(0))g(f^{n-1}(3)). \quad (2)
\end{aligned}$$

Using the induction hypothesis H_2 in Eq. (2), we get

$$g(f^{n+1}(3)) = g(f^n(3))g(f^{n-1}(0))g(f^n(3)).$$

This completes the proof.

Now we prove the following equivalence theorem about the words produced by the automaton in Fig. 1 and the word given by morphisms φ and τ.

Theorem 2. *The infinite words $\tau(\varphi^\omega(0))$ and $g(f^\omega(0))$ are equal.*

Proof. We prove this by a simultaneous induction on n with 3 hypotheses.

$$\tau(\varphi^k(0)) = g(f^k(0))g(f^{k-1}(3)) \tag{3}$$
$$\tau(\varphi^k(1)) = g(f^k(0))g(f^k(3)) \tag{4}$$
$$\tau(\varphi^k(2)) = g(f^k(0))g(f^{k+1}(3)) \tag{5}$$

The base case $k = 1$ can be checked by hand. Assume that the hypotheses hold for $k \leq n$. Next, we consider the following inductive steps using the definitions of φ and τ.

$$\begin{aligned}
\tau(\varphi^{n+1}(0)) &= \tau(\varphi^n(0))\tau(\varphi^n(1)) \\
&= g(f^n(0))g(f^{n-1}(3))g(f^n(0))g(f^n(3)) && \text{using (3,4)} \\
&= g(f^{n+1}(0))g(f^n(3)). && \text{using Lemma 1.}
\end{aligned}$$

$$\begin{aligned}
\tau(\varphi^{n+1}(1)) &= \tau(\varphi^n(0))\tau(\varphi^n(2)) \\
&= g(f^n(0))g(f^{n-1}(3))g(f^n(0))g(f^{n+1}(3)) && \text{using (3,5)} \\
&= g(f^{n+1}(0))g(f^{n+1}(3)) && \text{using Lemma 1.}
\end{aligned}$$

$$\begin{aligned}
\tau(\varphi^{n+1}(2)) &= \tau(\varphi^n(0))\tau(\varphi^n(2))\tau(\varphi^n(2)) \\
&= g(f^n(0))g(f^{n-1}(3))g(f^n(0))g(f^{n+1}(3))g(f^n(0))g(f^{n+1}(3)) \\
&= g(f^{n+1}(0))g(f^{n+2}(3)) && \text{using Lemmas 1, 2.}
\end{aligned}$$

This proves that the hypotheses are true. From Eq. (3), we have $\tau(\varphi^k(0)) = g(f^k(0))g(f^{k-1}(3))$. Letting $k \to \infty$, we get $\tau(\varphi^\omega(0)) = g(f^\omega(0))$. This completes the proof.

2.4 Proof of Palindromic Richness

We claim that the infinite word $\mathbf{r} = \mathbf{s} = g(f^\omega(0)) = 001001100100110\cdots$ is rich. The proof is carried out using Walnut by constructing a set of predicates based on Theorem 3, as done in [23]. We say that a word w has a *unioccurrent* suffix s if s is not a factor of any proper prefix of w. We recall that A and E are Walnut's way of expressing the universal and existential quantifiers, respectively. All of the computations we describe can be carried out in a matter of a few seconds on a Linux machine.

Theorem 3 *(Glen et al. [11]). A word w is rich if and only if every prefix of w has a unioccurrent palindromic suffix.*

In the following predicates, R denotes the automaton in Fig. 1. First, we introduce the fundamental predicates that form the building blocks for verification of the richness property.

1. The predicate FactorEq takes 3 parameters i, j, n and evaluates to true if the length-n factors of r starting at indices i and j are equal.
2. The predicate Occurs takes 4 parameters i, j, m, n and evaluates to true if the length-m factor of r starting at index i occurs in the length-n factor starting at index j, i.e., $R[i..i + m − 1]$ is a factor of $R[j..j + n − 1]$.
3. The predicate Palindrome takes 2 parameters i, n and evaluates to true if the length-n factor of r starting at index i is a palindrome.

```
1   def FactorEq "?msd_pell Ak (k < n) => (R[i + k] = R[j + k])";
2   def Occurs "?msd_pell (m <= n) &
3       (Ek (k + m <= n) & $FactorEq(i, j + k, m))";
4   def Palindrome "?msd_pell Aj,k ((k < n) & (j + k + 1 = n)) =>
5       (R[i + k] = R[i + j])";
```

By Theorem 3, for any finite word to be rich, it is sufficient to check if all its prefixes have a unioccurrent palindromic suffix. We use this property to construct the predicate RichFactor which takes two parameters i, n, and evaluates to true if the length-n factor of r starting at index i is rich. Figure 3 shows the representation of variables in the predicate.

```
1   def RichFactor "?msd_pell
2       Am ((m >= 1) & (m <= n)) =>
3           (Ej (i <= j) & (j < i + m) &
4           $Palindrome(j, i + m - j) &
5           ~$Occurs(j, i, i + m - j, m - 1))";
```

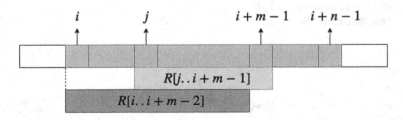

Fig. 3. Representation of variables i, j, m, n in the predicate RichFactor. It evaluates to true if the word $R[i..i + n − 1]$ is rich.

Now, we simply check that all prefixes of r are rich to show that the infinite word r is rich. The following predicate, R_Is_Rich evaluates to true, which completes the proof.

```
1   eval R_Is_Rich "?msd_pell An $RichFactor(0, n)";
```

2.5   Determining the Critical Exponent

First we observe that the critical exponent of **r** is < 3. This can be checked in Walnut as follows:

```
1   eval CheckCritExp "?msd_pell ~(E i, p (p >= 1) &
2       Aj (j < 2*p) => R[i + j] = R[i + j + p])";
```

Next, we compute the periods p such that a repetition with exponent $\geq 5/2$ and period p occurs in **r**.

```
1   eval HighPowPeriods "?msd_pell (p >= 1) &
2       (Ei Aj (2*j <= 3*p) => R[i + j] = R[i + j + p])";
```

The language accepted by the produced automaton is 0*1100*, which is the Pell-base representation of numbers of the form $P_t + P_{t-1}$, for $t \geq 3$. Next, we compute pairs of integers (n, p) such that **r** has a factor of length $n + p$ with period p, and this factor cannot be extended to a longer factor beginning at the same position, of length $n + p + 1$ with the same period.

```
1   def MaximalReps "?msd_pell Ei
2       (Aj (j < n) => R[i + j] = R[i + j + p]) &
3       (R[i + n] != R[i + n + p])";
```

Finally, we compute the pairs (n, p) where p matches the regular expression 0*1100* in the Pell base representation, and $n + p$ is the maximum possible length of any factor with period p. (Such an n exists for each p because we showed the critical exponent is < 3.)

```
1   eval HighestPowers "?msd_pell
2       $HighPowPeriods(p) &
3       $MaximalReps(n, p) &
4       (Am $MaximalReps(m, p) => m <= n)";
```

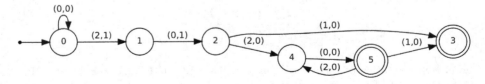

Fig. 4. Pairs (n, p) satisfying the predicate HighestPowers.

Figure 4 shows the automaton produced by the predicate `HighestPowers`. It accepts pairs (n, p) of the following forms:

$$\begin{pmatrix}0\\0\end{pmatrix}^*\begin{pmatrix}2\\1\end{pmatrix}\begin{pmatrix}0\\1\end{pmatrix}\begin{pmatrix}1\\0\end{pmatrix}, \tag{6}$$

$$\begin{pmatrix}0\\0\end{pmatrix}^*\begin{pmatrix}2\\1\end{pmatrix}\begin{pmatrix}0\\1\end{pmatrix}\begin{pmatrix}2\\0\end{pmatrix}\begin{pmatrix}0\\0\end{pmatrix}\left\{\begin{pmatrix}2\\0\end{pmatrix}\begin{pmatrix}0\\0\end{pmatrix}\right\}^*, \text{ or} \tag{7}$$

$$\begin{pmatrix}0\\0\end{pmatrix}^*\begin{pmatrix}2\\1\end{pmatrix}\begin{pmatrix}0\\1\end{pmatrix}\begin{pmatrix}2\\0\end{pmatrix}\begin{pmatrix}0\\0\end{pmatrix}\left\{\begin{pmatrix}2\\0\end{pmatrix}\begin{pmatrix}0\\0\end{pmatrix}\right\}^*\begin{pmatrix}1\\0\end{pmatrix}. \tag{8}$$

Here, the length of the words is $l = n + p$ and the period is p. Equation (6) corresponds to $n = (201)_P = 11$ and $p = (110)_P = 7$. Thus we have

$$e = \frac{l}{p} = \frac{n+p}{p} = \frac{18}{7} \approx 2.57.$$

Equation (7) corresponds to

$$n = \sum_{1 \le i \le k} 2P_{2k} = P_{2k+1} - 1, \; p = P_{2k} + P_{2k-1}.$$

Equation (8) corresponds to

$$n = 1 + \sum_{1 \le i \le k} 2P_{2k+1} = P_{2k+2} - 1, \; p = P_{2k+1} + P_{2k}.$$

Putting $m = 2k-1$ for (7), and $m = 2k$ for (8), we notice that the expressions for n and p coincide.

$$e = \frac{P_{m+2} + P_{m+1} + P_m - 1}{P_{m+1} + P_m}$$
$$= 2 + \frac{P_{m+1} - 1}{P_{m+1} + P_m}.$$

Since Pell numbers are the convergents of $\sqrt{2} - 1$, and the ratio P_{m+1}/P_m converges to $\sqrt{2} + 1$, we have that

$$e = 2 + \frac{P_{m+1} - 1}{P_{m+1} + P_m} \tag{9}$$

$$< 2 + \frac{\sqrt{2} + 1 + 1/P_m^2 - 1/P_m}{\sqrt{2} + 2 - 1/P_m^2}. \tag{10}$$

For $m \ge 4$, as $m \to \infty$, the value in Eq. (10) is increasing, and tends to $2 + \sqrt{2}/2$. Thus, the critical exponent of the word $\mathbf{r}$ is $2 + \sqrt{2}/2$. The `Walnut` commands for verifying richness and computing the critical exponent are available on GitHub.[2]

---

[2] URL: https://github.com/aseemrb/Walnut/blob/master/CommandFiles/rich2.txt.

2.6 Optimality of the Critical Exponent

A backtracking computation shows that the longest rich binary word with critical exponent <2.700 is of length 1339. Combining this with the result above, we obtain the following bounds.

$$2.700 \leq RT(2) \leq 2 + \frac{\sqrt{2}}{2} = 2.7071\ldots$$

3 Faster Backtracking

In this section, we discuss some methods to optimize our backtracking algorithm. The most obvious optimization is to consider the following.

1. Without loss of generality, we assume that the word starts with a 0.
2. We impose the restriction that the first occurrence of the symbol a occurs before the first occurrence of symbol b if $a < b$.

3.1 Lyndon Method

Since our goal is to check if there is an infinite rich word with critical exponent less than a preset threshold, we can utilize the Lyndon method to prune certain branches of the backtracking search tree. A Lyndon word is a primitive nonempty word that is strictly smaller in lexicographic order than all of its rotations. If a word satisfies the properties of richness and the critical exponent being less than some threshold, then all factors of the word also satisfy these properties. This fact helps us by pruning those paths in the search tree that lead to a suffix that is lexicographically smaller than the word itself.

3.2 Counting Palindromes

To check for richness, Groult et al. give a linear time algorithm to count the number of distinct palindromes in a word [12]. Their algorithm is based on two major ideas: a linear-time algorithm by Gusfield to compute all maximal palindromes in a word [14], and a linear-time algorithm by Crochemore and Ilie to compute the LPF (longest previous factor) array [7]. However, their approach is not helpful to our problem since it requires linear pre-processing time.

What we require is a fast online algorithm such that given the number of distinct palindromes for a word w over an alphabet Σ, we can find the number of distinct palindromes in the word wa for all $a \in \Sigma$ in constant amortized time. Such an algorithm is given by Rubinchik and Shur [22]. Their primary idea is to construct a graph where each node represents a unique palindrome. There are two types of edges in this graph:

1. **Border edge**: This is a directed edge from p to q labeled a, if $q = apa$ for some $a \in \Sigma$.

2. **Suffix edge**: This is an unlabeled directed edge from p to q, if q is the longest proper palindromic suffix of p.

Whenever we append a new symbol to an already processed word, it takes amortized constant time to maintain this graph. The C++ implementation of the algorithm can be found on GitHub.[3]

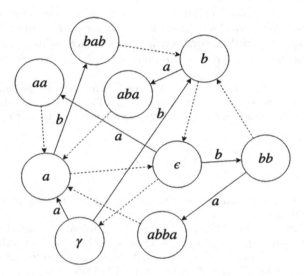

Fig. 5. The graph of palindromes for the word $w = aababba$. Here ϵ is the empty word and γ is the imaginary palindrome word of length -1 [22].

Example 2. Figure 5 shows the graph construction for the rich word $aababba$. The number of nonempty palindromes is equal to 7. Note that we have an imaginary word γ that has length -1 and is a palindrome. The suffix edges are shown by dashed lines, while the border edges are shown with solid lines having labels. We say that a palindrome consisting of a single symbol borders γ, which makes the implementation of the algorithm easy.

3.3   Computing Maximal Runs

In [6], Chen et al. present a survey of fast space-efficient algorithms for computing all maximal runs in a string. They also propose some new and faster algorithms for this task. In future work, we aim to understand and implement these algorithms in our backtracking search. Faster computation of maximal runs will help us to efficiently reject those paths in the backtracking search that violate the critical exponent threshold. Thus, we may be able to compute tighter lower bounds on the repetition threshold more efficiently.

---

[3] URL: https://github.com/aseemrb/research-scripts/blob/master/palindromes/palin.cpp.

4   Future Prospects

For an alphabet of size $k = 3$, backtracking search shows that $RT(3) \geq 9/4$. The longest word that has a critical exponent $<9/4$ is of length 114. For $k = 4$ and the exponent threshold $11/5$, our search program has reached words of length 3800 and has not terminated.

An obvious direction for further research is to develop novel ideas and methods that may help us prove lower bounds on the repetition threshold of infinite rich words. Another possible direction is to construct infinite rich words over larger alphabets that may serve as candidates for the repetition threshold.

Acknowledgments. We are grateful to the referees for their suggestions.

After our paper was submitted, we learned from Edita Pelantová that our word **r** is a complementary symmetric Rote word [21], and hence by [3,18] it follows that **r** is rich.

References

1. Angluin, D.: Learning regular sets from queries and counterexamples. Inf. Comput. **75**(2), 87–106 (1987)
2. Baranwal, A.R., Shallit, J.: Critical exponent of infinite balanced words via the Pell number system. Preprint: https://arxiv.org/abs/1902.00503 (2019)
3. Blondin Massé, A., Brlek, S., Labbé, S., Vuillon, L.: Palindromic complexity of codings of rotations. Theoret. Comput. Sci. **412**, 6455–6463 (2011)
4. Brlek, S., Hamel, S., Nivat, M., Reutenauer, C.: On the palindromic complexity of infinite words. Int. J. Found. Comput. Sci. **15**, 293–306 (2004)
5. Bucci, M., De Luca, A., Glen, A., Zamboni, L.Q.: A new characteristic property of rich words. Theoret. Comput. Sci. **410**, 2860–2863 (2009)
6. Chen, G., Puglisi, S.J., Smyth, W.F.: Fast & practical algorithms for computing all the runs in a string. In: Ma, B., Zhang, K. (eds.) CPM 2007, LNCS, vol. 4580, pp. 307–315. Springer, Heidelberg (2007). https://doi.org/10.1007/978-3-540-73437-6_31
7. Crochemore, M., Ilie, L.: Computing longest previous factor in linear time and applications. Inform. Process. Lett. **106**(2), 75–80 (2008)
8. Currie, J., Rampersad, N.: A proof of Dejean's conjecture. Math. Comput. **80**(274), 1063–1070 (2011)
9. de Luca, A., Glen, A., Zamboni, L.Q.: Rich, Sturmian, and trapezoidal words. Theoret. Comput. Sci. **407**, 569–573 (2008)
10. Dejean, F.: Sur un théorème de Thue. J. Combin. Theory. Ser. A **13**(1), 90–99 (1972)
11. Glen, A., Justin, J., Widmer, S., Zamboni, L.Q.: Palindromic richness. Eur. J. Comb. **30**, 510–531 (2009)
12. Groult, R., Prieur, E., Richomme, G.: Counting distinct palindromes in a word in linear time. Inform. Process. Lett. **110**, 908–912 (2010)
13. Guo, C., Shallit, J., Shur, A.M.: Palindromic rich words and run-length encodings. Inform. Process. Lett. **116**, 735–738 (2016)
14. Gusfield, D.: Algorithms on Strings, Trees, and Sequences: Computer Science and Computational Biology. Cambridge University Press, Cambridge (1997)

15. Mousavi, H.: Automatic theorem proving in Walnut. Preprint: https://arxiv.org/abs/1603.06017 (2016)
16. Ostrowski, A.: Bemerkungen zur Theorie der diophantischen Approximationen. Abh. Math. Semin. Univ. Hamburg **1**(1), 77–98 (1922)
17. Pelantová, E., Starosta, Š.: Languages invariant under more symmetries: overlapping factors versus palindromic richness. Discrete Math. **313**, 2432–2445 (2013)
18. Pelantová, E., Starosta, Š.: Constructions of words rich in palindromes and pseudopalindromes. Discrete Math. Theoret. Comput. Sci. 18, Paper #16 (2016). https://dmtcs.episciences.org/2202
19. Rampersad, N., Shallit, J., Vandomme, E.: Critical exponents of infinite balanced words. Theoret. Comput, Sci. **777**, 454–463 (2018)
20. Rao, M.: Last cases of Dejean's conjecture. Theoret. Comput. Sci. **412**(27), 3010–3018 (2011)
21. Rote, G.: Sequences with subword complexity $2n$. J. Number Theory **46**, 196–213 (1994)
22. Rubinchik, M., Shur, A.M.: EERTREE: an efficient data structure for processing palindromes in strings. In: Lipták, Z., Smyth, W.F. (eds.) IWOCA 2015. LNCS, vol. 9538, pp. 321–333. Springer, Cham (2016). https://doi.org/10.1007/978-3-319-29516-9_27
23. Schaeffer, L., Shallit, J.: Closed, palindromic, rich, privileged, trapezoidal, and balanced words in automatic sequences. Electronic J. Combinatorics **23**, 1–25 (2016)
24. Vesti, J.: Extensions of rich words. Theoret. Comput. Sci. **548**, 14–24 (2014)
25. Vesti, J.: Rich square-free words. Theoret. Comput. Sci. **687**, 48–61 (2017)

Generalized Lyndon Factorizations
of Infinite Words

Amanda Burcroff[(⊠)] and Eric Winsor

University of Michigan, Ann Arbor, MI 48109, USA
{burcroff,rcwnsr}@umich.edu

Abstract. A generalized lexicographic order on words is a lexicographic order where the total order of the alphabet depends on the position of the comparison. A generalized Lyndon word is a finite word which is strictly smallest among its class of rotations with respect to a generalized lexicographic order. This notion can be extended to infinite words: an infinite generalized Lyndon word is an infinite word which is strictly smallest among its class of suffixes. We prove a conjecture of Dolce, Restivo, and Reutenauer: every infinite word has a unique nonincreasing factorization into finite and infinite generalized Lyndon words. When this factorization has finitely many terms, we characterize the last term of the factorization. Our methods also show that the infinite generalized Lyndon words are precisely the words with infinitely many generalized Lyndon prefixes.

Keywords: Generalized lexicographic order ·
Infinite generalized Lyndon word ·
Unique nonincreasing Lyndon factorization

1 Introduction

A *rotation* of a finite word w is a word of the form vu, where $w = uv$ is a factorization of w. A finite word is called *Lyndon* if it is strictly smallest among its class of rotations with respect to the standard lexicographic order. In particular, every finite word is a conjugate of some power of a Lyndon word. Lyndon words were introduced in 1953 by Shirshov in [12] and studied by Lyndon in [8]. Lyndon words have been given various names throughout their history, including *standard lexicographic sequences*, *regular words*, and *prime words*. These names hint at their significant role in the factorization of words.

Let A^* denote the free monoid on a totally ordered (possibly infinite) alphabet A, where A^* is ordered lexicographically. The Chen-Fox-Lyndon factorization theorem for words states that the Lyndon words form a basis for A^* [2]. Put more concretely, any finite word on A can be written uniquely as a product of nonincreasing Lyndon words.

About 40 years later, infinite Lyndon words were introduced in [13]. There are several equivalent definitions, but we use the definition which focuses on the

R. Mercaş and D. Reidenbach (Eds.): WORDS 2019, LNCS 11682, pp. 106–118, 2019.
https://doi.org/10.1007/978-3-030-28796-2_8

idea of rotation. An infinite word is called *Lyndon* if it is strictly smallest among its suffixes with respect to the standard lexicographic order. If w is an infinite word with a nontrivial factorization uv, the suffix v can be viewed as the rotation with respect to this factorization. Let A^ω denote the set of sequences, or infinite words, over A. These too yielded deep factorization properties; Siromoney et al. showed that every sequence in A^ω has a unique factorization as a nonincreasing product of finite and infinite Lyndon words.

The extension of the Lyndon property to generalized lexicographic orders came about 10 years later by Reutenauer [11]. A *generalized lexicographic order* is a modified lexicographic order where the total order of the alphabet depends on the index of comparison. This naturally induces a notion of finite and infinite *generalized Lyndon words* under a generalized lexicographic order. (See Sect. 2.) Reutenauer showed that the finite generalized Lyndon words form a basis for A^* using Hall set theory, and Dolce et al. provided a combinatorial proof in 2018 [3,11]. Generalized Lyndon words are studied further by Dolce et al. in [4].

An example of a generalized lexicographic order is the *alternating order* $\leq_{\mathrm{alt}}$, where the alphabet is given its standard order when the index of comparison is odd and its opposite order when the index is even. This order can be connected with continued fractions by noting that the map $\phi : \mathbb{N}^\omega \to \mathbb{R}$ defined by

$$\phi(x_1 x_2 \cdots) = x_1 + \cfrac{1}{x_2 + \cfrac{1}{\ddots}}$$

satisfies $u \leq_{\mathrm{alt}} v$ in $\mathbb{N}^\omega$ if and only if $\phi(u) \leq \phi(v)$ in $\mathbb{R}$. Generalized Lyndon words with respect to the alternating order are called *Galois words*, and Galois factorizations were given further characterization in [4]. Another special case are the anti-Lyndon words, introduced in [5], which are generalized Lyndon words with respect to the opposite lexicographic order.

Dolce et al. conjectured that the finite and infinite generalized Lyndon words provide a unique nonincreasing (with respect to ω-powers) factorization of all infinite words. Our main result is to show that this is indeed the case.

In Sect. 3, we focus on words with a generalized Lyndon suffix. Theorem 3 shows that these are precisely the words with finitely many terms in their nonincreasing generalized Lyndon factorization. Moreover, we characterize the last term as the first generalized Lyndon suffix (with respect to the index).

Sections 4 and 5 focus on the existence and uniqueness, respectively, of nonincreasing generalized Lyndon factorizations for words which have no generalized Lyndon suffix. In the process we develop powerful machinery to take advantage of the strong properties of these factorizations. A product of this machinery is presented briefly in Sect. 6, where we show that an infinite word is generalized Lyndon if and only if it has infinitely many generalized Lyndon prefixes. This is the generalized analogue of the result of Siromoney et al. showing that infinite Lyndon words are precisely the words with infinitely many Lyndon prefixes.

2   Preliminaries

Let $\mathbb{N} = \{1, 2, 3, \dots\}$. Words are finite or infinite (to the right) sequences of letters from a fixed alphabet A. For $i < j$, the contiguous substring beginning at the ith letter and ending with the jth (inclusive on both ends) is denoted $x[i, j]$. A word v is a *factor* of x if $x = uvw$ for (possibly empty) words u and w. In the case that u is empty, v is a *prefix* of x, and if w is empty, then v is a *suffix* of x. If in addition w (resp. u) is nonempty, we say that the prefix (resp. suffix) is *proper*. If x is an infinite word, the suffix of x beginning at the jth index of x is denoted $x[j, \infty)$. The length of a finite word w is denoted by $|w|$.

Let $A^\infty = A^* \cup A^\omega$. Given a total order on an alphabet A, the *lexicographic ordering* $<_{\text{lex}}$ on A^∞ is defined such that $x <_{\text{lex}} y$ if and only if x is a proper prefix of y or $x = pas$ and $y = pbs'$ for words p, s, s' and letters $a < b$. We are primarily interested in a generalization of this order.

For each $n \in \mathbb{N}$, let $<_n$ be a total order on A. The *generalized lexicographic order* $<$ induced by $(<_n)_{n \in \mathbb{N}}$ is defined such that $x < y$ if and only if x is a proper prefix of y or $x = pas$ and $y = pbs'$ for words p, s, s' and letters $a <_{|p|+1} b$.

If u is a prefix of v or v is a prefix of u, we write $u \sim v$. Note that if $|u| = |v|$, then $u \sim v$ implies $u = v$. We will use the $\sim$ operator "transitively", where the expression $w_1 \sim w_2 \sim \cdots \sim w_n$ implies that the shortest of the n words is a prefix of the rest. We also define a modified comparison operator $\lesssim$ such that $w_1 \lesssim \cdots \lesssim w_n$ if the prefixes p_i of w_i having length $\min\{|w_1|, \dots, |w_n|\}$ satisfy $p_1 \leq \cdots \leq p_n$, where $\leq$ is the generalized lexicographic order. The same property of only comparing the prefixes up to the length of the shortest word in a chain also applies when the operators $\sim$ and $\lesssim$ are applied together in a chain.

A finite word v is called a *power* of a finite word u if $v = u^k$ for some integer $k \geq 2$. Let the ω-power of u, denoted by u^ω, be the infinite word $\prod_{i=1}^{\infty} u$. An infinite word v is called a *power* of a finite word u if $v = u^\omega$; we also say that v is *periodic*. If u is infinite, we use the convention $u^\omega = u$. An infinite word with a periodic suffix is called *eventually periodic*, and an infinite word which is not eventually periodic is called *aperiodic*. A word which is not a power is called *primitive*. A finite word w is called a *fractional power* of a finite word u if $w \sim u^\omega$. We write $w = u^{|w|/|u|}$, e.g., $01 = (0111)^{1/2}$. See [6], [7], and [10] for more on the combinatorics of words.

A word w is a *finite generalized Lyndon word* if it is strictly smallest among its class of rotations with respect to a generalized lexicographic order. That is, for any nontrivial factorization $w = uv$, we have $uv < vu$. An infinite word w is an *infinite generalized Lyndon word* if it is strictly smallest among its class of suffixes with respect to a generalized lexicographic order. A nonincreasing generalized Lyndon factorization of a word w is a product of the form $w = \prod_{i=1}^{n} \ell_i$ where $n \in \mathbb{N} \cup \{\infty\}$, each ℓ_i is generalized Lyndon, and $\ell_i^\omega \geq \ell_{i+1}^\omega$ for all $i \in [1, n)$.

3   Existence and Uniqueness of Finite Factorizations

In this section, we show that the words admitting a unique finite generalized Lyndon factorization are precisely the words with a generalized Lyndon suffix.

Lemma 1 ([3], Lemma 31). *Let u, v be nonempty finite words. Then the following four conditions are equivalent:*

(1) $u^\omega < v^\omega$ (2) $(uv)^\omega < v^\omega$ (3) $u^\omega < (vu)^\omega$ (4) $(uv)^\omega < (vu)^\omega$.

We will also make use of a result by Lyndon and Schützenberger concerning commuting words, which can easily be strengthened when one of the words is generalized Lyndon.

Lemma 2 [9]. *Two finite words commute if and only if they are powers of a common word.*

Corollary 1. *Suppose u is a finite generalized Lyndon word, v is any finite word, and $uv = vu$. Then v is a power of u.*

Proof. This follows from Lemma 2 and the fact that generalized Lyndon words are primitive. ∎

Lemma 3. *Suppose u and v are finite words satisfying $u^\omega \lesssim v \lesssim u^n v$ (resp. $u^\omega \gtrsim v \gtrsim u^n v$) for some $n \in \mathbb{N}$. Then $v \sim u^\omega$.*

Proof. Suppose there exists a maximum nonnegative integer, m, such that $u^m \sim v$; note that $|u^m| < |v|$. Then

$$u^{m+1} \leq u^\omega \lesssim v \lesssim u^n v \sim u^{m+n}.$$

Thus $v \sim u^{m+n}$, a contradiction to our choice of m. The proof proceeds analogously for the case where the inequalities are reversed. ∎

Theorem 1. *Suppose w is an infinite word. If w is a nonincreasing product of finite generalized Lyndon words, then w has no generalized Lyndon suffixes.*

Proof. Suppose w has a generalized Lyndon suffix ℓ. Without loss of generality, we can assume $w = \ell_0 \ell_1 \ell_2 \cdots$ and $\ell = u\ell_1 \ell_2 \cdots$, where each ℓ_i is a generalized Lyndon word, $\ell_0^\omega \geq \ell_1^\omega \geq \cdots$, and u is a suffix of ℓ_0. Since ℓ_0 is generalized Lyndon, Lemma 1 implies $u^\omega \geq \ell_0^\omega$. Furthermore, since ℓ is generalized Lyndon, we have $\ell_r \gtrsim u$ for all $r \in \mathbb{N}$. Thus, for all $r \in \mathbb{N}$ we have $u^\omega \geq \ell_0^\omega \geq \ell_r^\omega \sim \ell_r \gtrsim u$, hence $\ell_r \sim u$.

Suppose that there exists some $r \in \mathbb{N}$ such that $|\ell_r| < |u|$. Note that each such ℓ_r is a prefix of u. By the nonincreasing property of the generalized Lyndon factors, either there exist finitely many such r, or there exists some $n \in \mathbb{N}$ and $\alpha \in (0, 1)$ such that for all $r \geq n$, we have $\ell_r = u^\alpha$. The latter case holds because there are only finitely many prefixes of u, so one prefix must appear infinitely many times. By the nonincreasing property of the factorization, this means that all terms in the factorization after the first term equal to this prefix must also equal this prefix. Observe that in the latter case, we have

$$(u^\alpha)^\omega = \ell_n^\omega \leq u^\omega \sim u \lesssim \ell_n \ell_{n+1} \cdots = (u^\alpha)^\omega,$$

hence $u \sim (u^\alpha)^\omega$.

We conclude there exists a minimal $k \in \mathbb{N}$ such that $\ell_k = u^\alpha$ for some $\alpha \in (0,1)$ and $u \sim \ell_{k+1}\ell_{k+2}\cdots$. Since $u^\alpha u \sim \ell_k \ell_{k+1}\cdots$, then $u^\alpha u \gtrsim u\ell_1$. Thus, $u^\alpha u \gtrsim u \sim u^\omega \geq \ell_k^\omega = (u^\alpha)^\omega$, so Lemma 3 implies $u \sim (u^\alpha)^\omega$. Suppose $|\ell_1| \geq |u|$. Then uu^α is a prefix of w, so

$$(u^\alpha)^\omega \sim u^\alpha u \geq uu^\alpha \sim u^\omega \geq (u^\alpha)^\omega,$$

hence u is a power of u^α by Corollary 1. Thus, u is not generalized Lyndon, so $u^\omega > \ell_0^\omega \geq \ell_k^\omega = (u^\alpha)^\omega = u^\omega$, a contradiction.

Thus, we must have that $|\ell_1| < |u|$. By the minimality of k, we have that $|\ell_r| < |u|$ for $1 \leq r \leq k$, which implies that $\ell_r \sim u \sim (u^\alpha)^\omega = \ell_k^\omega$ for $1 \leq r \leq k$. As $u \sim \ell_k^\omega$, we have that $u \sim \ell_k u \sim \ell_k \ell_{k+1}$. Hence

$$\ell_{k-1}^\omega \leq u^\omega \sim u \lesssim \ell_{k-1}\ell_k\ell_{k+1}\cdots \sim \ell_{k-1}u.$$

In particular, by Lemma 3, we have $u \sim \ell_{k-1}^\omega$ and $u \sim \ell_{k-1}\ell_k \cdots$. We repeat this process, showing that $u \sim \ell_i^\omega$ and $u \sim \ell_i\ell_{i+1}\cdots$ for all $1 \leq i \leq k$. Hence $\ell_1 u \sim \ell_1\ell_2 \cdots \gtrsim u\ell_1$. However, since $\ell_1^\omega \leq u^\omega$, Lemma 1 implies $\ell_1 u \leq u\ell_1$. Thus, u and ℓ_1 commute, so Corollary 1 implies u is a power of ℓ_1. In particular, ℓ_1 is a proper suffix of ℓ_0, so $\ell_1^\omega > \ell_0^\omega$, contradicting our nonincreasing assumption.

Thus, we must have $|\ell_r| \geq |u|$ for all $r \in \mathbb{N}$. w has a generalized Lyndon suffix, so it cannot be periodic. We can fix s to be the smallest index such that $\ell_s \neq u$. By Lemma 3, the inequality

$$u^\omega \geq \ell_s^\omega \sim \ell_s \gtrsim u\ell_1 \cdots \ell_s = u^s\ell_s,$$

implies that $\ell_s \sim u^\omega$. Hence $\ell_s = u^{n+\beta}$ for some $n \in \mathbb{N}$ and $\beta \in [0,1)$. On the one hand, we have $u^\omega \geq \ell_s^\omega = (u^{n+\beta})^\omega$, hence $u^n uu^\beta \geq u^n u^\beta u$. On the other hand, since $u^n uu^\beta \sim u^{s+n+\beta}$ is a prefix of ℓ and $u^n u^\beta u \sim \ell_s\ell_{s+1}$ is a factor, we have $u^n uu^\beta \leq u^n u^\beta u$ because ℓ is generalized Lyndon. Hence, Lemma 2 implies $(u^\beta)^\omega = u^\omega = \ell_s^\omega$, contradicting that $(u^\beta)^\omega > \ell_s^\omega$ by the generalized Lyndon property of ℓ_s.

Lemma 4. *If u is a finite word and v is an infinite word, then $u^\omega > v$ (resp. $u^\omega < v$) if and only if $uv > v$ (resp. $uv < v$).*

Proof. Suppose $u^\omega > v$. Let j be the largest integer such that $u^j \sim v$. Hence $v = u^j v'$ for some infinite word $v' \not\sim u$. Thus, the comparison between uv and v happens between index $j|u| + 1$ and index $(j+1)|u|$, inclusive. In particular, $uv \sim u^{j+1} > v$.

Now suppose $uv > v$. Let k be the largest index such that $u^k \sim v$. Thus, the comparison between uv and v happens between index $k|u| + 1$ and index $(k+1)|u|$, inclusive. In particular, $u^\omega \sim u^{k+1} \sim uv > v$.

The proof with the reverse inequalities proceeds analogously.

In order to show the existence and uniqueness of generalized Lyndon factorizations of infinite words, we will invoke a theorem of Reutenauer which gives the analogous result for finite words [11].

Theorem 2 [3,11]. *Any finite word has a unique nonincreasing factorization into generalized Lyndon words.*

Theorem 3. *An infinite word with an infinite generalized Lyndon suffix has a unique factorization into generalized Lyndon words, and this factorization is finite. Furthermore, the last term in this factorization is the first generalized Lyndon suffix by index.*

Proof. We first show existence. Let ℓ be the first generalized Lyndon suffix of w by index, that is, $w = v\ell$ where the length of v is minimum such that ℓ is generalized Lyndon. Let $\ell_1, \ldots, \ell_n$ be the unique nonincreasing factorization of v from Theorem 2. It is enough to show that $\ell_n^\omega \geq \ell$, as this will yield $\ell_1, \ldots, \ell_n, \ell$ as a nonincreasing generalized Lyndon factorization of w.

Suppose that $\ell_n^\omega < \ell$. By Lemma 4, this implies $\ell_n\ell < \ell$. Let s be the shortest (not necessarily proper) suffix of ℓ_n such that $s\ell$ is minimal. Note that we have $s\ell \leq \ell_n\ell < \ell$, so s is nonempty. However, by construction we have $s\ell \leq s'\ell$ for every suffix s' of s. Notably, $s\ell \leq \ell \leq \ell'$ for any suffix ℓ' of ℓ because ℓ is generalized Lyndon. Thus $s\ell$ is generalized Lyndon. This contradicts our choice of ℓ to be the first generalized Lyndon suffix of w. Therefore $\ell_n^\omega \geq \ell$, so we have produced a nonincreasing factorization of w.

By Theorem 1, any factorization of w must have only finitely many terms. Let $\ell_1, \ldots, \ell_n\ell$ be a nonincreasing factorization of w into generalized Lyndon words. Suppose, seeking a contradiction, that ℓ is not the longest generalized Lyndon suffix of w, i.e., there is a suffix s of w of the form $u\ell_{j+1} \cdots \ell_n\ell$ where u is a suffix of ℓ_j. From the nonincreasing property of the factorization w and the generalized Lyndon property of ℓ_j, we know $u^\omega \geq \ell_j^\omega \geq \cdots \geq \ell_n^\omega \geq \ell$. By Lemma 1, $(u\ell_j)^\omega \geq \ell_j^\omega$. Inductively, we find $(u\ell_j \cdots \ell_n)^\omega \geq \ell_n^\omega \geq \ell$. Thus, by Lemma 4, we have $s = u\ell_j \cdots \ell_n\ell \geq \ell$, contradicting that s is generalized Lyndon.

Now that we have uniquely determined ℓ, the other factors $\ell_1, \ldots, \ell_n$ are uniquely determined. This follows because the prefix $w[1, |\ell_1| + \cdots + |\ell_n|]$ of w has a unique nonincreasing factorization into generalized Lyndon words. Thus by our initial assumption that $\ell_1, \ldots, \ell_n, \ell$ is a nonincreasing factorization of w, the unique factorization of $w[1, |\ell_1| + \cdots + |\ell_n|]$ must be $\ell_1, \ldots, \ell_n$.

4   Existence of Infinite Factorizations

In this section, we describe a method to construct an infinite factorization of a word with no generalized Lyndon suffix by taking a limit of the finite factorizations of some of its prefixes.

Lemma 5. *If a primitive infinite word has infinitely many generalized Lyndon prefixes, then it is a generalized Lyndon word.*

Proof. Let w be a primitive word which is not infinite generalized Lyndon, and let $m \in \mathbb{N}$ be minimal such that $w[m, \infty) < w$. Let i be the index of comparison between $w[m, \infty)$ and w. Then for any $n \geq m+i$, we have $w[m, n] \sim w[m, \infty) <$

$w \sim w[1,n]$ with a comparison at index i. Thus $w[1,n]$ is not generalized Lyndon for any $n \geq m + i$, so we can conclude that w has finitely many generalized Lyndon prefixes.

Lemma 6. *If ℓ is a finite word that is not generalized Lyndon, then ℓ^ω has finitely many generalized Lyndon prefixes.*

Proof. If ℓ is not generalized Lyndon, then we can write $\ell = uv$ where $vu < uv$ for some prefix u. Observe that vu will be a factor of any prefix of ℓ^ω having length at least $|\ell| + |u|$, and uv will be a prefix of any such prefix of ℓ^ω. Thus any prefix of ℓ^ω having length at least $|\ell| + |u|$ is not generalized Lyndon.

Theorem 4. *An infinite word has a nonincreasing factorization into generalized Lyndon words.*

Proof. Fix an infinite word w. Theorem 3 completes the proof in the case that w has an infinite generalized Lyndon suffix. So we can assume that w has no infinite generalized Lyndon suffix. In particular, w is not generalized Lyndon.

We will first consider the case where w is not eventually periodic. Since w is not generalized Lyndon, Lemma 5 implies that w has finitely many generalized Lyndon prefixes. Thus one of its generalized Lyndon prefixes must appear in the factorization of $w[1,n]$ yielded by Theorem 2 for infinitely many $n \in \mathbb{N}$. Let ℓ_1 be such a prefix, and let $w = \ell_1 w_1$.

We will now inductively construct a factorization of w. Suppose we can write $w = \ell_1 \cdots \ell_k w_k$ such that each ℓ_j is a finite generalized Lyndon word, $\ell_1^\omega \geq \cdots \geq \ell_k^\omega$, and w has infinitely many prefixes whose factorizations begin with $\ell_1, \ldots, \ell_k$. Since w has no generalized Lyndon suffixes, w_k is not generalized Lyndon, so it must have finitely many generalized Lyndon prefixes. Since infinitely many prefixes of w have factorizations beginning with $\ell_1, \ldots, \ell_k$, one of the generalized Lyndon prefixes of w_k, which we label ℓ_{k+1}, must be such that infinitely many prefixes of w have factorizations beginning with $\ell_1, \ldots, \ell_k, \ell_{k+1}$. We can then write $w = \ell_1 \cdots \ell_{k+1} w_{k+1}$. Note that by construction, $\ell_k^\omega \geq \ell_{k+1}^\omega$. By induction, we get a nonincreasing generalized Lyndon factorization $\ell_1, \ell_2, \ldots$ of w.

Now suppose that w is eventually periodic. If w is a power of a generalized Lyndon word ℓ, we can use the factorization $w = \ell^\omega$. Otherwise, w is a power of a finite word that is not generalized Lyndon or w is primitive. In either case, Lemmas 5 and 6 imply that w has finitely many generalized Lyndon prefixes. We can thus apply the construction from the previous paragraph, in each step yielding a factorization of w starting with $\ell_1, \cdots, \ell_k$. This process will halt only if w_k has infinitely many generalized Lyndon prefixes p_i such that $\ell_1, \ldots, \ell_k, p_i$ is the factorization of a prefix of w. By Lemmas 5 and 6, this implies that w_k is a power of a generalized Lyndon word ℓ. Moreover, since the p_i's have unbounded length and $\ell_k^\omega \geq p_i^\omega$, we must have $\ell_k^\omega \geq \ell^\omega$. Therefore $\ell_1, \ldots, \ell_k, \ell^\omega$ is a factorization of w. Thus, in any case, this construction yields a nonincreasing factorization of w into generalized Lyndon words.

5 Uniqueness of Infinite Factorizations

We will determine the uniqueness of the factorization constructed in Sect. 4, handling first the eventually periodic words and then aperiodic words with no generalized Lyndon suffix.

Theorem 5. *An eventually periodic infinite word has a unique nonincreasing factorization into generalized Lyndon words.*

Proof. Fix an infinite word w with a periodic suffix. Observe that this implies we can write w as $u\ell^\omega$ where u is a (possibly empty) finite word and ℓ is a nonempty finite generalized Lyndon word. We may assume w has no generalized Lyndon suffix, as this case is handled by Theorem 3.

We first claim that the factorization (from Theorem 4) of $w = \prod_{i=1}^\infty \ell_i$ must terminate with ℓ^ω. Since ℓ is generalized Lyndon hence not equal to any of its rotations, we have that $\ell^\omega[i, \infty) = \ell^\omega$ if and only if $i - 1$ is an integer multiple of $|\ell|$. Moreover, if $i - 1$ is not a multiple of $|\ell|$, then $\ell^\omega[i, \infty)$ is a power of a word which is not generalized Lyndon and hence has finitely many generalized Lyndon prefixes by Lemma 6. If one of these generalized Lyndon prefixes, ℓ', appears infinitely many times in the factorization of w, then $(\ell')^\omega$ is a suffix of w. Since $(\ell')^\omega$ and ℓ^ω are suffixes of w, they are powers of rotations of ℓ and ℓ', respectively. Because ℓ and ℓ' are generalized Lyndon, this means that $\ell = \ell'$. That is, only finitely many terms of the factorization are not equal to ℓ. Thus, we can conclude $\ell_i = \ell$ for sufficiently large i.

Now suppose $\ell_1, \ldots, \ell_n, \ell^\omega$ and $h_1, \ldots, h_m, \ell^\omega$ are two distinct factorizations of w. Note that $|\ell_1 \cdots \ell_n| - |h_1 \cdots h_m|$ must be an integer multiple of $|\ell|$, as ℓ is a generalized Lyndon word and hence not equal to any of its rotations. Without loss of generality, assume $|\ell_1 \cdots \ell_n| - |h_1 \cdots h_m| > 0$. In this case, there exists $k \in \mathbb{N}$ such that $\ell_1 \cdots \ell_n = h_1 \cdots h_m \ell^k$, which violates the uniqueness of the nonincreasing generalized Lyndon factorization for finite words from Theorem 2. Thus, the nonincreasing factorization of w into generalized Lyndon words is unique.

Lemma 7. *Let $w = v\ell_1\ell_2 \cdots \ell_n u$ be a finite generalized Lyndon word where $n \in \mathbb{Z}_{\geq 0}$, ℓ_i is a finite generalized Lyndon word for all $i \in \{1, \ldots, n\}$, v is a suffix of a finite generalized Lyndon word ℓ_0, u is a prefix of a finite generalized Lyndon word ℓ_{n+1}, and $\ell_0^\omega \geq \ell_1^\omega \geq \cdots \geq \ell_{n+1}^\omega$. Then $u \sim v$ and $u \sim v \sim \ell_i \cdots \ell_n u$ for all $i \in \{1, \ldots, n\}$.*

Proof. The generalized Lyndon property of w implies $u \gtrsim v$. The nonincreasing property of the factors implies $v \sim v^\omega \geq \ell_0^\omega \geq \ell_{n+1}^\omega \sim u$. Combining these inequalities, we have $u \sim v$.

Suppose $|u| \leq |v|$. The generalized Lyndon property of w and the nonincreasing property furthermore implies

$$\ell_n u \gtrsim v \sim u \sim v \sim v^\omega \geq \ell_0^\omega \geq \ell_n^\omega.$$

Hence Lemma 3 implies that $u \sim \ell_n^\omega$, so $\ell_n u \sim u$. Repeating this process, we can conclude $u \sim v \sim \ell_n u \sim \ell_{n-1}\ell_n u \sim \cdots \sim \ell_1 \cdots \ell_n u$.

Similarly, suppose $|u| > |v|$. The generalized Lyndon property of w and the nonincreasing property implies

$$\ell_n v \sim \ell_n u \gtrsim v \sim v^\omega \geq \ell_0^\omega \geq \ell_n^\omega.$$

Hence Lemma 3 implies that $v \sim \ell_n^\omega$, so $v \sim \ell_n v \sim \ell_n u$. Repeating this process, we can conclude $v \sim u \sim \ell_n u \sim \ell_{n-1}\ell_n u \sim \cdots \sim \ell_1 \cdots \ell_n u$.

Lemma 8. *Let w satisfy the hypotheses of Lemma 7. If $|u| \geq |v|$, then there exists some m with $0 \leq m \leq n$ such that*

$$\ell_j = \begin{cases} v & \text{if } 1 \leq j \leq m \\ v^{\alpha_j} \text{ for some } \alpha_j \in (0,1) & \text{if } m < j \leq n. \end{cases}$$

Proof. We assume that $\ell_j = v$ for $1 \leq j \leq k$ with $0 \leq k \leq n-1$ and proceed by induction on k. Note that the base case of $k = 0$ is automatic. Furthermore, we suppose there exists m with $0 \leq m \leq n$ that satisfies the property of Lemma 8 when we restrict to considering ℓ_j with $j \leq k$. Note that we can have $m \geq k$. By Lemma 7, we have $\ell_{k+1} \sim u \sim v$ and $|v| \leq |u|$, hence $\ell_{k+1} \sim v$. Thus, if $|\ell_{k+1}| < |v|$, then we are done.

Suppose $m < k$ and $|\ell_{k+1}| \geq |v|$, so v is a prefix of ℓ_{k+1}. Let $\ell_{m+i} = v^{\alpha_i}$ for $1 \leq i \leq k - m$, where each $\alpha_i \in (0,1)$. Let $t = k - m$. Thus $v\ell_1 \cdots \ell_k \ell_{k+1} \sim v^k v^{\alpha_1} \cdots v^{\alpha_t} v$. By Lemma 7, we have $v \sim u \sim v^{\alpha_2} \cdots v^{\alpha_{k-m}} v$, so $v^{\alpha_1} v$ is a factor of w. Since vv^{α_1} is a prefix of w, by the generalized Lyndon property we have $v^{\alpha_1} v \geq vv^{\alpha_1}$. On the other hand, we have $v^\omega \geq \ell_0^\omega \geq \ell_{m+1}^\omega = (v^{\alpha_1})^\omega$, which implies $v^\omega \geq (v^{\alpha_1} v)^\omega$ by Lemma 1. In particular, we have $vv^{\alpha_1} \geq v^{\alpha_1} v$. Combining inequalities yields $vv^{\alpha_1} = v^{\alpha_1} v$, implying v is a power of v^{α_1} by Corollary 1. Thus

$$(v^{\alpha_1})^\omega = v^\omega \geq \ell_0^\omega \geq \ell_{m+1}^\omega = (v^{\alpha_1})^\omega,$$

so Corollary 1 and Lemma 1 imply $\ell_0 = v^{\alpha_1} = v$. This contradicts our choice of m, hence $|\ell_{k+1}| < |v|$ and $\ell_{k+1} \sim v$, as desired.

In the other case, we need to consider is $|\ell_{k+1}| > |v|$ and $m \geq k$. Let $\ell_{k+1} = v^r v^\alpha$ for $r \in \mathbb{N}$ and $\alpha \in (0,1)$, noting that $r + \alpha \notin \mathbb{N}$ since ℓ_{k+1} is generalized Lyndon, and hence primitive. If $k + 1 = n$ or if $|\ell_{k+2}| \geq v$, then $\ell_{k+1} v$ is a factor of w. Note by our inductive hypothesis that $v^r vv^\alpha$ is a prefix of w. By the generalized Lyndon property, we have $v^r v^\alpha v \geq v^r vv^\alpha$. However, we also have

$$v^r v^\alpha v \sim (v^r v^\alpha)^\omega = \ell_{k+1}^\omega \leq \ell_0^\omega \leq v^\omega \sim v^r vv^\alpha.$$

Combining inequalities yields $v^r v^\alpha v = v^r vv^\alpha$, implying v is a power of v^α by Corollary 1. This means $\ell_{k+1} = v^r v^\alpha$ is a power of v^α, contradicting the primitiveness of ℓ_{k+1}, so we must have $k < n - 1$ and $|\ell_{k+2}| < |v|$. Since we assume $|u| > |v|$, there must exist some $q \in \{k + 2, \ldots, n\}$ such that $|\ell_q| < n$ and

$\ell_q \ell_{q+1} \cdots \ell_n u \sim \ell_q v$. Notably, the largest value of q such that $|\ell_q| < n$ works. Then $v\ell_q$ is a prefix of w, and $\ell_q v$ is a factor of w. By the generalized Lyndon property of w and our inductive hypothesis, we have $\ell_q v \geq v\ell_q$. However, we also have $\ell_q^\omega \leq \ell_0^\omega \leq v^\omega$, hence $(\ell_q v)^\omega \leq v^\omega$ by Lemma 1. In particular, $\ell_q v \lesssim vv \sim v\ell_q$. Again, we combine inequalities and use Corollary 1 and Lemma 1 to conclude $\ell_q = v$, our final contradiction.

Lemma 9. *If w satisfies the hypotheses of Lemma 7, then $\ell_1 = \cdots = \ell_n = v$.*

Proof. It is enough to show that $\ell_n = v$, since $v^\omega \geq \ell_0^\omega \geq \ell_i^\omega \geq \ell_n^\omega = v^\omega$. If $\ell_n \neq v$, then by Lemma 8, we have $\ell_n = v^\alpha$ for some $\alpha \in (0,1)$. Moreover, by Lemma 7 and our assumption $|u| \geq |v|$, we have $\ell_1 \ldots \ell_n u \sim v$, hence $w = v\ell_1 \ldots \ell_n u \sim vv$. Thus, $v^\alpha v$ is a factor of w and vv^α is a prefix, so the generalized Lyndon property of w implies $vv^\alpha \leq v^\alpha v$. Since we have $(v^\alpha)^\omega = \ell_n^\omega \leq \ell_0^\omega \leq v^\omega \sim v \lesssim v^\alpha v$, Lemma 3 implies $v \sim (v^\alpha)^\omega$. In particular, $v^\alpha v \sim (v^\alpha)^\omega \leq v^\omega \sim vv^\alpha$. Combining inequalities, we have $vv^\alpha \leq v^\alpha v$. This implies v is a power of v^α by Corollary 1. Thus $(v^\alpha)^\omega = v^\omega \geq \ell_0^\omega \geq \ell_n^\omega = (v^\alpha)^\omega$. In particular $\ell_0^\omega = v^\omega$, which implies $v = \ell_0$ by the generalized Lyndon property of ℓ_0. Therefore $v = \ell_0 = v^\alpha$, contradicting our choice of α.

Corollary 2. *Let w be as in the statement of Lemma 9. If we additionally assume that v is a proper suffix of ℓ_0 and $n \geq 1$, then $|u| < |v|$.*

Proof. By Lemma 9, we have $\ell_1 = \ell_2 = \cdots = \ell_n = v$. Since v is a proper suffix of ℓ_0, we have $v^\omega > \ell_0^\omega$. Thus $v^\omega > \ell_0^\omega \geq \ell_1^\omega = v^\omega$, which is a contradiction.

Theorem 6. *An aperiodic infinite word with no generalized Lyndon suffix has a unique nonincreasing factorization into finite generalized Lyndon words.*

Proof. Suppose w is an aperiodic word with no generalized Lyndon suffix such that w has two distinct nonincreasing factorizations into generalized Lyndon words. Note that each factor in both factorizations must be finite. We can remove any initial common factors, so without loss of generality $w = \prod_{i=0}^\infty w_i = \prod_{j=0}^\infty \ell_j$ where $w_i^\omega \geq w_{i+1}^\omega$ for all $i \in \mathbb{Z}_{\geq 0}$, $\ell_j^\omega \geq \ell_{j+1}^\omega$ for all $j \in \mathbb{Z}_{\geq 0}$, and $|w_0| > |\ell_0|$. Since we know finite words have unique nonincreasing generalized Lyndon factorizations from Theorem 2, we have $\prod_{i=0}^x w_i \neq \prod_{j=0}^y \ell_j$ for any $x, y \in \mathbb{Z}_{\geq 0}$.

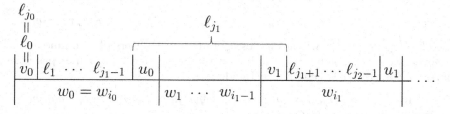

Fig. 1. The construction of v_k, ℓ_{j_k}, w_{i_k}, and u_k

Define $v_0 = \ell_0$, $\ell_{j_0} = \ell_0$, and $w_{i_0} = w_0$. We define j_{k+1} to be the unique integer such that w_{i_k} can be written as $v_k \ell_{j_k+1} \cdots \ell_{j_{k+1}-1} u_k$, where u_k is a prefix of $\ell_{j_{k+1}}$. We define i_{k+1} to be the unique integer such that $\ell_{j_{k+1}}$ can be written as $u_k w_{j_k+1} \cdots w_{j_{k+1}-1} v_{k+1}$, where v_{k+1} is a prefix of $w_{j_{k+1}}$. This construction is illustrated in Fig. 1. Observe that for each $k \in \mathbb{Z}_{\geq 0}$ we have that v_k is a proper prefix of w_{i_k}, u_k is a proper suffix of w_{i_k}, u_k is a proper prefix of $\ell_{j_{k+1}}$, and v_{k+1} is a proper suffix of $\ell_{j_{k+1}}$.

We aim to show that $|v_{k+1}| < |v_k|$ for each $k \in \mathbb{Z}_{\geq 0}$, and since this reduction can only be applied finitely many times, we will reach a contradiction. Assume not, that $|v_{k+1}| \geq |v_k|$ for a certain $k \in \mathbb{Z}_{\geq 0}$.

First suppose that $|u_k| \geq |v_k|$, and note that we cannot have u_k be a power of v_k or $u_k = v_k$, or else w_{i_k} is not primitive by Lemma 9. Thus Lemmas 9 and 7 imply that $u_k = v_k^r v_k^\alpha$ for some $r \in \mathbb{N}$ and $\alpha \in (0,1)$. Moreover, Corollary 2 implies that $w_{i_k} = v_k u_k$. Lemma 7 implies that $v_k \sim u_k$ and $u_k \sim v_{k+1}$, hence $|v_k| \leq |u_k|, |v_{k+1}|$ implies $v_k \sim v_{k+1}$. Furthermore, Lemma 7 yields $v_k \sim v_{k+1} \sim w_{i_k+1} \cdots w_{i_{k+1}-1} v_{k+1}$, so $u_k w_{i_k+1} \cdots w_{i_{k+1}-1} v_{k+1} \sim v_k^r v_k^\alpha v_k$. Thus

$$v_k^r v_k v_k^\alpha \sim v_k^\omega \geq \ell_{j_k}^\omega = (u_k w_{i_k+1} \cdots w_{i_{k+1}-1} v_{k+1})^\omega \sim v_k^r v_k^\alpha v_k.$$

However, by the generalized Lyndon property of w_{i_k} and Lemma 8, we also have

$$v_k^r v_k^\alpha v_k \sim (v_k^r v_k^\alpha)^\omega = u_k^\omega \geq (v_k \ell_{j_k+1} \cdots \ell_{j_{k+1}-1})^\omega = v_k^\omega \sim v_k^r v_k v_k^\alpha.$$

Combining inequalities yields $v_k^r v_k^\alpha v_k = v_k^r v_k v_k^\alpha$, which implies that v_k and v_k^α are powers of a common word by Lemma 2. Thus w_{i_k} is not primitive, contradicting that it is generalized Lyndon. So in this case we have $|v_{k+1}| < |v_k|$.

Now suppose $|u_k| < |v_k|$. By the generalized Lyndon property of w_k, we have $v_k \lesssim u_k$. The nonincreasing property of our factors implies $u_k \sim \ell_{j_{k+1}}^\omega \leq \ell_{j_k}^\omega \leq v_k^\omega \sim v_k$. Hence $v_k \sim u_k$, so $u_k = v_k^\alpha$ for some $\alpha \in (0,1)$. Since $|u_k| < |v_k| \leq |v_{k+1}|$, Corollary 2 applies to ℓ_{j_k}. In particular, $\ell_{j_{k+1}} = u_k v_{k+1}$. By the generalized Lyndon property of w_{i_k} and $\ell_{j_{k+1}}$ along with Lemma 1, we have

$$(v_k^\alpha)^\omega = u_k^\omega \geq w_{i_k}^\omega \geq w_{i_{k+1}}^\omega \sim v_{k+1} \sim v_{k+1}^\omega \geq u_k^\omega = (v_k^\alpha)^\omega.$$

Hence $v_{k+1} \sim (v_k^\alpha)^\omega$. Note that $v_k \sim w_{i_k}^\omega$, so we also have $v_k \sim v_{k+1} \sim (v_k^\alpha)^\omega$. On the one hand, we have

$$v_k^\alpha v_k \sim u_k v_{k+1} = \ell_{j_{k+1}} \sim \ell_{j_{k+1}}^\omega \leq \ell_{j_k}^\omega \leq v_k^\omega \sim v_k v_k^\alpha.$$

However, the generalized Lyndon property of w_1 and Lemma 7 imply

$$v_k^\alpha v_k = u_k v_k \sim u_k v_k \ell_{j_k+1} \cdots \ell_{j_{k+1}-1} \gtrsim v_k \ell_{j_k+1} \cdots \ell_{j_{k+1}-1} u_k \sim v_k u_k = v_k^\alpha v_k.$$

Therefore $v_k^\alpha v_k = v_k v_k^\alpha$, which implies v_k and v_k^α are powers of a common word. However, we reach our final contradiction by noting that $\ell_{j_{k+1}}$ is not primitive, contradicting that it is generalized Lyndon.

Theorem 7. *Every infinite word has a unique factorization into a nonincreasing product of generalized Lyndon words.*

Proof. This follows directly from Theorems 3, 5, and 6.

6   Characterization of Infinite Generalized Lyndon Words

Siromoney et al. showed in [13] that the infinite Lyndon words are precisely the limits of prefix-preserving increasing sequences of finite Lyndon words. We show that this result still holds when Lyndon words are replaced with generalized Lyndon words provided that the infinite word is primitive.

Theorem 8. *A primitive infinite word is generalized Lyndon if and only if it has infinitely many generalized Lyndon prefixes.*

Proof. Lemma 5 handles the reverse direction. Suppose that there exists an infinite generalized Lyndon word w with finitely many generalized Lyndon prefixes. Since w has infinitely many prefixes, one of its generalized Lyndon prefixes must appear in the unique nonincreasing generalized Lyndon factorizations (from Theorem 2) of infinitely many of the prefixes of w.

We will now use the method presented in the proof of Theorem 4 to construct a nontrivial factorization of w, contradicting the result of Theorem 3. Suppose that $w = \ell_1 \ldots \ell_n w_n$ where ℓ_j is a finite generalized Lyndon word and $\ell_1^\omega \geq \ell_2^\omega \geq \ell_n^\omega > w_n$. Further suppose that w has infinitely many prefixes with factorizations beginning with $\ell_1, \ldots, \ell_n$. If w_n is not generalized Lyndon, the process proceeds as in Theorem 4.

Suppose w_n is generalized Lyndon. If we can choose a generalized Lyndon prefix ℓ_{n+1} of w_n such that infinitely many prefixes of w have factorizations beginning with $\ell_1, \ldots, \ell_n, \ell_{n+1}$, then the process can continue. Otherwise, there must be infinitely many prefixes p of w_n such that $\ell_1, \ldots, \ell_n, p$ is a factorization of a prefix of w. In particular, we have that $p^\omega \leq \ell_n^\omega$ for infinitely many prefixes of p. Taking the limit of these prefixes, we find that $w_n \leq \ell_n^\omega$. Thus, $\ell_1, \ldots, \ell_n, w_n$ is a nontrivial factorization of w, contradicting Theorem 3.

Therefore either the process terminates and produces a nontrivial finite generalized Lyndon factorization of w, or it continues indefinitely and produces a nonincreasing generalized Lyndon factorization of w. Either case contradicts Theorem 3, so w must have infinitely many generalized Lyndon prefixes.

We cannot hope this result extends to the case where the infinite word is not primitive. For example, consider $(01)^\omega$ under the alternating order. It has infinitely many Galois prefixes, namely the prefixes of the form $(01)^k 0$ for any $k \in \mathbb{N}$, but 01 is not Galois.

7   Further Directions

Given that every finite and infinite word has a unique nonincreasing factorization into generalized Lyndon words, one may wish to characterize or compute this factorization. For example, given a simple (e.g. a finite expression of products and powers) representation of an infinite word and a generalized lexicographical ordering, one may wish to compute the factorization of the word in polynomial time. In a different direction, [1] proved the existence and uniqueness of a factorization of a general transfinite (ordinally indexed) word into Lyndon words. It remains to be seen whether this factorization theorem still holds when using generalized Lyndon words. Lastly, one may seek a general characterization of the first factor in a generalized Lyndon factorization along the lines of [14]. While simple characterizations such as longest generalized Lyndon prefix fail, there may be a more clever characterization lurking in the background.

References

1. Boasson, L., Carton, O.: Transfinite Lyndon words. In: Potapov, I. (ed.) DLT 2015. LNCS, vol. 9168, pp. 179–190. Springer, Cham (2015). https://doi.org/10. 1007/978-3-319-21500-6_14
2. Chen, K.-T., Fox, R.H., Lyndon, R.C.: Free differential calculus, IV. Ann. Math. **68**, 81–95 (1958)
3. Dolce, F., Restivo, A., Reutenauer, C.: On generalized Lyndon words. Theoret. Comput. Sci. (2018). https://doi.org/10.1016/j.tcs.2018.12.015
4. Dolce, F., Restivo, A., Reutenauer, C.: Some variations on Lyndon words. arXiv:1904.00954 [math.DM] (2019)
5. Gewurz, D.A., Merola, F.: Numeration and enumeration. Eur. J. Comb. **33**(7), 1547–1556 (2012)
6. Lothaire, M.: Combinatorics on Words. Cambridge Mathematical Library. Cambridge University Press, Cambridge (1997)
7. Lothaire, M.: Algebraic Combinatorics on Words. Encyclopedia of Mathematics and Its Applications, vol. 90. Cambridge University Press, Cambridge (2002)
8. Lyndon, R.C.: On Burnside's problem. Trans. Am. Math. Soc. **77**, 202–215 (1954)
9. Lyndon, R.C., Schützenberger, M.P.: The equation $a^M = b^N c^P$ in a free group. Michigan Math. J. **9**(4), 289–298 (1962)
10. Perrin, D., Restivo, A.: Words. In: Handbook of Enumerative Combinatorics, pp. 509–564. CRC Press (2015)
11. Reutenauer, C.: Mots de Lyndon généralisés. Sém. Lothar. Combin. **54**, B54h (2006)
12. Shirshov, A.I.: Subalgebras of free Lie algebras. Matematicheskii Sbornik **75**(2), 441–452 (1953)
13. Siromoney, R., Mathew, L., Dare, V.R., Subramanian, K.G.: Infinite Lyndon words. Inf. Process. Lett. **50**(2), 101–104 (1994)
14. Ufnarovskij, V.A.: Combinatorial and asymptotic methods in algebra. In: Kostrikin, A.I., Shafarevich, I.R. (eds.) Algebra, VI, Encyclopedia of Mathematical Sciences, vol. 57, pp. 1–196. Springer, Heidelberg (1995)

On the Commutative Equivalence
of Bounded Semi-linear Codes

Arturo Carpi[1] and Flavio D'Alessandro[2,3(✉)]

[1] Dipartimento di Matematica e Informatica, Università degli Studi di Perugia,
via Vanvitelli 1, 06123 Perugia, Italy
[2] Dipartimento di Matematica, Università di Roma "La Sapienza",
Piazzale Aldo Moro 2, 00185 Rome, Italy
dalessan@mat.uniroma1.it
[3] Department of Mathematics, Boğaziçi University, 34342 Bebek, Istanbul, Turkey

Abstract. The problem of the commutative equivalence of semigroups
generated by semi-linear languages is studied. In particular conditions
ensuring that the Kleene closure of a bounded semi-linear code is commutatively equivalent to a regular language are investigated.

Keywords: Commutative equivalence ·
Bounded semi-linear language · Uniquely decipherable code ·
Kleene closure · Exponential growth

1 Introduction

In this paper, we study the commutative equivalence of context-free and regular
languages. Two words are said to be *commutatively equivalent* if one is obtained
from the other by rearranging the letters of the word. Two languages L_1 and L_2
are said to be *commutatively equivalent* if there exists a bijection $f : L_1 \to L_2$
such that every word $u \in L_1$ is commutatively equivalent to $f(u)$. This notion
plays an important role in the study of several problems of Theoretical Computer
Science such as, for instance, in the Theory of Codes, where it is involved in the
celebrated Schützenberger conjecture about the commutative equivalence of a
maximal finite code with a prefix one (see, e.g, [4]). The question of our interest
can be formulated as follows:

Commutative Equivalence Problem. *Given a context-free language L_1, does
there exist a regular language L_2 which is commutatively equivalent to L_1?*

In the sequel, for short, we refer to it as *CE Problem*. A language which is
commutatively equivalent to a regular one will be called *commutatively regular*.

It is worth noticing that commutatively equivalent languages share the same
alphabet and their generating series are equal. In particular, the generating series

The research of F. D'Alessandro was partially supported by TUBITAK Project 2221
(Scientific and Technological Research Council of Turkey) and by the National Group
for Algebraic and Geometric Structures, and their Applications (GNSAGA–INdAM).

R. Mercaş and D. Reidenbach (Eds.): WORDS 2019, LNCS 11682, pp. 119–132, 2019.
https://doi.org/10.1007/978-3-030-28796-2_9

of a commutatively regular language must be rational. This remark leads us to recall that a conceptually related study was conducted by Béal and Perrin in [1], where the generating series of regular languages on alphabets of prescribed size are studied. Béal and Perrin provided a characterization of such series and this remarkable contribution thus defines the theoretical setting in which the CE Problem can be naturally fitted in.

For our discussion, the following notions are useful. Given a language L, the *growth function* g_L returns, for any non-negative integer n, the number of the words of L whose length is less than or equal to n. A language L is called *sparse* if its growth function is polynomially upper bounded. A language L is said to be *of exponential growth* if there exists a real number $k > 1$ such that $g_L(n) > k^n$ for all sufficiently large n. Two results are relevant in this context. The first proved in [5, 26, 29] states that every context-free language is either sparse or of exponential growth. The second, proved in [23, 27], states that the class of sparse context-free languages coincides with that of *bounded languages*. We recall that a language L is termed *bounded* if there exist k words $u_1, \ldots, u_k$ such that $L \subseteq u_1^* \cdots u_k^*$. Bounded context-free languages play a meaningful role in Computer Science and in Mathematics and have been widely investigated in the past so that their structure has been characterized by several theorems [2, 7, 9, 10, 13–15, 18, 19, 21–25, 27, 28]. A characterisation of regular bounded sets, based upon a combinatorial property of the factors of the words of the language, has been obtained by Restivo [28] and, subsequently, extended to context-free bounded languages by Boasson and Restivo [2].

Very recently, results on the counting functions of context-free languages, based upon the notion of strongly counting-regularity, have been obtained in [24]. An excellent survey on the relationships between bounded languages and semigroups has been given by de Luca and Varricchio in [15].

Another theorem that is central in this setting has been proved by Ginsburg and Spanier [18, 19]. This theorem allows one to represent, in a canonical way, bounded context-free languages by means of sets of vectors. For this purpose, let us first introduce a notion. Let $L \subseteq u_1^* \cdots u_k^*$ be a bounded language where, for every $i = 1, \ldots, k$, u_i is a word over the alphabet A. Let $\varphi : \mathbb{N}^k \longrightarrow u_1^* \cdots u_k^*$ be the map defined as: for every tuple $(\ell_1, \ldots, \ell_k) \in \mathbb{N}^k$,

$$\varphi(\ell_1, \ldots, \ell_k) = u_1^{\ell_1} \cdots u_k^{\ell_k}.$$

The map φ is called the *Ginsburg map*. Ginsburg and Spanier proved that L is context-free if and only if the subset $\varphi^{-1}(L)$ of $\mathbb{N}^k$ is a finite union of linear sets, each having a stratified set of periods. Roughly speaking, a stratified set of periods corresponds to a system of well-formed parentheses.

In view of the Ginsburg and Spanier theorem, bounded context-free languages are special instances of a broader class of languages called *bounded semi-linear*. A language $L \subseteq u_1^* \cdots u_k^*$ is called bounded semi-linear if $L = \varphi(B)$, where B is a semi-linear set, that is, a finite union:

$$\bigcup_{i=1}^{n} B_i, \tag{1}$$

of linear subsets B_i of $\mathbb{N}^k$, $1 \leq i \leq n$, of dimension $k_i \geq 0$:

$$B_i = \{\mathbf{b}_0^{(i)} + x_1\mathbf{b}_1^{(i)} + \cdots + x_{k_i}\mathbf{b}_{k_i}^{(i)} \mid x_1, \ldots, x_{k_i} \in \mathbb{N}\}. \tag{2}$$

In [11–13] the solution (in the affirmative) of the CE Problem was given for sparse languages: *Every bounded context-free language L_1 is commutatively equivalent to a regular language L_2. Moreover the language L_2 can be effectively constructed starting from an effective presentation of L_1.* It is also shown that the CE Problem can be solved in the affermative for the wider class of bounded semi-linear languages.

In view of the last theorem and of the results mentioned above, the CE Problem remains open for the class of context-free languages of exponential growth.

A relevant fact in this context, is that the generating series of a commutatively regular language L is always rational. This implies that the answer to the CE Problem is not affirmative in general. Indeed, there exist context-free languages whose generating series are algebraic but not rational. It is worth noting that Flajolet even provided remarkable examples of linear unambiguous context-free languages with a transcendental generating series (*cf* [17], Theorem 3).

In [6], the study of the CE Problem has been investigated in connection with languages of exponential growth generated by unambiguous non-expansive grammars [3] and unambiguous minimal linear grammars [8,20], respectively. In particular conditions ensuring that such languages are commutatively regular have been provided. It is worth noting that an unambiguous minimal linear grammar gives an interesting generalization of the concept of unique-factorization code since it inherits several properties of this structure.

As a continuation of this work, we investigate the CE Problem with respect to the Kleene closure of languages. This operation is of interest for this study since it preserves the property of context-freeness of languages, while it does not preserve the property of boundedness: finite sets of words are the simplest example of bounded languages whose Kleene closure is not bounded. It is also interesting to note that Dyck and semi-Dyck languages provide another natural class of monoids that are not commutatively regular. In view of the classical theorem by Chomsky and Schützenberger for the representation of context-free languages, such monoids can be considered very general.

It is useful to observe that a Dyck monoid is a free monoid whose minimal set of generators, i.e. the set of Dyck prime words, is a context-free bifix code.

In contrast with the previous situation, the main result of this paper shows, up to a technical restriction, that the monoid generated by a prefix (or suffix) bounded semi-linear language, is commutatively regular. Precisely, the main contribution of the paper is the following.

Theorem 1. *Let L be a bounded semi-linear language and $L = \varphi(B)$ where B is the semi-linear set of Eq. (1) associated with L.*

Suppose that, for every $i = 1, \ldots, n$ and for every $j = 0, \ldots, k_i$, the vector $\mathbf{b}_j^{(i)}$ of Eq. (2) is such that its corresponding word $\varphi(\mathbf{b}_j^{(i)})$ contains two distinct letters.

If L is a prefix (or suffix) code, then there exists a regular code L′ which is commutatively equivalent to L. Consequently, L* is commutatively equivalent to (L′)*. Moreover, L′ can be effectively constructed starting from an effective presentation of L.

In order to prove Theorem 1, we use two arguments: the first concerns codes and equations of words and makes it possible to separate the languages that represent the simple sets B_i of the decomposition of B. The second one is a technique of an algebro-geometrical nature for the decomposition into parallelepipeds of simple sets. Among the rational operations on languages, the Kleene closure is the sole one not preserving the property of boundedness and for which therefore the study of the commutative regularity becomes an interesting issue. In view of Theorem 1, we conjecture that the last property can be extended to all monoids generated by bounded semi-linear languages. The result of this paper could be the first step along this direction.

2   Preliminaries

The aim of this section is to introduce some preliminary results on semi-linear sets and bounded context-free languages. We assume that the reader is familiar with the basic notions of context-free languages (see [4, 19] for a reference).

The free abelian monoid on k generators is identified with $\mathbb{N}^k$ with the usual additive structure. Let $B = \{\mathbf{b}_1, \ldots, \mathbf{b}_m\}$, $m \geq 0$, be a finite subset of $\mathbb{N}^k$. Then we denote by $B^\oplus$ the submonoid of $\mathbb{N}^k$ generated by B, that is

$$B^\oplus = \mathbf{b}_1^\oplus + \cdots + \mathbf{b}_m^\oplus = \{x_1\mathbf{b}_1 + \cdots + x_m\mathbf{b}_m \mid x_i \in \mathbb{N}, \ 1 \leq i \leq m\}.$$

Definition 2. *Let X be a subset of $\mathbb{N}^k$. Then*

1. *X is* linear *if $X = \mathbf{b}_0 + \{\mathbf{b}_1, \ldots, \mathbf{b}_m\}^\oplus$, where $\mathbf{b}_0, \mathbf{b}_1, \ldots, \mathbf{b}_m \in \mathbb{N}^k$,*
2. *X is* simple *if the vectors $\mathbf{b}_1, \mathbf{b}_2, \ldots, \mathbf{b}_m$ are linearly independent in $\mathbb{Q}^k$,*
3. *X is* semi-linear *if X is a finite union of linear sets in $\mathbb{N}^k$,*
4. *X is* semi-simple *if X is a finite disjoint union of simple sets in $\mathbb{N}^k$.*

If $X = \mathbf{b}_0 + \{\mathbf{b}_1, \ldots, \mathbf{b}_m\}^\oplus$ is a simple set, then we say that the vectors $\mathbf{b}_0, \mathbf{b}_1, \ldots, \mathbf{b}_m$, form the *(unambiguous) representation* of X. The vector $\mathbf{b}_0$ is called *constant* and $\mathbf{b}_1, \ldots, \mathbf{b}_m$ are called *generators* of the representation, respectively. One can prove that the representation of a simple set is unique. The number m is called the *dimension* of B. Obviously one has $m \leq k$. The following is an important characterization of semi-linear sets.

Theorem 3 *(Eilenberg and Schützenberger, [16]). Let X be a subset of $\mathbb{N}^k$. Then X is semi-linear in $\mathbb{N}^k$ if and only if X is semi-simple in $\mathbb{N}^k$.*

Let $A = \{a_1, \ldots, a_t\}$ be an alphabet of t letters and A^* the free monoid generated by A. The empty word of A^* is denoted by ε. The length of every word u is denoted by $|u|$. For every $a \in A$ and $u \in A^*$, the number of occurrences of a in u will be denoted by $|u|_a$. We let $\psi \colon A^* \to \mathbb{N}^t$ denoting the *Parikh map over A*, defined, for each $u \in A^*$ as $\psi(u) \coloneqq (|u|_{a_1}, |u|_{a_2}, \ldots, |u|_{a_t})$.

A subset L of words of A^+ is said to be a *code (over A)* if every word of L^+ admits a unique factorization in term of words of L. A set L over the alphabet A is said to be a *prefix set* if $LA^+ \cap L = \emptyset$.

Let L_1, L_2 be two languages over A. We say that L_1 is *commutatively equivalent* to L_2 if a bijection $f \colon L_1 \to L_2$ exists such that, for every $u \in L_1$, one has $\psi(u) = \psi(f(u))$. In the sequel, by simplicity, if L_1 and L_2 are so, we write $L_1 \sim L_2$. If $u_1, \ldots, u_k$ are k words of A^+, the *Ginsburg map*

$$\varphi \colon \mathbb{N}^k \to u_1^* \cdots u_k^*, \tag{3}$$

is the map defined, for each tuple $(\ell_1, \ldots, \ell_k) \in \mathbb{N}^k$, as $\varphi(\ell_1, \ldots, \ell_k) = u_1^{\ell_1} \cdots u_k^{\ell_k}$. A remarkable observation in [21] (*cf* Lemma 2.1) allows one to prove that the representation of Eq. (1) of a bounded semi-linear language is faithful. Moreover a previously mentioned theorem by Ginsburg and Spanier [18] provides a fundamental tool to represent, in terms of semi-linear sets, the bounded context-free languages. For our purposes, these two results can be stated as follows.

Theorem 4. *Let $L \subseteq u_1^* \cdots u_k^*$ be a bounded semi-linear language. Then there exists a semi-simple set B of $\mathbb{N}^k$ such that $\varphi(B) = L$ and φ is injective on B. Moreover, B can be effectively constructed.*

In particular, the condition above holds for bounded context-free languages.

3   Some Results of Combinatorics on Words

From now on, in the sequel of the paper, we assume that $u_1, \ldots, u_k$, is a list of k words over A, fixed once for all.

In the sequel we assume that $B = \{\mathbf{b}_0 + x_1\mathbf{b}_1 + \cdots + x_m\mathbf{b}_m \mid x_i \in \mathbb{N}, 1 \leq i \leq m\}$ is a simple set satisfying the following:

Assumption 5. *For every $i = 0, \ldots, m$, the word $\varphi(\mathbf{b}_i)$ contains two distinct letters.*

Remark 6. One could prove that Assumption 5 is equivalent to the following statement:

There exists a real number $\rho < 1$ such that, for every $\mathcal{X} \in \varphi(B)$ and $a \in A$, one has

$$|\mathcal{X}|_a < \rho|\mathcal{X}|. \tag{4}$$

Thus, Assumption 5 implies, in particular, that no word of $\varphi(B)$ can be a power of a single letter.

Lemma 7. *Let* $L = \varphi(B) \subseteq u_1^* \cdots u_k^*$ *be the bounded semi-linear language defined by the simple set B above. There exists a positive constant γ such that for every pair of distinct letters $a, b \in A$ the word*

$$u = (a^\gamma b)^k \tag{5}$$

is not a prefix of any word in LA^*.

In the sequel, γ will denote the minimum constant specified by Lemma 7.

Let $\mathbf{v} = (v_1, \ldots, v_t) \in \mathbb{N}^t$ be a vector. We denote by $|\mathbf{v}|$ the non-negative integer $|\mathbf{v}| = v_1 + \cdots + v_t$. Let

$$z_1, \ldots, z_m, \tag{6}$$

be a list of (not necessarily pairwise distinct) words of A^+. We associate with the list (6) its multiset of Parikh vectors $\{(\alpha_1, \mathbf{v}_1), \ldots, (\alpha_\ell, \mathbf{v}_\ell)\}$, where, for every $i = 1, \ldots, \ell$, α_i is the multiplicity of $\mathbf{v}_i$.

Lemma 8. *Let us consider the list of words (6) together with its multiset of Parikh vectors. Suppose that:*

(i) for every $j = 1, \ldots, m$, z_j contains, at least, two different letters;
(ii) for every $j = 1, \ldots, \ell$, $|\mathbf{v}_j| = \beta$, where β is a constant not depending on j.

Let N_j be the greatest integer such that $\mathbf{v}_j$ has the form $\mathbf{v}_j = N_j \bar{\mathbf{v}}_j$ with $\bar{\mathbf{v}}_j \in \mathbb{N}^t$, for every $j = 1, \ldots, \ell$. If, for every $j = 1, \ldots, \ell$,

$$N_j > mk(\gamma + 1),$$

then there exists a code $\mathcal{W}$, over the alphabet A, of m (distinct) words such that, each word has length β, and

$$\forall\, i = 1, \ldots, \ell, \quad \mathrm{Card}(\{w \in \mathcal{W} \mid \psi(w) = \mathbf{v}_i\}) = \alpha_i. \tag{7}$$

Moreover the word $(a^\gamma b)^k$ of Lemma 7 is a prefix of each word of $\mathcal{W}$.

A simple argument based on code composition allows one to verify the following lemma which will be useful in the sequel.

Lemma 9. *Let $X = X_1 \cup X_2$ be a partition of a code $X \subseteq A^*$ in two subsets X_1 and X_2. Then, $X_1 X_2^*$ is a code.*

4   The CE Problem for Bounded Semi-linear Languages

Here we recall details of a construction (see [11]) of an algebro-geometrical nature, that we will use in the proof of Theorem 1.

4.1   A Geometrical Decomposition of a Simple Set of $\mathbb{N}^k$

Let B be a simple set of $\mathbb{N}^k$ of dimension $m > 0$:

$$B = \mathbf{b}_0 + \mathbf{b}_1^{\oplus} + \cdots + \mathbf{b}_m^{\oplus} = \{\mathbf{b}_0 + x_1\mathbf{b}_1 + \cdots + x_m\mathbf{b}_m \mid x_i \in \mathbb{N}, \ 1 \le i \le m\}, \quad (8)$$

where the vectors $\mathbf{b}_0, \mathbf{b}_1, \ldots, \mathbf{b}_m$ form the representation of B. Let

$$(N_1, \ldots, N_m) \quad (9)$$

be a sequence of m positive integers. From an intuitive point of view, our goal is to define, according to this sequence of m positive integers, a suitable decomposition of B into parallelepipeds of dimension lower than or equal to m. For this purpose, let $\{+, -\}$ be an alphabet of two symbols and let $\mathcal{E}$ be the set

$$\mathcal{E} = \{(\epsilon_1, \ldots, \epsilon_m) \mid \epsilon_i \in \{+, -\}, \ i = 1, \ldots, m\},$$

of all sequences of length m with elements in the set $\{+, -\}$. With every sequence $(\epsilon_1, \ldots, \epsilon_m) \in \mathcal{E}$, we associate the set of vectors $B^{(\epsilon_1, \ldots, \epsilon_m)}$ defined as:

$$B^{(\epsilon_1, \ldots, \epsilon_m)} = \{\mathbf{b}_0 + x_1\mathbf{b}_1 + \cdots + x_m\mathbf{b}_m \mid x_i \in R_i, \ 1 \le i \le m\}, \quad (10)$$

where, for every $i = 1, \ldots, m$, one has:

$$R_i = \{x \in \mathbb{N} \mid x \ge N_i\} \text{ if } \epsilon_i = +, \qquad R_i = \{x \in \mathbb{N} \mid x < N_i\} \text{ if } \epsilon_i = -.$$

One can verify that the family (10) is a finite set of pairwise disjoint semi-simple sets that gives a partition of B. The partition (10) can be refined as follows. Let $(\epsilon_1, \ldots, \epsilon_m) \in \mathcal{E} \setminus \{(-, -, \ldots, -)\}$, that is, there exists i, with $1 \le i \le m$, where $\epsilon_i = +$. Then there exists a non-negative integer p, depending on $(\epsilon_1, \ldots, \epsilon_m)$, such that the set of indices i, with $i = 1, \ldots, m$ is partitioned in two sets:

$$I_{\epsilon_1 \cdots \epsilon_m}^{-} = \{i_1, \ldots, i_p\}, \quad I_{\epsilon_1 \cdots \epsilon_m}^{+} = \{i_{p+1}, \ldots, i_m\}, \quad (11)$$

where $\epsilon_{i_\ell} = -$ for $\ell = 1, \ldots, p$, and $\epsilon_{i_\ell} = +$ for $\ell = p+1, \ldots, m$.

It is worth remarking that:

- If, for every $i = 1, \ldots, m$, $\epsilon_i = +$, then $I_{\epsilon_1 \cdots \epsilon_m}^{-} = \emptyset$;
- the integer p depends upon the sequence $(\epsilon_1, \ldots, \epsilon_m)$.

Denote by $\mathcal{C}_{\epsilon_1 \cdots \epsilon_m}^{+}$ the set of all sequences

$$(r_{i_{p+1}}, \ldots, r_{i_m}). \quad (12)$$

with $0 \le r_{i_\ell} < N_{i_\ell}$, $i_\ell \in I_{\epsilon_1 \cdots \epsilon_m}^{+}$. For every sequence $(r_{i_{p+1}}, \ldots, r_{i_m})$ in $\mathcal{C}_{\epsilon_1 \cdots \epsilon_m}^{+}$, define the set of vectors $B_{r_{i_{p+1}} \cdots r_{i_m}}^{(\epsilon_1, \ldots, \epsilon_m)}$ as:

$$\mathbf{b}_0 + \sum_{\ell=1}^{p} \{c_{i_\ell}\mathbf{b}_{i_\ell} \mid 0 \le c_{i_\ell} < N_{i_\ell}\} + \sum_{\ell=p+1}^{m} \{(r_{i_\ell} + N_{i_\ell}x_{i_\ell})\mathbf{b}_{i_\ell} \mid x_{i_\ell} \ge 1\}. \quad (13)$$

One can verify that the sets $B_{r_{i_{p+1}}\cdots r_{i_m}}^{(\epsilon_1,\ldots,\epsilon_m)}$ are pairwise disjoint, semi-simple and they give a partition of $B^{(\epsilon_1,\ldots,\epsilon_m)}$. Thus, by the last remarks, the simple set B admits the partition into semi-simple sets

$$B^- \cup \bigcup_{(\epsilon_1,\ldots,\epsilon_m)\in\mathcal{E}\setminus\{(-,-,\ldots,-)\}} \bigcup_{(r_{i_{p+1}},\ldots,r_{i_m})\in\mathcal{C}_{\epsilon_1\cdots\epsilon_m}^+} B_{r_{i_{p+1}}\cdots r_{i_m}}^{(\epsilon_1,\ldots,\epsilon_m)}, \tag{14}$$

where B^- denotes the set $B^{(\epsilon_1,\ldots,\epsilon_m)}$ such that, for every $i = 1,\ldots,m$, $\epsilon_i = -$.
   Let $L^- = \varphi(B^-)$ and let

$$L_{r_{i_{p+1}}\cdots r_{i_m}}^{(\epsilon_1,\ldots,\epsilon_m)} = \varphi(B_{r_{i_{p+1}}\cdots r_{i_m}}^{(\epsilon_1,\ldots,\epsilon_m)}), \tag{15}$$

be the languages obtained as the images, under the map φ, of the sets of the family (13). In view of the last remarks and Theorem 4, we get

Theorem 10. *The family (15), together with L^-, is a partition of L.*

4.2   The Construction of the Regular Language

Let $L \subseteq u_1^* \cdots u_k^*$ be a bounded semi-linear language. By Theorem 4 there exists a semi-simple set B of $\mathbb{N}^k$ such that $\varphi(B) = L$ and φ is injective on B.
   Now we assume that $B = \{\mathbf{b}_0 + x_1\mathbf{b}_1 + \cdots + x_m\mathbf{b}_m \mid x_i \in \mathbb{N}, \ 1 \leq i \leq m\}$ is a simple set satisfying Assumption 5, already introduced in Sect. 3:

For every $i = 0,\ldots,m$, the word $\varphi(\mathbf{b}_i)$ contains two distinct letters.

   Let c be a non-negative integer and $\beta(c) = \Pi_{\ell=1}^m |\varphi(\mathbf{b}_\ell)|c$.
For every $i = 1,\ldots,m$, let $N_i(c)$ be the number defined as $N_i(c) = \beta(c)/|\varphi(\mathbf{b}_i)|$. From now on, we will assume that c is a positive integer such that, for every $i = 1,\ldots,m$, $N_i(c) > mk(\gamma+1)$, where m is the number of the vectors of the representation of B and γ is the fixed constant of Lemma 7. The list of (possibly equal) words

$$\varphi(N_1(c)\mathbf{b}_1),\ldots,\varphi(N_m(c)\mathbf{b}_m) \tag{16}$$

satisfy the hypotheses of Lemma 8. Hence by applying this lemma to the list (16), one gets the existence of a code of m distinct words $\mathcal{W} = \{w_1,\ldots,w_m\}$, each of length $\beta(c)$, where, for every $i = 1,\ldots,m$, $\psi(w_i) = \psi(\varphi(N_i(c)\mathbf{b}_i))$. Thus we can define the bijective function such that, for every $i = 1,\ldots,m$,

$$N_i(c)\mathbf{b}_i \longrightarrow w_i, \tag{17}$$

maps the i-th vector $N_i(c)\mathbf{b}_i$ into the i-th word of $\mathcal{W}$.
   For the sake of simplicity, from now on, the numbers $N_1(c),\ldots,N_m(c)$ and $\beta(c)$ will be denoted respectively as:

$$N_1,\ldots,N_m, \ \beta. \tag{18}$$

Let us now consider the partition (14) of B, given in Subsect. 4.1, with respect to the sequence $(N_1,\ldots,N_m)$.

In order to simplify the notation, we let $\mathcal{B}_1, \ldots, \mathcal{B}_s$, $s \geq 1$, denote an enumeration of the sets of (13). Hence, for every $(\epsilon_1, \ldots, \epsilon_m) \in \mathcal{E} \setminus (-,-,\cdots,-)$, and for every $(r_{i_{p+1}}, \ldots, r_{i_{m-p}}) \in \mathcal{C}^+_{\epsilon_1 \cdots \epsilon_m}$, there exists exactly one index i, with $1 \leq i \leq s$ such that $\mathcal{B}_i = B^{(\epsilon_1, \ldots, \epsilon_m)}_{r_{i_{p+1}} \cdots r_{i_m}}$.

Let us associate with every set $\mathcal{B}_i$, with $i = 1, \ldots, s$, the regular language:

$$L'_i = \varphi \left(\mathbf{b}_0 + \left\{ \sum_{\ell=1}^{p} c_{i_\ell} \mathbf{b}_{i_\ell} \,\middle|\, 0 \leq c_{i_\ell} < N_{i_\ell} \right\} + \sum_{\ell=p+1}^{m} r_{i_\ell} \mathbf{b}_{i_\ell} \right) w^+_{i_{p+1}} w^+_{i_{p+2}} \cdots w^+_{i_m},$$

(19)

where, for every $\ell = p+1, \ldots, m$, w_{i_ℓ} is the word of the code $\mathcal{W}$ associated by the coding (17) to the index i_ℓ.

In order to keep our notation uniform, we denote the languages of the family (15) as $L_i = \varphi(\mathcal{B}_i)$, for every $i = 1, \ldots, s$.

Let $u \in L_i$. By Theorem 4 there exists exactly one vector $\mathbf{b} \in \mathcal{B}_i = B^{(\epsilon_1, \ldots, \epsilon_m)}_{r_{i_{p+1}} \cdots r_{i_m}}$ such that $u = \varphi(\mathbf{b})$. Moreover by the fact that $\mathbf{b}_0, \ldots, \mathbf{b}_m$ form the representation of B, there exist exactly a tuple $x_{i_\ell} \geq 1$, with $i_\ell = p+1, \ldots, m$ and a tuple c_{i_ℓ}, with $i_\ell = 1, \ldots, p$ where $0 \leq c_{i_\ell} < N_{i_\ell}$ such that

$$\mathbf{b} = \mathbf{v} + \sum_{\ell=p+1}^{m} N_{i_\ell} x_{i_\ell} \mathbf{b}_{i_\ell}, \quad x_{i_\ell} > 0,$$

with $\mathbf{v} = \mathbf{b}_0 + \sum_{\ell=1}^{p} c_{i_\ell} \mathbf{b}_{i_\ell} + \sum_{\ell=p+1}^{m} r_{i_\ell} \mathbf{b}_{i_\ell}$.

Let us define the map: $f_i : L_i \longrightarrow L'_i$ such that, for every $u \in L_i$,

$$f_i(u) = \varphi(\mathbf{v}) w^{x_{i_{p+1}}}_{i_{p+1}} \cdots w^{x_{i_m}}_{i_m}.$$

(20)

One can prove that f_i is a bijection from L_i to L'_i that preserves the Parikh vectors of words of L_i (cf [11], Lemma 10). Let L' be the union of all the languages (19) together with L^-:

$$L' = L^- \cup \bigcup_{j=1}^{s} L'_j.$$

(21)

It is worth noticing that the construction of L' can be effectively done, starting from an effective presentation of L. Let

$$f : L \to L',$$

(22)

be the map defined as follows: on the set L^-, f coincides with the identity and, on the set L_i, for every $i = 1, \ldots, s$, f coincides with the map f_i defined in (20).

In view of Theorem 10, f is well defined as a map from L into L'. Since every map f_i of (20) is a surjection from L_i onto L'_i, that preserves the Parikh vectors, then f does the same from L into L'.

The injectivity of f comes immediately from the fact that the languages (19) and $L^- = \varphi(B^-)$ are all pairwise disjoint (cf [11], Lemma 9). Hence L and L' are commutatively regular. More precisely one has

Theorem 11. *The map f is a bijection of L and L', preserving Parikh vectors.*

5   The Main Result

Let $L \subseteq u_1^* \cdots u_k^*$ be a bounded language such that $\varphi(B) = L$ for a semi-simple set B of $\mathbb{N}^k$ and φ is injective on B. For the sake of simplicity, we first prove Theorem 1 in the case that B is a simple set.

5.1   The Simple Case

Let $B = \{\mathbf{b}_0 + x_1 \mathbf{b}_1 + \cdots + x_m \mathbf{b}_m \mid x_i \in \mathbb{N},\ 1 \leq i \leq m\}$ be the simple set such that $L = \varphi(B)$. Moreover we suppose that B satisfies Assumption 5.

In order to prove the theorem, the following lemma is needed.

Lemma 12. *The language L' is a code over A.*

Proof. Since, by construction, $L' \subseteq LW^*$, it is sufficient to show that the set LW^* is a code. In view of Lemma 9, this result will be achieved if we show that $L \cup W$ is a code and $L \cap W = \emptyset$. We notice that

$$LA^* \cap WA^* = \emptyset. \tag{23}$$

Indeed, by construction, the word u of Eq. (5) is a prefix of all words of WA^*, while by Lemma 7, u is not a prefix of any word in LA^*. Taking into account that both L and W are prefix codes, from Eq. (23) one easily derives that also $L \cup W$ is a prefix code and, moreover, $L \cap W = \emptyset$. This completes the proof. $\square$

Proof (of Theorem 1, for simple sets). By construction and by the latter lemma, L' is a regular code such that $L \sim L'$. Let us now show that $L^* \sim (L')^*$. Let $f : L \to L'$ be the map (22) between L and the regular language L'. In view of Theorem 11, f is a bijection preserving the Parikh vectors. Let $g : L^* \to L'^*$ be the morphism generated by the map f on L: for every $u \in L^*$

$$g(u) = \begin{cases} \varepsilon & \text{if } u = \varepsilon \\ f(v_1) \cdots f(v_n) & \text{if } u = v_1 \cdots v_n,\ v_i \in L,\ n \geq 1 \end{cases} \tag{24}$$

Since L is a code, every word of L^+ admits a unique factorization over L, thus implying that g is well defined as a map. Since f preserves the Parikh vectors, one easily verifies that g does it as well, that is, for every $u \in L^*$, one has $\psi(g(u)) = \psi(u)$. By construction, g is a surjective mapping from L^* onto L'^*.

Let us now verify that g is injective. Assume that $g(u) = g(u')$, for $u, u' \in L^*$. One can suppose that $u, u' \neq \varepsilon$, so that $u = v_1 \cdots v_n$, $v_i \in L, n \geq 1$, and $u' = v_1' \cdots v_\ell'$, $v_i' \in L$, $\ell \geq 1$. From the latter we get $f(v_1) \cdots f(v_n) = f(v_1') \cdots f(v_\ell')$, where $f(v_i), f(v_j') \in L', 1 \leq i \leq n, 1 \leq j \leq \ell$. By Lemma 12, one has $n = \ell$ and for every $i = 1, \ldots, n, f(v_i) = f(v_i')$. By the injectivity of f, from the latter one has $v_i = v_i'$, for every $i = 1, \ldots, n$, so $u = u'$. Thus g is a bijection between L^* and L'^*, preserving the Parikh vectors, so $L^* \sim L'^*$. $\square$

The following example clarifies some of the basic ideas underlying the proof of Theorem 1.

Example 13. Let $L = \{a^n b^m c^m d^n : n, m \geq 0\}$ be the bounded semi-linear language over the alphabet $A = \{a, b, c, d\}$. The Ginsburg map is defined as $\varphi : \mathbb{N}^4 \to a^* b^* c^* d^*$, and $L = \varphi(B)$, where B is the simple subset of $\mathbb{N}^4$ $B = \{\mathbf{v}_0 + n\mathbf{v}_1 + m\mathbf{v}_2 : n, m \geq 0\}$, where $\mathbf{v}_1 = (1, 0, 0, 1)$, $\mathbf{v}_2 = (0, 1, 1, 0)$ and $\mathbf{v}_0 = (0, 0, 0, 0)$. Notice that $\varphi(\mathbf{v}_1) = ad$, $\varphi(\mathbf{v}_2) = bc$ and $\varphi(\mathbf{v}_0) = \varepsilon$, so that B satisfies a lighter variant of Assumption 5. It is easily verified that L is a prefix set and $L \cap A^+$ is a prefix code.

The structure of the set B is sufficiently simple not to use the partition procedure of Sect. 4, in order to construct the regular language L' such that $L \sim L'$. However, the decomposition of L, on which the construction of L' is based upon, follows a (kind of) similar argument.

Since φ is injective on $\mathbb{N}^4$, every vector $\mathbf{v} = (n, m, m, n)$ of B can be faithfully represented by the word $a^n b^m c^m d^n$, so that we may directly work on the language L and get, for this language, the partition $L = L_0 \cup L_1 \cup L_2 \cup L_3 \cup L_4 \cup L_5$, where

- $L_0 = \{\varepsilon, abcd\}$, $L_1 = \{a^n d^n : n \geq 1\}$,
- $L_2 = \{b^m c^m : m \geq 1\}$, $L_3 = \{a^{2+n} b^{2+m} c^{2+m} d^{2+n} : n, m \geq 0\}$,
- $L_4 = \{a^{2+n} bcd^{2+n} : n \geq 0\}$, $L_5 = \{ab^{2+m} c^{2+m} d : m \geq 0\}$.

Let L' be the regular language $L' = L_0 \cup L_1' \cup L_2' \cup L_3' \cup L_4' \cup L_5'$, where the L_i''s are the regular languages defined as:

- $L_1' = \{(ad)^{n-1} da : n \geq 1\}$, $L_2' = \{(bc)^{m-1} cb : m \geq 1\}$,
- $L_3' = \{a^2 b^2 (ad)^n (bc)^m c^2 d^2 : n, m \geq 0\}$, $L_4' = \{a^2 (ad)^n (bc) d^2 : n \geq 0\}$,
- $L_5' = \{b^2 (ad)(bc)^m c^2 : m \geq 0\}$.

Note that all the languages of the partition of L' are pairwise disjoint and that L' is a prefix set. Let $f : L \to L'$ be the map defined as follows:

- f is the identity on L_0,
- $\forall v = a^n d^n \in L_1$, $f(v) = (ad)^{n-1} da \in L_1'$,
- $\forall v = b^m c^m \in L_2$, $f(v) = (bc)^{m-1} cb \in L_2'$,
- $\forall v = a^{2+n} b^{2+m} c^{2+m} d^{2+n} \in L_3$, $f(v) = a^2 b^2 (ad)^n (bc)^m c^2 d^2 \in L_3'$,
- $\forall v = a^{2+n} bcd^{2+n} \in L_4$, $f(v) = a^2 (ad)^n (bc) d^2 \in L_4'$,
- $\forall v = ab^{2+m} c^{2+m} d \in L_5$, $f(v) = b^2 (ad)(bc)^m c^2 \in L_5'$.

One easily verifies that the restriction of f on each L_i is a bijection between L_i and L_i' that preserves the Parikh vectors, so that $L_i \sim L_i'$. Moreover, since the languages L_i's provide a partition of L, and the languages L_i''s provide a partition of L', one easily gets $L \sim L'$.

5.2   The Semi-simple Case

We outline the proof of Theorem 1 in its full generality. The proof essentially reuses the arguments of the previous subsection. In the general case we can decompose L as the disjoint union $L = \bigcup_{i=1}^{n} L_i$, where, for $i = 1, \ldots, n$, $L_i = \varphi(B_i)$ for a suitable simple set B_i. Moreover all sets B_i satisfy Assumption 5. By the results of Sect. 4 and Subsect. 5.1, for all $i = 1, \ldots, n$, one can find a regular code L_i', a uniform code $\mathcal{W}_i$ and an integer γ_i such that

1. $L_i \sim L_i'$,
2. $L_i' \subseteq L_i \mathcal{W}_i^*$,
3. $\mathcal{W}_i A^* \cap L A^* = \emptyset$,
4. the word $a^{\gamma_i} b$ is a prefix of all the words of $\mathcal{W}_i$.

Moreover, as shown in the proof of Lemma 7, the integers $\gamma_1, \ldots, \gamma_n$ can be chosen arbitrarily, provided that they are sufficiently large. Thus, with no loss of generality, we may assume that the integers above are pairwise distinct. Since the sets L_i, $i = 1, \ldots, n$, are pairwise disjoint subsets of the prefix code L, one easily derives that the sets $L_i \mathcal{W}_i^*$ are pairwise disjoint too. Thus, by Condition 2, the sets L_i' are pairwise disjoint, too. This, together with Condition 1, implies that the set $L' = \bigcup_{i=1}^n L_i'$ is commutatively equivalent to L.

Now, to complete the proof, it is sufficient to verify that L' is a code. First, let us verify that

$$\mathcal{W} = \mathcal{W}_1 \cup \cdots \cup \mathcal{W}_n,$$

is a prefix code. Indeed, let $w \in \mathcal{W}_i$, $1 \le i \le n$. The word w cannot be a prefix of another word of $\mathcal{W}_i$ because $\mathcal{W}_i$ is a uniform code and it cannot be a prefix of a word $w' \in \mathcal{W}_j$, with $j \neq i$, because $a^{\gamma_i} b$ is a prefix of w while $a^{\gamma_j} b$ is a prefix of w' and $\gamma_i \neq \gamma_j$. We conclude that $\mathcal{W}$ is a prefix code.

In view of Condition 3, one derives that also the set $L \cup \mathcal{W}$ is a prefix code and, moreover, $L \cap \mathcal{W} = \emptyset$. Thus, by Lemma 9, $L\mathcal{W}^*$ is a code. Observing that, in view of Condition 2, $L' \subseteq L\mathcal{W}^*$, we conclude that L' is a regular code. Note that all the steps of the construction of L' are constructive.

Finally, by using the very same argument of the proof of Theorem 1 for simple sets, one constructs a bijection between L^* and L'^*, preserving the Parikh vectors, thus implying $L^* \sim L'^*$. This completes the proof. $\square$

References

1. Béal, M.-P., Perrin, D.: On the generating sequences of regular languages on k symbols. J. ACM **50**, 955–980 (2003)
2. Boasson, L., Restivo, A.: Une Caractérisation des Langages Algébriques Bornés. ITA **11**, 203–205 (1977)
3. Baron, G., Kuich, W.: The characterization of nonexpansive grammars by rational power series. Inf. Control **48**, 109–118 (1981)
4. Berstel, J., Perrin, D., Reutenauer, C.: Codes and Automata Encyclopedia of Mathematics and its Applications, vol. 129. Cambridge University Press, Cambridge (2009)
5. Bridson, M.R., Gilman, R.H.: Context-free languages of sub-exponential growth. J. Comput. Syst. Sci. **64**, 308–310 (1999)
6. Carpi, A., D'Alessandro, F.: On the commutative equivalence of context-free languages. In: Hoshi, M., Seki, S. (eds.) DLT 2018. LNCS, vol. 11088, pp. 169–181. Springer, Cham (2018). https://doi.org/10.1007/978-3-319-98654-8_14
7. Choffrut, C., D'Alessandro, F., Varricchio, S.: On bounded rational trace languages. Theory Comput. Syst. **46**, 351–369 (2010)

8. Chomsky, N., Schützenberger, M.-P.: The algebraic theory of context-free languages. In: Braffort, P., Hirschberg, D. (eds.) Computer Programming and Formal Systems, pp. 118–161. North Holland Publishing Company, Amsterdam (1963)
9. D'Alessandro, F., Intrigila, B., Varricchio, S.: The Parikh counting functions of sparse context-free languages are quasi-polynomials. Theoret. Comput. Sci. **410**, 5158–5181 (2009)
10. D'Alessandro, F., Intrigila, B., Varricchio, S.: Quasi-polynomials, linear Diophantine equations and semi-linear sets. Theoret. Comput. Sci. **416**, 1–16 (2012)
11. D'Alessandro, F., Intrigila, B.: On the commutative equivalence of bounded context-free and regular languages: the code case. Theoret. Comput. Sci. **562**, 304–319 (2015)
12. D'Alessandro, F., Intrigila, B.: On the commutative equivalence of semi-linear sets of $\mathbb{N}^k$. Theoret. Comput. Sci. **562**, 476–495 (2015)
13. D'Alessandro, F., Intrigila, B.: On the commutative equivalence of bounded context-free and regular languages: the semi-linear case. Theoret. Comput. Sci. **572**, 1–24 (2015)
14. D'Alessandro, F., Ibarra, O.H., McQuillan, I.: On finite-index indexed grammars and their restrictions. In: Drewes, F., Martín-Vide, C., Truthe, B. (eds.) LATA 2017. LNCS, vol. 10168, pp. 287–298. Springer, Cham (2017). https://doi.org/10.1007/978-3-319-53733-7_21
15. de Luca, A., Varricchio, S.: Finiteness and Regularity in Semigroups and Formal Languages. Springer, Berlin (1999). https://doi.org/10.1007/978-3-642-59849-4
16. Eilenberg, S., Schützenberger, M.-P.: Rational sets in commutative monoids. J. Algebra **13**, 173–191 (1969)
17. Flajolet, P.: Analytic models and ambiguity of context-free languages. Theoret. Comput. Sci. **49**, 283–309 (1987)
18. Ginsburg, S., Spanier, E.H.: Semigroups, presburger formulas, and languages. Pacific J. Math. **16**, 285–296 (1966)
19. Ginsburg, S.: The Mathematical Theory of Context-Free Languages. Mc Graw-Hill, New York (1966)
20. Gross, M.: Inherent ambiguity of minimal linear grammars. Inf. Control **7**, 366–368 (1964)
21. Honkala, J.: Decision problems concerning thinness and slenderness of formal languages. Acta Inf. **35**, 625–636 (1998)
22. Honkala, J.: On Parikh slender context-free languages. Theoret. Comput. Sci. **255**, 667–677 (2001)
23. H.Ibarra, O., Ravikumar, B.: On sparseness, ambiguity and other decision problems for acceptors and transducers. In: Monien, B., Vidal-Naquet, G. (eds.) STACS 1986. LNCS, vol. 210, pp. 171–179. Springer, Heidelberg (1986). https://doi.org/10.1007/3-540-16078-7_74
24. Ibarra, O.H., McQuillan, I., Ravikumar, B.: On counting functions of languages. In: Hoshi, M., Seki, S. (eds.) DLT 2018. LNCS, vol. 11088, pp. 429–440. Springer, Cham (2018). https://doi.org/10.1007/978-3-319-98654-8_35
25. Ilie, L., Rozenberg, G., Salomaa, A.: A characterization of poly-slender context-free languages. RAIRO Inform. Théor. Appl. **34**, 77–86 (2000)
26. Incitti, R.: The growth function of context-free languages. Theoret. Comput. Sci. **255**, 601–605 (2001)
27. Latteux, M., Thierrin, G.: On bounded context-free languages. Elektron. Inform. Verarb. u. Kybern. **20**, 3–8 (1984)

28. Restivo, A.: A characterization of bounded regular sets. In: Brakhage, H. (ed.) GI-Fachtagung 1975. LNCS, vol. 33, pp. 239–244. Springer, Heidelberg (1975). https://doi.org/10.1007/3-540-07407-4_26
29. Trofimov, V.I.: Growth functions of some classes of languages. Kibernetika (Kiev), (6), 9–12, 149 (1981)

Circularly Squarefree Words
and Unbordered Conjugates:
A New Approach

Trevor Clokie, Daniel Gabric, and Jeffrey Shallit[✉]

School of Computer Science, University of Waterloo, Waterloo, ON N2L 3G1, Canada
{trevor.clokie,dgabric,shallit}@uwaterloo.ca

Abstract. Using a new approach based on automatic sequences, logic, and a decision procedure, we reprove some old theorems about circularly squarefree words and unbordered conjugates in a new and simpler way. Furthermore, we prove two new results about unbordered conjugates: we complete the classification, due to Harju and Nowotka, of binary words with the maximum number of unbordered conjugates, and we prove that for every possible number, up to the maximum, there exists a word having that number of unbordered conjugates.

1 Introduction

Throughout this paper, Σ_k denotes the alphabet $\{0, 1, \ldots, k-1\}$.

Two finite words are said to be *conjugate* if one is a cyclic shift of the other, as in the English words `enlist` and `listen`.

A finite word w has a *border* x if $x \notin \{\epsilon, w\}$ and x is both a prefix and suffix of w; the two occurrences of x are allowed to overlap each other. For example, `alfa` is a border of `alfalfa`. A finite word w is said to be *bordered* if it has a border, and otherwise, it is *unbordered*. A finite word w if bordered iff it has a border of length $\leq |w|/2$, for if a word has a longer border y, then the nonempty overlap of the two occurrences of y—one as prefix and one as suffix—provides a shorter border. For example, `alfalfa` is also bordered by `a`.

A finite word w is said to be a *square* if $w = xx$ for some nonempty word x. An example in French is the word `couscous`. A word (finite or infinite) is *squarefree* if no nonempty factor is a square. Let μ be the *Thue-Morse morphism*, defined by $\mu(0) = 01$ and $\mu(1) = 10$. The Thue-Morse word $\mathbf{t} = 01101001\cdots$ is the infinite fixed point, starting with 0, of μ. Thue [4,12,13] proved that there exist infinite squarefree words over a three-letter alphabet; also see [2]. A famous example of such a word can be obtained from the Thue-Morse word as follows: count the number of 1's between two consecutive 0's in $\mathbf{t}$. This gives the so-called *ternary Thue-Morse word*

$$\mathbf{c} = 210201\cdots,$$

© Springer Nature Switzerland AG 2019
R. Mercaş and D. Reidenbach (Eds.): WORDS 2019, LNCS 11682, pp. 133–144, 2019.
https://doi.org/10.1007/978-3-030-28796-2_10

and is squarefree. An alternative description of **c** is as follows [3]: it is the image, under τ of the fixed point of the morphism φ defined below:

$$\varphi(0) = 01 \qquad\qquad \tau(0) = 2$$
$$\varphi(1) = 20 \qquad\qquad \tau(1) = 1$$
$$\varphi(2) = 23 \qquad\qquad \tau(2) = 0$$
$$\varphi(3) = 02 \qquad\qquad \tau(3) = 1$$

A word w is *circularly squarefree* if every one of its conjugates is squarefree. For example, outshout is squarefree, but not circularly squarefree. Clearly we have

Proposition 1. *A word is circularly squarefree iff all its conjugates are unbordered.*

We now turn to a description of what we do in this paper. Using a complicated case-based argument, Currie [6] proved that there exist circularly squarefree ternary words of every length n, except for $\{5, 7, 9, 10, 14, 17\}$. The first of our main results is a new proof of Currie's theorem, based on the following result:

Theorem 1. *For all natural numbers $n > 3$, except $5, 7, 9, 10, 14, 17, 21$, and 28, there exists a factor $x = x(n)$ of the ternary Thue-Morse word **c** that is either*

(a) of length $n - 3$, and $x021$ is circularly squarefree;
(b) of length $n - 4$, and $x2120$ is circularly squarefree.

The virtues of our proof are (a) it requires very little work—just setting up the appropriate logical predicates—and (b) it gives specific examples of the desired words that are very easy to describe and compute. For a completely different approach, which has the virtue of allowing one to give a good estimate for the number of circularly squarefree words of length n, see Shur [11].

We now turn to unbordered conjugates. In two fundamental papers, Harju and Nowotka [7,8] studied the unbordered conjugates of a word. In particular, letting $\mathrm{nuc}(w)$ denote the number of unbordered conjugates of w, and $\mathrm{mnuc}_k(n)$ denote the maximum number of unbordered conjugates of a length-n word over a k-letter alphabet, they proved that

(a) for binary words w of length $n \geq 4$ we have $\mathrm{nuc}(w) \leq n/2$;
(b) for $n > 2$ even, there exists a binary word of length n having $n/2$ unbordered conjugates iff $n = 2^k$ or $n = 3 \cdot 2^k$ for some $k \geq 1$.

In other words, they explicitly computed $\mathrm{mnuc}_2(n)$ for all even n and bounded it above for odd n. We complete the understanding of $\mathrm{mnuc}_2(n)$ by proving that $\mathrm{mnuc}_2(n) = \lfloor n/2 \rfloor$ for all odd $n > 3$. Our strategy is to show that the maximum of $\mathrm{nuc}(w)$, over all words of length n, is actually achieved by a factor of the Thue-Morse word.

More precisely, we prove

Theorem 2. *For all $n \geq 1$, there exists a length-n factor w of the Thue-Morse word **t** with $\mathrm{nuc}(w) = \mathrm{mnuc}_2(n)$. Furthermore, such a factor is guaranteed to occur starting at a position $\leq n$ in **t**.*

2 Circularly Squarefree Ternary Words via Walnut

Since the ternary Thue-Morse word **c** is squarefree, it is reasonable to hope its factors might be a good source of circularly squarefree words. Unfortunately, **c** contains circularly squarefree words of length n for only about $1/8$ of all natural numbers n, as the following result shows.

Theorem 3. *There is a length-n factor of **c** that is circularly squarefree iff $(n)_2$ is accepted by the automaton in Fig. 1.*

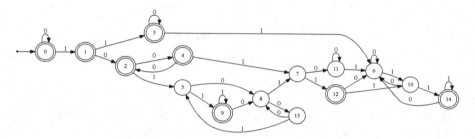

Fig. 1. Automaton accepting lengths $(n)_2$ of circularly squarefree words occurring in **c**

To prove this result, we make use of the fact that many first-order statements concerning claims about k-automatic sequences are decidable [5]. Furthermore, there is free software called `Walnut` available to decide these claims [9].

Let $(n)_k$ denote the canonical base-k representation of n, starting with the most significant digit, having no leading zeroes. A sequence $(a_n)_{n \geq 0}$ is k-*automatic* if there is a deterministic finite automaton with output (DFAO) taking $(n)_k$ as input, and reaching a state with a_n as output. For example, Fig. 2 illustrates the DFAO generating the sequence **c**. The notation q/a in a state means the name of the state is q and the output is a. For more about automatic sequences, see [1].

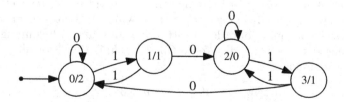

Fig. 2. DFAO computing the sequence **c**

Proof. We can use the ideas in [10], adapted for our case. We create first-order logical predicates crep, facge2, and circsf as follows:

- crep(i, m, p, n, s) evaluates to **true** iff in the length-n word (considered circularly) starting at position s of the word **c**, there is a factor w of length m and (not necessarily least) period $p \geq 1$ starting at position i;
- facge2(n, s) evaluates to **true** iff in the length-n word (considered circularly) starting at position s of the word **c** there is a square or higher power;
- circsf(n) evaluates to **true** iff some length-n factor (considered circularly) of the word **c** has no squares.

$$\text{crep}(i, m, n, p, s) := \exists j \, ((j \geq i) \wedge (j + p < s + n) \wedge (j + p < i + m)) \implies \mathbf{c}[j] = \mathbf{c}[j + p]) \wedge$$
$$(\forall j \, ((j \geq i) \wedge (j < s + n) \wedge (j + p \geq s + n) \wedge (j + p < i + m)) \implies \mathbf{c}[j] = \mathbf{c}[j + p - n]) \wedge$$
$$(\forall j \, ((j \geq i) \wedge (j \geq s + n) \wedge (j + p < i + m)) \implies \mathbf{c}[j - n] = \mathbf{c}[j + p - n])$$

$$\text{facge2}(n, s) := \exists i, m, p \, (p \geq 1) \wedge (m \leq n) \wedge (i \geq s) \wedge (i < s + n) \wedge (m \geq 2p) \wedge \text{crep}(i, m, n, p, s)$$

$$\text{circsf}(n) := \exists s \, \neg \, \text{facge2}(n, s)$$

When we evaluate these predicates in `Walnut`, we get the automaton depicted in Fig. 1. It accepts those $(n)_2$ for which `circsf` evaluates to **true**.

Remark 1. All the `Walnut` code for the theorems in this paper is available at https://cs.uwaterloo.ca/~shallit/papers.html. The reader can therefore verify our results.

Corollary 1. *The number of lengths ℓ, with $2^n \leq \ell < 2^{n+1}$ and $n \geq 4$, such that **c** contains a circularly squarefree factor of length ℓ, is $2^{n-3} - F_{n-3} + 2$, where F_n is the n'th Fibonacci number.*

Proof. By standard techniques, by determining the roots of the characteristic polynomial of the 15×15 matrix encoding transitions of the automaton in Fig. 1.

So while the factors of the ternary Thue-Morse word alone do not suffice for our purpose, it turns out that a small modification of them do. We now give the proof of our first main result, Theorem 1.

Proof (of Theorem 1). Let $n \geq 4$ and $w \in \{x021, y2120\}$, where x, y are factors of the ternary Thue-Morse word **c** of lengths $n - 3$ and $n - 4$, respectively.

First, we create a predicate sq021(i, n, p, s) which evaluates to **true** if $w' := x021x02$ contains a square of order p with $p \geq 1$ and $2p \leq n$ beginning at index $i - s$, where $x = \mathbf{c}[s..s + n - 4]$. We do this by defining $w[j]$ for all j such that $i \leq j < i + p$ as follows:

$$w[j] = \begin{cases} \mathbf{c}[j], & \text{if } j < s + n - 3; \\ 0, & \text{if } j \in \{s + n - 3, \ s + 2n - 3\}; \\ 2, & \text{if } j \in \{s + n - 2, \ s + 2n - 2\}; \\ 1, & \text{if } j = s + n - 1; \\ \mathbf{c}[j - n], & \text{if } s + n \leq j < s + 2n - 3. \end{cases}$$

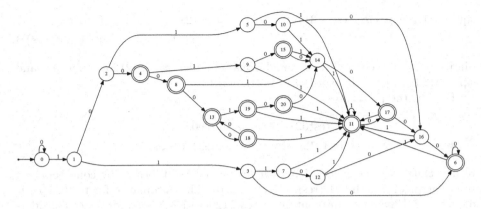

Fig. 3. DFA computing $\exists s$ sqfree021(n, s)

The goal is that sq021 should represent the implication

$$\forall j\left((i \leq j) \wedge (j < i + p)\right) \implies w[j] = w[j + p].$$

It is formed by constructing the conjunction of the predicates

$$\forall j\left((i \leq j) \wedge (j < i + p) \wedge (w[j] = \alpha) \wedge (w[j + p] = \beta)\right) \implies \alpha = \beta$$

for each possible combination j and $j + p$, and simplifying.

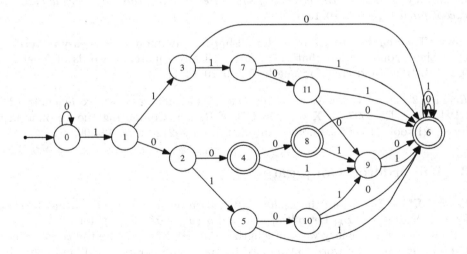

Fig. 4. DFA computing acceptable n

Next, we create a second predicate sqfree021(n, s), which evaluates to **true** if there exists x where $w = x021$ is circularly squarefree, for the given values of n and s:

$$\text{sqfree021}(i, n, p, s) := (n > 3) \wedge (\forall i, p\, ((1 \leq p) \wedge (2p \leq n) \wedge (s \leq i) \wedge (i < s + n))$$
$$\implies \neg(\text{sq021}(i, n, p, s))).$$

Similarly, we create the analogous predicates $\text{sq2120}(i, n, p, s)$ and $\text{sqfree2120}(n, s)$ for the word $w' := y2120y212$.

Finally, the predicates

$$\text{test021}(n) := \exists s\ \text{sqfree021}(n, s)$$
$$\text{test2120}(n) := \exists s\ \text{sqfree2120}(n, s)$$

return **true** if there exists a length-n squarefree word formed by concatenating some factor of **c** with 021 (respectively, 2120). The automaton for $\text{test021}(n)$ is depicted in Fig. 3; the automaton for $\text{test2120}(n)$ is omitted for space considerations.

When we now evaluate the predicate

$$\text{currie}(n) := \text{test021}(n) \vee \text{test2120}(n)$$

with **Walnut**, we get the automaton depicted in Fig. 4.

By inspection we easily see that the automaton in Fig. 4 accepts the base-2 representation of all n except $0, 1, 2, 3, 5, 7, 9, 10, 14, 17, 21, 28$.

As a consequence we now get Currie's theorem:

Corollary 2. *There exist circularly squarefree ternary words of every length n, except for $n \in \{5, 7, 9, 10, 14, 17\}$.*

Proof. Theorem 1 gives the result for all but finitely many n. It is easy to verify by a short computation that there are cyclically squarefree words of lengths $0, 1, 2, 3, 21, 28$, and none for lengths $5, 7, 9, 10, 14, 17$.

Remark 2. These calculations were done in **Walnut** on a Linux machine (2 CPU—Intel E5-2697 v3 Xeon, 256 GB of RAM). Computing the automaton for sq021 took 115.505 s, and the automaton for sq2120 took 124.908 s.

3   Unbordered Conjugates

Let $\sigma : \Sigma_k^* \to \Sigma_k^*$ denote the *cyclic shift function*, where $\sigma(\epsilon) = \epsilon$, $\sigma(cw) = wc$ for $w \in \Sigma_k^*$ and $c \in \Sigma_k$. Let $\sigma^0(w) = w$ and $\sigma^i(w) = \sigma^{i-1}(\sigma(w))$ for $i \geq 1$.

Suppose w is a binary word of length n. Let $\beta : \Sigma_k^* \to \Sigma_k^*$ be the *border correlation function* of a word (introduced by Harju and Nowotka [7]), and defined as follows: $\beta(w) = a_0 a_1 \cdots a_{n-1}$, where

$$a_i = \begin{cases} u, & \text{if } \sigma^i(w) \text{ is unbordered;} \\ b, & \text{if } \sigma^i(w) \text{ is bordered.} \end{cases}$$

For example, $\beta(0001) = ubbu$ since 0001 is unbordered, while 0010, and 0100 are both bordered, and 1000 is unbordered. Let $u, v \in \Sigma_k^*$. We say u is the i'th *cyclic shift* of v if $\sigma^i(v) = u$.

A result from Harju and Nowotka [7] shows that a binary word has no two consecutive cyclic shifts that are unbordered. This result immediately tells us that a binary word of length n can have at most $\lfloor n/2 \rfloor$ unbordered conjugates. For a binary word w of even length to achieve this bound, every other cyclic shift must be unbordered, or, in other words either $\beta(w) = (ub)^{|w|/2}$ or $\beta(w) = (bu)^{|w|/2}$. Harju and Nowotka [7] showed that the only words of even length that achieve this bound are the circularly overlap-free words, which are of length $3 \cdot 2^i$ and 2^i for $i \geq 1$.

Let w be a binary word. Suppose w is of even length and is not circularly overlap-free. Clearly w cannot have $|w|/2$ unbordered conjugates, but it could potentially have $|w|/2 - 1$ unbordered conjugates. Then $\beta(w) = (ub)^i b(ub)^{|w|/2-i-1} b$ for some $i \geq 0$, up to conjugation. Now suppose w is of odd length. No circularly overlap-free words exist of odd length, so it makes sense to think that w could contain a maximum of $\lfloor |w|/2 \rfloor$ unbordered conjugates. Then $\beta(w) = (ub)^{\lfloor |w|/2 \rfloor} b$, up to conjugation.

Let w be a bordered binary word. Then $w = uvu$ for some words u and v. By the *left border* of w we mean the occurrence of u that begins at position 1 of w, and by the *right border* we mean the occurrence of u that begins at position $|w| - |u| + 1$ of w.

Now we prove Theorem 2.

Proof. When $n = 1, 2, 3$ the maximum number of unbordered conjugates $\mathrm{mnuc}_2(n)$ is achieved by the words 0, 01, and 011 respectively. Specifically we have that $\mathrm{mnuc}_2(1) = 1$, $\mathrm{mnuc}_2(2) = 2$, and $\mathrm{mnuc}_2(3) = 2$. It is readily verified that each of these words occur as a factor of the Thue-Morse word at position $\leq n$.

Let w be a length-n word at position m of the Thue-Morse word. The first step is to create a first-order predicate isBorder(l, m, n) that asserts that a cyclic shift of w has a border of a certain length. More specifically, we want to know whether the l'th cyclic shift of w has a border of length k. There are three cases to consider.

1. When a prefix of the right border is a suffix of w and a suffix of the right border is a prefix of w. In other words, $w = yuvx$ for words u, v, x, y where $xy = u$, $|y| = l$, and $|u| = k$. This predicate is denoted by isBorderC1(k, l, m, n).
2. When both borders are completely contained inside of w. In other words, $w = yuux$ for words y, u, x where $|yu| = l$, and $|u| = k$. This predicate is denoted by isBorderC2(k, l, m, n).
3. When a prefix of the left border is a suffix of w and a suffix of the left border is a prefix of w. In other words, $w = yvux$ for words u, v, x, y where $xy = u$, $|yvu| = l$, and $|u| = k$. This predicate is denoted by isBorderC3(k, l, m, n).

$\text{isBorderC1}(k, l, m, n) := ((k + l > n) \Rightarrow ((\forall i(i < n - l) \Rightarrow T[m + l + i] = T[m + l - k + i])$
$\qquad \land (\forall i(i < k + l - n) \Rightarrow T[m + i] = T[m + n - k + i])))$

$\text{isBorderC2}(k, l, m, n) := (((k + l \leq n) \land (l \geq k)) \Rightarrow (\forall i (i < k) \Rightarrow$
$\qquad T[m + l + i] = T[m + l - k + i]))$

$\text{isBorderC3}(k, l, m, n) := (((k + l \leq n) \land (l < k)) \Rightarrow ((\forall i (i < k - l) \Rightarrow T[m + n - k + l + i]$
$\qquad = T[m + l + i]) \land (\forall i (i < l) \Rightarrow T[m + i] = T[m + k + i])))$

$\text{isBorder}(k, l, m, n) := \text{isBorderC1}(k, l, m, n) \land \text{isBorderC2}(k, l, m, n) \land \text{isBorderC3}(k, l, m, n).$

We define the predicate isBordered(l, m, n) that asserts that the l'th cyclic shift of a length-n word at position m in the Thue-Morse word is bordered. We can create this predicate by checking whether this word has a border of size $\leq n/2$.

$$\text{isBordered}(l, m, n) := \exists i(2i \leq n \land i \geq 1 \land \text{isBorder}(i, l, m, n)).$$

Recall that when $|w|$ is odd and w has a maximum number of unbordered conjugates, we have that $\beta(w) = (ub)^{\lfloor |w|/2 \rfloor} b$, up to conjugation. So we have exactly one pair of adjacent bordered cyclic shifts, and the rest of the cyclic shifts of w alternate between bordered and unbordered. The predicate isAlternating0(l, m, n) asserts that all of the cyclic shifts of a length-n word at position m in the Thue-Morse word alternate between unbordered and bordered, except for the l'th and $l + 1$'th cyclic shifts, which are both bordered.

$\text{isAlternating0}(l, m, n) :=$
$\forall i(((i \neq l \land i < n - 1) \Rightarrow (\text{isBordered}(i, m, n) = \neg \text{isBordered}(i + 1, m, n)))) \land$
$(((i \neq l) \land (i = n - 1)) \Rightarrow (\text{isBordered}(n - 1, m, n) = \neg \text{isBordered}(0, m, n))).$

Now we create a predicate hasMNUCO(m, n) that asserts that a length-n word at position m in the Thue-Morse word achieves the maximum number of unbordered conjugates.

$\text{hasMNUCO}(m, n) := \exists i(((i < n - 1 \land \text{isBordered}(i, m, n) \land \text{isBordered}(i + 1, m, n)) \lor$
$(i = n - 1 \land \text{isBordered}(n - 1, m, n) \land \text{isBordered}(0, m, n))) \land \text{isAlternating0}(i, m, n)).$

Similarly, recall that when $|w|$ is even and w has a maximum number of unbordered conjugates, we have that $\beta(w) = (ub)^i b(ub)^{|w|/2 - i - 1} b$ for some $i \geq 0$ or $\beta(w) = (ub)^{|w|/2}$, up to conjugation. So we have that either all of the cyclic shifts of w alternate between bordered and unbordered, or there are exactly two pairs of adjacent bordered cyclic shifts, and the rest of the cyclic shifts of w alternate between bordered and unbordered. The predicate

$$\text{isAlternatingE}(e, l, m, n)$$

asserts that all of the cyclic shifts of a length-n word at position m in the Thue-Morse word alternate between unbordered and bordered, except for the

l'th, $l + 1$'th, e'th, and $e + 1$'th cyclic shifts, which are all bordered. Note that isAlternatingE(n, n, m, n) asserts that all of the cyclic shifts of a length n word at position m in the Thue-Morse word alternate between unbordered and bordered.

$$\text{isAlternatingE}(e, l, m, n) := (\forall i \, (((i \neq l \wedge i \neq e \wedge i < n - 1) \Rightarrow (\text{isBordered}(i, m, n) \Leftrightarrow$$
$$\neg \, \text{isBordered}(i + 1, m, n)))) \wedge (((i \neq l) \wedge (i \neq e) \wedge (i = n - 1)) \Rightarrow$$
$$(\text{isBordered}(n - 1, m, n) \Leftrightarrow \neg \, \text{isBordered}(0, m, n))))$$

Now we create a predicate hasMNUCE(m, n) that asserts that a length-n word at position m in the Thue-Morse word achieves the maximum number of unbordered conjugates.

$$\text{hasMNUCE}(m, n) := (\exists i, j \, ((i < j) \wedge (i < n - 1 \wedge \text{isBordered}(i, m, n) \wedge \text{isBordered}(i + 1, m, n)) \wedge$$
$$((j = n - 1 \wedge \text{isBordered}(n - 1, m, n) \wedge \text{isBordered}(0, m, n)) \vee ((j < n - 1) \wedge$$
$$\text{isBordered}(j, m, n) \wedge \text{isBordered}(j + 1, m, n))) \wedge \text{isAlternatingE}(i, j, m, n))) \vee$$
$$\text{isAlternatingE}(n, n, m, n).$$

With these predicates we can write a predicate asserting that the Thue-Morse word contains factors of every length $n > 3$ that are maximally unbordered and occur at position $\leq n$. We split the computation into cases, one for even length words, and one for odd:

$$\forall n \, ((n \geq 2) \implies (\exists i \, \text{hasMNUCE}(i, 2n)) \wedge i \leq 2n)$$
$$\forall n \, ((n \geq 2) \implies (\exists i \, \text{hasMNUCO}(i, 2n + 1)) \wedge i \leq 2n + 1),$$

and `Walnut` evaluates these predicates to be true.

Thus we have that

$$\text{mnuc}_2(n) = \begin{cases} 1, & \text{if } n = 1; \\ 2, & \text{if } n = 2 \text{ or } n = 3; \\ n/2, & \text{if } n \in \{2^{i+1}, 3 \cdot 2^i : i \geq 1\}; \\ n/2 - 1, & \text{if } n > 3 \text{ even and } n \notin \{2^i, 3 \cdot 2^i : i \geq 1\}; \\ \lfloor n/2 \rfloor, & \text{if } n > 3 \text{ odd.} \end{cases}$$

As a corollary, we easily get the following.

Corollary 3. *Let $f(n) = \text{mnuc}_2(n) - \lfloor n/2 \rfloor$. Then f is a 2-automatic sequence.*

4   More About Unbordered Conjugates

In this section we show that there exist binary words of length n that have exactly i unbordered conjugates where $1 < i \leq \text{mnuc}_2(n)$.

The general idea behind the proof is to pick some $i > 1$ and then pick a word w of odd length such that $\text{nuc}(w) = i$ and $\text{mnuc}_2(|w|) = i$. Furthermore we only consider such words w such that one of w's conjugates contain 000 as a factor. Then we keep adding 0's to w precisely where 000 first occurs. This keeps the number of unbordered conjugates the same. Then we can keep increasing the size of w in this way until we hit the length we want.

Lemma 1. *For* $n > 4$ *odd, there exists a word* $w \in \Sigma_2^n$ *such that* $\mathrm{nuc}(w) = \lfloor n/2 \rfloor$ *and* 000 *is a factor of some conjugate of* w.

Proof. By Theorem 2, such a word w exists as a factor of the Thue-Morse word. It is well known that the Thue-Morse word is overlap-free. So 000 cannot be a factor of such a word w. But it is possible that $w = 0u00$, or $w = 00u0$ for some word u. We can check whether this is the case for all odd $n > 4$ by modifying our predicate from the proof of Theorem 2:

$$\forall n \, ((n \geq 2) \implies (\exists i \; \mathrm{hasMNUCO}(i, 2n+1)) \wedge ((T[i] = 0 \wedge T[i+1] = 0 \wedge T[2n+i] = 0)$$
$$\vee \, (T[i] = 0 \wedge T[2n-1+i] = 0 \wedge T[2n+i] = 0))),$$

which evaluates to true.

Lemma 2. *Let* $n > 4$ *be odd and* w *be a binary word of length* n *such that a conjugate of* w *has* 000 *as a factor and* $\mathrm{nuc}(w) = \lfloor n/2 \rfloor$. *Then every conjugate of* w *contains at most one distinct occurrence of* 000 *as a factor.*

Proof. Suppose, contrary to what we want to prove that a conjugate of w contains at least two distinct occurrences of 000 as a factor. Call this conjugate w'.

If the two occurrences of 000 overlap, then we can write $w' = s0000t$ for some words s, t. Then the cyclic shifts $0ts000$, $00ts00$, and $0ts000$ are bordered. This means that only $\lfloor |ts|/2 \rfloor + 1$ of the remaining cyclic shifts of w can be unbordered since any unbordered cyclic shift must be followed by a bordered one. But $\lfloor |ts|/2 \rfloor + 1 = \lfloor (n-4)/2 \rfloor + 1 < \lfloor n/2 \rfloor$, so the two occurrences of 000 cannot overlap.

If the two occurrences of 000 do not overlap, then we can write $w' = s000t000$ for some words s, t where s, and t are non-empty. Then the conjugates $00t000s0$, $0t000s00$, $00s000t0$, and $0s000t00$ are bordered. By the same argument as above, of the remaining cyclic shifts, a maximum of $\lfloor |st|/2 \rfloor + 2$ of them can be unbordered. But $\lfloor |st|/2 \rfloor + 2 = \lfloor (n-6)/2) \rfloor + 2 < \lfloor n/2 \rfloor$, a contradiction.

Lemma 3. *Let* $n > 4$ *be odd and* w *be a binary word of length* n *such that a conjugate* w' *of* w *has* 000 *as a prefix and* $\mathrm{nuc}(w) = \lfloor n/2 \rfloor$. *Then* $\mathrm{nuc}(w) = \mathrm{nuc}(w') = \mathrm{nuc}(0^i w')$ *for all* $i \geq 0$.

Proof. Let $i \geq 0$ be an integer. We can write $w' = 000u$ for some word u. It is clear that $0^j u 0^{i+3-j}$ is bordered for all $1 \leq j \leq i+2$. Therefore, it suffices to prove that $s000t$ is bordered if and only if $s0^{i+3}t$ is bordered where $u = ts$.

First we prove the forward direction. Suppose $s000t$ is bordered. By Lemma 2 we have that $s000t$ contains only one occurrence of 000 as a factor. So 000 is neither a prefix of $s00$ nor a suffix of $00t$. Thus, any border of $s000t$ must of length $\leq \min\{|s|, |t|\} + 2$. But such a border would also be a border of $s0^{i+3}t$.

A similar argument works for the reverse direction. Therefore $\mathrm{nuc}(w) = \mathrm{nuc}(w') = \mathrm{nuc}(0^i w')$ for all $i \geq 0$.

Theorem 4. *For all $1 < i \leq \mathrm{mnuc}_k(n)$ there exists $w \in \Sigma_k^n$ such that $\mathrm{nuc}(w) = i$.*

Proof. Let $C = \{5, 7, 9, 10, 14, 17\}$. For $k \geq 4$, Harju and Nowotka [8] showed that for all integers i with $1 < i \leq n$ there exists a word $w \in \Sigma_k^n$ such that $\mathrm{nuc}(w) = i$. For $k = 3$, Harju and Nowotka [8] showed that if $n \notin C$ then for all integers i with $1 < i \leq n$ there exists a word $w \in \Sigma_k^n$ such that $\mathrm{nuc}(w) = i$, and if $n \in C$ then for all integers i with $1 < i < n$ there exists a word $w \in \Sigma_k^n$ such that $\mathrm{nuc}(w) = i$.

To the best of the authors' knowledge, there is no known proof of the existence of such words for $k = 2$. Suppose $k = 2$. By Theorem 2 there exists a $w \in \Sigma_2^n$ such that w is a factor of the Thue-Morse word and $\mathrm{mnuc}_2(n) = \mathrm{nuc}(w)$. So assume $i < \mathrm{mnuc}_2(n)$. By Lemma 1 there exists a binary word u of odd length m such that $\mathrm{nuc}(u) = i = \lfloor m/2 \rfloor$ and 000 is a factor of some conjugate of u. Let u' be the conjugate of u such that 000 is a prefix of u'. Lemma 3 tells us $\mathrm{nuc}(u) = \mathrm{nuc}(u') = \mathrm{nuc}(0^{n-m} u')$. Since $\mathrm{nuc}(0^{n-m} u') = i$ and $|0^{n-m} u'| = n$, we have that for all $1 < i \leq \mathrm{mnuc}_2(n)$, there exists a $w \in \Sigma_2^n$ such that $\mathrm{nuc}(w) = i$.

5 Conclusions

We want to emphasize that our experience shows that rephrasing problems in combinatorics on words using the first-order logical theory of automatic sequences can be a useful tool in solving these problems. We encourage others to adopt this approach.

Acknowledgments. We thank Dirk Nowotka for helpful discussions. We are very grateful to the referees for helpful comments, and for reminding us about the paper [11].

References

1. Allouche, J.P., Shallit, J.: Automatic Sequences: Theory, Applications, Generalizations. Cambridge University Press, Cambridge (2003)
2. Allouche, J.P., Shallit, J.O.: The ubiquitous Prouhet-Thue-Morse sequence. In: Ding, C., Helleseth, T., Niederreiter, H. (eds.) Sequences and Their Applications. DISCMATH, pp. 1–16. Springer, London (1999)
3. Berstel, J.: Sur la construction de mots sans carré. Séminaire de Théorie des Nombres, pp. 18.01–18.15 (1978–1979)
4. Berstel, J.: Axel Thue's Papers on Repetitions in Words: a Translation. No. 20 in Publications du Laboratoire de Combinatoire et d'Informatique Mathématique, Université du Québec à Montréal, February 1995

5. Bruyère, V., Hansel, G., Michaux, C., Villemaire, R.: Logic and p-recognizable sets of integers. Bull. Belgian Math. Soc. **1**, 191–238 (1994). corrigendum, Bull. Belg. Math. Soc. **1**, 577 (1994)
6. Currie, J.: There are ternary circular square-free words of length n for $n \geq 18$. Electron. J. Combin. **9**, #N10 (2002). https://www.combinatorics.org/v9i1n10
7. Harju, T., Nowotka, D.: Border correlation of binary words. J. Combin. Theory Ser. A **108**, 331–341 (2004)
8. Harju, T., Nowotka, D.: Bordered conjugates of words over large alphabets. Electron. J. Combin. **15**, #N41 (2008). https://www.combinatorics.org/ojs/index. php/eljc/article/view/v15i1n41
9. Mousavi, H.: Automatic theorem proving in Walnut (2016), arxiv preprint arXiv:1603.06017. Software available at https://github.com/hamousavi/Walnut
10. Shallit, J., Zarifi, R.: Circular critical exponents for Thue-Morse factors. RAIRO Inform. Théor. App. **53**, 37–49 (2019)
11. Shur, A.M.: On ternary square-free circular words. Electron. J. Combin. **17**, #R140 (2010). https://www.combinatorics.org/ojs/index.php/eljc/article/view/ v17i1r140
12. Thue, A.: Über unendliche Zeichenreihen. Norske vid. Selsk. Skr. Mat. Nat. Kl. **7**, 1–22 (1906). reprinted in Selected Mathematical Papers of Axel Thue, T. Nagell, editor, Universitetsforlaget, Oslo, pp. 139–158 (1977)
13. Thue, A.: Über die gegenseitige Lage gleicher Teile gewisser Zeichenreihen. Norske vid. Selsk. Skr. Mat. Nat. Kl. **1**, 1–67 (1912). reprinted in Selected Mathematical Papers of Axel Thue, T. Nagell, editor, Universitetsforlaget, Oslo, pp. 413–478 (1977)

The Undirected Repetition Threshold

James D. Currie[iD] and Lucas Mol[(⊠)][iD]

The University of Winnipeg, Winnipeg, MB R3B 2E9, Canada
{j.currie,l.mol}@uwinnipeg.ca

Abstract. For rational $1 < r \leq 2$, an *undirected r-power* is a word of the form xyx', where x is nonempty, $x' \in \{x, x^R\}$, and $|xyx'|/|xy| = r$. The *undirected repetition threshold* for k letters, denoted $\mathrm{URT}(k)$, is the infimum of the set of all r such that undirected r-powers are avoidable on k letters. We first demonstrate that $\mathrm{URT}(3) = \frac{7}{4}$. Then we show that $\mathrm{URT}(k) \geq \frac{k-1}{k-2}$ for all $k \geq 4$. We conjecture that $\mathrm{URT}(k) = \frac{k-1}{k-2}$ for all $k \geq 4$, and we confirm this conjecture for $k \in \{4, 8, 12\}$.

Keywords: Repetition thresholds · Gapped repeats ·
Gapped palindromes · Pattern avoidance · Patterns with reversal

1 Introduction

A *square* is a word of the form xx, where x is a nonempty word. An *Abelian square* is a word of the form $x\tilde{x}$, where $\tilde{x}$ is an anagram (or permutation) of x. The notions of square and Abelian square can be extended to fractional powers in a natural way. Let $1 < r \leq 2$ be a rational number. An *(ordinary) r-power* is a word of the form xyx, where x is a nonempty word, and $|xyx|/|xy| = r$. An *Abelian r-power* is a word of the form $xy\tilde{x}$, where x is a nonempty word, $\tilde{x}$ is an anagram of x, and $|xy\tilde{x}|/|xy| = r$.[1]

In general, if $\sim$ is an equivalence relation on words that respects length (i.e., we have $|x| = |x'|$ whenever $x \sim x'$), then an *r-power up to* $\sim$ is a word of the form xyx', where x is nonempty, $x \sim x'$, and $|xyx'|/|xy| = r$. The notion of r-power up to $\sim$ generalizes ordinary r-powers and Abelian r-powers, where the equivalence relations are equality and "is an anagram of", respectively.

Let $\sim$ be an equivalence relation on words that respects length. For a real number $1 < \alpha \leq 2$, a word w is called *α-free up to* $\sim$ if no factor of w is an r-power up to $\sim$ for $r \geq \alpha$. Moreover, the word w is called *α⁺-free up to* $\sim$ if no factor of w is an r-power up to $\sim$ for $r > \alpha$. For every integer $k \geq 2$, we say that α-powers up to $\sim$ are *k-avoidable* if there is an infinite word on k letters

---

[1] We use the definition of Abelian r-power of Cassaigne and Currie [5]. We note that several distinct definitions exist (see [17,28], for example).

J. D. Currie—Supported by the Natural Sciences and Engineering Research Council of Canada (NSERC), [funding reference number 2017-03901].

R. Merçaş and D. Reidenbach (Eds.): WORDS 2019, LNCS 11682, pp. 145–158, 2019.
https://doi.org/10.1007/978-3-030-28796-2_11

that is α-free up to $\sim$, and k-*unavoidable* otherwise. For every integer $k \geq 2$, the *repetition threshold up to* $\sim$ for k letters, denoted $\mathrm{RT}_\sim(k)$, is defined as

$$\mathrm{RT}_\sim(k) = \inf\{r \colon r\text{-powers up to } \sim \text{ are } k\text{-avoidable}\}.$$

Since we have only defined r-powers for $r \leq 2$, it follows that $\mathrm{RT}_\sim(k) \leq 2$ or $\mathrm{RT}_\sim(k) = \infty$ for any particular value of k.

It is well-known that squares are 3-avoidable [1]. Thus, for $k \geq 3$, we have that $RT_=(k)$ is the usual *repetition threshold*, denoted simply $\mathrm{RT}(k)$. Dejean [14] proved that $\mathrm{RT}(3) = 7/4$, and conjectured that $\mathrm{RT}(4) = 7/5$ and $\mathrm{RT}(k) = k/(k-1)$ for all $k \geq 5$. This conjecture has been confirmed through the work of many authors [3,12–14,22,23,25,26].

It is also known that Abelian squares are 4-avoidable [21]. Let $\approx$ denote the equivalence relation "is an anagram of". Thus, for all $k \geq 4$, we see that $\mathrm{RT}_\approx(k)$ is equal to the *Abelian repetition threshold* (or *commutative repetition threshold*) for k letters, introduced by Cassaigne and Currie [5], and denoted $\mathrm{ART}(k)$. Relatively less is known about the Abelian repetition threshold. Cassaigne and Currie [5] give (weak) upper bounds on $\mathrm{ART}(k)$ in demonstrating that $\lim_{k\to\infty} \mathrm{ART}(k) = 1$. Samsonov and Shur [28] conjecture that $\mathrm{ART}(4) = 9/5$ and $\mathrm{ART}(k) = (k-2)/(k-3)$ for all $k \geq 5$, and give a lower bound matching this conjecture.[2]

For every word $x = x_1 x_2 \cdots x_n$, where the x_i are letters, we let x^R denote the *reversal of* x, defined by $x^R = x_n \cdots x_2 x_1$. For example, if $x = \texttt{time}$ then $x^R = \texttt{emit}$. Let $\simeq$ be the equivalence relation on words defined by $x \simeq x'$ if $x' = x$ or $x' = x^R$. In this article, we focus on determining $\mathrm{RT}_\simeq(k)$. We simplify our notation and terminology as follows. We refer to r-powers up to $\simeq$ as *undirected r-powers*. These come in two types: words of the form xyx are ordinary r-powers, while we refer to words of the from xyx^R as *reverse r-powers*.[3] For example, the English words $\texttt{edited}$ and $\texttt{render}$ are undirected $\frac{3}{2}$-powers; $\texttt{edited}$ is an ordinary $\frac{3}{2}$-power, while $\texttt{render}$ is a reverse r-power.

We say that a word w is *undirected α-free* if it is α-free up to $\simeq$. The definition of an *undirected α^+-free* word is analogous. We let $\mathrm{URT}(k) = \mathrm{RT}_\simeq(k)$, and refer to this as the *undirected repetition threshold* for k letters.

---

[2] Samsonov and Shur define weak, semi-strong, and strong Abelian α-powers for all real numbers $\alpha > 1$. For rational $1 < r \leq 2$, their definitions of semi-strong Abelian r-power and strong Abelian r-power are both equivalent to our definition of Abelian r-power.

[3] We note that words of the form xyx are sometimes referred to as *gapped repeats*, and that words of the form xyx^R are sometimes referred to as *gapped palindromes*. In particular, an ordinary (reverse, respectively) r-power satisfying $r \geq 1 + 1/\alpha$ is called an α-*gapped repeat* (α-*gapped palindrome*, respectively). Algorithmic questions concerning the identification and enumeration of α-gapped repeats and palindromes in a given word, along with some related questions, have recently received considerable attention; see [6,16,19,20] and the references therein. Gapped repeats and palindromes are important in the context of DNA and RNA structures, and this has been the primary motivation for their study.

It is clear that $\simeq$ is coarser than $=$ and finer than $\approx$. Thus, for every rational $1 < r \leq 2$, an r-power is an undirected r-power, and an undirected r-power is an Abelian r-power. As a result, we immediately have

$$\mathrm{RT}(k) \leq \mathrm{URT}(k) \leq \mathrm{ART}(k)$$

for all $k \geq 2$.

Since only a weak upper bound on $\mathrm{ART}(k)$ is currently known, we provide an alternate upper bound on $\mathrm{URT}(k)$ for large enough k. For words $u = u_0 u_1 \cdots$ and $v = v_0 v_1 \cdots$ of the same length (possibly infinite) over alphabets A and B, respectively, the *direct product* of u and v, denoted $u \otimes v$, is the word on alphabet $A \times B$ defined by

$$u \otimes v = (u_0, v_0)(u_1, v_1) \cdots .$$

A word x is called a *reversible factor* of w if both x and x^R are factors of w.

Theorem 1. *For every* $k \geq 9$, *we have* $\mathrm{URT}(k) \leq \mathrm{RT}(\lfloor k/3 \rfloor)$.

Proof. Fix $k \geq 9$, and let $\ell = \lfloor k/3 \rfloor$. Evidently, we have $\ell \geq 3$ and thus $\mathrm{RT}(\ell) < 2$. Let $\boldsymbol{u}$ be an infinite $\mathrm{RT}(\ell)^+$-free word on ℓ letters. We claim that the word $\boldsymbol{u} \otimes (123)^\omega$ on $3\ell < k$ letters is undirected $\mathrm{RT}(\ell)^+$-free, from which the theorem follows. Since the only reversible factors of $(123)^\omega$ have length at most 1, any reverse r-power xyx^R in $\boldsymbol{u} \otimes (123)^\omega$ satisfies $x = x^R$, and hence is an ordinary r-power as well. Since $\boldsymbol{u}$ is ordinary $\mathrm{RT}(\ell)^+$-free, so is $\boldsymbol{u} \otimes (123)^\omega$. This completes the proof of the claim. □

We now describe the layout of the remainder of the article. In Sect. 2, we discuss related problems in pattern avoidance, and give some implications of our main results in that setting. In Sect. 3, we show that $\mathrm{URT}(3) = 7/4$ using a standard morphic construction. In Sect. 4, we demonstrate that $\mathrm{URT}(k) \geq (k-1)/(k-2)$ for all $k \geq 4$. In Sect. 5, we use a variation of the encoding introduced by Pansiot [25] to prove that $\mathrm{URT}(k) = (k-1)/(k-2)$ for $k \in \{4, 8, 12\}$. In light of our results, we propose the following.

Conjecture 2. For all $k \geq 4$, we have $\mathrm{URT}(k) = (k-1)/(k-2)$.

We briefly place this conjecture in context. We know that $\mathrm{RT}(k) = k/(k-1)$ for all $k \geq 5$, we conjecture that $\mathrm{URT}(k) = (k-1)/(k-2)$ for all $k \geq 4$, and Samsonov and Shur [28] conjecture that $\mathrm{ART}(k) = (k-1)/(k-2)$ for all $k \geq 5$. Let us fix $k \geq 5$. In [29], Shur proposes splitting all exponents greater than $\mathrm{RT}(k)$ into levels as follows[4]:

$$
\begin{array}{cccc}
\text{1st level} & \text{2nd level} & \text{3rd level} & \cdots \\
\left[\frac{k}{k-1}^+, \frac{k-1}{k-2}\right] & \left[\frac{k-1}{k-2}^+, \frac{k-2}{k-3}\right] & \left[\frac{k-2}{k-3}^+, \frac{k-3}{k-4}\right] & \cdots
\end{array}
$$

---

[4] Shur considers exponents belonging to the "extended rationals". This set includes all rational numbers and all such numbers with a $+$, where x^+ covers x, and the inequalities $y \leq x$ and $y < x^+$ are equivalent.

For $\alpha, \beta \in \left[\frac{k}{k-1}^+, \frac{k-3}{k-4}\right]$, Shur provides evidence that the language of α-free k-ary words and the language of β-free k-ary words exhibit similar behaviour (e.g., with respect to growth) if α and β are in the same level, and quite different behaviour otherwise; see [29,30]. If the conjectured values of $\mathrm{URT}(k)$ and $\mathrm{ART}(k)$ are correct, then the undirected repetition threshold and the Abelian repetition threshold provide further evidence of the distinction between levels.

We now introduce some terminology that will be used in the sequel. Let A and B be alphabets, and let $h\colon A^* \to B^*$ be a morphism. Using the standard notation for images of sets, we have $h(A) = \{h(a)\colon a \in A\}$, which we refer to as the set of *blocks* of h. A set of words $P \subseteq A^*$ is called a *prefix code* if no element of P is a prefix of another. If P is a prefix code and w is a nonempty factor of some element of P^+, a *cut* of w over P is a pair (x, y) such that (i) $w = xy$; and (ii) for every pair of words p, s with $pws \in P^+$, we have $px \in P^*$. We use vertical bars to denote cuts. For example, over the prefix code $\{01, 10\}$, the word 11 has cut $1|1$. The prefix code that we work over will always be the set of blocks of a given morphism, and should be clear from context.

2 Related Problems in Pattern Avoidance

Let $p = p_1 p_2 \cdots p_n$ be a word over alphabet V, where the p_i are letters called *variables*. In this context, the word p is called a *pattern*. If $\sim$ is an equivalence relation on words, then we say that the word w *encounters* p *up to* $\sim$ if w contains a factor of the form $X_1 X_2 \cdots X_n$, where each word X_i is nonempty and $X_i \sim X_j$ whenever $p_i = p_j$. Otherwise, we say that w *avoids* p *up to* $\sim$. A pattern p is k-*avoidable up to* $\sim$ if there is an infinite word on a k-letter alphabet that avoids p up to $\sim$. Otherwise, the pattern p is k-*unavoidable up to* $\sim$. Finally, the pattern p is *avoidable up to* $\sim$ if it is k-avoidable for some k, and *unavoidable up to* $\sim$ otherwise.

When $\sim$ is equality, we recover the ordinary notion of *pattern avoidance* (see [4]). When $\sim$ is $\approx$ (i.e., "is an anagram of"), we recover the notion of *Abelian pattern avoidance* (see [8,9,27], for example). One could also explore pattern avoidance up to $\simeq$, or *undirected pattern avoidance*. We discuss some initial results in this direction. While there are patterns that are avoidable in the ordinary sense but not in the Abelian sense [8, Lemma 3], every avoidable pattern is in fact avoidable up to $\simeq$, as we show below.

Theorem 3. *Let p be a pattern. Then p is avoidable in the ordinary sense if and only if p is avoidable up to $\simeq$.*

Proof. If p is unavoidable in the ordinary sense, then clearly p is unavoidable up to $\simeq$. If p is avoidable in the ordinary sense, then let $\boldsymbol{u}$ be an ω-word avoiding p. The direct product $\boldsymbol{u} \otimes (123)^\omega$ avoids p up to $\simeq$ by an argument similar to the one used in Theorem 1. $\qquad\square$

Questions concerning the k-avoidability of patterns up to $\simeq$ appear to be more interesting. The *avoidability index* of a pattern p up to $\sim$, denoted $\lambda_\sim(p)$, is the least positive integer k such that p is k-avoidable up to $\sim$, or ∞ if p is unavoidable. In general, for any pattern p, we have

$$\lambda_=(p) \leq \lambda_\simeq(p) \leq \lambda_\approx(p).$$

The construction of Theorem 3 can be used to show that $\lambda_\simeq(p) \leq 3\lambda_=(p)$, though we suspect that this bound is not tight.

The study of the undirected repetition threshold will have immediate implications on avoiding patterns up to $\simeq$. For example, we can easily resolve the avoidability index of unary patterns up to $\simeq$ using known results along with a result proven later in this article.

Theorem 4. $\lambda_\simeq(x^k) = \begin{cases} 3, & \text{if } k \in \{2,3\}; \\ 2, & \text{if } k \geq 4. \end{cases}$

Proof. We prove that $\mathrm{URT}(3) = 7/4$ in Sect. 3, from which it follows that $\lambda_\simeq(xx) = 3$. Backtracking by computer, one finds that the longest binary word avoiding xxx in the undirected sense has length 9, so $\lambda_\simeq(xxx) \geq 3$. Since $\lambda_\approx(xxx) = 3$ [15], we conclude that $\lambda_\simeq(xxx) = 3$. Finally, since $\lambda_\approx(x^4) = 2$ [15], we have $\lambda_\simeq(x^k) = 2$ for all $k \geq 4$. $\qquad\square$

We plan to determine the avoidability index of all binary patterns up to $\simeq$ in a future work.

Finally, we remark that the study of k-avoidability of patterns up to $\simeq$ has implications for k-avoidability of *patterns with reversal* (see [7,10,11] for definitions and examples). In particular, if pattern p is k-avoidable up to $\simeq$, then all patterns with reversal that are obtained by swapping any number of letters in p with their mirror images are *simultaneously k-avoidable*; that is, there is an infinite word on k letters avoiding all such "decorations" of p.

3 URT$(3) = \frac{7}{4}$

Dejean [14] demonstrated that $\mathrm{RT}(3) = 7/4$, and hence we must have $\mathrm{URT}(3) \geq 7/4$. In order to show that $\mathrm{URT}(3) = 7/4$, it suffices to find an infinite ternary word that is undirected $\frac{7}{4}^+$-free. We provide a morphic construction of such a word. Let f be the 24-uniform morphism defined by

$$0 \mapsto 012\,021\,201\,021\,012\,102\,120\,210$$
$$1 \mapsto 120\,102\,012\,102\,120\,210\,201\,021$$
$$2 \mapsto 201\,210\,120\,210\,201\,021\,012\,102.$$

The morphism f is similar in structure to the morphism of Dejean [14] whose fixed point avoids ordinary $7/4^+$-powers (but not undirected $7/4^+$-powers). Note, in particular, that f is "symmetric" in the sense of [18].

The following theorem was verified by one of the anonymous reviewers using the automatic theorem proving software `Walnut` [24]. In light of this fact (and in order to meet space constraints), we omit our original proof.

Theorem 5. *The word $f^\omega(0)$ is undirected $\frac{7}{4}^+$-free.*

Thus, we conclude that $\mathrm{URT}(3) = \mathrm{RT}(3) = \frac{7}{4}$. We will see in the next section that $\mathrm{URT}(k)$ is strictly greater than $\mathrm{RT}(k)$ for every $k \geq 4$.

4   A Lower Bound on URT(k) for $k \geq 4$

Here, we prove that $\mathrm{URT}(k) \geq (k-1)/(k-2)$ for $k \geq 4$.

Theorem 6. *If $k \geq 4$, then $\mathrm{URT}(k) \geq \frac{k-1}{k-2}$, and the longest k-ary word that is undirected $(k-1)/(k-2)$-free has length $k+3$.*

Proof. For $k \in \{4, 5\}$, the statement is checked by a standard backtracking algorithm, which we performed by computer. We now provide a general backtracking argument for all $k \geq 6$.

Fix $k \geq 6$, and suppose that w is a k-ary word of length $k + 4$ that is undirected $(k-1)/(k-2)$-free. It follows that at least $k - 2$ letters must appear between any two repeated occurrences of the same letter in w, so that any length $k - 1$ factor of w must contain $k - 1$ distinct letters. So we may assume that w has prefix $12\cdots(\mathtt{k\text{-}1})$. Further, given any prefix u of w of length at least $k - 1$, there are only two possibilities for the next letter in w, as it must be distinct from the $k - 2$ distinct letters preceding it. These possibilities are enumerated in the tree of Fig. 1.

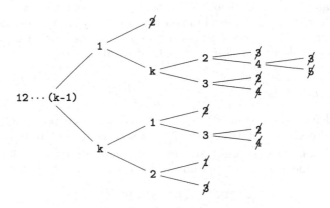

Fig. 1. The tree of undirected $(k-1)/(k-2)$-power free words on k letters.

We now explain why each word corresponding to a leaf of the tree contains an undirected r-power for some $r \geq (k-1)/(k-2)$. We examine the leaves from top to bottom, and use the fact that $(k+1)/(k-1) > (k+2)/k > (k-1)/(k-2)$ when $k \geq 6$.

– The factor $12\cdots(\mathtt{k\text{-}1})\mathtt{12}$ is an ordinary $(k+1)/(k-1)$-power.

- The factor $23\cdots(\text{k-1})\text{1k23}$ is an ordinary $(k+2)/k$-power.
- The factor $34\cdots(\text{k-1})\text{1k243}$ is a reverse $(k+2)/k$-power.
- The factor $45\cdots(\text{k-1})\text{1k245}$ is an ordinary $(k+1)/(k-1)$-power.
- The factor $23\cdots(\text{k-1})\text{1k32}$ is an ordinary $(k+2)/k$-power.
- The factor $34\cdots(\text{k-1})\text{1k34}$ is an ordinary $(k+1)/(k-1)$-power.
- The factor $12\cdots(\text{k-1})\text{k12}$ is an ordinary $(k+2)/k$-power.
- The factor $23\cdots(\text{k-1})\text{k132}$ is a reverse $(k+2)/k$-power.
- The factor $34\cdots(\text{k-1})\text{k134}$ is an ordinary $(k+1)/(k-1)$-power.
- The factor $12\cdots(\text{k-1})\text{k21}$ is a reverse $(k+2)/k$-power.
- The factor $23\cdots(\text{k-1})\text{k23}$ is an ordinary $(k+1)/(k-1)$-power.   □

Conjecture 2 proposes that the value of $\mathrm{URT}(k)$ matches the lower bound of Theorem 6 for all $k \geq 4$. In the next section, we confirm Conjecture 2 for several values of k.

5 $\mathrm{URT}(k) = \frac{k-1}{k-2}$ for $k \in \{4, 8, 12\}$

First we explain why we rely on a different technique than in Sect. 3. Fix $k \geq 4$, and let $\Sigma_k = \{1, 2, \ldots, \text{k}\}$. A morphism $h : A^* \to B^*$ is called α-free (α^+-free, respectively) if it maps every α-free (α^+-free, respectively) word in A^* to an α-free (α^+-free, respectively) word in B^*. The morphism h is called *growing* if $h(a) > 1$ for all $a \in A^*$. Brandenburg [2] demonstrated that for every $k \geq 4$, there is no growing $\mathrm{RT}(k)^+$-free morphism from Σ_k^* to Σ_k^*. By a minor modification of his proof, one can show that there is no growing $(k-1)/(k-2)^+$-free morphism from Σ_k^* to Σ_k^*. While this does not entirely rule out the possibility that there is a morphism from Σ_k^* to Σ_k^* whose fixed point is $(k-1)/(k-2)^+$-free, it suggests that different techniques may be required. Our technique relies on an encoding similar to the one introduced by Pansiot [25] in showing that $\mathrm{RT}(4) = 7/5$. Pansiot's encoding was later used in all subsequent work on Dejean's Conjecture.

5.1   A Ternary Encoding

We first describe an alternate definition of ordinary r-powers which will be useful in this section. A word $w = w_1 \cdots w_n$, where the w_i are letters, is *periodic* if for some positive integer q, we have $w_{i+q} = w_i$ for all $1 \leq i \leq n - q$. In this case, the integer q is called a *period* of w. The *exponent* of w, denoted $\exp(w)$, is the ratio between its length and its minimal period. If $r = \exp(w)$, then w is an r-power.[5] We can write any r-power w as $w = pe$, where $|pe|/|p| = r$ and e is a prefix of pe. In this case, we say that e is the *excess* of the r-power w.

Suppose that $w \in \Sigma_k^*$ is an undirected $(k-1)/(k-2)^+$-free word that contains at least $k - 1$ distinct letters. Write $w = w_1 w_2 \cdots w_n$ with $w_i \in \Sigma_k$. Certainly,

---

[5] If $r \leq 2$, then w is an r-power as we have defined it in Sect. 1. If $r > 2$, then we take this as the definition of an (ordinary) r-power. For example, the English word `alfalfa` has minimal period 3 and exponent $\frac{7}{3}$, so it is a $\frac{7}{3}$-power.

every length $k - 2$ factor of w contains $k - 2$ distinct letters, and it is easily checked that every length k factor of w contains at least $k - 1$ distinct letters.

Now let $w \in \Sigma_k^*$ be any word containing at least $k - 1$ distinct letters and satisfying these two properties:

- Every length $k - 2$ factor of w contains $k - 2$ distinct letters; and
- Every length k factor of w contains at least $k - 1$ distinct letters.

Let u be the shortest prefix of w containing $k - 1$ distinct letters. We see immediately that u has length $k - 1$ or k. Write $w = uv$, where $v = v_1 v_2 \cdots v_n$ with $v_i \in \Sigma_k$. Define $p_0 = u$ and $p_i = uv_1 \cdots v_i$ for all $i \in \{1, \ldots, n\}$. For all $i \in \{0, 1, \ldots, n\}$, the prefix p_i determines a permutation

$$r_i = \begin{pmatrix} 1 & 2 & \ldots & k \\ r_i[1] & r_i[2] & \ldots & r_i[k] \end{pmatrix},$$

of the letters of Σ_k, which ranks the letters of Σ_k by the index of their final appearance in p_i. In other words, the word $r_i[3] \cdots r_i[k]$ is the length $k - 2$ suffix of p_i, and of the two letters in $\Sigma_k \backslash \{r_i[3], \ldots, r_i[k]\}$, the letter $r_i[2]$ is the one that appears last in p_i. Note that the final letter $r_i[1]$ may not even appear in p_i. For example, on Σ_6, the prefix **123416** gives rise to the permutation

$$\begin{pmatrix} 1\,2\,3\,4\,5\,6 \\ 5\,2\,3\,4\,1\,6 \end{pmatrix}.$$

Since every factor of length $k - 2$ in w contains $k - 2$ distinct letters, for any $i \in \{1, \ldots, n\}$, the letter v_i must belong to the set $\{r_{i-1}[1], r_{i-1}[2], r_{i-1}[3]\}$. This allows us to encode the word w over a ternary alphabet, as described explicitly below.

For $1 \leq i \leq n$, define $t(w) = t_1 \cdots t_n$, where for all $1 \leq i \leq n$, we have

$$t_i = \begin{cases} 1, & \text{if } v_i = r_{i-1}[1]; \\ 2, & \text{if } v_i = r_{i-1}[2]; \\ 3, & \text{if } v_i = r_{i-1}[3]. \end{cases}$$

For example, on Σ_5, for the word $w = $ **12342541243**, the shortest prefix containing 4 distinct letters is **1234**, and w has encoding $t(w) = $ **3131231**. Given the shortest prefix of w containing $k - 1$ distinct letters, and the encoding $t(w)$, we can recover w. Moreover, if w has period $q < n$, then so does $t(w)$. The exponent $|w|/q$ of w corresponds to an exponent $|v|/q$ of $t(w)$.

Let S_k denote the symmetric group on Σ_k with left multiplication. Define a morphism $\sigma : \Sigma_3^* \to S_k$ by

$$\sigma(1) = \begin{pmatrix} 1\,2\,3\,4 \ldots k-1\,k \\ 2\,3\,4\,5 \ldots \quad k \quad 1 \end{pmatrix}$$

$$\sigma(2) = \begin{pmatrix} 1\,2\,3\,4 \ldots k-1\,k \\ 1\,3\,4\,5 \ldots \quad k \quad 2 \end{pmatrix}$$

$$\sigma(3) = \begin{pmatrix} 1\,2\,3\,4 \ldots k-1\,k \\ 1\,2\,4\,5 \ldots \quad k \quad 3 \end{pmatrix}.$$

One proves by induction that $r_0\sigma(t(p_i)) = r_i$. It follows that if $w = pe$ has period $|p|$, and e contains at least $k-1$ distinct letters, then the length $|p|$ prefix of $t(w)$ lies in the kernel of σ. In this case, the word $t(w)$ is called a *kernel repetition*. For example, over Σ_4, the word

$$w = 123243414212324$$

has period 10, and excess 12324. Hence, the encoding $t(w) = 312313123131$ is a kernel repetition; one verifies that $\sigma(3123131231) = \text{id}$.

Suppose that k is even. Then $\sigma(1)$ and $\sigma(3)$ are odd, while $\sigma(2)$ is even. It follows that $\sigma(31)$ is even, and hence the subgroup of S_k generated by $\sigma(2)$ and $\sigma(31)$ is a subgroup of the alternating group A_k. This simple observation leads to the following important lemma, which will be used to bound the length of reversible factors in the words we construct.

Lemma 7. *Let $k \geq 4$ satisfy $k \equiv 0 \pmod 4$. Let $w \in \Sigma_k^*$ be a word with prefix $12 \cdots (k-1)$ and encoding $t(w) \in \{31, 2\}^*$. Suppose that $u = u_1 u_2 \cdots u_{k-1}$ is a factor of w, where $u_1, u_2, \ldots, u_{k-1} \in \Sigma_k$ are distinct letters. Then u^R is not a factor of w.*

Proof. Suppose towards a contradiction that u and u^R are both factors of w. Assume without loss of generality that u appears before u^R in w. Then w contains a factor x with prefix u and suffix u^R. Consider the encoding $t(x)$, which is a factor of $t(w)$.

Immediately after reading u, the ranking of the letters in Σ_k is

$$\begin{pmatrix} 1 & 2 & 3 & \ldots & k \\ u_k & u_1 & u_2 & \ldots & u_{k-1} \end{pmatrix},$$

where u_k is the unique letter in $\Sigma_k \setminus \{u_1, u_2, \ldots, u_{k-1}\}$. Immediately after reading u^R, the ranking of the letters in Σ_k is

$$\begin{pmatrix} 1 & 2 & 3 & \ldots & k \\ u_k & u_{k-1} & u_{k-2} & \ldots & u_1 \end{pmatrix}.$$

Evidently, we have

$$\sigma(t(x)) = \begin{pmatrix} 1 & 2 & 3 & \ldots & \text{k-1} & \text{k} \\ 1 & \text{k} & \text{k-1} & \ldots & 3 & 2 \end{pmatrix}.$$

Since $k \equiv 0 \pmod 4$, we observe that $\sigma(t(x))$ is an odd permutation. We claim that $t(x)$ does not begin in 1 or end in 3, so that $t(x) \in \{31, 2\}^*$. But $\sigma(31)$ and $\sigma(2)$ are both even, which contradicts the fact that $\sigma(t(x))$ is odd.

The fact that $t(x)$ does not end in 3 follows immediately from the fact that $u_{k-1} \neq u_1$. It remains to show that $t(x)$ does not begin with 1. If x is a prefix of w, then $t(x)$ begins in 3 or 2, so we may assume that $w = yxz$ with $y \neq \varepsilon$. Then $t(w)$ has prefix $t(yu)t(x)$. If $t(x)$ began in 1, then $t(yu)$ would necessarily end in 3, and this is impossible since $u_1 \neq u_{k-1}$. This completes the proof of the claim, and the lemma. $\square$

5.2   Constructions

Define morphisms $f_4, f_8, f_{12} : \Sigma_2^* \to \Sigma_2^*$ as follows:

$$f_4(1) = 121$$
$$f_4(2) = 122$$

$$f_8(1) = 121212112122121$$
$$f_8(2) = 211212122122112$$

$$f_{12}(1) = 121212121211212122121$$
$$f_{12}(2) = 212122112121121212212.$$

Define $g : \Sigma_2^* \to \Sigma_3^*$ by

$$g(1) = 31$$
$$g(2) = 312.$$

A key property of each of the morphisms f_4, f_8, f_{12}, and g is that the images of 1 and 2 end in different letters.

Theorem 8. *Fix $k \in \{4, 8, 12\}$, and let $f = f_k$. Let $\boldsymbol{w}$ be the word over Σ_k with prefix $12 \cdots (\mathbf{k} - 1)$ and encoding $g(f^\omega(1))$. Then $\boldsymbol{w}$ is undirected $(k-1)/(k-2)^+$-free.*

The remainder of this section is devoted to proving Theorem 8. Essentially, we adapt the technique first used by Moulin-Ollagnier [23]. A simplified version of Moulin-Ollagnier's technique, which we follow fairly closely, is exhibited by Currie and Rampersad [13]. For the remainder of this section, we use notation as in Theorem 8. We let $r = |f(1)|$, i.e., we say that f is r-uniform.

We first discuss kernel repetitions appearing in $g(f^\omega(1))$. Let factor $v = pe$ of $g(f^\omega(1))$ be a kernel repetition with period q; say $g(f^\omega(1)) = xvy$. Let $V = x'vy'$ be the maximal period q extension of the occurrence xvy of v. Write $x = Xx'$ and $\boldsymbol{y} = y'\boldsymbol{Y}$, so that $g(f^\omega(1)) = XVY$. Write $V = PE = EP'$, where $|P| = q$. By the periodicity of PE, the factor P is conjugate to p, and hence P is in the kernel of σ. Write $P = \pi''g(\pi)\pi'$ where π'' is a proper suffix of $g(1)$ or $g(2)$, and π' is a prefix of $g(1)$ or $g(2)$. Analogously, write $E = \eta''g(\eta)\eta'$. Since $g(1)$ and $g(2)$ end in different letters, it follows from the maximality of V that $\pi'' = \eta'' = \varepsilon$. In particular, the word P begins in 3. It follows that E begins in 3, and thus we may assume that $\pi' = \varepsilon$. Finally, by the maximality of V, we have $\eta' = 31$, the longest common prefix of $g(1)$ and $g(2)$. Altogether, we can write

$$PE = g(\pi\eta)31,$$

where $g(\pi) = P$ and η is a prefix of π. We see that $|P| \geq 2|\pi|$ and $|E| \leq 3|\eta| + 2$.

Let $\tau : \Sigma_2^* \to S_k$ be the composite morphism $\sigma \circ g$. Evidently, we have

$$\tau(1) = \sigma(g(1)) = \sigma(31), \text{ and}$$
$$\tau(2) = \sigma(g(2)) = \sigma(312).$$

Since P was in the kernel of σ, we see that

$$\tau(\pi) = \sigma(g(\pi)) = \sigma(P) = \mathrm{id},$$

i.e., the word π is in the kernel of τ.

Now set $\pi_0 = \pi$ and $\eta_0 = \eta$. By the maximality of PE, the repetition $\pi\eta = \pi_0\eta_0$ must be a maximal repetition with period $|\pi_0|$ (i.e., it cannot be extended). If η_0 has a cut, then it follows by arguments similar to those used above that $\pi_0\eta_0 = f(\pi_1\eta_1)\eta'$, where η_1 is a prefix of π_1 and η' is the longest common prefix of $f(1)$ and $f(2)$. One checks that there is an element $\phi \in S_k$ such that

$$\phi \cdot \tau(f(\mathbf{a})) \cdot \phi^{-1} = \tau(\mathbf{a})$$

for every $\mathbf{a} \in \{1, 2\}$, i.e., the morphism τ satisfies the "algebraic property" described by Moulin-Ollagnier [23]. It follows that π_1 is in the kernel of τ. We can repeat this process until we reach a repetition $\pi_s\eta_s$ whose excess η_s has no cut. Recalling that f is an r-uniform morphism, we have

$$|\pi_0| = r^s|\pi_s|$$

and

$$|\eta_0| = r^s|\eta_s| + |\eta'| \sum_{i=0}^{s-1} r^i.$$

Note that $|\eta'| = 2$ if $k = 4$, while $|\eta'| = 0$ if $k \in \{8, 12\}$. Thus, we have

$$|\eta_0| = \begin{cases} r^s|\eta_s| + r^s - 1, & \text{if } k = 4; \\ r^s|\eta_s|, & \text{if } k \in \{8, 12\}. \end{cases}$$

It follows that $|\eta_0| \leq r^s|\eta_s| + r^s - 1$.

Proof of Theorem 8. We first show that w contains no reverse α-power with $\alpha > (k-1)/(k-2)$. Since 33 is not a factor of $g(f^\omega(1))$, every factor of length k in w contains a factor of the form $u = u_1u_2 \cdots u_{k-1}$, where $u_1, u_2, \ldots, u_{k-1}$ are distinct letters. Thus, by Lemma 7, if xyx^R is a factor of w with $|xyx^R|/|xy| > (k-1)/(k-2)$, then $|x| \leq k-1$. In turn, we have $|xyx^R| < (k-1)^2$. Therefore, we conclude by a finite check that w contains no reverse α-power with $\alpha > (k-1)/(k-2)$.

It remains to show that w is ordinary $(k-1)/(k-2)^+$-free. Suppose to the contrary that pe is a factor of w such that e is a prefix of pe and $|pe|/|p| > (k-1)/(k-2)$. We may assume that pe is maximal with respect to having period $|p|$. If e has less than $k-1$ distinct letters, then $|e| \leq k-1$. In turn, we have $|pe| < (k-1)^2$. By a finite check, the word w has no such factors.

So we may assume that e has at least $k-1$ distinct letters. Let $V = t(pe)$, and let P be the length $|P|$ prefix of V. So $V = PE$, where E is a prefix of P. Hence V is a kernel repetition, i.e., the word P is in the kernel of σ. By the maximality of pe, we see that P begins in 3. Hence, the length $k-1$ prefix of p

contains $k-1$ distinct letters, and $|e| = |E| + k - 1$. We can find a factor $\pi_s \eta_s$ of $f^\omega(1)$ as described above, such that η_s is a prefix of $\pi_s \eta_s$, the word π_s is in the kernel of τ, and η_s does not contain a cut. Now

$$
\frac{1}{k-2} < \frac{|e|}{|p|}
$$
$$
= \frac{|E| + k - 1}{|P|}
$$
$$
\leq \frac{3|\eta_0| + k + 1}{2|\pi_0|}
$$
$$
= \frac{3\left(r^s|\eta_s| + r^s - 1\right) + k + 1}{2r^s|\pi_s|}
$$
$$
= \frac{3r^s\left(|\eta_s| + 1\right) + k - 2}{2 \cdot r^s|\pi_s|}
$$
$$
= \frac{3(|\eta_s| + 1) + (k-2)r^{-s}}{2|\pi_s|}
$$
$$
\leq \frac{3|\eta_s| + k + 1}{2|\pi_s|}.
$$

Thus, we have

$$
|\pi_s| < \frac{(k-2)(3|\eta_s| + k + 1)}{2}. \tag{1}
$$

By exhaustive check, every factor of length r in $f^\omega(1)$ contains a cut, so we must have $|\eta_s| < r$, and we can list all possibilities for η_s. For each possible value of η_s, we can enumerate all possibilities for π_s using (1). At this point, our argument depends on the value of k.

If $k \in \{8, 12\}$, then we find that no such factor $\pi_s \eta_s$ exists in $f^\omega(1)$. On the other hand, if $k = 4$, then we find only the following two pairs satisfying (1):

- $\pi_s = 2121$ and $\eta_s = \varepsilon$;
- $\pi_s = 2112112212$ and $\eta_s = 21$.

Note, however, that for each pair, we have $|\pi_s|_1 = |\pi_s|_2$ and $|\eta_s|_1 = |\eta_s|_2$. Since $f_4(1) = 121$ and $f_4(2) = 122$, it follows that $|\pi_0|_1 = |\pi_0|_2$ and $|\eta_0|_1 = |\eta_0|_2$. In this case, we have $|P| = \frac{5}{2}|\pi_0|$ and $|E| = \frac{5}{2}|\eta_0| + 2$. By adapting the string of inequalities leading to (1), we find that we must in fact have

$$
|\pi_s| < 2|\eta_s| + 4,
$$

and this does not hold in either case.

Thus, we conclude that $\boldsymbol{w}$ is undirected $(k-1)/(k-2)^+$-free. $\square$

Acknowledgements. We thank the anonymous reviewers, whose comments helped to improve the article.

References

1. Berstel, J.: Axel Thue's papers on repetitions in words: a translation. In: Publications du LaCIM, vol. 20. Université du Québec à Montréal (1995)
2. Brandenburg, F.J.: Uniformly growing k-th power-free homomorphisms. Theoret. Comput. Sci. **23**(1), 69–82 (1983)
3. Carpi, A.: On Dejean's conjecture over large alphabets. Theoret. Comput. Sci. **385**(1–3), 137–151 (2007)
4. Cassaigne, J.: Unavoidable patterns. In: Lothaire, M. (ed.) Algebraic Combinatorics on Words, pp. 111–134. Cambridge University Press, Cambridge (2002)
5. Cassaigne, J., Currie, J.D.: Words strongly avoiding fractional powers. Eur. J. Combin. **20**(8), 725–737 (1999)
6. Crochemore, M., Kolpakov, R., Kucherov, G.: Optimal bounds for computing α-gapped repeats. In: Dediu, A.-H., Janoušek, J., Martín-Vide, C., Truthe, B. (eds.) LATA 2016. LNCS, vol. 9618, pp. 245–255. Springer, Cham (2016). https://doi.org/10.1007/978-3-319-30000-9_19
7. Currie, J.D., Lafrance, P.: Avoidability index for binary patterns with reversal. Electron. J. Combin. **23**(1), P1.36 (2016)
8. Currie, J.D., Linek, V.: Avoiding patterns in the Abelian sense. Canad. J. Math. **53**(4), 696–714 (2001)
9. Currie, J.D., Visentin, T.I.: Long binary patterns are Abelian 2-avoidable. Theor. Comput. Sci. **409**(3), 432–437 (2008)
10. Currie, J.D., Mol, L., Rampersad, N.: A family of formulas with reversal of high avoidability index. Int. J. Algebra Comput. **27**(5), 477–493 (2017)
11. Currie, J.D., Mol, L., Rampersad, N.: Avoidance bases for formulas with reversal. Theor. Comput. Sci. **738**, 25–41 (2018)
12. Currie, J.D., Rampersad, N.: Dejean's conjecture holds for $n \geq 27$. RAIRO - Theor. Inform. Appl. **43**(4), 775–778 (2009)
13. Currie, J.D., Rampersad, N.: A proof of Dejean's conjecture. Math. Comput. **80**(274), 1063–1070 (2011)
14. Dejean, F.: Sur un théorème de Thue. J. Combin. Theory Ser. A **13**, 90–99 (1972)
15. Dekking, F.M.: Strongly non-repetitive sequences and progression-free sets. J. Combin. Theory Ser. A **27**(2), 181–185 (1979)
16. Duchon, P., Nicaud, C., Pivoteau, C.: Gapped pattern statistics. In: 28th Annual Symposium on Combinatorial Pattern Matching (CPM 2017), pp. 21:1–21:12 (2017)
17. Fici, G., Langiu, A., Lecroq, T., Lefebvre, A., Mignosi, F., Peltomäki, J., Prieur-Gaston, É.: Abelian powers and repetitions in Sturmian words. Theor. Comput. Sci. **635**, 16–34 (2016)
18. Frid, A.E.: Overlap-free symmetric D0L words. Discret. Math. Theor. Comput. Sci. **4**(2), 357–362 (2001)
19. Gawrychowski, P., Manea, F.: Longest α-gapped repeat and palindrome. In: Kosowski, A., Walukiewicz, I. (eds.) FCT 2015. LNCS, vol. 9210, pp. 27–40. Springer, Cham (2015). https://doi.org/10.1007/978-3-319-22177-9_3
20. Tomohiro, I., Köppl, D.: Improved upper bounds on all maximal α-gapped repeats and palindromes. Theor. Comput. Sci. **753**, 1–15 (2019)
21. Keränen, V.: Abelian squares are avoidable on 4 letters. In: Kuich, W. (ed.) ICALP 1992. LNCS, vol. 623, pp. 41–52. Springer, Heidelberg (1992). https://doi.org/10.1007/3-540-55719-9_62

22. Mohammad-Noori, M., Currie, J.D.: Dejean's conjecture and Sturmian words. Eur. J. Combin. **28**(3), 876–890 (2007)

23. Moulin-Ollagnier, J.: Proof of Dejean's conjecture for alphabets with 5, 6, 7, 8, 9, 10, and 11 letters. Theor. Comput. Sci. **95**(2), 187–205 (1992)

24. Mousavi, H.: Automatic theorem proving in Walnut. Preprint, arXiv: 1603.06017 [cs.FL] (2016)

25. Pansiot, J.J.: A propos d'une conjecture de F. Dejean sur les répétitions dans les mots. Discret. Appl. Math. **7**(3), 297–311 (1984)

26. Rao, M.: Last cases of Dejean's conjecture. Theor. Comput. Sci. **412**(27), 3010–3018 (2011)

27. Rosenfeld, M.: Every binary pattern of length greater than 14 is Abelian-2-avoidable. In: 41st International Symposium on Mathematical Foundations of Computer Science (MFCS 2016), pp. 81:1–81:11 (2016)

28. Samsonov, A.V., Shur, A.M.: On Abelian repetition threshold. RAIRO - Theor. Inform. Appl. **46**(1), 147–163 (2012)

29. Shur, A.M.: On the existence of minimal β-powers. Int. J. Found. Comput. Sci. **22**(7), 1683–1696 (2011)

30. Shur, A.M.: Growth of power-free languages over large alphabets. Theor. Comput. Syst. **54**(2), 224–243 (2014)

Characteristic Parameters and Special Trapezoidal Words

Alma D'Aniello[1] and Alessandro De Luca[2]($\boxtimes$)

[1] Dipartimento di Matematica e Applicazioni "R. Caccioppoli",
Università degli Studi di Napoli Federico II, Naples, Italy
`alma.daniello@unina.it`
[2] DIETI, Università degli Studi di Napoli Federico II,
via Claudio 21, 80125 Naples, Italy
`alessandro.deluca@unina.it`

Abstract. Following earlier work by Aldo de Luca and others, we study trapezoidal words and their prefixes, with respect to their characteristic parameters K and R (length of shortest unrepeated suffix, and shortest length without right special factors, respectively), as well as their symmetric versions H and L. We consider the distinction between closed (i.e., periodic-like) and open prefixes, and between Sturmian and non-Sturmian ones. Our main results characterize right special and strictly bispecial trapezoidal words, as done by de Luca and Mignosi for Sturmian words.

Keywords: Trapezoidal word · Closed word · Periodic-like word

1 Introduction

Sturmian words are certainly among the most studied objects in combinatorics on words, thanks to their natural definition, interesting characterizations, and numerous applications in several fields; see [1,11] for surveys. An infinite word is Sturmian if it has exactly $n + 1$ distinct *factors* (blocks of consecutive letters) for each length $n \geq 0$.

Trapezoidal words, introduced in [4,8], are a natural finite analogue of Sturmian words. They have at most $n + 1$ factors of each length n, so that the graph of their *factor complexity* function is in the shape of an isosceles trapezoid (or triangle), whence their name.

The original definition of trapezoidal words, however, uses *characteristic parameters* K and R. For a finite word w, K_w denotes the length of the shortest unrepeated suffix of w, whereas R_w denotes the smallest integer $n \geq 0$ such that w has no *right special* factor of length n. A word w is then trapezoidal if and only if its length $|w|$ verifies $|w| = K_w + R_w$ (with $|w| \geq K_w + R_w$ true in general, see [8]). Finite Sturmian words, i.e., factors of Sturmian words, are trapezoidal, but there exist non-Sturmian trapezoidal words such as *aabb*.

© Springer Nature Switzerland AG 2019
R. Mercaş and D. Reidenbach (Eds.): WORDS 2019, LNCS 11682, pp. 159–166, 2019.
https://doi.org/10.1007/978-3-030-28796-2_12

In [2], the property of being closed (aka *periodic-like*) or open is considered for trapezoidal words. In particular, the case of (prefixes of) the Fibonacci word was completely characterized, and this was later (cf. [5]) extended to all characteristic Sturmian words. Our first aim, explored in Sect. 3, is to extend some of those arguments to the general trapezoidal case, with respect to the values of characteristic parameters for closed and open prefixes.

In [7], a characterization was given for right (resp. left) special Sturmian words, i.e., finite words w over $A = \{a, b\}$ such that both extensions wa, wb (resp. aw, bw) are Sturmian. Previously, in [9], *strictly bispecial* ones (that is, words w such that awa, awb, bwa, bwb are all Sturmian) had been characterized; in particular, these turn out to be the noteworthy family of *central* words. In [6], bispecial (i.e., simultaneously right and left special) Sturmian words were characterized. Our main objective in Sect. 4 is to give similar characterizations for special trapezoidal words. Special words in a language are often useful for dealing with enumerative and structural questions.

2   Notation and Preliminaries

Let $A = \{a, b\}$ be an alphabet. The free monoid of all words over A under concatenation is denoted by A^*; its neutral element is the *empty word* ε. For $w \in A^*$ and $x \in A$, $|w|_x$ denotes the number of occurrences of x in w.

If $w = pvs \in A^*$, we say that v is a *factor* of w, p is a *prefix*, and s is a suffix. A *border* of w is a word that is simultaneously a proper prefix and a suffix of w. The definitions of factor and prefix also apply to (right-)*infinite words* over A. A word u is a *right* (resp. *left*) *special* factor of a finite or infinite word w over A if ua and ub (resp. au, bu) are both factors of w.

As anticipated above, an infinite word is *Sturmian* if it has exactly $n + 1$ distinct factors of each length $n \geq 0$. Therefore Sturmian words are the simplest aperiodic words in terms of *factor complexity*, and this is one of the reasons of interest in their study (see [1,11]). Equivalently, a binary infinite word is Sturmian if and only if it has exactly one right (resp. left) special factor of each length. In particular, a Sturmian word is called *standard* or *characteristic* if all its left special factors occur as prefixes.

Among the many known characterizations of factors of Sturmian words, or *finite Sturmian words*, perhaps the most famous and widely used one deals with *balance*; $w \in A^*$ is Sturmian if and only if there is no word u such that aua and bub are both factors of w. Such a pair of factors for a non-Sturmian word is called a *pathological pair* (cf. [2,4]).

Central words are palindromic prefixes of characteristic Sturmian words. They enjoy many equivalent definitions and interesting properties (cf. [1,7]). In particular, a word is central if and only if it can be written as a^n, b^n, or $uabv = vbau$ for some integer $n \geq 0$ and $u, v \in A^*$; in the latter case, u and v are central words themselves.

The parameters K_w and R_w defined in the previous section were introduced in [8], along with their "left" counterparts H_w (length of the shortest unrepeated prefix) and L_w (smallest $n \geq 0$ such that w has no left special factor of length n). As already stated, a finite word is *trapezoidal* if $|w| = K_w + R_w$, or equivalently if $|w| = H_w + L_w$ (cf. [4,8]).

Another noteworthy parameter is the (minimal) *period* π_w of a word w, which can be defined by $\pi_w = |w| - |v|$ where v is the longest border of w. Central words can also be characterized in terms of periods; $w \in A^*$ is central if and only if $|w| = \pi_{wa} + \pi_{wb} - 2$ (cf. [7]). It is known (see for example [10]) that *periodic extensions* of a finite Sturmian word w, i.e., words w' such that w is a factor of w' and $\pi_{w'} = \pi_w$, are still Sturmian; however, this property does not extend to trapezoidal words.

Example 1. Let $w = aaababa$. Then $H_w = R_w = 3$ and $L_w = K_w = 4$, so that w is trapezoidal. Its period is $\pi_w = 6$, but the periodic extension $w' = abaaababa = abw$ is not trapezoidal, as $|w'| = 9 > 4 + 4 = K_{w'} + R_{w'}$.

The following theorem is essentially a restatement of [4, Theorem 5]. It characterizes non-Sturmian trapezoidal words as products of two periodic extensions of the elements in a pathological pair.

Theorem 2. *A word $w \in A^*$ is trapezoidal non-Sturmian if and only if it can be written as*

$$w = pxux \cdot yuyq$$

where u is a central word, $A = \{x, y\}$, $\pi_{pxux} = \pi_{ux}$, and $\pi_{yuyq} = \pi_{yu}$. Furthermore, $R_w = |pxux|$ and $K_w = |yuyq|$.

Note that for such a word, $\{xux, yuy\}$ is actually the *shortest* pathological pair (cf. [2]).

Example 3. The trapezoidal word $w = aaababa$ considered in Example 1 can be written as $aaa \cdot baba$, where aaa and $baba$ are periodic extensions (to the left and to the right, respectively) of the elements of the pathological pair $\{aaa, bab\}$.

The non-trapezoidal word $w' = abaaababa$ does not verify the condition in Theorem 2. Indeed, its only pathological pair is (aaa, bab), and writing $w' = abaaa \cdot baba$ we obtain $\pi_{abaaa} = 4 \neq 1 = \pi_{aa}$.

3 Closed and Open Trapezoidal Words

A finite, nonempty word w is said to be *closed* (or periodic-like in earlier works) if it has a border u with no internal occurrences, that is, a factor occurring exclusively as a prefix and as a suffix; another common terminology for describing this situation is that w is a *complete (first) return* to u. In particular, single letters are closed, their border being the empty word.

A non-closed word is said to be *open*. Equivalently, w is open if and only if its longest repeated prefix (resp. suffix) is a right (resp. left) special factor (cf. [8]).

The following result was proved in [3, Proposition 3.6].

Proposition 4. *All closed trapezoidal words are Sturmian.*

The following result, showing a basic connection between the behavior of H and the property of being closed or open, is essentially known (see [5, Lemma 6 and Remark 8]). We report a proof for the sake of completeness.

Lemma 5. *Let $w \in A^*$ and $x \in A$. Then $H_{wx} = H_w + 1$ if wx is closed, and $H_{wx} = H_w$ if wx is open.*

Proof. Trivial if $w = \varepsilon$. Let then w be nonempty, and hy ($y \in A$) be its shortest unrepeated prefix, so that $H_w = |hy|$. If wx is closed, then its longest border has to be longer than h since h has internal occurrences in wx, and not longer than hy since otherwise hy would reoccur in w. Hence $x = y$ and $H_{wx} = |hy|+1 = H_w+1$ as desired. If wx is open, then hy cannot have internal occurrences in wx, since it is unrepeated in w, and it cannot be a suffix either, otherwise wx would be closed. Hence hy is unrepeated in wx, i.e., $H_{wx} = H_w$. □

Lemma 6. *Let wx be a trapezoidal word, $x \in A$. Then $L_{wx} = L_w$ if wx is closed, and $L_{wx} = L_w + 1$ if wx is open.*

Proof. Follows from Lemma 5 as $|w| = H_w + L_w$ and $|wx| = H_{wx} + L_{wx}$. □

Clearly, the following symmetric statement holds for *left* extensions xw.

Lemma 7. *Let $w \in A^*$ and $x \in A$. Then $K_{xw} = K_w + 1$ if xw is closed, and $K_{xw} = K_w$ if xw is open. Moreover, if xw is trapezoidal, then $R_{xw} = R_w$ if xw is closed, and $R_{xw} = R_w + 1$ otherwise.*

For a trapezoidal word w, the equality $\{H_w, L_w\} = \{K_w, R_w\}$ holds (cf. [8]). The following theorem, proved in [2, Proposition 4.4], is more precise.

Theorem 8. *Let w be a trapezoidal word. Then $H_w = K_w$ and $L_w = R_w$ if w is closed, whereas $H_w = R_w$ and $L_w = K_w$ if w is open.*

Corollary 9. *Let wx be a trapezoidal word and $x \in A$. Then $K_{wx} = K_w + 1$ and $R_{wx} = R_w$, unless*

– w is closed and wx is open, or
– w is open and wx is closed,

in which cases we have $K_{wx} = R_w + 1$ and $R_{wx} = K_w$ instead.

Proof. Consequence of Lemmas 5, 6, and Theorem 8.

Proposition 10. *Let $w \in A^*$, $y \in A$. If wy is trapezoidal but not Sturmian, then w is open.*

Proof. If w is not Sturmian, by Proposition 4 we are done. Let then w be Sturmian and assume it is closed, by contradiction. By Proposition 4, wy is open. Writing $wy = pxux \cdot yuyq$ as in Theorem 2, we have $q = \varepsilon$ as w is Sturmian, and $H_w = H_{wy} = R_{wy} = |pxux|$ by Lemma 5 and Theorems 8 and 2. It follows that pxu is the longest border of w. As $x \neq y$ and w ends in yu, this is clearly absurd. □

Let $w_{[n]}$ denote the prefix of w of length n. The *oc-sequence* of a word w is the characteristic sequence of its closed prefixes. In other terms, it is the binary word OC_w such that

$$OC_w(n) = \begin{cases} 1 & \text{if } w_{[n]} \text{ is closed,} \\ 0 & \text{if } w_{[n]} \text{ is open.} \end{cases}$$

The oc-sequence is a useful tool in studying the structure of finite and infinite words. For example, in [5], the following was proved:

Theorem 11. *Let w be an infinite word, and let*

$$OC_w = \prod_{n=0}^{\infty} 1^{k_n} 0^{k'_n}$$

for suitable positive integers k_n, k'_n, with $n \geq 0$. Then $k_n \leq k'_n$ for all $n \geq 0$, with equality holding for all n if and only if w is a characteristic Sturmian word.

In terms of oc-sequences, an immediate consequence of Lemmas 5 and 6 is the following (see also [5, Remark 8]):

Proposition 12. *For any word w, H_w is the number of closed nonempty prefixes of w, i.e., $H_w = |OC_w|_1$. If w is trapezoidal, then $L_w = |OC_w|_0$.*

The following two results show the behavior of characteristic parameters H and L at the end of runs of 1 and 0 in the oc-sequence.

Proposition 13. *Let $w \in A^*$ and $x \in A$ be such that wx is an open trapezoidal word, while w is closed. Then $L_w < H_w$.*

Proof. Since $L_{wx} = L_w + 1$ by Lemma 6, the longest left special factor ℓx of wx occurs as a suffix, and ℓ is the longest left special factor of w. Clearly, the suffix ℓ has internal occurrences in w, so that it is strictly shorter than the longest border v. This proves $L_w - 1 = |\ell| < |v| = H_w - 1$, whence the assertion. $\square$

Proposition 14. *Let $w \in A^*$ and $x \in A$ be such that wx is a closed trapezoidal word, while w is open. Then $H_w \leq L_w$.*

Proof. Since $H_{wx} = H_w + 1$ by Lemma 5, the longest repeated prefix v of w occurs as a suffix, so that $H_w \leq K_w$. The assertion follows by Theorem 8. $\square$

While Theorem 11 gives local constraints for an oc-sequence (namely, each run of 1s is followed by a longer or equal run of 0s), our last three results can be viewed as more global constraints in the case of trapezoidal words. Considering the integer parameter

$$D_w := H_w - L_w = |OC_w|_1 - |OC_w|_0$$

gives an interesting way to picture this situation. Indeed by Proposition 12, if w is trapezoidal then $D_w = |OC_w|_1 - |OC_w|_0$, so that D increases or decreases by 1 at each subsequent prefix, depending on whether it is closed or open; moreover by Propositions 13–14, D is necessarily positive (resp. non-positive) when encountering the last closed (resp. open) prefix in a run.

Example 15. Let $w = baabaababab$. Then $w_{[n]}$ is closed for $n = 1$ and $4 \leq n \leq 8$, and open otherwise; that is, $OC_w = 10011111000$. As predicted by Propositions 13–14, D reaches its (positive) local maxima, respectively 1 and 4, at $n = 1$ and $n = 8$, and its (non-positive) local minimum of -1 for $n = 3$. Since w is not Sturmian, by Proposition 4 any subsequent trapezoidal right extension will be open, leading to an indefinite decrease of D.

4   Special Trapezoidal Words

In analogy with the case of finite Sturmian words (cf. [7,9]), we say that a trapezoidal word $w \in A^*$ is *right* (resp. *left*) *special* if wa and wb (resp. aw, bw) are both trapezoidal, and that w is strictly bispecial if awa, awb, bwa, and bwb are all trapezoidal.

Proposition 16. *A right special trapezoidal word is Sturmian.*

Proof. Let w be a non-Sturmian trapezoidal word, then open by Proposition 4. If $z \in A$ and wz is trapezoidal, then it is also not Sturmian (like w) and hence open. By Corollary 9, $R_{wz} = R_w$ and $K_{wz} = K_w + 1$. By Theorem 2, it follows $wz = pxux \cdot yuyqz$ with $\pi_{yuyqz} = \pi_{yu} = \pi_{yuyq}$. This shows that z is uniquely determined, so that w cannot be right special.   □

Symmetrically, one can prove that

Proposition 17. *A left special trapezoidal word is Sturmian.*

Theorem 18. *A trapezoidal word w is right special if and only if either of the following holds:*

1. *w is a suffix of a central word, or*
2. *$w = pxuxyu$ for a central word u, distinct letters x, y, and a word p such that $\pi_{pxux} = \pi_{ux}$.*

Symmetrically, w is a left special trapezoidal word if and only if it is either a prefix of a central word, or written as $w = uxyuyq$ for $x, y \in A$, $x \neq y$, and $\pi_{yuyq} = \pi_{yu}$.

Proof. As is well known (cf. [7]), both extensions wa, wb of a word w are Sturmian if and only if w is a suffix of a central word. Let now w be right special and such that one extension is not Sturmian. By Proposition 16, w is Sturmian. As a consequence of Theorem 2, we must have $w = pxux \cdot yu$ where $A = \{x, y\}$, u is some central word, and p is such that $\pi_{pxux} = \pi_{ux}$.

Conversely, if $w = pxuxyu$ with $A = \{x, y\}$, u central and $\pi_{pxux} = \pi_{ux}$, then w is Sturmian as $\{xux, yuy\}$ is the only pathological pair in the trapezoidal non-Sturmian word wy; therefore, wx must be Sturmian (and then trapezoidal) too.

The left special case is similar.   □

The following theorem is a restatement of results in [6, 7]; it characterizes Sturmian words that are bispecial (*as Sturmian words*).

Theorem 19. *Let $w \in A^*$. Then wa, wb, aw, bw are all Sturmian if and only if $w = (uxy)^n u$ for some central word u, $\{x, y\} = A$ and a nonnegative integer n. Furthermore, awa, awb, bwa, bwb are all Sturmian if and only if w is central, i.e., $n = 0$, whereas for $n > 0$ exactly one such bilateral extension is not Sturmian, namely xwy.*

Semicentral words were defined in [2] by the property of having their longest repeated prefix, longest repeated suffix, longest left special factor, and longest right special factor coincide. In the same paper, they were characterized as words w such that $w = uxyu$ for some central word u over $A = \{x, y\}$. Hence, they correspond to the case $n = 1$ in the previous theorem.

Our final result is a characterization of strictly bispecial trapezoidal words.

Theorem 20. *A trapezoidal word is strictly bispecial if and only if it is central or semicentral.*

Proof. By Theorem 19, central words are strictly bispecial. Moreover, by the same theorem all bilateral extensions of a semicentral word $uxyu$ are Sturmian, except for $xuxyuy$ which is trapezoidal non-Sturmian by Theorem 2.

Conversely, if w is a strictly bispecial trapezoidal word, then either all bilateral extensions are Sturmian, in which case w is central by Theorem 19 and we are done, or at least one is not.

Assume, for instance, that cwa is trapezoidal non-Sturmian, the other cases being similar. By Proposition 16, cw is Sturmian, so that cwb must be too. Symmetrically, as a consequence of Proposition 17, dwa must be Sturmian as well (where $\{c, d\} = A$). In all cases, wa, wb, aw, bw are all Sturmian. By Theorem 19, it follows $w = (uxy)^n u$ for some $n > 0$; as a consequence of Theorem 2, we must have $n = 1$. $\square$

5   Concluding Remarks

A few related problems remain open. In particular, in [5] the oc-sequence for (prefixes of) characteristic Sturmian words was characterized, see Theorem 11. The general trapezoidal case, and even the non-standard Sturmian one, is still open. We believe our results may shed some light on the matter, as illustrated at the end of Sect. 3.

Regarding the preceding section, a simple and elegant characterization of (not necessarily strictly) bispecial trapezoidal words, such as Theorem 19 is for the Sturmian case, is still missing. Theorem 18 might be an ingredient for such a result.

Acknowledgments. We thank the anonymous referees for their many helpful comments. This paper is dedicated to the memory of our dear colleague Aldo de Luca (1941–2018).

References

1. Berstel, J., Séébold, P.: Sturmian words. In: Lothaire, M. (ed.) Algebraic Combinatorics on Words, chap. 2. Cambridge University Press, Cambridge (2002)
2. Bucci, M., De Luca, A., Fici, G.: Enumeration and structure of trapezoidal words. Theoret. Comput. Sci. **468**, 12–22 (2013). https://doi.org/10.1016/j.tcs.2012.11.007
3. Bucci, M., de Luca, A., De Luca, A.: Rich and periodic-like words. In: Diekert, V., Nowotka, D. (eds.) DLT 2009. LNCS, vol. 5583, pp. 145–155. Springer, Heidelberg (2009). https://doi.org/10.1007/978-3-642-02737-6_11
4. D'Alessandro, F.: A combinatorial problem on trapezoidal words. Theoret. Comput. Sci. **273**, 11–33 (2002). https://doi.org/10.1016/S0304-3975(00)00431-X
5. De Luca, A., Fici, G., Zamboni, L.Q.: The sequence of open and closed prefixes of a Sturmian word. Adv. Appl. Math. **90**, 27–45 (2017). https://doi.org/10.1016/j.aam.2017.04.007
6. Fici, G.: On the structure of bispecial Sturmian words. J. Comput. Syst. Sci. **80**(4), 711–719 (2014). https://doi.org/10.1016/j.jcss.2013.11.001
7. de Luca, A.: Sturmian words: structure, combinatorics, and their arithmetics. Theoret. Comput. Sci. **183**, 45–82 (1997). https://doi.org/10.1016/S0304-3975(96)00310-6
8. de Luca, A.: On the combinatorics of finite words. Theoret. Comput. Sci. **218**, 13–39 (1999). https://doi.org/10.1016/S0304-3975(98)00248-5
9. de Luca, A., Mignosi, F.: Some combinatorial properties of Sturmian words. Theoret. Comput. Sci. **136**, 361–385 (1994). https://doi.org/10.1016/0304-3975(94)00035-H
10. de Luca, A., De Luca, A.: Some characterizations of finite Sturmian words. Theoret. Comput. Sci. **356**, 118–125 (2006). https://doi.org/10.1016/j.tcs.2006.01.036
11. Rigo, M.: Formal Languages, Automata and Numeration Systems: Introduction to Combinatorics on Words. Wiley, New York (2014). https://doi.org/10.1002/9781119008200

Return Words and Bifix Codes in Eventually Dendric Sets

Francesco Dolce[1]([✉]) and Dominique Perrin[2]

[1] IRIF, Université Paris Diderot, Paris, France
dolce@irif.fr
[2] LIGM, Université Paris-Est Marne-la-Vallée, Champs-sur-Marne, France

Abstract. A shift space (or its set of factors) is eventually dendric if the possible extensions of all long enough factors are described by a graph which is a tree. We prove two results on eventually dendric shifts. First, all sets of return words to long enough words have the same cardinality. Next, this class of shifts is closed under complete bifix decoding.

Keywords: Formal languages · Symbolic dynamics · Neutral words

1 Introduction

A shift space X can be defined as the set of two-sided infinite words with all their factors in a given extendable factorial set, called the *language* of the shift, and denoted $\mathcal{L}(X)$. Thus shift spaces and extendable factorial sets are two aspects of the same notion. The traditional hierarchy of classes of languages translates into a hierarchy of shift spaces. In particular, a shift space X is called sofic when its language $\mathcal{L}(X)$ is a regular language. It is called of finite type when its language is the complement of a finitely generated ideal.

The complexity of a shift space is the function $n \mapsto p(n)$ where $p(n)$ is the number of factors of length n of the shift. In this paper, we are interested in shift spaces of at most linear complexity. This class is important for many reasons and includes the class of Sturmian shifts which are by definition those of complexity $n + 1$, and which play a role as binary codings of discrete lines. Such shifts arise in many other contexts (see, e.g., [15] or [17]). A shift space X is recurrent if for every $u, v \in \mathcal{L}(X)$ there exists some w such that $uwv \in \mathcal{L}(X)$. It is uniformly recurrent if for every element $w \in \mathcal{L}(X)$ there is an integer n_w such that w occurs as a factor in each elements of $\mathcal{L}(X)$ longer than n_w. Thus, the notion of recurrence expresses the property that every factor has a second occurrence. Uniform recurrence correspond to the appearance of the second occurrence after bounded time. It is known that all uniformly recurrent factorial extendable sets of at most linear complexity have a finite $\mathcal{S}$-adic representation (i.e., a generalization to several morphisms of a fixed point of a morphism) [14,16]. Conversely, it is an open problem, known as the $\mathcal{S}$-*adic conjecture*, to characterise the $\mathcal{S}$-adic representations of uniformly recurrent sets of at most linear complexity (see [15,

© Springer Nature Switzerland AG 2019
R. Mercaş and D. Reidenbach (Eds.): WORDS 2019, LNCS 11682, pp. 167–179, 2019.
https://doi.org/10.1007/978-3-030-28796-2_13

Chap. 12]). Note that all substitutive shifts defined by a primitive morphism are both uniformly recurrent and of at most linear complexity ([17]).

In this contribution we study a class of shift spaces of at most linear complexity, called eventually dendric, recently introduced in [12]. We also call eventually dendric the language of an eventually dendric shift. This class extends the class of dendric sets introduced in [5] (under the name of *tree sets*) which themselves extend naturally episturmian sets (also called Arnoux-Rauzy sets) and interval exchange sets. It is known that the class of eventually dendric shifts is closed under the natural equivalence on shifts called conjugacy (see [12]). We prove here that it is closed under a second transformation, namely complete bifix decoding, which is important because it includes coding by non overlapping blocks of fixed length. These two results show the robustness of the class of eventually dendric sets, giving a strong motivation for its introduction.

A dendric set S is defined by introducing the extension graph of a word and by requiring that this graph is a tree for every word in S. It has many interesting properties which involve free groups. In particular, in a dendric set S on an alphabet A, the group generated by the set of return words (see Sect. 4) to some word in S is the free group on the alphabet and, in particular, has Card A free generators. This generalizes a property known for Sturmian sets whose link with automorphisms of the free group was noted by Arnoux and Rauzy. The class of eventually dendric sets, studied in this paper, is defined by the property that the extension graph of every long enough word in the set is a tree (for short words the graphs may be arbitrary). These sets are contained in the class of eventually neutral sets, where a weaker hypothesis on the extensions is required (see Sect. 3). Our main results are that: all sets of return words to a (long enough) word in a recurrent eventually neutral set S have the same cardinality (Theorem 1); the class of eventually dendric sets is closed under complete bifix decoding (Theorem 4). An interesting consequence of Theorem 1 is the equivalence of the notions of recurrence and uniform recurrence for eventually neutral (and thus eventually dendric) sets (Theorem 2).

The paper is organized as follows. In Sect. 2, we introduce the definition of extension graphs and of eventually dendric sets. In Sect. 3, we recall some known properties on the complexity of a factorial extendable set of words and of special words. In Sect. 4 we focus on (uniformly) recurrent eventually neutral sets and on return words in such sets. In particular we prove that for every word w the set of return words on w is finite and that when w is long enough, all these sets have the same cardinality (Theorem 1). In the same section, and actually as a consequence of Theorem 1, we also prove that an eventually neutral set is recurrent if and only if it is uniformly recurrent (Theorem 2), a property already known for neutral sets ([2,11]). In Sect. 5 we introduce generalized extension graphs in which extension by words over particular sets replaces extension by letters. We prove that one obtains an equivalent definition of eventually dendric shifts using these generalized extension graphs (Theorem 3). In Sect. 6, we use generalized extension graphs to prove that the class of recurrent eventually dendric sets is closed under complete bifix decoding (Theorem 4), a result already known for dendric sets. We conclude with some open questions.

2   Eventually Dendric Sets

A set of words S on the alphabet A is *factorial* if it contains the factors of its elements. It is called *extendable* if for any $w \in S$ there are letters $a, b \in A$ such that $awb \in S$. For any set X of two-sided infinite words the *language* of X is the set $S = \mathcal{L}(X)$ of finite factors of the elements of X. It is a factorial and extendable set. In the following we suppose that the alphabet is minimal, i.e., that $A \subseteq S$.

Given a factorial set S and an integer $n \geq 0$ we denote $S_n = S \cap A^n$ and $S_{\geq n} = \bigcup_{m \geq n} S_m$. For $w \in S$ and $n \geq 1$, we denote $L_n(w, S) = \{u \in S_n \mid uw \in S\}$, $R_n(w, S) = \{v \in S_n \mid wv \in S\}$ and $E_n(w, S) = \{(u, v) \in L_n(w, S) \times R_n(w, S) \mid uwv \in S\}$. The *extension graph* of order n of w, denoted $\mathcal{E}_n(w, S)$, is the undirected bipartite graph whose set of vertices the disjoint union of $L_n(w, S)$ and $R_n(w, S)$ and with edges the elements of $E_n(w, S)$. When the context is clear, we denote $L_n(w), R_n(w), E_n(w)$ and $\mathcal{E}_n(w)$ instead of $L_n(w, S), R_n(w, S)$, $E_n(w, S)$ and $\mathcal{E}_n(w, S)$. A path in an undirected graph is *reduced* if it does not contain successive equal edges. For any $w \in S$, since any vertex of $L_n(w)$ is connected to at least one vertex of $R_n(w)$, the bipartite graph $\mathcal{E}_n(w)$ is a tree if and only if there is a unique reduced path in $\mathcal{E}_n(w)$ between every pair of vertices of $L_n(w)$ (resp. $R_n(w)$). A factorial and extendable set S is said to be *eventually dendric* with *threshold* $m \geq 0$ if $\mathcal{E}_1(w)$ is a tree for every word $w \in S_{\geq m}$. It is said to be (purely) *dendric* if we can choose $m = 0$. Dendric sets were introduced in [5] under the name of tree sets. An important example of dendric sets is formed by *strict episturmian sets* (also called Arnoux-Rauzy sets), which are by definition factorial extendable sets closed by reversal and such that for every n there exists a unique $w_n \in S_n(X)$ such that $\mathrm{Card}(R_1(w_n)) = \mathrm{Card}(A)$ and such that for every $w \in S_n \setminus \{w_n\}$ one has $\mathrm{Card}(R_1(w)) = 1$ (see [1,5]).

Example 1. Let F be the *Fibonacci set*, which is the set of factors of the words $\varphi^n(a)$, where φ is the morphism $a \mapsto ab, b \mapsto a$. It is also the set of factors of the one-sided infinite word $\mathbf{x}$ having all $\varphi^n(a)$ as prefixes, called a *fixed point* of φ, since $\varphi(\mathbf{x}) = \mathbf{x}$. It is well known that F is a Sturmian set (see [15]). The graph $\mathcal{E}_1(a)$ is shown in Fig. 1 on the left. The graph $\mathcal{E}_3(a)$ is shown on the right.

Fig. 1. The graphs $\mathcal{E}_1(a)$ and $\mathcal{E}_3(a)$.

The tree sets of *characteristic* $c \geq 1$ introduced in [4,11] give an example of eventually dendric sets of threshold 1 (while $\mathcal{E}_1(\varepsilon)$ is a forest of c trees).

Example 2. Let S be the language of the infinite word obtained as fixed point of the morphism $\psi : a \mapsto ab, b \mapsto cda, c \mapsto cd, d \mapsto abc$. Its language is a tree set of characteristic 2 and it is actually a specular set ([4, Example 4.2]). The extension graph $\mathcal{E}_1(\varepsilon)$ is shown in Fig. 2. Since the extension graphs of all nonempty words are trees, the set is eventually dendric with threshold 1.

Fig. 2. The extension graph $\mathcal{E}_1(\varepsilon)$.

Example 3. Let S be the *Tribonacci set*, which is the set of factors of the fixed point of the morphism $\psi : a \mapsto ab, b \mapsto ac, c \mapsto a$. S is an Arnoux-Rauzy set and a dendric set (see [5]). Let α be the morphism $\alpha : a \mapsto a, b \mapsto a, c \mapsto c$. The set $\alpha(S)$ is eventually dendric with threshold 4 (see [12]).

3   Complexity of Eventually Dendric Sets

Let S be a factorial extendable set. For a word $w \in S$, we denote $\ell_k(w) = \operatorname{Card} L_k(w), e_k(w) = \operatorname{Card} E_k(w)$, and $r_k(w) = \operatorname{Card} R_k(w)$. For any $w \in S$, we have $1 \leq \ell_k(w), r_k(w) \leq e_k(w)$. The word w is *left-k-special* if $\ell_k(w) > 1$, *right-k*-special if $r_k(w) > 1$ and *k-bispecial* if it is both left-*k*-special and right-*k*-special. For $k = 1$, we use ℓ, r, e and we simply say special instead of *k*-special. We define the *multiplicity* of w as $m(w) = e(w) - \ell(w) - r(w) + 1$. We say that w is *strong* if $m(w) \geq 0$, *weak* if $m(w) \leq 0$ and *neutral* if $m(w) = 0$ (see [9]). It is clear that if $\mathcal{E}_1(w)$ is acyclic (resp., connected, a tree), then w is weak (resp., strong, neutral). The following proposition is easily verified.

Proposition 1. *Let S be a factorial extendable set and let $w \in S$. If w is neutral, then*

$$\ell(w) - 1 = \sum_{b \in R_1(w)} (\ell(wb) - 1) \tag{1}$$

A factorial and extendable set S is said to be *eventually neutral* with *threshold* $m \geq 0$ if w is neutral for every word $w \in S_{\geq m}$. It is said to be (purely) *neutral* if we can choose $m = 0$. Set further $p_n(S) = \operatorname{Card} S_n, s_n(S) = p_{n+1}(S) - p_n(S)$ and $b_n(S) = s_{n+1}(S) - s_n(S)$. The sequence $p_n(S)$ is called the *complexity* of S.

The following result is from [8] (see also [5] and [9, Theorem 4.5.4]).

Proposition 2. *We have for all $n \geq 0$,*

$$s_n(S) = \sum_{w \in S_n} (\ell(w) - 1) = \sum_{w \in S_n} (r(w) - 1) \quad \text{and} \quad b_n(S) = \sum_{w \in S_n} m(w).$$

In particular, the number of left-special (resp. right-special) words of length n is bounded by $s_n(S)$.

We will use the following easy consequence of Proposition 2.

Proposition 3. *Let S be a factorial extendable set. If S is eventually dendric, then the sequence $s_n(S)$ is eventually constant.*

The previous result implies that eventually dendric sets have eventual linear complexity. The converse of Proposition 3 is not true.

Example 4. Let C be the *Chacon ternary set*, which is the set of factors of the fixed point of the morphism $\varphi : a \mapsto aabc, b \mapsto bc, c \mapsto abc$. It is well known that the complexity of C is $p_n(C) = 2n + 1$ and thus that $s_n(C) = 2$ for all $n \geq 0$ (see [15, Sect. 5.5.2]). The extension graphs of abc and bca are shown in Fig. 3. Thus $m(abc) = 1$ and $m(bca) = -1$. Let now α be the map on words defined by $\alpha(x) = abc\varphi(x)$. Let us verify that if the extension graph of x is the graph of Fig. 3 on the left, the same holds for the extension graph of $y = \alpha(x)$. Indeed, since $axa \in C$, the word $\varphi(axa) = aabc\varphi(x)aabc = ayaabc$ is also in C and thus $(a, a) \in \mathcal{E}_1(y)$. Since $cxa \in C$ and since a letter c is always preceded by a letter b, we have $bcxa \in C$. Thus $\varphi(bcxa) = bcyaabc \in C$ and thus $(c, a) \in \mathcal{E}_1(y)$. The proof of the other cases is similar. The same property holds for a word x with the extension graph on the right of Fig. 3. This shows that there is an infinity of words whose extension graph is not a tree and thus the Chacon set is not eventually dendric.

Fig. 3. The extension graphs of abc and bca.

4 Recurrent Eventually Dendric Sets

A factorial set S is *recurrent* if for any $u, v \in S$ there is a word w such that $uwv \in S$. A set is *uniformly recurrent* whenever for all $w \in S$ there exists an $n \geq 0$ such that w is a factor of any word in S_n. This last property is called *minimality* in the context of dynamical systems. If S is uniformly recurrent and infinite, then either there exists for every $w \in S$ an integer $n \geq 1$ such that

$w^n \notin S$ or S is equal to the set of factors of an infinite periodic word $uuu \cdots$. A uniformly recurrent set is recurrent but the converse is false, since for example the A^* is recurrent but not uniformly recurrent as soon as A has at least two elements.

Let S be a factorial extendable set. The set of *complete return words* to a word $w \in S$ is the set $\mathcal{CR}_S(w)$ of words having exactly two factors equal to w, one as a proper prefix and the other one as a proper suffix. It is clear that S is uniformly recurrent if and only if it is recurrent and if for every word w the set of complete return words to w is finite. If wu is a complete return word to w, then u is called a (right) *return word* to w. We denote by $\mathcal{R}_S(w)$ the set of return words to w. Clearly $\mathrm{Card}(\mathcal{CR}_S(w)) = \mathrm{Card}(\mathcal{R}_S(w))$.

Example 5. Let S be the Tribonacci set (see Example 3). Then $\mathcal{R}_S(a) = \{a, ba, ca\}$ and $\mathcal{R}_S(c) = \{abac, ababac, abaabac\}$.

By a result of [2], if S is uniformly recurrent and neutral (a fortiori, if S is dendric) the set $\mathcal{R}_S(w)$ has $\mathrm{Card}(A)$ elements for every $w \in S$. This is not true anymore for eventually dendric sets, as shown in the following example.

Example 6. Let S be the Tribonacci set and let $Y = \alpha(S)$ be, as in Example 3 the image of S under the morphism $\alpha : a, b \to a, c \to c$. Then, using Example 5, we find $\mathcal{R}_Y(a) = \{a, ca\}$ while $\mathcal{R}_Y(c) = \{aaac, aaaaac, aaaaaac\}$.

We will prove that for eventually dendric sets, a weaker property is true. It implies that the cardinality of sets of return words is eventually constant. For $w \in S$, set $\rho_S(w) = r_1(w) - 1$ and for a set $W \subseteq S$, let $\rho_S(W) = \sum_{w \in W} \rho_S(w)$ (if W is infinite, $\rho_S(W)$ is the supremum of the values of $\rho_S(U)$ on the finite subsets U of W).

By the symmetric of Proposition 1, for every neutral word $w \in S$, we have

$$\rho_S(w) = \sum_{a \in L_1(w)} \rho_S(aw). \tag{2}$$

Theorem 1. *Let S be a recurrent set which is eventually neutral with threshold m. For every $w \in S$, the set $\mathcal{R}_S(w)$ is finite. Moreover, for every $w \in S_{\geq m}$, we have*

$$\mathrm{Card}(\mathcal{R}_S(w)) = 1 + \rho_S(S_m). \tag{3}$$

Note that for $m = 0$ we have $\mathrm{Card}(\mathcal{R}_S(w)) = \mathrm{Card}(A)$ since $\rho_S(\varepsilon) = \mathrm{Card}(A) - 1$.

A *prefix code* (resp. a *suffix code*) is a set X of words such that none of them is a prefix (resp. a suffix) of another one. A prefix code (resp. a suffix code) $U \subseteq S$ is called *S-maximal* if it is not properly contained in a prefix code (resp. suffix code) $Y \subseteq S$ (see, for instance, [3]).

Proposition 4. *Let S be an eventually neutral set with threshold m. Then $\rho_S(U)$ is finite for every suffix code $U \subseteq S$. If U is a finite S-maximal suffix code with $U \subseteq S_{\geq m}$, then*

$$\rho_S(U) = \rho_S(S_m). \tag{4}$$

Proof. For any suffix code $U \subseteq S$, let us set $U_m = (U \cap S_{<m}) \cup (T \cap S_m)$, where T is the set of words which are suffixes of some words of U. Note that U_m is a finite suffix code. It is equal to S_m if U is S-maximal and contained in $S_{\geq m}$.

Assume first that $U \subseteq S$ is a finite S-maximal suffix code. We prove by induction on the sum $\ell(U)$ of the lengths of the words of U that

$$\rho_S(U) = \rho_S(U_m). \tag{5}$$

If all words of U are of length at most m, then $U = U_m$ and thus Eq. (5) holds. Otherwise, let $u \in U$ be of maximal length. Set $u = av$ with $a \in A$. Then $Av \cap S \subseteq U$. Set $U' = (U \setminus Av) \cup \{v\}$. Thus U' is reduced to the empty word or is an S-maximal suffix code with $\ell(U') < \ell(U)$. We have $U = (U' \setminus v) \cup L_1(v)v$. Since $|v| \geq m$, v is neutral, we have, by Eq. (2), $\rho_S(U) = \rho_S(U') - \rho_S(v) + \sum_{a \in L_1(v)} \rho_S(av) = \rho_S(U')$. By induction hypothesis, Eq. (5) holds for U'. Since $U_m = U'_m$, we have $\rho_S(U) = \rho_S(U') = \rho_S(U'_m) = \rho_S(U_m)$. Thus Eq. (5) is proved.

If U is infinite, then $\rho_S(U)$ is the supremum of the values of $\rho_S(V)$ on the finite subsets V of U. Any finite suffix code $V \subset S$ is contained in a finite S-maximal suffix code W and $\rho_S(V) \leq \rho_S(W)$. By Eq. (5), this implies $\rho_S(V) \leq \rho_S(W_m)$. The number of possible W_m is finite, therefore $\rho_S(V)$ is bounded. We conclude that $\rho_S(U)$ is finite.

Proof (of Theorem 1). Consider a word $w \in S$ and let P be the set of proper prefixes of $\mathcal{CR}_S(w)$. For $p \in P$, denote $\alpha(p) = \mathrm{Card}\{a \in A \mid pa \in P \cup \mathcal{CR}_S(w)\} - 1$. Then $\mathcal{CR}_S(w)$ is finite if and only if P is finite. Moreover in this case, since $\mathcal{CR}_S(w)$ is a prefix code, we have by a well known property of trees (one can see $\mathcal{CR}_S(w)$ as set of leaves and P as set of internal nodes)

$$\mathrm{Card}(\mathcal{CR}_S(w)) = \alpha(P) + 1, \tag{6}$$

where $\alpha(P) = \sum_{p \in P} \alpha(p)$. Let U be the set of words in P which are not proper prefixes of w. We claim that U is an S-maximal suffix code. Indeed, if $u, vu \in U$, then w is a proper prefix of u and thus is an internal factor of vu, a contradiction unless $v = \varepsilon$. Thus U is suffix. Consider $r \in S$. Either r has a suffix in U or r is a suffix of a word in U. Indeed, let us suppose that r has no suffixes in U. Then, since S is recurrent, there is some $s \in S$ such that $wsr \in S$. Let u be the shortest suffix of wsr which has w for proper prefix. Then $u \in U$. This shows that U is an S-maximal suffix code. We have $\alpha(p) = 0$ for any proper prefix p of w since any word in $\mathcal{CR}_S(w)$ has w as a proper prefix. Next we have $\alpha(p) = \rho_S(p)$ for any $p \in U$. Indeed, if $ua \in S$ for $u \in P$ and $a \in A$, then $ua \in \mathcal{CR}_S(w) \cup P$ since S is recurrent. Thus we have $\alpha(P) = \rho_S(U)$. By Proposition 4, $\rho_S(U)$ is finite. Therefore, Eq. 6 shows that $\mathrm{Card}(\mathcal{CR}_S(w)) = \mathrm{Card}(\mathcal{R}_S(w))$ is finite.

Assume finally that $|w| \geq m$. Then $U \subseteq S_{\geq m}$ and thus, by Proposition 4, we have $\rho_S(U) = \rho_S(S_m)$. Thus we have $\alpha(P) = \rho_S(S_m)$. By Eq. (6), this implies Eq. (3).

It is known that for neutral set recurrence is enough to guarantee uniform recurrence [11]. We obtain as a direct corollary of Theorem 1 the following:

Theorem 2. *An eventually neutral set is recurrent if and only if it is uniformly recurrent.*

Proof. Let S be a recurrent eventually neutral set. By Theorem 1, the set $\mathcal{R}_S(w)$ is finite for every $w \in S$. Thus S is uniformly recurrent. $\qquad\square$

Theorem 2 shows also that in a recurrent eventually neutral set the cardinality of sets of complete return words is bounded. There exist (uniformly) recurrent sets which do not have this property (see [13, Example 3.17]).

5   Generalized Extension Graphs

We will now see how the conditions on extension graphs can be generalized to graphs expressing the extension by words having different length.

Proposition 5. *For every $n \geq 1$ and $m \geq 0$, the graph $\mathcal{E}_n(w)$ is a tree for all $w \in S_{\geq m}$ if and only if $\mathcal{E}_{n+1}(w)$ is a tree for all words $w \in S_{\geq m}$.*

To prove Proposition 5 we need some preliminary result as well as the following notions. Let S be a factorial extendable set of words over an alphabet A.

For $U, V \subseteq A^*$ and $w \in S$, let $L_U(w) = \{u \in U \mid uw \in S\}$ and $R_V(w) = \{v \in V \mid wv \in S\}$.

Let $U \subseteq A^*$ (resp. $V \subseteq A^*$) be a suffix code (resp. prefix code) and $w \in S$ be such that $L_U(w)$ is an S-maximal suffix code (resp. $R_V(w)$ is an S-maximal prefix code). The *generalized extension graph* of w relative to U, V is the following undirected bipartite graph $\mathcal{E}_{U,V}(w)$. The set of vertices is the disjoint union of $L_U(w)$ and $R_V(w)$. The edges are the pairs $(u, v) \in L_U(w) \times R_V(w)$ such that $uwv \in S$. In particular $\mathcal{E}_n(w) = \mathcal{E}_{S_n,S_n}(w)$. The only if part of the next result is [5, Lemmas 3.8 and 3.10].

Lemma 1. *Let S be a factorial extendable set and let $w \in S$. Let $U \subseteq S$ be a finite S-maximal suffix code and let $V \subseteq S$ be finite S-maximal prefix code. Let $\ell \in S$ be such that $A\ell \cap S \subseteq U$ and such that $\mathcal{E}_{A,V}(\ell w)$ is a tree. Set $U' = (U \setminus A\ell) \cup \{\ell\}$. The graph $\mathcal{E}_{U',V}(w)$ is a tree if and only if the graph $\mathcal{E}_{U,V}(w)$ is a tree.*

Proof. We need only to prove the if part. First, note that the hypothesis that $\mathcal{E}_{A,V}(\ell w)$ is a tree guarantees that the left vertices $A\ell$ in $\mathcal{E}_{U,V}(w)$ are clusterized: for any pair of vertices $a\ell, b\ell$ there exists a unique reduced path from $a\ell$ to $b\ell$ in $\mathcal{E}_{U,V}(w)$ using as left vertices only elements of $A\ell$. Indeed, such a path exists since the subgraph $\mathcal{E}_{A\ell,V}(w)$ of $\mathcal{E}_{U,V}(w)$ is isomorphic to $\mathcal{E}_{A,V}(\ell w)$ that is connected. Since $\mathcal{E}_{U,V}(w)$ is a tree, this path is unique. Let $v, v' \in R_V(w)$ be two distinct vertices and let π be the unique reduced path from v to v' in $\mathcal{E}_{U,V}(w)$. We show that we can find a unique reduced path π' from v to v' in $\mathcal{E}_{U',V}(w)$. If π does not pass by $A\ell$, we can simply define by π' a path passing by the same vertices than π. Otherwise, we can decompose π in a unique way as a concatenation of a path

π_1 from v to a vertex in $A\ell$ not passing by $A\ell$ before, followed by a path from $A\ell$ to $A\ell$ (using on the left only vertices from $A\ell$) and a path π_2 from $A\ell$ to v' without passing in $A\ell$ again. We consider in $\mathcal{E}_{U',V}(w)$ the unique path π_1' from v to ℓ obtained by replacing the last vertex of π_1 by ℓ and the unique reduced path π_2' from ℓ to v' obtained by replacing the first vertex of π_2 by ℓ. In this case we define π' as the concatenation of π_1' and π_2'. The reduced path π' is unique. Indeed, let us suppose that we have a different path π^* from v to v' in $\mathcal{E}_{U',V}(w)$. If π^* does not pass (on the left) by ℓ, we would find a path having the same vertices in $\mathcal{E}_{U,V}(w)$ which is impossible since the graph is acyclic. Let us suppose that both π' and π^* pass by ℓ. Without loss of generality let us suppose that we have a cycle in $\mathcal{E}_{U',V}(w)$ passing by ℓ and v (the case with v' being symmetric). Let us define by π_0' and π_0^* the two distinct subpaths of π' and π^* respectively going from v to ℓ. Since S is extendable, we can find $a\ell, b\ell \in U$, with $a, b \in A$ not necessarily distinct, and two reduced paths π_1 from v to $a\ell$ and and π_2 from v to $b\ell$ in $\mathcal{E}_{U,V}(w)$ obtained from π_0' and π_0^* by replacing the vertex ℓ by $a\ell$ and $b\ell$ respectively. From the remark at the beginning of the proof we know that we can find a reduced path in $\mathcal{E}_{U,V}(w)$ from $a\ell$ to $b\ell$. Thus we can find a nontrivial cycle in $\mathcal{E}_{U,V}(w)$, which contradicts the acyclicity of the graph.

A symmetric statement holds for $r \in S$ such that $rA \cap S \subseteq V$ and $\mathcal{E}_{U,A}(wr)$ is a tree, with $V' = (V \setminus rA) \cup \{r\}$: the graph $\mathcal{E}_{U,V}(w)$ is a tree if and only if $\mathcal{E}_{U,V'}(w)$ is a tree.

Lemma 2. *Let $n \geq 1$, let $m \geq 0$ and let V be a finite S-maximal prefix code. If $\mathcal{E}_{S_n,V}(w)$ is a tree for every $w \in S_{\geq m}$ then for each word $u \in S_{\geq n+m-1}$, the graph $\mathcal{E}_{A,V}(u)$ is a tree.*

Proof. The graph $\mathcal{E}_{A,V}(u)$ is obtained from $\mathcal{E}_{S_n,V}(u)$ by identifying the vertices of $L_n(u)$ ending with the same letter. Since $\mathcal{E}_{S_n,V}(u)$ is connected, $\mathcal{E}_{A,V}(u)$ is also connected. Set $u = \ell u'$ with $|\ell| = n - 1$. The graph $\mathcal{E}_{A,V}(u)$ is isomorphic to $\mathcal{E}_{A\ell,V}(u')$ which is a subgraph of $\mathcal{E}_n(u')$ and thus it is acyclic.

A symmetric statement holds for $n \geq 1$ and U a finite S-maximal suffix code: if $\mathcal{E}_{U,S_n}(w)$ is a tree for every $w \in S_{\geq m}$ then so is $\mathcal{E}_{U,A}(u)$ for every $u \in S_{n+m-1}$.

Proof (of Proposition 5). Assume first that $\mathcal{E}_n(w)$ is tree for every word $w \in S_{\geq m}$. We fix some $w \in S_{\geq m}$. We claim that for any finite S-maximal suffix code U formed of words of length n or $n + 1$, the graph $\mathcal{E}_{U,S_n}(w)$ is a tree. The proof is done by induction on $\gamma_{n+1}(U) = \text{Card}(L_U(w) \cap A^{n+1})$. The property is true for $\gamma_{n+1}(U) = 0$, since then $\mathcal{E}_{U,S_n}(w) = \mathcal{E}_n(w)$. Assume now that $\gamma_{n+1}(U) > 0$. Let $a\ell$ with $a \in A$ be a word of length $n + 1$ in $L_U(w)$. Since U is an S-maximal suffix code with words of length n or $n + 1$, we have $A\ell \cap S \subseteq U$. Let us consider $U' = (U \setminus A\ell) \cup \{\ell\}$. Since $\gamma_{n+1}(U') < \gamma_{n+1}(U)$, by induction hypothesis the graph $\mathcal{E}_{U',S_n}(w)$ is a tree. Moreover, by Lemma 2, the graph $\mathcal{E}_{A,S_n}(\ell w)$ is a tree. Thus, by Lemma 1, the graph $\mathcal{E}_{U,S_n}(w)$ is a tree. This proves the claim. We now claim that for any finite S-maximal prefix code V formed of words of length n or $n+1$, the graph $\mathcal{E}_{S_{n+1},V}(w)$ is a tree by induction on $\delta_{n+1}(V) = \text{Card}(R_V(w) \cap A^{n+1})$.

The property is true for $\delta_{n+1}(V) = 0$, since the graph $\mathcal{E}_{S_{n+1},V}(w) = \mathcal{E}_{S_{n+1},S_n}(w)$, is a tree. Assume now that $\delta_{n+1}(V) > 0$. Let ra with $a \in A$ be a word of length $n + 1$ in $R_V(w)$. Since V is an S-maximal prefix code with words of length n or $n + 1$, we have $rA \cap S \subseteq U$. Let us consider $V' = (V \setminus rA) \cup \{r\}$. Since $\delta_{n+1}(V') < \delta_{n+1}(V)$, by induction hypothesis the graph $\mathcal{E}_{S_{n+1},V'}(w)$ is a tree. Moreover, by the symmetric version of Lemma 2, the graph $\mathcal{E}_{S_{n+1},A}(wr)$ is a tree. This proves the claim. Since $\mathcal{E}_{n+1}(w) = \mathcal{E}_{S_{n+1},S_{n+1}}(w)$, we conclude that $\mathcal{E}_{n+1}(w)$ is a tree.

Assume now that $\mathcal{E}_{n+1}(w)$ is a tree for every $w \in S_{\geq m}$. Fix some $w \in S_{\geq m}$. We first claim that $\mathcal{E}_{U,S_{n+1}}(w)$ is a tree for every S-maximal suffix code U formed of words of length n or $n + 1$ by induction on $\gamma_n(U) = \mathrm{Card}(L_U(w) \cap A^n)$. The property is true if $\gamma_n(U) = 0$, since then $\mathcal{E}_{U,S_{n+1}}(w) = \mathcal{E}_{n+1}(w)$. Assume next that $\gamma_n(U) > 0$. Let $\ell \in L_U(w) \cap A^n$. Set $W = (U \setminus \{\ell\}) \cup A\ell$ or equivalently $U = (W \setminus A\ell) \cup \{\ell\}$. Then $\delta_n(W) < \delta_n(U)$ and consequently $\mathcal{E}_{W,S_{n+1}}(w)$ is a tree by induction hypothesis. On the other hand, by Lemma 2, the graph $\mathcal{E}_{A,S_{n+1}}(\ell w)$ is also a tree. By Lemma 1, the graph $\mathcal{E}_{U,S_{n+1}}(w)$ is a tree and thus the claim is proved. We now claim that $\mathcal{E}_{S_n,V}(w)$ is a tree for every S-maximal prefix code V formed of words of length n or $n + 1$ by induction on $\delta_n(V) = \mathrm{Card}(R_V(w) \cap A^n)$. The property is true if $\delta_n(V) = 0$. Assume now that $\delta_n(V) > 0$. Let $r \in R_V(w) \cap A^n$ and let $T = (V \setminus \{r\}) \cup rA$ or equivalently $V = (T \setminus rA) \cup \{r\}$. Then $\delta_n(T) < \delta_n(V)$ and thus $\mathcal{E}_{S_n,T}(w)$ is a tree by induction hypothesis. On the other hand, by the symmetric version of Lemma 2, the graph $\mathcal{E}_{S_n,A}(wr)$ is also a tree. By Lemma 1, the graph $\mathcal{E}_{S_n,T}(w)$ is a tree and thus the claim is proved. Since $\mathcal{E}_n(w) = \mathcal{E}_{U,V}(w)$ for $U = V = S_n$, it follows from the claim that $\mathcal{E}_n(w)$ is a tree.

The following result shows that in the definition of eventually dendric sets, one can replace the graphs $\mathcal{E}_1(w)$ by $\mathcal{E}_n(w)$ with the same threshold.

Theorem 3. *Let S be a factorial extendable set. For every $m \geq 1$, the following conditions are equivalent.*

  (i) *S is eventually dendric with threshold m,*
  (ii) *the graph $\mathcal{E}_n(w)$ is a tree for every $n \geq 1$ and every word $w \in S_{\geq m}$,*
  (iii) *there is an integer $n \geq 1$ such that $\mathcal{E}_n(w)$ is a tree for every word $w \in S_{\geq m}$.*

Proof. (i) $\Rightarrow$ (ii) and (iii) $\Rightarrow$ (i) follows from Proposition 5 using respectively ascending and descending induction on n. Finally, (ii) clearly implies $\Rightarrow$ (iii).

6   Complete Bifix Decoding

Let S be a factorial extendable set of words over an alphabet A. A set $U \subseteq S$ is said to be *right S-complete* (resp. *left S-complete*) if any long enough word of S has a prefix (resp. suffix) in U. It is *two-sided S-complete* if it is both left and right S-complete. A *bifix code* is a set of words that is both a prefix code and a suffix code. Similarly to what seen in Sect. 5, we say that a bifix code $U \subseteq S$ is

S-maximal if it is not properly contained in a bifix code $V \subseteq S$. If a bifix code $U \subseteq S$ is right S-complete (resp. left S-complete), it is an S-maximal bifix code since it is already an S-maximal prefix code (resp. suffix code). It can be proved conversely that if S is recurrent, a finite bifix code is S-maximal if and only if it is two-sided S-complete (see [3, Theorem 4.2.2]). This is not true in general, as shown by the following example.

Example 7. Let $S = a^*b^*$. The set $U = \{aa, b\}$ is an S-maximal bifix code. Indeed, it is a bifix code and it is left S-complete as one may verify. However it is not right S-complete since no word in ab^* has a prefix in U.

Let $S \subseteq A^*$ be a factorial extendable set and let U be a two-sided S-complete finite bifix code. Let $\varphi : B \to U$ be a coding morphism for U, that is, a bijection from an alphabet B onto U extended to a morphism from B^* into A^*. Then $\varphi^{-1}(S)$ is factorial and, since U is two-sided complete, it is extendable. The set $\varphi^{-1}(S)$ is called the *complete bifix decoding* of S with respect to U. For example, for any $n \geq 1$, the set S_n is a two-sided complete bifix code and the corresponding complete bifix decoding is the decoding of S by n-blocks. In [5, Theorem 3.13] it is proved that the maximal bifix decoding of a recurrent dendric set is a dendric set. Actually, the hypothesis that S is recurrent is only used to guarantee that the S-maximal bifix code used for the decoding is also an S-maximal prefix code and an S-maximal suffix code. In the definition used here of complete bifix decoding, we do not need this hypothesis. Note, however, that when S is recurrent, the two notions of complete and maximal bifix decoding coincide.

Theorem 4. *Any complete bifix decoding of an eventually dendric set is an eventually dendric set having the same threshold.*

Note that for any S-maximal suffix code U one has $\mathrm{Card}(U) \geq \mathrm{Card}(A)$. Indeed, every $a \in A$ appears as a suffix of (at least) an element of S.

Lemma 3. *A set S is an eventually dendric set with threshold n if and only if for any $w \in S_{\geq n}$, for any S-maximal suffix code U and for any S-maximal prefix code V, the graph $\mathcal{E}_{U,V}(w)$ is a tree.*

Proof. The "if" part is trivial. To prove the other direction, we use an induction on the sum of the lengths of the words in U, V. The property is true if the sum is equal to $2\,\mathrm{Card}(A)$. Indeed, for every $w \in S_{\geq n}$ one has $U = L(w)$ and $V = R(w)$ and thus $\mathcal{E}_{U,V}(w) = \mathcal{E}_1(w)$ is a tree. Otherwise, assume that U contains words of length at least 2 (the case with V being symmetrical). Let $u \in U$ be of maximal length. Set $u = a\ell$ with $a \in A$. Since U is an S-maximal suffix code, we have $A\ell \cap S \subseteq U$. Set $U' = (U \setminus A\ell) \cup \{\ell\}$. By induction hypothesis, both $\mathcal{E}_{U',V}(w)$ and $\mathcal{E}_{A,V}(\ell w)$ are trees. Thus, by Lemma 1, $\mathcal{E}_{U,V}(w)$ is also a tree.

Proof (of Theorem 4). Assume that S is eventually dendric with threshold n. Let $\varphi : B \to U$ be a coding morphism for U and let T be the decoding of S corresponding to U. Consider a word w of T of length at least n. By Lemma 3, and since $|\varphi(w)| \geq n$, the graph $\mathcal{E}_{U,U}(\varphi(w))$ is a tree. But for $b, c \in B$, one has $bwc \in T$ if and only if $\varphi(bwc) \in S$, that is, if and only if $(\varphi(b), \varphi(c)) \in E_1(\varphi(w))$. Thus $\mathcal{E}_1(w)$ is isomorphic to $\mathcal{E}_{U,U}(\varphi(w))$ and thus $\mathcal{E}_1(w)$ is a tree. This shows that T is eventually dendric with threshold n.

Example 8. Let S be the Fibonacci set. Then $U = \{aa, aba, b\}$ is an S-maximal bifix code. Let $\varphi : \{u, v, w\} \to U$ be the coding morphism for U defined by $\varphi : u \mapsto aa, v \mapsto aba, w \mapsto b$. The complete bifix decoding T of S with respect to U is a purely dendric set. It is actually the natural coding of an interval exchange transformation on three intervals (see [6]). The extension graphs $\mathcal{E}_1(\varepsilon, T)$ and $\mathcal{E}_1(v, T)$ are shown in Fig. 4.

Fig. 4. The graphs $\mathcal{E}_1(\varepsilon, T)$ and $\mathcal{E}_1(w, T)$.

A particular case of complete bifix decoding is related to the skew product of two dynamical systems, a notion which is well-known in topological dynamics (see [10]). Indeed, assume that we start with a shift space X, a transitive permutation group G on a set Q and a morphism $f : A^* \to G$. We denote by $q \mapsto q \cdot w$ the result of the action of the permutation $f(w)$ on the point $q \in Q$. The skew product of X and (G, Q) is the shift space Y on the alphabet $A \times Q$ formed by the bi-infinite words (a_i, q_i) such that $(a_i) \in X$ and $p_{i+1} = p_i \cdot f(a_i)$ for all $i \in \mathbb{Z}$. Fix a point $i \in Q$. The set of words w such that $i \cdot w = i$ is a submonoid generated by a bifix code U which is two-sided complete. The decoding of $S = \mathcal{L}(X)$ with respect to $U \cap S$ is the language of the dynamical system induced by Y on the set of $y \in Y$ such that $y_0 = (a, i)$ for some $a \in A$ (see [6] for more details).

Example 9. Let X be the Fibonacci shift, i.e., the shift whose language is the Fibonacci set. Let $Q = \{1, 2\}$, $G = \mathbb{Z}/2\mathbb{Z}$ and $f : A^* \to G$ defined by $a \mapsto (12), b \mapsto (1)$. Choosing $i = 1$, the bifix code U built as above is $U = \{aa, aba, b\}$ as in Example 8.

7   Conclusion

We have seen that the class of eventually dendric shifts is closed under complete bifix decoding. It is also known to be closed under conjugacy (see [12]), and thus it has strong closure properties. It would be interesting to know how properties which are known to hold for dendric sets (or language of dendric shifts) extend to this more general class. For instance, to which extent the properties of return words proved for recurrent dendric sets extend to eventually dendric ones? More precisely, what can we say about the subgroup of the free group generated by return words to a given word? In [5] it is proved that for recurrent dendric sets, every set of return words to a fixed word is a basis of the free group, while in the case of specular sets, the set of return words to a fixed word is a basis of a particular subgroup called the even subgroup (see [4]). Also, is there a finite S-adic representation for all recurrent eventually dendric sets? There is one for recurrent dendric sets [7].

Acknowledgements. We thank Valérie Berthé, Paulina Cecchi, Fabien Durand and Samuel Petite for useful conversations on this subject.

References

1. Arnoux, P., Rauzy, G.: Représentation géométrique de suites de complexité 2n + 1. Bull. Soc. Math. France **119**(2), 199–215 (1991)
2. Balková, L., Pelantová, E., Steiner, W.: Sequences with constant number of return words. Monatsh. Math. **155**(3–4), 251–263 (2008)
3. Berstel, J., De Felice, C., Perrin, D., Reutenauer, C., Rindone, G.: Bifix codes and Sturmian words. J. Algebra **369**, 146–202 (2012)
4. Berthé, V., et al.: Specular sets. Theor. Comput. Sci. **684**, 3–28 (2017)
5. Berthé, V., et al.: Acyclic, connected and tree sets. Monatsh. Math. **176**(4), 521–550 (2015)
6. Berthé, V., et al.: Bifix codes and interval exchanges. J. Pure Appl. Algebra **219**(7), 2781–2798 (2015)
7. Berthé, V., et al.: Maximal bifix decoding. Dicrete Math. **338**, 725–742 (2015)
8. Cassaigne, J.: Complexité et facteurs spéciaux. Bull. Belg. Math. Soc. Simon Stevin **4**(1), 67–88 (1997). Journées Montoises (Mons, 1994)
9. Cassaigne, J., Nicolas, F.: Factor complexity. In: Combinatorics, Automata and Number Theory. Encyclopedia of Mathematics and its Applications, vol. 135, pp. 163–247. Cambridge University Press, Cambridge (2010)
10. Cornfeld, I.P., Fomin, S.V., Sinai, Y.G.: Ergodic Theory. Grundlehren der Mathematischen Wissenschaften (Fundamental Principles of Mathematical Sciences), vol. 245. Springer-Verlag, New York (1982). https://doi.org/10.1007/978-1-4615-6927-5. Translated from the Russian by A. B. Sosinskii
11. Dolce, F., Perrin, D.: Neutral and tree sets of arbitrary characteristic. Theor. Comput. Sci. **658**(Part A), 159–174 (2017)
12. Dolce, F., Perrin, D.: Eventually dendric shifts. In: van Bevern, R., Kucherov, G. (eds.) CSR 2019. LNCS, vol. 11532, pp. 106–118. Springer, Cham (2019). https://doi.org/10.1007/978-3-030-19955-5_10
13. Durand, F., Leroy, J., Richomme, G.: Do the properties of an S-adic representation determine factor complexity? J. Integer Seq. **16**(2), 30 (2013). Article 13.2.6
14. Ferenczi, S.: Rank and symbolic complexity. Ergodic Theory Dyn. Syst. **16**(4), 663–682 (1996)
15. Fogg, N.P., Berthé, V., Ferenczi, S., Mauduit, C., Siegel, A.: Substitutions in Dynamics, Arithmetics and Combinatorics. LNM, vol. 1794. Springer, Berlin (2002)
16. Leroy, J.: Some improvements of the S-adic conjecture. Adv. Appl. Math. **48**(1), 79–98 (2012)
17. Queffélec, M.: Substitution Dynamical Systems-Spectral Analysis. LNM, vol. 1294, 2nd edn. Springer, Berlin (2010)

Enumeration and Extensions
of Word-Representants

Marisa Gaetz[1]($\boxtimes$) and Caleb Ji[2]($\boxtimes$)

[1] Massachusetts Institute of Technology,
77 Massachusetts Ave, Cambridge, MA 02139, USA
mgaetz@mit.edu
[2] Washington University in St. Louis,
1 Brookings Dr, St. Louis, MO 63130, USA
caleb.ji@wustl.edu

Abstract. Given a finite word w over a finite alphabet V, we may construct a graph with vertex set V and an edge between to elements of V if and only if they alternate in the word w. This is the notion of word-representability of graphs. In this paper, we first study minimal length words which represent graphs, giving an explicit formula for both the length and the number of such words in the case of trees and cycles. Then we extend this notion to study the graphs representable with other patterns in words, proving in all cases aside from one (still unknown to us), all graphs are representable by all other patterns. Finally, we pose a few open problems for further work.

1 Introduction

The theory of word-representable graphs gives a way to associate a graph with a word. Motivated by the study of Perkins semigroups [1], this topic has been the subject of much research since its inception. A major theme of this research has been on the classification of word-representable graphs [3,6,7]. Other papers have studied variants and extensions to the original notion of word-representability [4,5].

In this paper, we begin by giving a brief review of the basic definitions and results in this field. For a more thorough treatment, we refer the reader to [2]. Then we go on to study minimal length representants of word-representable graphs, first studied in [3]. We obtain an explicit formula for the number of words of minimal length representing a given tree or a cycle. In the following section, we extend the ideas from [5] to study the graphs representable with other patterns in words. Next, we consider the word-representability of hypergraphs through a natural generalization of the original notion. We will end with some open questions and avenues for further research.

Definition 1 ([2]). *A simple, undirected graph $G = (V, E)$ is word-representable if there exists a word w over the alphabet V such that any $x, y \in V$*

© Springer Nature Switzerland AG 2019
R. Mercaş and D. Reidenbach (Eds.): WORDS 2019, LNCS 11682, pp. 180–192, 2019.
https://doi.org/10.1007/978-3-030-28796-2_14

alternate in w if and only if xy ∈ E; that is, if there are no two consecutive instances of x or y without an occurrence of the other in between. We require that w contains each letter of V at least once. We say that w represents G or that w is a word-representant, or simply representant, for G.

A word w is called *k-uniform* if each letter in w occurs exactly k times. For example, 12332414 is a 2-uniform word, while the 1-uniform words are precisely the permutations.

Definition 2 ([2]). *A graph G is k-word-representable if it has a k-uniform representant w.*

In fact, it has been shown that Definitions 1 and 2 are equivalent.

Theorem 1 ([3]). *A graph is word-representable if and only if it is k-word-representable for some k.*

For a word w, we call the permutation obtained by removing all but the leftmost occurrence of each letter in w the *initial permutation* of w. Let $\pi(w)$ denote the initial permutation of w. Similarly, we call the permutation obtained by removing all but the rightmost occurrence of each letter in w the *final permutation* of w. Let $\sigma(w)$ denote the final permutation of w.

Henceforth, all graphs will be taken to be simple and undirected.

In addition to the initial and final permutations of a word, it will often be useful to consider the restriction of a word to some subset of the vertices it is defined on. To denote a word w restricted so some letters $x_1, \ldots, x_m$, we write $w|_{x_1 \cdots x_m}$. For example, if $w = 132435213$, then $w|_{12} = 1221$.

2   Enumeration of Minimal Length Representants

By Theorem 1, the following notion introduced in [2] by Kitaev is well-defined.

Definition 3 ([2]). *Let G be a word-representable graph. The representation number of G is the least k such that G is k-word-representable.*

In [2], Kitaev goes on to study the class of graphs with representation number two. Rather than following Kitaev in investigating minimal length *uniform* representants of graphs, we here choose to study *absolute* minimal length representants.

Definition 4. *For a word-representable graph G, define ℓ(G) as the minimal length of a word w that represents G.*

It is important to note that, following Kitaev's definition of "word-representant" (stated above as Definition 1), we require that a word-representant w for a graph $G = (V, E)$ contains each letter of V at least once. As a result of this requirement, we do not need to worry about whether it makes sense for letters x and y to "alternate" in w when one (or both) of x and y do not appear in w.

Definition 5. *For a word-representable graph G, define n(G) as the number of words of length ℓ(G) that represent G.*

2.1   General Results

We obtain the following bound for the $\ell(G)$ of a general graph.

Theorem 2. *Let $G = (V, E)$ be a word-representable graph with connected components $\{G_i = (V_i, E_i)\}_{i=1}^k$. Then*

$$\ell(G) \leq \sum_{i=1}^{k} (\ell(G_i) + |V_i|) - \max_{1 \leq j \leq k} |V_j|.$$

Proof. For each $i \in \{1, 2, \ldots, k\}$, let w_i be a minimal length representant of G_i. We claim that the word $w = w_1 \sigma(w_1) \ w_2 \sigma(w_2) \cdots w_{k-1} \sigma(w_{k-1}) \ w_k$ represents G and has length $\sum_{i=1}^{k} (\ell(G_i) + |V_i|) - |V_k|$. For each $i \in \{1, 2, \ldots, k\}$, we have that w_i represents G_i over the alphabet V_i and that $w_i \sigma(w_i)$ represents G_i over the alphabet V_i (since appending $\sigma(w_i)$ to w_i does not affect which letters alternate). Noting that the sets $V_1, V_2, \ldots, V_k$ are pairwise disjoint, we have that $w|_{V_i}$ represents G_i for all $1 \leq i \leq k$. Furthermore, for all pairs (i, j) satisfying $1 \leq i < j \leq k$, we have that every vertex in V_i occurs at least twice in w before any vertex in V_j appears. Therefore, w accurately encodes each connected component of G as well as the information that there are no edges between different connected components of G. It follows that w represents G, as desired. Finally, it is clear by construction that w has length $\sum_{i=1}^{k} (\ell(G_i) + |V_i|) - |V_k|$. $\square$

2.2   Trees

In the case of trees, we find a precise value for $\ell(G)$.

Theorem 3. *Let $T = (V, E)$ be a tree, and let $n := |V|$. Then $\ell(T) = 2n - 2$.*

Proof. We first prove that $\ell(T) \geq 2n - 2$. Let w be a minimal length word-representant of T. By definition of word-representant (see Definition 1), we have that all elements of V occur at least once in w. We claim that there are at most two elements of V that occur only once in w. Suppose, for the sake of contradiction, that $x, y, z \in V$ each occur exactly once in w. Then any pair of letters chosen from $\{x, y, z\}$ must alternate in w. Consequently, there are edges $xy, yz, xz \in E$ forming a triangle in T, contradicting the fact that T is a tree. Therefore, at most two elements of V occur only once in w, meaning $\ell(T) \geq 2(n - 2) + 2 = 2n - 2$.

We now show by induction on n that there is a word of length $2n - 2$ that represents T. Observe that the word 12 represents the unique tree on two vertices (namely, the tree $T_2 = (V_2, E_2)$ defined by $V_2 = \{1, 2\}$ and $E_2 = \{12\}$). In other words, the result holds for $n = 2$. Assume that the result holds up to $n = k - 1$, where $k \geq 3$. Let $T_k = (V_k, E_k)$ be any tree on k vertices. Let a denote a leaf of T_k, and let b denote the parent of a. By the inductive hypothesis, there is a word w' of length $2k - 4$ that represents the tree $T_k \setminus \{a\}$ obtained from T_k by removing the vertex a and the edge ab. Now, replace one instance of b in w' with aba, and let w denote the resulting word.

We claim that w represents T_k. Recall from the above argument that at most two elements of $V_k \setminus \{a\}$ occur only once in w'. Since w' has length $2k-4$, all of the other $k-3$ elements of $V_k \setminus \{a\}$ must occur exactly twice in w'. In particular, there are at most two instances of b in w' (and hence in w). It follows that a and b alternate in w. Furthermore, a clearly does not alternate with any other letters in w. Consequently, we see that w is a length $2k-2$ word-representant of T. The theorem follows by induction.

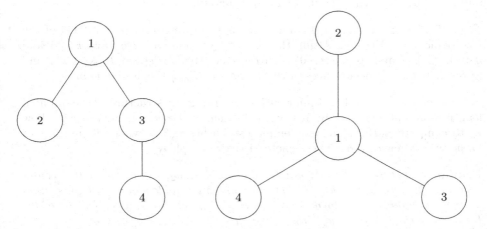

Fig. 1. A tree on four vertices.   **Fig. 2.** The star graph S_3.

Example 1. Figure 1 depicts a tree on four vertices. According to Theorem 3, a minimal length representant for this tree has length $2 \cdot 4 - 2 = 6$. It is straightforward to check that 212434 is a minimal length representant of the depicted tree.

For a tree $T = (V, E)$, we would now like to count the number $n(T)$ of minimal length representants for T. By Theorem 3, such representants have length $2|V|-2$ and are such that two vertices occur once and all other vertices occur twice. Note that the two vertices appearing once alternate and thus are connected in the graph. Although this gives us some idea as to the structure of minimal length representants of T, we would like a more detailed picture. To this end, we establish the following notation and lemma.

For a tree $T = (V, E)$ and an edge $xy \in E$, let $T_{x,xy}$ be the subtree of T obtained by deleting the edge xy and taking the connected component containing x. Define $T_{y,xy}$ similarly.

Lemma 1. *Let T be a tree and w a minimal length representant for T. Let $x, y \in V(T)$ be the vertices of T that occur only once in w; without loss of generality, let x occur before y in w. Then w is of the form $w = w_{x_1} w_{x_2} \cdots w_{x_m}\, w_{y_1} w_{y_2} \cdots w_{y_n}$, where $w_{x_1}, w_{x_2}, \ldots, w_{x_m} \in T_{x,xy}$ and $w_{y_1}, w_{y_2}, \ldots, w_{y_n} \in T_{y,xy}$.*

Proof. For sake of contradiction, assume that there are vertices $x_i \in T_{x,xy}$ and $y_j \in T_{y,xy}$ such that a copy of $T|_{x_i y_j} = y_j x_i x_i y_j$. Let $xx_1 \ldots, x_i$ and $yy_1 \ldots, y_j$ be paths in T. Since x appears before y and x_1 and y_1 alternate with x and y respectively but not with each other, we have that every occurrence of x_1 appears before every appearance of y_1. Then inductively it is clear that not both copies of y_j appear before x, so they both appear after x. Since the x_ks must alternate with each other but not with y_j, we see that all of them must appear between the two y_js as well. This implies that both x_1s are after one of the y_js and thus after x, giving the desired contradiction.

Example 2. Consider again the minimal length representant $w = 212434$ of the tree depicted in Figure 1. Using the notation of Lemma 1, we have $x = 1$ and $y = 3$. As Lemma 1 predicts, all occurrences of the vertices of $T_{x,xy}$ (i.e. 1 and 2) occur before all occurrences of the vertices of $T_{y,xy}$ (i.e. 3 and 4) in w.

Lemma 1 gives us a lot of information regarding the structure of the minimal length representants of T. In fact, with the help of the next lemma, it will allow us to compute $n(T)$. This next lemma establishes $n(S_k)$, where S_k is the *star graph* with k leaves (i.e. the complete bipartite graph $K_{1,k}$).

Lemma 2. *Let S_k be the star graph with leaves $u_1, u_2, \ldots, u_k$ and with a center vertex c. Then $n(S_k) = 2 \cdot k!$. Furthermore, any minimal length word-representant w of S_k is either of the form $w = u_k\, u_1 u_2 \cdots u_{k-1}\, c\, u_{k-1} u_{k-2} \cdots u_1$ or of the form $w = u_1 u_2 \cdots u_{k-1}\, c\, u_{k-1} u_{k-2} \cdots u_1\, u_k$.*

Proof. Note that K_n is representable by the word $123 \cdots n$. Now say the word w that represents a graph G. Say ij is an edge in G and G' is the graph G after deleting the edge ij.

Take $w' = w\sigma(w)t[i,j]$. It suffices to show that w' t-represents G'.

Indeed, because w' contains w, only the edges in G will be included, and because w' includes $t[i,j]$, there will be no edge between i and j. Thus it suffices to show that no other edges are removed.

Assume there exists an instance of the pattern $c^k d^l$ in w' that isn't in w, where $\{c,d\} \neq \{i,j\}$. If $c^k d^l$ begins in the w section, then the entirety of c^k must be contained in w, or else a d will appear before all k cs appear in a row. Now if the ds after c^k begins at the end of w, then c comes before d in $\sigma(w)$, which doesn't work either. Thus $c^k d^l$ has to appear after w in w'. Then it has to begin in $\sigma(w)$, but there is only one appearance of c in $\sigma(w)$. Thus both c and d appear in $t[i,j]$, and thus $\{c,d\} = \{i,j\}$, contradiction. Thus there is no such instance of $c^k d^l$ in w', so no new edges other than ij are removed. Thus w' t-represents G, as desired.

Example 3. Figure 2 shows the star graph S_3. Using the notation of Lemma 2, $c = 1$ for this graph. According to Lemma 2, the depicted graph has the 12 minimal length representants given in the following Table 1:

Table 1. Minimal length representants of the star graph S_3 (see Fig. 2).

$u_k = 2$	$u_k = 3$	$u_k = 4$
234143	324142	432123
341432	241423	321234
243134	342124	423132
431342	421243	231324

With Lemmas 1 and 2, we are now ready to compute $n(T)$.

Theorem 4. *Let $T = (V, E)$ be a tree on at least two vertices. Then*

$$n(T) = 2 \prod_{v \in V} \deg(v)! \sum_{xy \in E} \frac{1}{\deg(x) \deg(y)},$$

where $\deg(v)$ denotes the degree of v in T.

We briefly present the idea of the proof below.

Proof (Idea). We loop over all edges xy and sum up the number of representants which have exactly one occurrence of x and y. We use Lemma 1 to break up the representant into halves and inductively apply Lemma 2. Some calculation gives the desired result.

Example 4. Consider the tree $T = (V, E)$ on four vertices depicted in Fig. 1. According to Theorem 4,

$$n(T) = 2 \prod_{v \in V} \deg(v)! \sum_{xy \in E} \frac{1}{\deg(x) \deg(y)}$$

$$= 2 \left(2! \cdot 1! \cdot 2! \cdot 1! \right) \left(\frac{1}{1 \cdot 2} + \frac{1}{2 \cdot 2} + \frac{1}{2 \cdot 1} \right) = 8 \left(\frac{5}{4} \right) = 10.$$

Indeed, it is straightforward to verify that the ten representants shown in Table 2 are precisely the minimal word-representants of T, where x and y are the vertices appearing once.

Table 2. Minimal length representants of the tree shown in Fig. 1.

$\{x, y\}$	Corresponding minimal length representants of T
$\{1, 2\}$	231434, 314342, 243413, 434132
$\{1, 3\}$	212434, 434212
$\{3, 4\}$	212314, 132124, 421231, 413212

Theorem 4 has the following easy corollary.

Corollary 1. *Let P_k be the path on k vertices. Then for $k \geq 3$, we have*

$$n(P_k) = (k + 1) \cdot 2^{k-3}.$$

2.3   Cycles

In this subsection, we consider cycle graph C_n on $n \geq 3$ vertices. The minimal representants of C_3 are quite easy to understand.

Proposition 1. $\ell(C_3) = 3$ and $n(C_3) = 6$.

Proof. Since a representant by definition contains each vertex at least once, $\ell(C_3) \geq 3$. It is then straightforward to see that the six permutations 123, 132, 213, 231, 312, 321 are the only length-three word-representants of C_3, verifying that $\ell(C_3) = 3$ and that $n(C_3) = 6$. ∎

For $n \geq 4$, the minimal representants of C_n are more complicated.

Theorem 5. *Let* $C_n = (V, E)$ *be the cycle graph on* $n \geq 4$ *vertices. Then* $\ell(C_n) = 2n - 2$.

Proof. Let w be a minimal length word-representant for C_n. We first show that $\ell(C_n) \geq 2n - 2$. Recall that by definition of word-representant (see Definition 1), every element of V occurs at least once in w. Therefore, to show that $\ell(C_n) \geq 2n - 2$, it suffices to show that no more than two vertices can appear only once in w. Suppose, for the sake of contradiction, that three vertices $x, y, z \in V$ occur exactly once in w. Then $xy, xz, yz \in E$, and these edges form a triangle in C_n. This contradicts the fact that C_n is the cycle graph on $n \geq 4$ vertices.

We now show that $\ell(C_n) \leq 2n - 2$ by constructing a length $2n - 2$ word-representant for C_n. Suppose that the vertices of C_n are labeled $1, 2, \ldots, n$ such that there are edges between vertices labeled with consecutive integers, as well as between n and 1. Consider the word $w' = n1\ (n-1)n\ (n-2)(n-1)\ (n-3)(n-2)\ \cdots\ 45\ 34\ 23$. Here, we format w' by adding space between every pair of letters to illustrate the structure of the word; the first letters in each pair form the decreasing sequence $n, n-1, n-2, n-3, \ldots, 4, 3, 2$, while the second letters in each pair form the sequence $1, n, n-1, n-2, \ldots, 5, 4, 3$. In this construction, the letters 1 and 2 appear exactly once, while the letters $3, 4, \ldots, n$ appear exactly twice. Moreover, for every vertex $v \in V \setminus \{1, 2\}$, the letters appearing between the two occurrences of v in w' are precisely $v + 1$ and $v - 1$. Consequently, v alternates only with $v + 1$ and $v - 1$ in w'. Additionally, 1 alternates only with 2 and n, and 2 alternates only with 1 and 3. ∎

Example 5. Consider the cycle graph C_5 shown in Fig. 3. According to Theorem 5, $\ell(C_5) = 2 \cdot 5 - 2 = 8$. It is straightforward to check that 51453423 is therefore a minimal length representant of C_5.

Having established that $\ell(C_3) = 3$ and that $\ell(C_n) = 2n - 2$ for $n \geq 4$, we would now like to establish $n(C_n)$. We have already shown that $n(C_3) = 6$. In fact, $n(C_n) = 2n$ for all $n \geq 3$.

Theorem 6. *Let* $C_n = (V, E)$ *be the cycle graph on* n *vertices. For* $n \geq 3$, $n(C_n) = 2n$.

Proof. Let w be a minimal length word-representant for C_n. Since $\ell(C_n) = 2n-2$, two elements of V will occur exactly once in w, while the remaining vertices will occur exactly twice.

There are n choices to choose the pair of vertices which appear only once, since they alternate and thus there must be an edge between them. After this pair of vertices is chosen, there are two orders they can occur in a minimal length representant. Without loss of generality, suppose the pair we have chosen is $(1, 2)$ and that 1 occurs before 2.

We will now consider the ways in which two instances of each of $3, 4, \ldots, n$ can be placed in this minimal word-representant w (which is assumed to be such that $w|_{12} = 12$. Certainly, the two 3's must surround the 2 but not the 1. Moreover, the two 4's must alternate with the two 3's without alternating with the 1 or the 2. We claim that the 4's must be between 1 and 2. Indeed, otherwise both 4's will be to the right of the 2. If this is the case, then every successive number must have at least one instance to the right of the 2, and hence both instances to the right of the 2 (so as to avoid alternating with 2). This would make it impossible for n to alternate with 1. Therefore, our representant must satisfy $w|_{1234} = 143423$. Similarly, the two 5's must surround the leftmost 4 without alternating with 3, giving $w|_{12345} = 15453423$. We can continue in this manner and conclude with the two n's immediately surrounding 1 and the leftmost instance of $n - 1$.

Observe that after choosing the pair $(1, 2)$ and their relative locations, this process has involved no choices. In other words, once 1 and 2 were chosen and placed in that order, there was only one way to form a word representant of C_n. It follows that $n(C_n) = 2n$, as desired.

3 Graphs Representable from Pattern Avoidance in Words

So far in this paper we have exclusively considered word-representable graphs. In this section, we define a more general notion of representability for graphs, motivated by Kitaev in [5]. To this end, we first establish two preliminary definitions.

Definition 6. *Two words are said to be isomorphic if there is a bijective, pattern-preserving correspondence between their letters.*

Example 6. The words 112134, 332378, and *aabacd* are all isomorphic.

Definition 7. *Let w be a word defined on an alphabet V, and let u be a word defined on an alphabet U. Then w avoids u if the set*

$$\big\{\{x_1, x_2, \ldots, x_{|U|}\} \subseteq V \; : \; w|_{x_1, \ldots, x_n} \text{ has a contiguous subword isomorphic to } u\big\}$$

is empty. If w does not avoid u, we say w contains u.

Example 7. Let $w = 121223$ and let $u = 112$. Then w contains u, since $w|_{23} = 2223$ has a subword that is isomorphic to u (namely, 223).

With these definitions in mind, we can now introduce a generalized notion of graph representability.

Definition 8. *Let t be a word on two letters. A t-representable graph is one where there is an edge between x and y if and only if the subword induced by x and y avoids the pattern given by t.*

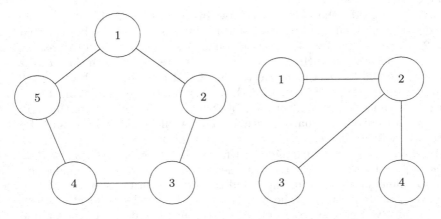

Fig. 3. The cycle graph C_5.   **Fig. 4.** A 112-representable graph.

Example 8. The graph shown in Fig. 4 is 112-representable with $w = 121334$ serving as a 112-representant.

Remark 1. The word-representable graphs are precisely the 11-representable graphs. This can be seen by noting that "alternating" is equivalent to avoiding the word 11. In general, a word 1^k-represents a graph if and only if it a^k-represents it.

Remark 2. In [5], Kitaev defines "u-representability", an alternate way of defining representability from a word u. However, his definition depends on the ordering of the vertices comprising u. For instance, 112-representants are distinct from 221-representants in his definition. Our version is consistent with the traditional notion of pattern avoidance and does not distinguish between these.

Example 9. Let $w = 2123$. Then w aba-represents the graph with vertex set $\{1, 2, 3\}$ with edges between 1 and 3 and between 2 and 3. However, w 121-represents the complete graph on the vertices $\{1, 2, 3\}$.

We now prove several results on the representability of graphs for various t. In [5], Kitaev shows all graphs can be u-represented if the length of u is at least 3 in the following way. If w u-represents some graph G, he constructs a word w' from w that u-represents G', where G' is the graph G after deleting any edge. We use this same approach to show that all graphs are t-representable for certain t.

Note that if t is of the form 1^k for some $k > 1$, then t-representations are the same as u-representations. Thus we only consider the cases in which t has two distinct letters.

Theorem 7. *If t is of the form $a^k b^l a \cdots$ for positive integers k and l, then every graph is t-representable.*

Proof. Note that K_n is representable by the word $123 \cdots n$. Now say the word w that represents a graph G. Say ij is an edge in G and G' is the graph G after deleting the edge ij.

Let $t[i,j]$ be the word t after the substitution $a \to i, b \to j$ has been made. Take $w' = wr(\sigma(w))_1^{l+1} r(\sigma(w))_2^{l+1} \cdots r(\sigma(w))_n^{l+1} t[i,j]$. We claim that w' t-represents G'.

Indeed, because w' contains w, only the edges in G will be included, and because w' includes $t[i,j]$, there will be no edge between i and j. Thus it suffices to show that no other edges are removed.

Assume there exists an instance of the pattern $c^k d^l c \cdots$ in w' that isn't in w, where $\{c,d\} \neq \{i,j\}$. Then consider where the d^l can be placed. It cannot be a part of in the section $r(\sigma(w))_1^{l+1} r(\sigma(w))_2^{l+1} \cdots r(\sigma(w))_n^{l+1}$, or else there would be $l+1$ consecutive appearances of d. It cannot come in $t[i,j]$, or else $\{c,d\} = \{i,j\}$. Therefore it must be completely contained in w. Then we know that there can be no instance of c in w that comes after those l bs, or else the pattern would have already existed in w. Then c must appear before d in w' after w. But by our construction, the opposite happens. Thus there is no such instance of $c^k d^l c \cdots$ in w', so no new edges other than ij are removed. Thus w' t-represents G, as desired.

In this way, we can construct a word t-representing G for any graph G by removing edges from K_n to reach G one at a time.

Theorem 8. *If t is of the form $a^k b^l$ where $k, l \geq 2$, then every graph is t-representable.*

Proof. Note that K_n is representable by the word $123 \cdots n$. Now say the word w that represents a graph G. Say ij is an edge in G and G' is the graph G after deleting the edge ij.

Take $w' = w\sigma(w)t[i,j]$. It suffices to show that w' t-represents G'.

Indeed, because w' contains w, only the edges in G will be included, and because w' includes $t[i,j]$, there will be no edge between i and j. Thus it suffices to show that no other edges are removed.

Assume there exists an instance of the pattern $c^k d^l$ in w' that isn't in w, where $\{c,d\} \neq \{i,j\}$. If $c^k d^l$ begins in the w section, then the entirety of c^k must be contained in w, or else a d will appear before all k cs appear in a row. Now if the ds after c^k begins at the end of w, then c comes before d in $\sigma(w)$, which doesn't work either. Thus $c^k d^l$ has to appear after w in w'. Then it has to begin in $\sigma(w)$, but there is only one appearance of c in $\sigma(w)$. Thus both c and d appear in $t[i,j]$, and thus $\{c,d\} = \{i,j\}$, contradiction. Thus there is no such instance of $c^k d^l$ in w', so no new edges other than ij are removed. Thus w' t-represents G, as desired.

The remaining cases are of the form $t = a^k b$ and $t = ab^k$, where $k \geq 1$. First, note that the representable graphs in these cases are the same. Indeed, if w $a^k b$-represents some graph G, then the reverse of w ab^k-represents G and vice versa.

For the case of $k = 1$, because every vertex must appear at least once in a representant, there will always be an appearance of the pattern ab for any two vertices. Thus, only the graph with no edges is ab-representable.

We now deal with the case $k \geq 2$.

Theorem 9. Let $k \geq 3$ and set $t = a^k b$. Then every graph is t-representable.

Proof. Note that K_n is representable by the word $123 \cdots n$. Now say the word w that represents a graph G. Say ij is an edge in G and G' is the graph G after deleting the edge ij.

Let v be a permutation of the elements of the set $V(G) \backslash \{i,j\}$. Take $w' = i^{k-1} v i j \pi(w) w$. It suffices to show that w' t-represents G'.

Indeed, because w' contains w, only the edges in G will be included, and because w' includes an instance of $i^k j$ at the front, there will be no edge between i and j. Thus it suffices to show that no other edges are removed.

Assume there exists an instance of the pattern $c^k d^l$ in w' that isn't in w, where $\{c,d\} \neq \{i,j\}$. Clearly the c^k section cannot intersect with the i^{k-1} section, so it must begin in the section $v i j \pi(w)$. Note that this is a concatenation of two permutations of the elements of the set $V(G)$. Thus there can only be two occurrences of c in that section. But because $\pi(w)$ is the initial permutation of w, every other element will appear between the second occurrence of c and its first occurrence in w, so we cannot have an instance of $c^k d$. Thus there is no such instance of $c^k d^l$ in w', so no new edges other than ij are removed. Thus w' t-represents G, as desired.

For $k = 2$, we have not yet fully determined the $a^k b$-representable graphs. However, we have the following theorem.

Theorem 10. All aa-representable graphs are also aab-representable.

Proof. Let w aa-represent a graph G. We claim that $w\sigma(w)$ aab-represents G, where $\sigma(w)$ denotes the final permutation of w. Let x and y be any two distinct vertices of G.

Suppose first that $(x,y) \notin E(G)$. Then the subword $w|_{xy}$ of w induced by x and y contains at least one occurrence of the pattern aa. If $w|_{xy}$ also contains at least one occurrence of aab, then so does $w\sigma(w)$, and this direction of the proof is complete. Therefore, suppose that $w|_{xy}$ avoids aab. Then $w|_{xy}$ must end in the pattern aa. If $w|_{xy}$ ends in xx, then $\sigma(w)|_{xy} = yx$, meaning $(w\sigma(w))|_{xy}$ ends in $xxyx$, which contains 112. The case in which $w|_{xy}$ ends in yy follows similarly.

Next, suppose that $(x, y) \in E(G)$. Then $w|_{xy}$ avoids aa (and therefore avoids aab). Appending $\sigma(w)$ to w does not introduce any occurrences of the pattern aa. Therefore, $(w\sigma(w))|_{xy}$ avoids aab.

4 Further Work

In Sect. 2, we proved the following upper bound for $l(G)$ in terms of its connected components.

$$\ell(G) \leq \sum_{i=1}^{k}(\ell(G_i) + |V_i|) - \max_{1 \leq j \leq k} |V_j|. \tag{1}$$

It would be interesting to strengthen this bound. This leads to the following problem.

Problem 1. Find stricter bounds for $l(G)$ in terms of its connected components and classify all graphs for which equality in Eq. 1 holds.

We also proved that for both trees and cycles, $\ell(G) = 2|G| - 2$. For all triangle-free graphs G, it is clear that $\ell(G) \geq 2|G| - 2$, because no more than two vertices can appear only once. We ask if these observations can be extended to classify graphs with representation number 2 for which $\ell(G)$ is close to $2|G|$.

Problem 2. Classify all graphs G with representation number 2 for which

(a) $\ell(G) = 2|G| - 2$,
(b) $\ell(G) = 2|G| - 1$,
(c) $\ell(G) = 2|G|$.

Recall that in [2], Kitaev classifies all graphs with representation number 2. This gives a nice starting point from which to attack the problem above.

In Sect. 3, we show that all graphs can be t-represented as long as t as length at least 3, except for the cases where t is of the form ab^k and a^kb. Resolving this last case is an open problem.

Problem 3. Characterize the graphs which are t-representable for $t = ab^k$ or $t = a^kb$.

Acknowledgments. This research was conducted at the University of Minnesota Duluth REU, funded by NSF Grant 1650947 and NSA Grant H98230-18-1-0010. The authors would like to thank Joe Gallian for running the REU and suggesting the topic. The authors also thank their advisors Levent Alpoge, Aaron Berger, and Colin Defant for their support, as well as the anonymous reviewers for many helpful comments.

References

1. Kitaev, S., Seif, S.: Word problem of the Perkins semigroup via directed acyclic graphs. Order **25**(3), 177–194 (2008)
2. Kitaev, S.: A comprehensive introduction to the theory of word-representable graphs. In: Charlier, É., Leroy, J., Rigo, M. (eds.) DLT 2017. LNCS, vol. 10396, pp. 36–67. Springer, Cham (2017). https://doi.org/10.1007/978-3-319-62809-7_2
3. Kitaev, S., Pyatkin, A.: On representable graphs. J. Autom. Lang. Comb. **13**(1), 45–54 (2008)
4. Cheon, G., Kim, J., Kim, M., Kitaev, S., Pyatkin, A.: On k-11-representable graphs. arXiv preprint: https://arxiv.org/abs/1803.01055.pdf
5. Kitaev, S.: Existence of u-representation of graphs. J. Graph Theory **85**(3), 661–668 (2017)
6. Halldórsson, M.M., Kitaev, S., Pyatkin, A.: Graphs capturing alternations in words. In: Gao, Y., Lu, H., Seki, S., Yu, S. (eds.) DLT 2010. LNCS, vol. 6224, pp. 436–437. Springer, Heidelberg (2010). https://doi.org/10.1007/978-3-642-14455-4_41
7. Kitaev, S., Salimov, P., Severs, C., Ulfarsson, H.: Word-representability and line graphs. Open J. Discrete Math. **1**(2), 96–101 (2011)

Localisation-Resistant Random Words with Small Alphabets

Cyril Gavoille[1]([✉]), Ghazal Kachigar[1,2], and Gilles Zémor[2]

[1] LaBRI, University of Bordeaux, Bordeaux, France
{gavoille,ghazal.kachigar}@labri.fr
[2] IMB, University of Bordeaux, Bordeaux, France
zemor@math.u-bordeaux.fr

Abstract. We consider q-coloured words, that is words on $\{1, \ldots, q\}$ where no two consecutive letters are equal. Motivated by multipartite colouring games with nonsignalling resources, we are interested in random q-coloured words satisfying a k-*localisability* property. More precisely, the probability of containing any given pair of words as subwords spaced at least k letters apart can depend only on their lengths. We focus on the issue of the smallest alphabet size q for which a probability distribution for such random words can exist. For $k = 1$, we prove a lower bound of $q \geqslant 4$. The bound is optimal because there exists a suitable distribution for random 4-colourings that was constructed by Holroyd and Liggett in 2015. Our lower bound can be generalized to k-localisable random words where the letters of each subword of $k + 1$ letters must be pairwise different. We show that the alphabet size in this case must be at least $(k + 1) \cdot (1 + 1/k)^k$.

Keywords: Random words · Stochastic colouring process · Hard-core process · Colouring game

1 Introduction

Multipartite Colouring Game. Let us consider the following general multipartite graph colouring game. We are given a graph G with nodes $v_1, \ldots, v_n$, a colour bound q, and m players $P_1, \ldots, P_m$. The referee virtually places each player P_i at a node v_j and gives them a personalised input. This information depends on the variant of the game. For example, it can consist of the index j that the player P_i is placed on. Each player then has to output a colour for its node, i.e., an integer taken from $\{1, \ldots, q\}$. The players win if the resulting node colouring is a q-colouring of the coloured subgraph of G, i.e., the subgraph induced by the nodes hosting at least one player. More precisely, colours must differ for adjacent players and coincide for players that have been placed on the same node, if any.

Supported by the French ANR projects ANR-16-CE40-0023 (DESCARTES) and ANR-18-CE47-0010 (QUDATA).

R. Mercaş and D. Reidenbach (Eds.): WORDS 2019, LNCS 11682, pp. 193–206, 2019.
https://doi.org/10.1007/978-3-030-28796-2_15

Players are allowed to agree on a joint strategy beforehand, which may depend on G, q and m. Once placed on their node with the referee's input, players are not allowed to communicate in any way. So, the output colour can only depend on the joint strategy and on the referee's input. Unless specified by the referee's input, players are not aware whether other players stand at the same node, or which players are adjacent to them.

The main question is to understand how small the colour bound q can be so that the m players can still win the game for G, under given assumptions on their joint strategy and on the referee's inputs.

This question is related to fundamental problems in graph theory, distributed computing, and quantum information. To illustrate, assume that the referee's input consists of the node index where each player is placed. For $m = n$, the smallest q is precisely the chromatic number of G since the referee can force the players to output a q-colouring of the whole graph, and a strategy for the players consists of agreeing beforehand on any given q-colouring. For $m = 2$, and if the two players share quantum resources (materialised say by entangled particles), this leads to the notion of *quantum chromatic number*. Interestingly, this variant of chromatic number can be smaller than the classical one. For instance, there is a graph with 18 nodes and chromatic number 5 on which the two players Alice and Bob can win this colouring game with only 4 colours [8].

Two-partite games with quantum resources (sometimes called pseudo-telepathy games) are well studied in Computer Science and in Physics. However, multipartite quantum games with a large number m of players are much less understood. There are multipartite games where quantum superiority can be proved, and also outperformed (in terms of winning probability) by general *nonsignalling* resources [1,2,7,12]. Such exotic resources, which are not predicted to exist according to current physical theories, allow the players to use *any* non-local correlations in their outputs without any communication.

Links to Distributed Computing. Colouring a network with a minimal amount of communication is a fundamental symmetry-breaking problem studied in distributed computing (see [4,10,13,15,16] for recent breakthroughs, and [3] for a book dedicated to this field). In this setting and in brief, each player acts at a single node as a processor that can exchange messages with its neighbours in some underlying graph. They must output a colouring of the graph after a limited number of synchronous rounds of communication. In this model, a.k.a. the LOCAL model, there are distributed algorithms for q-colouring n-node paths[1] that require $O(\log^* n - \log^* q)$ rounds[2] of communication for $q > 2$, and this is tight. More precisely, [26,27] showed that after collecting the IDs of k neighbours around each node, i.e., $k = 2t$ numbers after t two-sided rounds of communication in the path, any possibly randomised q-colouring algorithm must satisfy $q = \Omega(\log^{(k)} n)$.

---

[1] This holds also for cycles, and more generally for graphs of maximum degree Δ with $q > \Delta$.

[2] We write $\log^* n = \min\{i : \log_2^{(i)} n \leqslant 1\}$, the inverse function of a power-2 tower.

Since after t rounds every node has been able to communicate with nodes at distance at most t, these t-round algorithms imply that information about nodes at distance t suffices to provide a q-colouring. Yet, the colouring problem in the LOCAL model can be viewed as a particular setting of the general multipartite colouring game where the referee's placement is a permutation ($m = n$) and where the input for each player is the t-neighbourhood of its node[3]. Whether the number of rounds t can be significantly reduced if quantum resources are available is a widely open question [14,24,25] even for path graphs.

Lastly, we notice that the colouring game as described above can be further extended to *locally checkable labelling* games where the goal for the players is to output a label taken from a predefined set and satisfying some local constraints in the graph. This captures not only colouring problems, but also maximal independent set, dominating set, weak 2-colouring[4], and many others [6,28].

Our Contribution. In this paper we consider the multipartite q-colouring game for the path $v_1 - v_2 - \cdots - v_n$ where the referee's placement is a permutation σ on $\{1, \ldots, n\}$ not revealed to the players. So player P_i is placed at position $\sigma(i)$ on the path, i.e., at the node $v_{\sigma(i)}$. The referee's input for P_i consists only of the index $\sigma^{-1}(\sigma(i)+1)$ of the player placed on its right neighbouring node[5], $v_{\sigma(i)+1}$. As a result, each player outputs a colour given its own index and the one of its right neighbour. The players win the game if they produce a q-colouring of the path. Thus in this game a player can coordinate only with its right neighbour and its colour cannot depend on its position $\sigma(i)$.

As explained above, this game can be seen as the q-colouring problem in the distributed LOCAL model where each processor has received information only from its right neighbour, after what may be called a half-round of communication. From the above lower bound with $k = 1$, we must have $q = \Omega(\log n)$ for every joint strategy based on classical resources including shared randomness.

The question we want to address is what is the minimum number of colours q that can be achieved if players are allowed to use quantum resources, and more generally any nonsignalling resources. It should be stressed that quantum resources allow each player to use non-local correlations that may in fact beat the previous $\Omega(\log n)$ lower bound on q. We prove in this paper that $q \geqslant 4$ for n large enough, and this is optimal.

To formally state our main theorem, we interpret the colouring resulting from a run of the game as a random word $X_1 X_2 \cdots X_n$, where each letter X_i is a random variable ranging in $\{1, \ldots, q\}$ and corresponding to the colour output by the player at node v_i. Here randomness may come from the kind of resource used by the players in their joint strategy (e.g., a quantum state) that is revealed at the time they output the colour (e.g., measurement).

---

[3] Say, the list of player's IDs and the edges list between them.

[4] In this problem, each player must produce one out two possible colours such that at least one of its adjacent node receives a different colour.

[5] Player at v_n receives index 0. Alternatively, we may assume that an extra player P_0 is always placed at a virtual node v_{n+1} and does not take part in the colouring game.

Given an interval $I = [a, b]$, we use the notation X_I for the subword $X_a X_{a+1} \cdots X_{b-1} X_b$. Define the *distance* between any two intervals I, J as $\inf \{|i - j| : i \in I, j \in J\}$. (The distance between I and $J = \varnothing$ is $+\infty$ by convention). Note that the two subwords X_I and X_J are separated by k letters in $X_1 \cdots X_n$ if and only if I and J are at distance $k + 1$. We say that a word is *coloured* if any two consecutive letters are distinct.

In order to lower bound q for any probability distribution for random q-coloured words coming from such games, we introduce the notion of k-localisability defined as follows:

Definition 1. *A probability distribution for a random word $X_1 \cdots X_n$ is k-localisable if, for all intervals $I, J \subseteq \{1, \ldots, n\}$ at distance more than k, the distribution of (X_I, X_J) can only depend on $\{|I|, |J|\}$.*

Informally, this means that the probability of having two given words S and T in a random word depends neither on their absolute positions, nor on their order, nor on their distance in the word, as long as the number of letters between them is at least k.

Coming back to our colouring game on the path where players are only aware of their immediate right neighbour, the word distribution resulting of any winning strategy based on nonsignalling resources must be 1-localisable. This is because otherwise two players at nodes v_i and v_j sufficiently far apart could collectively retrieve information about i, j or $|i - j|$ from their colour distribution. From the rules of the game this is not possible without *signals* (i.e., communication). This holds for any nonsignalling theory including quantum mechanics. Note however that a k-localisable colour distribution does not forbid non-local correlation.

Theorem 1. *Every 1-localisable probability distribution for random q-coloured words of length n requires $q \geqslant 4$ for n large enough.*

As we will see in the next paragraph, the lower bound of Theorem 1 is tight. This is actually a consequence of the random 4-colouring given in [20].

Our approach to prove Theorem 1 is to study random binary words $Y_1 \cdots Y_n$ obtained from a random q-coloured word $X_1 \cdots X_n$ by fixing any colour $c \in \{1, \ldots, q\}$ and by setting $Y_i = 1$ if $X_i = c$, and $Y_i = 0$ otherwise. Observe that $Y_1 \cdots Y_n$ codes an independent set of the n-node path, and let us call an *independent-set* word any binary word that does not contain any two consecutive ones. Such random words can also be seen as *hard-core* processes where the variable Y_i indicates the presence of a radius-1 hard-core particle at position i on the discrete line.

The lower bound of Theorem 1 is actually a corollary of our following main technical contribution. It gives a fine analysis of the marginal probabilities of having a given number of ones in fixed positions for 1-localisable random independent-set words, a result interesting in its own right. We let $c_n = \frac{1}{n+1}\binom{2n}{n}$ denote the n-th Catalan number.

Theorem 2. *Let p_i denote the probability of having i ones in the positions indexed by the odd integers $1, 3, \ldots, 2i - 1$, for a random independent-set word of length $n \geqslant 2i$. Let $\ell = \lfloor n/2 \rfloor$. Then, for every even n:*

i. *Every 1-localisable probability distribution for random independent-set words of length n satisfies, for each $i \in \{0, \ldots, \ell\}$, $p_i \leqslant c_{\ell-i+1}/c_{\ell+1}$.*

ii. *There exists a 1-localisable probability distribution for random independent-set words of length n such that, for each $i \in \{0, \ldots, \ell\}$, $p_i = c_{\ell-i+1}/c_{\ell+1}$.*

By marginalising, it is easy to derive from Theorem 2(i) that $p_i \leqslant c_{\lceil n/2 \rceil - i + 1}/c_{\lceil n/2 \rceil + 1}$ for every length n, and not only for even n.

Let us explain why Theorem 1 follows from Theorem 2. The first observation is that any letter transformation $Y_i = f(X_i)$ preserves the k-localisability of the distribution as long as f does not depend on i. Now, given any 1-localisable distribution for a random q-coloured word $X_1 \cdots X_n$, consider the most frequent colour c, so appearing with probability at least $1/q$ in the random word. The random independent-set word $Y_1 \cdots Y_n$ as defined above has a 1-localisable distribution. And the probability of having a one at any fixed position in $Y_1 \cdots Y_n$ is $p_1 \geqslant 1/q$. However, from Theorem 2(i) applied to $Y_1 \cdots Y_n$, we get that $p_1 \leqslant c_5/c_6 = 7/22$ whenever $n \geqslant 10$ noting that $c_\ell/c_{\ell+1} = (\ell + 2)/(4\ell + 2)$. Thus, we obtain $1/q \leqslant p_1 \leqslant 7/22$, implying that $q > 3$ as claimed in Theorem 1.

Related Works. The notion of k-localisability introduced in this paper is a natural notion for the study of multipartite colouring games on paths with quantum resources (and beyond). A related notion in probability theory is the well-known k-dependence of random variables [17,22] studied for more than seven decades. A probability distribution for random variables $X_1 \cdots X_n$ is k-*dependent* if, for all intervals $I, J \subseteq \{1, \ldots, n\}$ at distance more than k, the variables X_I and X_J are independent. Clearly, 0-dependence is the same as independence.

Recall that a probability distribution for a random word $X_1 \cdots X_n$ is *stationary* if, for every interval $I \subseteq \{1, \ldots, n\}$, the distribution of X_I can depend only on $|I|$. It is not difficult to see that any stationary k-dependent distribution is also k-localisable: for k-dependent distributions, the distribution of (X_I, X_J) is the product of the marginals which, by stationarity, can depend only on $|I|$ and on $|J|$. However, the reverse is false. Although every k-localisable distribution is stationary (setting $J = \varnothing$ in the definition), there exist k-localisable distributions that are not k-dependent. For instance $X_i = \sigma(i)$ for a uniform random permutation σ of $\{1, \ldots, n\}$ defines a 0-localisable distribution that is not k-dependent for every k. Indeed, $\mathbb{P}(X_I = S) = (n - |I|)!/n!$ and $\mathbb{P}(X_I = S, X_J = T) = (n - (|I| + |J|))!/n!$ for any two disjoint intervals I, J (so at distance more than k for some $k \geqslant 0$). However, $\mathbb{P}(X_I = S, X_J = T) \neq \mathbb{P}(X_I = S) \cdot \mathbb{P}(X_J = T)$ for every k. Furthermore, the random binary word defined by $Y_i = X_i \bmod 2$ is still 0-localisable and once again not k-dependent for every k.

Interestingly, the notion of 0-localisability corresponds to the notion of *exchangeability* [5,9], in connection with the celebrated de Finetti Theorem that explains the relationship between exchangeability and independence. Random variables are (finitely) exchangeable if they are invariant under permutations of their indices, i.e., if $\mathbb{P}(X_{\sigma(1)}, \cdots, X_{\sigma(n)}) = \mathbb{P}(X_1, \cdots, X_n)$ for any permutation σ on $\{1, \ldots, n\}$.

Until very recently, no stationary k-dependent distribution for q-coloured words of growing length n was known, even for large q. It is easy to see that $k \geqslant 1$ and $q \geqslant 3$ are required. Indeed, $k \geqslant 1$ since X_i and X_{i+1} cannot be independent. And $q \geqslant 3$, since for 2-colouring $\mathbb{P}(X_i = X_j)$ depends on the parity of $|i - j|$ that can be much larger than k. In fact, for large enough n, it has been proved in [21] that no stationary 1-dependent 3-colouring exists. This result is actually implied by our Theorem 2(i).

The relationship between k and q has been investigated in [18–20]. In particular, in [20] a stationary 1-dependent 4-colouring is constructed, as well as a 2-dependent 3-colouring of words of infinite length. The construction is based on recursion formulae extending a suitable colouring of a word of length n to a word of length $n + 1$. This stationary 1-dependent 4-colouring implies that a 1-localisable 4-colouring exists. Thus our lower bound in Theorem 1 is tight.

Overview. Let $\mathcal{I}_n \subset \{0, 1\}^n$ be the set of all independent-set words of length n, i.e., the binary words of length n with no two consecutive ones. As explained in the previous paragraph, Theorem 1 is a corollary of Theorem 2(i). So we focus on 1-localisable distributions for binary words of $\mathcal{I}_n$.

In a first step, we show that, for every 1-localisable probability distribution $\mathbb{P}$, the probability $\mathbb{P}(s)$ of every binary word s of length n can always be written as a linear combination with integral coefficients of the p_i's, i.e., the probabilities of a random word having i ones in positions $1, 3, \ldots, 2i - 1$. This leads to a system of linear inequalities with $O(n)$ variables p_i and with $O(|\mathcal{I}_n|)$ constraints. We can in principle find the maximum value of p_1 by solving such a linear programming problem. Unfortunately, $|\mathcal{I}_n|$ grows exponentially in n since it satisfies a Fibonacci recurrence. This approach may at first seem intractable.

However, we show that the $O(|\mathcal{I}_n|)$ constraints are highly redundant and that there is a subset of only $O(n)$ constraints strictly equivalent to the original ones. Hence, we end up with a much smaller linear programming problem with $n/2$ variables and $n/2$ constraints which moreover turns out to be sufficiently structured so as to admit a closed-form solution.

Section 2 is dedicated to deriving this structured linear programming problem with p_1 as the linear objective function that we are maximising. Section 3 addresses the problem of solving this linear program. We are first able to derive a feasible solution for the linear program involving a binomial formula for the Catalan numbers (namely Corollary 1). We then show that the feasible solution we found at the previous step is indeed the optimal one by using the duality theorem for linear programming. We also show that this particular solution maximises simultaneously all the p_i's, which will prove Theorem 2(i & ii).

Due to space limitations, proofs and intermediate lemmas will appear in the full version.

2   Localisable Distribution on Independent-Set Words

A small worked-out example will go a long way towards explaining what the present and following section are about. Let $\mathbb{P}$ be a 1-localisable probability distribution on $\mathcal{I}_4$, and let $X_1 X_2 X_3 X_4$ be a random word with this distribution. We define $p_1 = \mathbb{P}(X_1 = 1)$ and $p_2 = \mathbb{P}(X_1 = X_3 = 1)$. Let us now consider the probabilities of the 8 individual words of $\mathcal{I}_4$:

$$0000, 1000, 0100, 0010, 0001, 1010, 0101, 1001$$

We have $\mathbb{P}(1010) = p_2$ by definition, and 1-localisability tells us that $\mathbb{P}(X_1 = 1, X_3 = 1) = \mathbb{P}(X_1 = 1, X_4 = 1) = \mathbb{P}(X_2 = 1, X_4 = 1)$. Hence, $\mathbb{P}(1010) = \mathbb{P}(1001) = \mathbb{P}(0101) = p_2$. Now we also have: $p_1 = \mathbb{P}(1000) + \mathbb{P}(1010) + \mathbb{P}(1001)$. Hence the value of $\mathbb{P}(1000)$, which is readily seen to be the same as $\mathbb{P}(0001)$:

$$\mathbb{P}(1000) = \mathbb{P}(0001) = p_1 - 2p_2.$$

From $\mathbb{P}(X_2 = 1) = \mathbb{P}(0100) + \mathbb{P}(0101)$ and $\mathbb{P}(X_2 = 1) = \mathbb{P}(X_1 = 1)$ we get the value of $\mathbb{P}(0100)$ and similarly of $\mathbb{P}(0010)$:

$$\mathbb{P}(0100) = \mathbb{P}(0010) = p_1 - p_2.$$

The only probability of an individual word that is unaccounted for is $\mathbb{P}(0000)$. Writing that all probabilities of all individual words sum to 1, we get:

$$\mathbb{P}(0000) = 1 - 4p_1 + 3p_2.$$

We may now notice two things. Any 1-localisable distribution on $\mathcal{I}_4$ is entirely determined by the two values p_1 and p_2. Conversely, any probability distribution defined as above by the two values p_1 and p_2 is 1-localisable. Finally, given any two positive numbers p_1 and p_2, such a probability distribution is well-defined if and only if all the linear expressions in p_1, p_2 that we have just computed take positive values. In other words, the values of p_1, p_2 for which there exists a 1-localisable probability distribution on $\mathcal{I}_4$ such that $\mathbb{P}(X_1 = 1) = p_1$ and $\mathbb{P}(X_1 = X_3 = 1) = p_2$, are exactly the solutions of the system of linear inequalities:

$$p_1, p_2 \geqslant 0, \quad p_1 - 2p_2 \geqslant 0, \quad p_1 - p_2 \geqslant 0, \quad \text{and} \quad 1 - 4p_1 + 3p_2 \geqslant 0.$$

Determining the largest allowable value of p_1 consists therefore in solving the associated linear program for the objective function p_1. In the present example we find that the maximum value is $p_1 = 2/5$. Our goal is to prove that the phenomena that we observe on this small example carry over to the general case of 1-localisable distributions on $\mathcal{I}_n$. We will then solve the general linear program associated with the maximisation of p_1.

We will find it convenient to write expressions such as $\mathbb{P}(10 \star \star)$ for the value $\mathbb{P}(X_1 = 1) = p_1$. More generally, for a distribution $\mathbb{P}$ for binary words of length n, two words s, t, and an integer $i \geqslant 0$ such that $|s| + i + |t| = n$, we will write:

$$\mathbb{P}(s \star^i t) = \sum_{u \in \{0,1\}^i} \mathbb{P}(s\,u\,t).$$

We now focus on the case of even n, and set $n = 2\ell$. It will be useful to introduce an algebraic formalism that will enable us to manipulate the general linear program and identify redundant linear inequalities.

Consider ℓ variables $p_1, \ldots, p_\ell$. Consider a function $\Lambda_n : \{0,1\}^n \to \mathbb{Z}[p_1, \ldots, p_\ell]$, and define $p_0 = \sum_{s \in \{0,1\}^n} \Lambda_n(s)$. We define the following rule for extending the domain of Λ_n to $\{0,1,\star\}^n$:

(R0) $\Lambda_n(s\star t) = \Lambda_n(s0t) + \Lambda_n(s1t)$ for every s, t such that $|s| + |t| = n - 1$.

Repeated application of rule (R0) until only the symbol $\star$ remains on the left-hand side gives that $\Lambda_n(\star^n) = \sum_{s \in \{0,1\}^n} \Lambda_n(s) = p_0$.

We also define the following properties:

(R1) $\Lambda_n(s) = 0$ if $s \in \{0,1\}^n \setminus \mathcal{I}_n$.
(R2) $\Lambda_n(s\star t\star) = \Lambda_n(s\star\star t) = \Lambda_n(\star s\star t)$ for s, t such that $|s| + |t| = n - 2$.
(R3) $\Lambda_n(s\star\star t) = \Lambda_n(t\star\star s)$ for s, t such that $|s| + |t| = n - 2$.
(R4) $\Lambda_n((1\star)^i \star^{n-2i}) = p_i$ for $i \in \{1, \ldots, \ell\}$.

Lemma 1. *For every p_0, there is a unique function Λ_n on $\{0,1\}^n$ satisfying (R1), (R2) and (R4) and $p_0 = \sum_{s \in \{0,1\}^n} \Lambda_n(s)$. For every $s \in \mathcal{I}_{2\ell}$, $\Lambda_n(s)$ is a linear function of $p_1, \ldots, p_\ell$. Furthermore, Λ_n satisfies Property (R3).*

From now on, we consider only functions Λ_n that satisfy (R1) through (R4). We now introduce the system of linear inequalities:

System 1. $p_i \geqslant 0$ and $\Lambda_{2\ell}(s) \geqslant 0$, for all $i \in \{1, \ldots, \ell\}$ and $s \in \mathcal{I}_{2\ell}$.

We then have the relatively straightforward result:

Theorem 3. *Let $p_1, \ldots, p_\ell \in [0,1]$. There exists a 1-localisable probability distribution $\mathbb{P}$ on $\mathcal{I}_{2\ell}$ such that $\mathbb{P}((1\star)^i \star^{2\ell - 2i}) = p_i$ for all $i \in \{1, \ldots, \ell\}$ iff System 1 is satisfied with $p_0 = 1$. We then have $\Lambda_{2\ell}(s) = \mathbb{P}(s)$.*

Let $\mathcal{S}_n = \{(10)^k 0^{n-2k} : k \in \{0, \ldots, \ell\}\} \subset \mathcal{I}_n$. We define the following sub-system of System 1:

System 2. $p_i \geqslant 0$ and $\Lambda_{2\ell}(s) \geqslant 0$, for all $i \in \{1, \ldots, \ell\}$ and $s \in \mathcal{S}_{2\ell}$.

We have the following:

Lemma 2. *For every $s \in \mathcal{I}_n$, there is $(a_t)_{t \in \mathcal{S}_n}$, $a_t \in \mathbb{N}$, such that $\Lambda_n(s) = \sum_{t \in \mathcal{S}_n} a_t \Lambda_n(t)$.*

Using this, we prove that

Proposition 1. *System 1 is equivalent to System 2, i.e., any solution of one is also a solution of the other.*

Thus, one can focus on the much more manageable System 2. We have the following expressions for the Λ-values of the elements of $\mathcal{S}_n$:

Lemma 3. $\Lambda_n((10)^k 0^{n-2k}) = \sum_{i=0}^{\ell-k} (-1)^i \binom{2\ell - 2k + 1 - i}{i} p_{k+i}$, for $k \in \{0, \ldots, \ell\}$.

3   Solving the LP System

To summarise, we have shown so far that the existence of a 1-localisable probability distribution on $\mathcal{I}_{2\ell} \subset \{0,1\}^{2\ell}$ is equivalent to the solvability of a system of $O(|\mathcal{I}_n|) \sim \exp(\Omega(n))$ inequalities $\Lambda_{2\ell}(s) \geqslant 0$ for $s \in \mathcal{I}_{2\ell}$. Moreover, every $\Lambda_{2\ell}(s)$, for $s \in \mathcal{I}_{2\ell}$, is a linear function of $p_i = \Lambda_{2\ell}((1\star)^i \star^{n-2i})$ for $1 \leqslant i \leqslant \ell$. We obtain therefore a system of *linear* inequalities. We furthermore showed that there is a size-ℓ subset $\mathcal{S}_{2\ell}$ of $\mathcal{I}_{2\ell}$ such that the inequalities corresponding to its members imply all inequalities for all the members of $\mathcal{I}_{2\ell}$.

Since we are interested in the values that can be taken by $p_1, \ldots, p_\ell$, in particular p_1 and its maximum value, Lemma 3 tells us that we are now faced with the explicit linear programming problem defined by $p_0 = 1$ and:

$$\text{maximise} \quad p_1 \quad \text{subject to:}$$

$$\begin{cases} p_i \geqslant 0, & i \in \{1, \ldots, \ell\} \\ \sum_{i=0}^{\ell-k} (-1)^i \binom{2\ell-2k+1-i}{i} p_{k+i} \geqslant 0, & k \in \{0, \ldots, \ell-1\}. \end{cases} \quad (1)$$

Once we know this maximum value of p_1, we set the value of p_1 to be something less than or equal to this maximum value. It turns out that this gives rise to another linear programming problem which is very similar in form to the first one, and where the goal is now to maximise p_2. We repeat this procedure until we get the maximum value of every p_i when the values of p_j for $j < i$ are set to something less than or equal to their maximum possible value. Indeed, we will show that we have the following, which implies directly Theorem 2:

Theorem 4. *Any solution* $(p_1, \ldots, p_\ell) \in \mathbb{R}^\ell$ *to the system of inequalities* (1) *satisfies* $p_i \leqslant (\mathsf{c}_\ell/\mathsf{c}_{\ell+1}) \cdot p_{i-1} \leqslant (\mathsf{c}_{\ell-i+1}/\mathsf{c}_{\ell+1}) \cdot p_0$, *possibly with equality.*

We now need linear programming notation:

Definition 2. *Let* $m, n \in \mathbb{N}$, $c_i, b_j, a_{i,j} \in \mathbb{R}$ *for* $1 \leqslant i \leqslant m$ *and* $1 \leqslant j \leqslant n$. *Let* $\mathbf{c} = (c_1, \ldots, c_n)^\mathsf{T}$, $\mathbf{b} = (b_1, \ldots, b_m)^\mathsf{T}$, $\mathbf{x} = (x_1, \ldots, x_n)^\mathsf{T}$ *and* $\mathbf{A} = (a_{i,j})$ *be an* $m \times n$ *matrix. A problem of the form:*

$$\textit{Maximise } \mathbf{c}^\mathsf{T}\mathbf{x}, \textit{ subject to } \mathbf{A}\mathbf{x} \leqslant \mathbf{b} \textit{ and } \mathbf{x} \geqslant \mathbf{0},$$

is called an LP *problem in standard form. The linear expression* $\mathbf{c}^\mathsf{T}\mathbf{x}$ *is called the* objective function, $\mathbf{A}\mathbf{x} \leqslant \mathbf{b}$ *and* $\mathbf{x} \geqslant \mathbf{0}$ *are called the* constraints, *the latter being more specifically non-negativity constraints.*

The corresponding dual problem *is defined as the following problem on* m *variables* $(y_1, \ldots, y_m)^\mathsf{T} = \mathbf{y}$*:*

$$\textit{Minimise } \mathbf{b}^\mathsf{T}\mathbf{y}, \textit{ subject to } \mathbf{A}^\mathsf{T}\mathbf{y} \geqslant \mathbf{c} \textit{ and } \mathbf{y} \geqslant \mathbf{0}.$$

We will need the duality theorem, see for instance [11, Chap. 5].

Theorem 5 (Duality Theorem).   *If the primal problem has an optimal solution* $\mathbf{x}^* = (x_1^*, \ldots, x_n^*)^\mathsf{T}$, *then the dual problem has an optimal solution* $\mathbf{y}^* = (y_1^*, \ldots, y_m^*)^\mathsf{T}$ *such that* $\mathbf{c}^\mathsf{T}\mathbf{x}^* = \mathbf{b}^\mathsf{T}\mathbf{y}^*$. *Furthermore, if* $\mathbf{x}$ *and* $\mathbf{y}$ *are feasible solutions to the primal and the dual problem respectively, such that* $\mathbf{c}^\mathsf{T}\mathbf{x} = \mathbf{b}^\mathsf{T}\mathbf{y}$, *then this common value optimises both objective functions.*

The solution to the linear program (1) will be a consequence of the following:

Theorem 6.   *Consider an* $n \times n$ *matrix* $\mathbf{A_n}$ *which is of the form*

$$\begin{bmatrix} a_{1,1} & a_{1,2} & \ldots & \ldots & a_{1,n-1} & a_{1,n} \\ -1 & a_{2,2} & \ldots & \ldots & a_{2,n-1} & a_{2,n} \\ 0 & -1 & \ddots & \ddots & \vdots & \vdots \\ \vdots & \vdots & \ddots & \ddots & \vdots & \vdots \\ 0 & 0 & \ldots & -1 & a_{n-1,n-1} & a_{n-1,n} \\ 0 & 0 & \ldots & 0 & -1 & a_{n,n} \end{bmatrix}.$$

Consider the LP maximisation problem $\mathbf{P_n}$ *associated with* $(\mathbf{A_n}, \mathbf{c_n}, \mathbf{b_n}, \mathbf{x_n})$, *where* $\mathbf{b_n} = (b, 0, \ldots, 0)^\mathsf{T}$, $\mathbf{c_n} = (c, 0, \ldots, 0)^\mathsf{T}$ *and* $\mathbf{x_n} = (x_1, \ldots, x_n)^\mathsf{T}$ *are vectors of length* n. *Then, the optimal value of the objective function of* $\mathbf{P_n}$ *is obtained by solving the special case* $\mathbf{A_n}\mathbf{x_n} = \mathbf{b_n}$. *And this optimal value is* $\frac{u_n}{u_{n+1}}bc$, *where the sequence* $(u_k)_{k \geqslant 1}$ *is defined by* $u_1 = 1$ *and* $u_{k+1} = \sum_{i=1}^{k} a_{n-k+1, n-k+i}\, u_{k+1-i}$.

Some comments are in order: in matrix form, the linear program (1) is exactly of the form envisaged by Theorem 6 with $b = c = 1$. We will therefore obtain the maximum of p_1 predicted by Theorem 4 by applying Theorem 6 and by proving that the associated sequence $(u_n)_{n \geqslant 1}$ is the sequence of Catalan numbers.

Our goal is therefore to prove Theorem 6. In other words the aim is to solve the following LP maximisation problem $\mathbf{P_n}$ associated with $(\mathbf{A_n}, \mathbf{c_n}, \mathbf{b_n}, \mathbf{x_n})$.

Problem 1. *Maximise* $\mathbf{c_n}^\mathsf{T}\mathbf{x_n}$, *subject to* $\mathbf{A_n}\mathbf{x_n} \leqslant \mathbf{b_n}$ *and* $\mathbf{x_n} \geqslant \mathbf{0}$.

We will first compute the special solution given by $\mathbf{A_n}\mathbf{x_n} = \mathbf{b_n}$. We obtain:

Proposition 2. *The value of the objective function of Problem 1 in the case where* $\mathbf{A_n}\mathbf{x_n} = \mathbf{b_n}$ *is* $\frac{u_n}{u_{n+1}}bc$.

Proposition 2 follows from the following intermediate results.

Lemma 4. *In the case where* $\mathbf{A_n}\mathbf{x_n} = \mathbf{b_n}$, *there is* $(\mu_j)_{1 \leqslant j \leqslant n}$ *such that* $x_{n-j} = \mu_j x_{n-j+1}$ *for* $j \neq n$, $b = \mu_n x_1$.

Corollary 1. *Let* $u_1 = 1$ *and* $u_i = \prod_{j=1}^{i-1} \mu_j$ *for* $i \in \{2, \ldots, n+1\}$. *Then, we have the recurrence relation* $u_{k+1} = \sum_{i=1}^{k} a_{n-k+1, n-k+i}\, u_{k+1-i}$.

Corollary 2. *The sequence* $(u_j)_{1 \leqslant j \leqslant n+1}$ *as defined in Corollary 1 satisfies* $x_{n-j} = \frac{u_{j+1}}{u_j}x_{n-j+1}$ *for* $j \neq n$ *and* $b = \frac{u_{n+1}}{u_n}x_1$.

Proposition 2 now follows from Corollary 2 by remarking that the objective function is cx_1.

We now wish to prove that the value of p_1 given by Proposition 2 actually maximises p_1. To this end we consider the dual of Problem 1, namely:

Problem 2. *Minimise* $\mathbf{b_n}^\mathsf{T}\mathbf{y_n}$, *subject to* $\mathbf{A_n}^\mathsf{T}\mathbf{y_n} \geqslant \mathbf{c_n}$ *and* $\mathbf{y_n} \geqslant \mathbf{0}$.

Once again we solve a particular instance of this problem, namely $\mathbf{A_n}^\mathsf{T}\mathbf{y_n} = \mathbf{c_n}$. We will show that

Proposition 3. *The value of the objective function of Problem 2 in the case where* $\mathbf{A_n}^\mathsf{T}\mathbf{y_n} = \mathbf{c_n}$ *is* $\frac{u_n}{u_{n+1}}bc$.

It will be useful to now define a sequence of LP maximisation problems $(\mathbf{P_k})_{k \leqslant n}$ associated with $(\mathbf{A_k}, \mathbf{c_k}, \mathbf{b_k}, \mathbf{x_k})$, where $\mathbf{A_{k-1}}$ is the $(k-1) \times (k-1)$ submatrix at the bottom right of $\mathbf{A_k}$. In other words, $\mathbf{A_k} = (a_{i,j}^k)$ where:

$$a_{i,j}^n = a_{i,j} \quad \text{and} \quad a_{i,j}^{k-1} = a_{i+1,j+1}^k$$

and where $\mathbf{b_{k-1}} = (x_{n-k}, 0, \ldots, 0)^\mathsf{T}$ and $\mathbf{x_{k-1}} = (x_{n-k}, \ldots, x_n)^\mathsf{T}$. We will now write $a_{i,j}^n$ instead of $a_{i,j}$ because we shall need to modify the superscript n later.

We next prove the following results.

Proposition 4. *There are* $(U_{k,n})_{1 \leqslant k \leqslant n+1}$ *and* $(V_{k,n})_{1 \leqslant k \leqslant n+1}$ *such that*

$$y_k = U_{k,n}\,y_1 + V_{k,n} \quad for\ 1 \leqslant k \leqslant n$$
$$0 = U_{n+1,n}\,y_1 + V_{n+1,n}$$

Furthermore,

(1) $V_{k,n} = -cU_{k-1,n-1}$, *and* (2) $U_{k,n} = \displaystyle\sum_{\substack{0 \leqslant j \leqslant k-1 \\ 0=i_0 < \cdots < i_j = k-1}} \prod_{0 \leqslant u \leqslant j-1} a_{i_u+1, i_{u+1}}^n$, *for* $k \geqslant 2$.

Corollary 3. *We have* $y_1 = \frac{U_{n,n-1}}{U_{n+1,n}}c$ *in the case of equality in* $\mathbf{A_n}^\mathsf{T}\mathbf{y_n} = \mathbf{c_n}$.

Lemma 5. $U_{n+1,n} = u_{n+1}$, *where the sequence* $(u_n)_{n \geqslant 1}$ *is as in Corollary 1.*

Corollary 3 and Lemma 5 together prove Proposition 3, remarking that the objective function is by_1. Since we have found a solution to Problem 1 that gives the value $(u_n/u_{n+1})bc$ for its objective function and we have also found a solution to its dual problem, namely Problem 2, that gives the very same value for the dual objective function, the Duality Theorem implies that this common value maximises both objective functions. This proves therefore Theorem 6.

It remains to compute the specific value of the sequence (u_n) in the case of the LP problem (1). We have:

Proposition 5. *The sequence* (u_n) *is exactly the sequence of Catalan numbers.*

Thus, the maximum value that can be taken by p_1 is $c_\ell/c_{\ell+1}$.

The proof of Theorem 4 is completed by determining the optimal values of the remaining variables $p_2, \ldots, p_\ell$. This amounts to solving linear systems $\mathbf{P}_k$ for decreasing values of k, so that all the preceding techniques and results apply.

4 Generalisation and Conclusion

In this paper we have introduced k-localisable distributions for random words. They generalise the notion of exchangeability for random variables in much the same way as k-dependence generalises independence. Furthermore, we believe this notion is of great interest for the study of multipartite games with nonsignalling resources (capturing quantum resources). This raises fundamental questions in graph theory (through chromatic numbers) and distributed computing (through symmetry-breaking problems). We have given a fine-grained analysis of 1-localisable distributions for independent-set words, implying an optimal lower bound of $q \geqslant 4$ for 1-localisable random q-coloured words.

Using the same approach, we can extend Theorem 1 to d-distance q-coloured words in which $d + 1$ consecutive letters must receive pairwise distinct colours (Theorem 1 is for $d = 1$). This also corresponds to d-distance q-colouring of a path, a well-known notion in graph theory [23, 29]. As d-distance chromatic number of the path is $d + 1$, we must have $q \geqslant d + 1$. Observe that there is no k-localisable d-distance q-colouring for $k < d$ since $\mathbb{P}(X_i = X_j)$ depends on whether $|i - j| = d$ (it must be 0) or $|i - j| > d$ (it must be >0 for $q < n$). So, we investigate the case $k = d$, the d-localisable distributions for random d-distance q-coloured words of length n. We can show that the minimum number of colours must be $q \geqslant (d+1) \cdot (1 + 1/d)^d$ for n large enough, generalising Theorem 1. Using the same approach as for $d = 1$, we consider distance-d independent-set words (i.e., binary words with no two ones at distance $\leqslant d$). We show, using the same technique, that the probability of having a one in any fixed position is upper bounded by the ratio of two consecutive Fuss-Catalan numbers of parameters $n/(d + 1)$, a generalisation of Catalan numbers, whose limit is $d^d/(d + 1)^{d+1}$. This will appear in the full version of the paper.

A step further would be to extend the results to combinatorial structures other than words. The notion of k-localisability extends naturally to graphs as follows. Here each node v of a graph G gets a random variable X_v. Let X_S, for every subset S of nodes, denote the collection of random variables X_s with $s \in S$, and let G_S denote any graph isomorphic to $G[S]$, the subgraph of G induced by S. Then, a probability distribution for random variables (X_v) with support the nodes of G is k-localisable if, for every two subsets I, J at distance more than k in G such that G_I and G_J are connected, the distribution of (X_I, X_J) can depend only on $\{G_I, G_J\}$. The notion of independent-set word transfers also to binary variables encoding independent sets in G. The study of k-localisable q-colourings (or independent-sets) on graphs would have potential applications to understanding the possibilities of distributed quantum computing.

References

1. Almeida, M.L., Bancal, J.-D., Brunner, N., Acín, A., Gisin, N., Pironio, S.: Guess your neighbor's input: a multipartite nonlocal game with no quantum advantage. Phys. Rev. Lett. **104** (2010). https://doi.org/10.1103/PhysRevLett.104.230404

2. Arfaoui, H., Fraigniaud, P.: What can be computed without communications? In: Even, G., Halldórsson, M.M. (eds.) SIROCCO 2012. LNCS, vol. 7355, pp. 135–146. Springer, Berlin (2012). https://doi.org/10.1007/978-3-642-31104-8_12
3. Barenboim, L., Elkin, M.: Distributed graph coloring: fundamentals and recent developments. Synth. Lect. Distrib. Comput. Theory 4(1) (2013). https://doi.org/10.2200/S00520ED1V01Y201307DCT011
4. Barenboim, L., Elkin, M., Pettie, S., Schneider, J.: The locality of distributed symmetry breaking. In: 53rd Annual IEEE Symposium on Foundations of Computer Science (FOCS), pp. 321–330. IEEE Computer Society Press, October 2012. https://doi.org/10.1109/FOCS.2012.60
5. Brandão, F.G.S.L., Harrow, A.W.: Quantum de Finetti theorems under local measurements with applications. In: 45th Annual ACM Symposium on Theory of Computing (STOC), pp. 861–870. ACM Press, June (2013). https://doi.org/10.1145/2488608.2488718
6. Brandt, S., et al.: LCL problems on grids. In: 35th Annual ACM Symposium on Principles of Distributed Computing (PODC), pp. 101–110. ACM Press, July 2016. https://doi.org/10.1145/3087801.3087833
7. Brassard, G., Broadbent, A., Hänggi, E., Méthot, A.A., Wolf, S.: Classical, quantum and nonsignalling resources in bipartite games. Theor. Comput. Sci. **486**, 61–72 (2013). https://doi.org/10.1016/j.tcs.2012.12.017
8. Cameron, P.J., Montanaro, A., Newman, M.W., Severin, S., Winter, A.: On the quantum chromatic number of a graph. Electron. J. Comb. **14**, R81 (2007)
9. Caves, C.M., Fuchs, C.A., Schack, R.: Unknown quantum states: the quantum de Finetti representation. J. Math. Phys. **43**, 4537 (2001). https://doi.org/10.1063/1.1494475
10. Chang, Y.-J., Li, W., Pettie, S.: An optimal distributed $(\delta+1)$-coloring algorithm? In: 50th Annual ACM Symposium on Theory of Computing (STOC), pp. 445–456. ACM Press, June 2018. https://doi.org/10.1145/3188745.3188964
11. Chvátal, V.: Linear Programming. W. H. Freeman, New York (1983)
12. Czekaj, Ł., Pawłowski, M., Vértesi, T., Grudka, A., Horodecki, M., Horodecki, R.: Quantum advantage for distributed computing without communication. Phys. Rev. A **92**, 032122 (2015). https://doi.org/10.1103/PhysRevA.92.032122
13. Fraigniaud, P., Heinrich, M., Kosowski, A.: Local conflict coloring. In: 57th Annual IEEE Symposium on Foundations of Computer Science (FOCS), pp. 625–634. IEEE Computer Society Press, October 2016. https://doi.org/10.1109/FOCS.2016.73
14. Gavoille, C., Kosowski, A., Markiewicz, M.: What can be observed locally? In: Keidar, I. (ed.) DISC 2009. LNCS, vol. 5805, pp. 243–257. Springer, Heidelberg (2009). https://doi.org/10.1007/978-3-642-04355-0_26
15. Ghaffari, M., Kuhn, F., Maus, Y.: On the complexity of local distributed graph problems. In: 49th Annual ACM Symposium on Theory of Computing (STOC), pp. 784–797. ACM Press, June 2017. https://doi.org/10.1145/3055399.3055471
16. Harris, D.G., Schneider, J., Su, H.-H.: Distributed $(\Delta+1)$-coloring in sublogarithmic rounds. In: 48th Annual ACM Symposium on Theory of Computing (STOC), pp. 465–478. ACM Press, June 2016. https://doi.org/10.1145/2897518.2897533
17. Hoeffding, W., Robbins, H.: The central limit theorem for dependent variables. Duke Math. J. **15**, 773–780 (1948). https://doi.org/10.1215/S0012-7094-48-01568-3
18. Holroyd, A.E., Hutchcroft, T., Levy, A.: Finitely dependent cycle coloring. Electron. Commun. Probab. **23**, 1–8 (2018). https://doi.org/10.1214/18-ECP118

19. Holroyd, A.E., Liggett, T.M.: Symmetric 1-dependent colorings of the integers. Electron. Commun. Probab. **20**, 1–8 (2015). https://doi.org/10.1214/ECP.v20-4070
20. Holroyd, A.E., Liggett, T.M.: Finitely dependent coloring. Forum Math., Pi **4**, 1–43 (2016). https://doi.org/10.1017/fmp.2016.7
21. Holroyd, A.E., Schramm, O., Wilson, D.B.: Finitary coloring. Ann. Probab. **45**, 2867–2898 (2017). https://doi.org/10.1214/16-AOP1127
22. Işlak, U.: Asymptotic normality of random sums of m-dependent random variables. Stat. Probab. Lett. **109**, 22–29 (2016). https://doi.org/10.1016/j.spl.2015.10.015
23. Kramer, F., Kramer, H.: A survey on the distance-colouring of graphs. Discrete Math. **308**, 422–426 (2008). https://doi.org/10.1016/j.disc.2006.11.059
24. Le Gall, F., Magniez, F.: Sublinear-time quantum computation of the diameter in CONGEST networks. In: 37th Annual ACM Symposium on Principles of Distributed Computing (PODC), pp. 337–347. ACM Press, July 2018. https://doi.org/10.1145/3212734.3212744
25. Le Gall, F., Nishimura, H., Rosmanis, A.: Quantum advantage for the LOCAL model in distributed computing. Technical report, October 2018. arXiv:1810.10838v1 [quant-ph]
26. Linial, N.: Locality in distributed graphs algorithms. SIAM J. Comput. **21**, 193–201 (1992). https://doi.org/10.1137/0221015
27. Naor, M.: A lower bound on probabilistic algorithms for distributive ring coloring. SIAM J. Discrete Math. **4**, 409–412 (1991). https://doi.org/10.1137/0404036
28. Naor, M., Stockmeyer, L.: What can be computed locally. SIAM J. Comput. **24**, 1259–1277 (1995). https://doi.org/10.1137/S0097539793254571
29. Niranjan, P.K., Kola, S.R.: The k-distance chromatic number of trees and cycles. AKCE Int. J. Graphs Comb. (2017, in press). https://doi.org/10.1016/j.akcej.2017.11.007

On Codeword Lengths Guaranteeing Synchronization

Vladimir V. Gusev[1] and Elena V. Pribavkina[2(✉)]

[1] Leverhulme Research Centre for Functional Materials Design,
University of Liverpool, Liverpool, UK
`vladimir.gusev@liverpool.ac.uk`
[2] Institute of Natural Sciences and Mathematics, Ural Federal University,
Ekaterinburg, Russia
`elena.pribavkina@urfu.ru`

Abstract. Prefix codes such as Huffman codes are commonly used for loseless data compression. The class of synchronizing codes is often chosen to improve error resilience or to enable parallel decoding of data. Such codes have a special sequence whose occurrence realigns decoding process leading to recovery from errors in a data stream. In the present paper we identify a class of codes whose synchronizability depends only on the lengths of codewords. Namely, we show that every maximal finite prefix code with only two codeword lengths is synchronizing if and only if these lengths are coprime.

Keywords: Synchronizing codes · Totally synchronizing digraphs · Self-synchronizing Huffman code · Synchronizing automata

1 Introduction

Prefix codes are widely used for data compression and transmission, for example, they arise from the classical Huffman coding scheme. Error resilience of variable length codes is a major practical concern: a single bit error at the very beginning can propagate and make decoding of the whole message incorrect.

One of the key ideas to address this issue is the notion of synchronization [3, Chap. 3.6]. A word w is *synchronizing* for a code X if uw is in X^* for all words $u \in \Sigma^*$. Essentially, an occurrence of w synchronizes all possible decodings and blocks propagation of an error. A code X is called *synchronizing* if there exists a synchronizing word for X. Such codes happen to be not only more robust, but they also admit parallel decoding schemes described in [11,21].

Fortunately, almost all maximal binary prefix codes are synchronizing [6]. Furthermore, in many cases we can construct a synchronizing code given the

---

V. Gusev is supported by the Leverhulme Trust. E. Pribavkina was supported by Russian Ministry of Education and Science Project No. 1.3253.2017 and the Competitiveness Enhancement Program of Ural Federal University.

R. Mercaş and D. Reidenbach (Eds.): WORDS 2019, LNCS 11682, pp. 207–216, 2019.
https://doi.org/10.1007/978-3-030-28796-2_16

desired codeword lengths. Recall that the code is called *aperiodic* if the greatest common divisor of codeword lengths is equal to one. It is not hard to see that aperiodicity is necessary for synchronizing codes. A kind of reverse statement is true as well: for every aperiodic code there is a synchronizing code with the same multiset of codeword lengths [18]. In certain cases we can infer that a code is synchronizing from a very limited knowledge about the code. A great example of this is the following result: if the orders of letters are relatively prime, then the code is synchronizing [3, Theorem 3.6.10]. In the present paper we push this idea to an extreme: can we infer that a code is synchronizing by looking only at its codeword lengths? For a more thorough overview of a vast number of results about codes and their synchronizing properties we refer the reader to [3,4].

Synchronizing automata constitute another source of motivation for our paper. Recall that a deterministic finite automaton is *synchronizing* if there exists a word mapping all states to a fixed one independently of the starting state. Synchronization is a fundamental notion that arises naturally in different areas, e.g. in matrix theory [7], semigroup theory [16], group theory [1], etc., see [17,20] for a general overview of the topic. Also, synchronizing codes and synchronizing automata are very similar concepts [3]. A direct connection between them in the case of maximal prefix codes is described in Sect. 2, more recent works on the border between synchronizing automata and codes are presented in [2,5,14,15].

One of the early challenging questions about synchronizing automata, known as the *Road coloring problem*, is to characterize fixed out-degree digraphs that can be turned into synchronizing automata by assigning letters to their edges. Significant research effort has culminated in the following statement: a digraph has a synchronizing coloring if and only if it is *aperiodic*, i.e. the gcd of its cycle lengths is equal to one [19]. A brief overview of results connecting digraphs and synchronizing automata can be found in [10]. The Road coloring problem in the context of synchronizing codes was studied in [13].

The study of quantitative versions of the road coloring problem was recently initiated in [8,9]. It is conjectured that aperiodic digraphs always have a large number of synchronizing colorings, moreover, almost all aperiodic digraphs are *totally synchronizing*, i.e. all their colorings are synchronizing. At the moment, these conjectures seem to be quite hard. Moreover, no efficient characterization of totally synchronizing digraphs is known, i.e. can we check that a digraph is totally synchronizing in polynomial time?

In the present paper we put synchronizing codes in the context of total synchronizability. Our main contribution is that every aperiodic maximal finite prefix code with only two different codeword lengths is synchronizing. Since Huffman codes are maximal binary prefix codes, this statement applies to them as well. In the context of synchronizing automata, our result provides a new class of totally synchronizing digraphs. Our proof relies on a neat result of Perrin [12] stating that prefix codes form a free semigroup and a technical analysis of the automaton associated with a code.

2   Preliminaries

Let Σ be an alphabet. By Σ^* we denote the free monoid generated by Σ, and by Σ^+ we denote the set of all non-empty words on Σ.

A *deterministic finite automaton* $\mathscr{A}$ over the alphabet Σ is defined by a triple $(Q, \Sigma, .)$, where Q is a finite set of states, and $p.w$ is the result of applying a word w to a state p. An automaton $\mathscr{A}$ is said to be *synchronizing*, if there exists a word w such that $|Q.w| = 1$, i.e. the image of the state set under the action of w is a singleton.

A *prefix code* on Σ is a subset X of Σ^+ which contains no proper prefix of its elements. A prefix code is said to be *maximal* if it is maximal with respect to inclusion among the prefix codes on Σ. In this paper all the codes we consider are finite maximal prefix codes. A prefix code can be identified with a labeled tree so that the words of the code correspond bijectively to the leaves of the tree. Namely, a tree of a maximal prefix code X is a complete tree $T(X)$, where each node is either an internal node with $k = |\Sigma|$ children, or a leaf node with no children. Each outgoing edge is labeled by a letter of the alphabet. The *label* of a node q is the word $\pi(q)$ that labels the path from the root to q. The label of the root is ε. Then a maximal prefix code is the set of words X such that $X = \{\pi(q) \mid q \text{ is a leaf of } T\}$ for some tree T.

We may turn the tree $T(X)$ corresponding to a code X into an automaton $\mathscr{A}(X)$ recognizing X by identifying the leaves with the root, and setting the root as both initial and final state. It is easy to see that this automaton is deterministic since the code is maximal. Moreover, it is synchronizing iff the code X is synchronizing. We say that a state of the automaton $\mathscr{A}(X)$ has *depth* i if the corresponding vertex in $T(X)$ has depth i, i.e. it can be reached by reading a word of length i from the root. The root has depth 0. Let n be the maximal length of words in X. Thus, the depth ranges from 0 to $n - 1$.

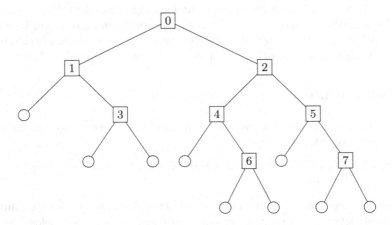

Fig. 1. Non-synchronizing code C with three codeword lengths

An example of a code and the corresponding tree is shown in Fig. 1. The code C consists of 9 words $\{a^2, aba, ab^2, ba^2, baba, bab^2, b^2a, b^3a, b^4\}$. It is sometimes easier to deal with the minimized automaton $\mathcal{M}(X)$ rather than with the automaton $\mathscr{A}(X)$. The minimized automaton of the code C is shown on Fig. 2.

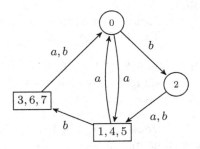

Fig. 2. Minimized automaton of the code C

In what follows we will make use of the following simple fact.

Lemma 1. *Let $\mathcal{M}(X)$ be the minimized automaton of a maximal prefix code X. The automaton $\mathcal{M}(X)$ is synchronizing iff the code X itself is synchronizing.*

Proof. If the minimized automaton is synchronizing, then it can be synchronized to the class containing the root, since the automaton is strongly connected. This class is a singleton, since the root is the unique final state. So the initial automaton $\mathscr{A}(X)$ can also be synchronized to the root. The other direction is trivial.

Consider an arbitrary maximal prefix code X. Note that all the states of depth $n-1$ in $\mathscr{A}(X)$ have the same right language (equal to Σ), so they can always be identified. Furthermore, if ℓ is the second to maximal length of codewords in X, then all the states of depths i where $\ell \leq i \leq n-1$ can be identified. Indeed, the right language of any such state of depth i is equal to Σ^{n-i}.

3   Main Results

Here we focus on the case of a maximal finite prefix code with only two codeword lengths $d < n$, $(d, n) = 1$.

Theorem 1. *A maximal finite prefix code X with two codeword lengths is synchronizing iff X is aperiodic.*

To prove the theorem we will make use of several lemmata. First consider in detail the structure of the automaton of such code. As observed before, we may identify all the states of depth i for $i \in \{d, \ldots, n-1\}$. For convenience let us denote the root by 0, and the states of depth i for $i \in \{d, \ldots, n-1\}$ by their depths. We will refer to the subset of states $\{d, \ldots, n-1\}$ as *chain-states*, and denote this subset by $\mathscr{C}$. Other states except the root are called *tree-states*, and the set of such states is denoted by $\mathscr{T}$. Thus, $Q = \{0\} \sqcup \mathscr{T} \sqcup \mathscr{C}$. A *level* is the set of states of the same depth.

We say that a state $p \in \mathscr{T}$ of depth i *has type* '$-$', if there is a word $u \in \Sigma^d$ such that $depth(p.u) = (d+i) \mod n$, $0.u = 0$; and has type '$+$', if there exists a word $u \in \Sigma^d$ such that $depth(p.u) = i$, $0.u = d$. Note that in principle, a state may have both types, or have no type at all, see Figs. 3 and 4.

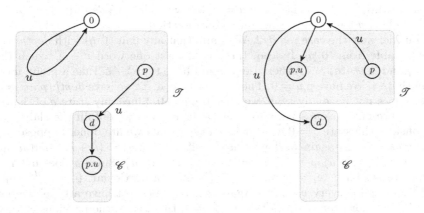

Fig. 3. Types '$-$' and '$+$'

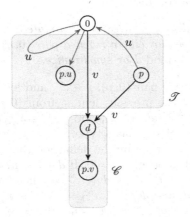

Fig. 4. No type

Lemma 2. *For every state $p \in \mathscr{T}$ either p has a type, or a pair $(0, q)$ is reachable from $(0, p)$ such that $depth(q) = depth(p)$, and q has a type.*

Proof. Suppose, there is a state $p \in \mathscr{T}$ having no type. This means that $0.u = d$ iff $depth(p.u) = (d + i) \mod n$, and $0.u = 0$ iff $depth(p.u) = i$, where $i = depth(p)$. In other words, the states 0 and p either go simultaneously on a cycle of length d or on the cycle of length n.

Assume first that $i \geq d/2$. Consider the word v such that $0.v = p$, and its prefix w of length $d - i$. We have $p.w = x \in \{0, d\}$. Then $0.wt = x$ for every $t \in \Sigma^i$. In particular, $p.s = x$ for every word $s \in \Sigma^{d-i}$. If $x = 0$, then we can choose a word $u \in \Sigma^d$ such that $0.u = d$, and we will have $depth(p.u) = i$. This means p actually has type '+' which is a contradiction. If $x = d$, then we can choose a word $u \in \Sigma^d$ such that $0.u = 0$, but we will have $depth(p.u) = (d + i) \mod n$. Thus, p has type '−', again, a contradiction.

Consider now the case $i < d/2$. We claim that any pair $(0, q)$ with $depth(q) = i$ is reachable from $(0, p)$. Indeed, there is at least one word $v \in \Sigma^d$ such that $p.v = q$ and $depth(q) = i$ (otherwise p would have type '−'). Take a prefix u of v of length $d-i$. We have $p.u = 0$. Then for every $w \in \Sigma^i$ we have $depth(p.uw) = i$. Since p does not have any type, we have $0.uw = 0$. Since any state q of depth i can be represented as $q = p.uw$ for a suitable $w \in \Sigma^i$, we obtain the claim.

Consider the state $q = 0.a^i$. If it has a type, then we are done. Suppose, it is not the case. Then again we have $0.u = d$ iff $depth(q.u) = (d + i) \mod n$, and $0.u = 0$ iff $depth(q.u) = i$. We have $q.a^{d-i} = x \in \{0, d\}$. Since q does not have a type, we get $0.a^{d-i}v_1 = x$ for every $v_1 \in \Sigma^i$. In particular, $(0.a^i).a^{d-2i}v_1 = q.a^{d-2i}v_1 = x$ for every $v_1 \in \Sigma^i$. Again, since q does not have a type, we have $0.a^{d-2i}v_2 = x$ for every $v_2 \in \Sigma^{2i}$. If $d - 2i \leq i$, then, as in the previous case, we have $q.u = x$ for every $u \in \Sigma^{d-i}$. If $x = 0$, then q has type '+', if $x = d$, then q has type '−'. A contradiction with the hypothesis. Otherwise, if $i \leq d/3$, we can continue with the induction argument until we find a positive integer k such that $d - ki \leq i$, and get a contradiction.

Lemma 3. *If every level $1 \leq i < d$ contains a state of type '−', then from any pair $(0, p)$ with $p \in \mathscr{T}$ we can either reach a pair $(0, q)$ such that $q \in \mathscr{T}$ and $depth(q) = (depth(p) - n) \mod d$ or synchronize.*

Proof. Consider an arbitrary pair $(0, p)$, $p \in \mathscr{T}$, and let $i = depth(p)$. If p does not have a type, then by the previous lemma from $(0, p)$ we can reach a pair $(0, p')$ such that $depth(p') = depth(p)$, and p' has a type. So without loss of generality we may assume that p has a type.

Case 1. If p has type '−', then by the definition there is a word $u \in \Sigma^d$ such that $depth(p.u) = (d + i) \mod n$, $0.u = 0$. If $d + i < n$, then let $\alpha \geq 1$ be such that $p.u^{\alpha-1} \in \mathscr{C}$, and $p.u^\alpha \in \mathscr{T}$. If $d + i \geq n$, then we set $\alpha = 1$. We have $d^\alpha + i = n + j$, $0 \leq j < d$, and $depth(p.u^\alpha) = j = (i - n) \mod d$. On the other hand, $0.u^\alpha = 0$. Thus, if $j \neq 0$, we are done. If $j = 0$, then the pair $(0, p)$ is synchronized by the word u^α.

Case 2. Suppose now that the only type the state p has is '+'. Consider the following subcases.

(i) There is a word $u \in \Sigma^d$ such that $0.u = d$, and $depth(p.u) = d+i \mod n$. If $d + i < n$, then from $(0, p)$ we can reach any pair $(0, q)$ with $depth(q) = i$.

Indeed, let q be an arbitrary state of depth i, and let $w \in \Sigma^i$ be the word such that $q = 0.w$. We have $(0,p).uvw = (0,q)$ for an arbitrary word v of length $n-d-i$. If $d+i > n$, then consider the prefix v of u of length $d-i$. We have $p.v = d$. Then $0.vw = d$ for any word $w \in \Sigma^i$, otherwise we would have a word $vw' \in \Sigma^d$ such that $0.vw' = 0$ and $depth(p.vw') = (d+i) \mod n$, meaning that the state p has type '$-$'. We have that any pair (d,p') with $depth(p') = d+i-n < i$ is reachable from $(0,p)$. Then after applying an arbitrary word of length $n - d$ we obtain an arbitrary pair $(0,q)$ with $depth(q) = i$. Since the level i contains a state q of type '$-$', from $(0,p)$ we can reach the pair $(0,q)$, and we enter the previous Case 1.

(ii) There is no word $u \in \Sigma^d$ such that $0.u = d$, and $depth(p.u) = d + i$ mod n. Since p does not have type '$-$', we get that $p.v = 0$ for every $v \in \Sigma^{d-i}$. Consider the word w_1 such that $0.w_1 = p$.

(ii).a. Suppose first, that $i \leq d/2$. Then $p.w_1v = 0$ for any word $v \in \Sigma^{d-2i}$. So for any state q of depth i we have $q = p.w_1vu$ for a suitable word $u \in \Sigma^i$. On the other hand, we have $0.w_1vu = p.vu = 0$. Thus, any pair $(0,q)$ such that $depth(q) = i$ is reachable from the pair $(0,p)$. Then we can reach such a pair that q has type '$-$', and we enter Case 1.

(ii).b. Suppose now that $i > d/2$. Then $p.w_1 = q_1$ and $depth(q_1) = 2i - d$. Let w_2 be the word such that $0.w_2 = q_1$, $|w_2| = 2i - d$. Then for any $u \in \Sigma^{d-i}$ we have

$$(0,p).w_1u = (0,q_1.u),$$

$$depth(q_1.u) = 2i - d + d - i = i.$$

If among the states $q_1.u$ there is a state of type '$-$', we enter Case 1, and we are done. Likewise, if among the states $q_1.u$ there is a state satisfying condition (i), then we are done as well. So we may assume condition (ii) holds for every state $q_1.u$, $u \in \Sigma^{d-i}$. This means that $q_1.v = 0$ for every $v \in \Sigma^{2(d-i)}$.

We claim that if $2(d - i) \geq i$, which is equivalent to $i \leq \frac{2}{3}d$, then every pair $(0,q)$ with $depth(q) = i$ is reachable from $(0,p)$. Indeed, let $q = 0.v$ be an arbitrary state of depth i, so $|v| = i$. Let v_2 be an arbitrary word of length $2d - 3i$. We have $0.w_2v_2v = q_1.v_2v = 0$, and $p.w_2v_2v = 0.v = q$. So among the states q we find such that its type is '$-$', and finish the proof. If $i > \frac{2}{3}d$, then we continue by induction: suppose on the $(k - 1)$-th step we obtain $i > \frac{k-1}{k}d$. On this step we have the word w_{k-1} of length $(k - 1)i - (k - 2)d$ such that $0.w_{k-1} = q_{k-2}$. Consider the k-th step. We have $|w_{k-1}| > d-i$, so we may assume $w_{k-1} = w'_k w_k$, where $|w'_k| = d - i$, and $|w_k| = ki - (k - 1)d$. Let $q_{k-1} = 0.w_k$. We have $p.w_{k-1} = p.w'_k w_k = 0.w_k = q_{k-1}$. For every word $u \in \Sigma^{(k-1)(d-i)}$ we have

$$(0,p).w_{k-1}u = (q_{k-2}, q_{k-1}).u = (0, q_{k-1}.u),$$

$$depth(q_{k-1}.u) = ki - (k - 1)d + (k - 1)(d - i) = i.$$

We apply the previous arguments and either obtain a reachable pair $(0,q)$ with $depth(q) = i$ and such that the type of q is '$-$', or we need to continue to the next step. But there will be only finite number of steps, namely $k \leq d$, since $i > d - 1$ cannot hold true.

Corollary 1. *If every level $1 \leq i < d$ contains a state of type '$-$', then every pair $(0, p)$, $p \in \mathscr{T}$ is synchronizing.*

Proof. By the previous lemma we can consider the sequence of reachable pairs

$$(0, p) = (0, p_0) \to (0, p_1) \to \cdots \to (0, p_k),$$

such that $depth(p_j) = (depth(p_{j-1}) - n) \mod d = (depth(p) - jn) \mod d$. Since $(d, n) = 1$, there is $0 < j < d$ such that $jn \mod d = depth(p)$, thus the pair $(0, p)$ synchronizes after at most $d - 1$ steps.

Lemma 4. *Every level $1 \leq i < d$ contains a state of type '$-$'.*

Proof. Suppose on the contrary, there is a depth $1 \leq i < d$ such that all the states of depth i either have type '$+$' or no type. Let X_d be the set of codewords of length d. By definition $u \in X_d$ iff $u \in \Sigma^d$ and $0.u = 0$. Our hypothesis means that for every state q of depth i, and for every $u \in X_d$ we have $depth(q.u) = i$. In other terms, for every word $v \in \Sigma^i$ and every $u \in X_d$ the prefix u' of the word vu of length d fixes the root 0, hence $u' \in X_d$. Thus, we obtain the inclusion $\Sigma^i X_d \subseteq X_d \Sigma^i$. Since both sets are finite having the same number of elements, we have the equality $\Sigma^i X_d = X_d \Sigma^i$. Note, that both Σ^i and X_d are prefix codes themselves. By a neat result by Perrin [12] the set of prefix codes is a free monoid under concatenation. It follows, that two elements commute iff they are a power of the same element. We have $\Sigma^i = (\Sigma^j)^r$, so $X_d = (\Sigma^j)^\ell$, which is a contradiction, since the set of codewords of length n would be empty in this case.

Proof (of Theorem 1). In order to prove the code is synchronizing, it is enough to prove that every pair (p, q) is synchronizing. By Corollary 1 and Lemma 4 every pair $(0, p)$ with $p \in \mathscr{T}$ is synchronizing. For a pair $(0, p)$ with $p \in \mathscr{C}$ we can apply a word $u \in X_d$ a suitable number of times (at most $[\frac{n}{d}]$) and reach a pair $(0, p')$ with $p' \in \mathscr{T}$. For a pair (p, q) with $p, q \in \mathscr{T} \cup \mathscr{C}$ we apply a word of length at most n to bring one of these states to the root.

The other direction is a well-known observation. We will present it here for completeness. Suppose that the code X is synchronizing, and let r be a common divisor of codeword lengths. Let w be a synchronizing word for the code X. By the definition, for any $u \in \Sigma^*$ we have $uw \in X^*$. In particular, for $u = \varepsilon$ we have $w \in X^*$, thus the length of the word w should be divisible by r. On the other hand, for a word u of length 1 we should also have $uw \in X^*$, thus its length, equal to $|w| + 1$, should also be divisible by r. Therefore, $r = 1$, so the code is aperiodic.

Note that Theorem 1 cannot be extended to codes of more than two codeword lengths: there exists a non-synchronizing code with codeword lengths $\{2, 3, 4\}$. The code is shown on Fig. 1. Its minimized automaton is shown on Fig. 2. It is easy to see that the pair of states $(0, \{1, 4, 5\})$ cannot be synchronized.

Theorem 2. *Given an aperiodic maximal prefix code X with two codeword lengths $d < n$, the shortest synchronizing word is bounded by $2n^2d$.*

Proof. Let us first show that every pair $(0, p)$ with $p \in \mathscr{T}$ can be mapped to a pair $(0, q)$ with $depth(q) = (depth(p) - n) \mod d$ by a word of length at most $2n$. Indeed, if p has type '$-$', then by Lemma 3 such a pair $(0, q)$ can be reached by applying a word u^α of length at most n. If p has no type, then we should first move it to the pair $(0, p')$ such that p' has type '$-$'. This is possible, since by Lemma 2 any pair $(0, p')$ with p' having the same depth as p can be reached from $(0, p)$ by a word of length d. By Lemma 4 we can reach such a pair with p' having type '$-$'. If p has type '$+$', then by Lemma 3 we can reach a pair $(0, p')$ with p' of type '$-$' by applying a word of length at most n. Overall, in order to obtain a pair $(0, q)$ with $depth(q) = (depth(p) - n) \mod d$ from a pair $(0, p)$ with $p \in \mathscr{T}$ a word of length at most $2n$ is sufficient.

In order to synchronize a pair $(0, p)$ with $p \in \mathscr{T}$ we need to apply this process of reducing depth at most $d - 1$ times according to Corollary 1. Thus, such a pair is synchronizing by a word of length at most $2n(d - 1)$. To synchronize a pair $(0, p)$, where $p \in \mathscr{C}$, we first apply a word of length at most n to reach a pair $(0, p')$ such that $p' \in \mathscr{T}$. Thus, such a pair can be synchronized by a word of length at most $n + 2n(d - 1)$. We also note that we always synchronize to the root.

Now to synchronize the automaton $\mathscr{A}(X)$ we first apply a word a^n for some letter $a \in \Sigma$. Then $Q.a^n = \{0, 0.a, 0.a^2, \ldots, 0.a^\ell\}$ with $\ell \in \{d - 1, n - 1\}$. Then we successively synchronize pairs of the form $(0, p)$. In the beginning there are at most $n - 1$ such pairs, and at each step their number is reduced at least by one. Thus, a rough estimate for the length of a synchronizing word is $n + (n + 2n(d - 1))(n - 1) \leq 2n^2d$.

References

1. Araújo, J., Cameron, P.J., Steinberg, B.: Between primitive and 2-transitive: synchronization and its friends. EMS Surv. Math. Sci. **4**(2), 101–184 (2017)
2. Berlinkov, M.V., Szykuła, M.: Algebraic synchronization criterion and computing reset words. Inf. Sci. **369**, 718–730 (2016)
3. Berstel, J., Perrin, D., Reutenauer, C.: Codes and Automata. Cambridge University Press, New York (2009)
4. Biskup, M.: Error resilience in compressed data - selected topics. Ph.D. thesis, University of Warsaw (2008)
5. Biskup, M.T., Plandowski, W.: Shortest synchronizing strings for Huffman codes. Theor. Comput. Sci. **410**(38), 3925–3941 (2009)
6. Freiling, C.F., Jungreis, D.S., Theberge, F., Zeger, K.: Almost all complete binary prefix codes have a self-synchronizing string. IEEE Trans. Inf. Theory **49**(9), 2219–2225 (2003)
7. Gerencsér, B., Gusev, V.V., Jungers, R.: Primitive sets of nonnegative matrices and synchronizing automata. SIAM J. Matrix Anal. Appl. **39**(1), 83–98 (2018)
8. Gusev, V.V., Szykuła, M.: On the number of synchronizing colorings of digraphs. In: Drewes, F. (ed.) CIAA 2015. LNCS, vol. 9223, pp. 127–139. Springer, Cham (2015). https://doi.org/10.1007/978-3-319-22360-5_11

9. Gusev, V.V., Pribavkina, E.V.: On synchronizing colorings and the eigenvectors of digraphs. In: Faliszewski, P., Muscholl, A., Niedermeier, R. (eds.) Mathematical Foundations of Computer Science (MFCS). Leibniz International Proceedings in Informatics (LIPIcs), vol. 58, pp. 48:1–48:14 (2016)
10. Gusev, V.V., Pribavkina, E.V., Szykuła, M.: Around the road coloring theorem. In: TUCS Proceedings of Russian Finnish Symposium on Discrete Mathematics, vol. 26, pp. 52–56 (2017)
11. Klein, S.T., Wiseman, Y.: Parallel Huffman decoding with applications to JPEG files. Comput. J. **46**(5), 487–497 (2003)
12. Perrin, D.: Codes conjugués. Inf. Control **20**(3), 222–231 (1972)
13. Perrin, D., Schützenberger, M.P.: Synchronizing prefix codes and automata and the road coloring problem. In: Symbolic dynamics and its applications, Contemporary Mathematics, vol. 135, pp. 295–318. American Mathematical Society (1992)
14. Pribavkina, E.V.: Slowly synchronizing automata with zero and noncomplete sets. Math. Notes **90**(3), 411–417 (2011)
15. Ryzhikov, A., Szykuła, M.: Finding short synchronizing words for prefix codes. In: Potapov, I., Spirakis, P., Worrell, J. (eds.) Mathematical Foundations of Computer Science (MFCS). Leibniz International Proceedings in Informatics (LIPIcs), vol. 117, pp. 21:1–21:14 (2018)
16. Salomaa, A.: Composition sequences for functions over a finite domain. Theor. Comput. Sci. **292**(1), 263–281 (2003)
17. Sandberg, S.: Homing and synchronizing sequences. In: Broy, M., Jonsson, B., Katoen, J.-P., Leucker, M., Pretschner, A. (eds.) Model-Based Testing of Reactive Systems. LNCS, vol. 3472, pp. 5–33. Springer, Heidelberg (2005). https://doi.org/10.1007/11498490_2
18. Schützenberger, M.P.: On synchronizing prefix codes. Inf. Control **11**(4), 396–401 (1967)
19. Trahtman, A.N.: The road coloring problem. Israel J. Math. **172**(1), 51–60 (2009)
20. Volkov, M.V.: Synchronizing automata and the Černý conjecture. In: Martín-Vide, C., Otto, F., Fernau, H. (eds.) LATA 2008. LNCS, vol. 5196, pp. 11–27. Springer, Heidelberg (2008). https://doi.org/10.1007/978-3-540-88282-4_4
21. Weißenberger, A., Schmidt, B.: Massively parallel Huffman decoding on GPUs. In: International Conference on Parallel Processing (ICPP), pp. 27:1–27:10 (2018)

Binary Intersection Revisited

Štěpán Holub[(✉)]

Department of Algebra, Charles University,
Sokolovská 83, 175 86 Praha, Czech Republic
holub@karlin.mff.cuni.cz

Abstract. We reformulate the classical result by Juhani Karhumäki characterizing intersections of two languages of the form $\{x,y\}^* \cap \{u,v\}^*$. We use the terminology of morphisms which allows to formulate the result in a shorter and more transparent way.

1 Introduction

One of the classical results that deserve to be better known is the description Karhumäki gave in [6] for the intersection of two free monoids of rank two, that is, for languages of the form $\{x,y\}^* \cap \{u,v\}^*$ where x and y, as well as u and v, do not commute. We reformulate here the result in terms of morphisms which allows an exposition that is much shorter, and hopefully also more transparent.

It is well known that an intersection of two free submonoids of a free monoid is free. On the other hand, the intersection $\{x,y\}^* \cap \{u,v\}^*$ can have infinite rank. The Theorem 2 in [6] gives two possible forms: $\{\beta,\gamma\}^*$ and $(\beta_0 + \beta(\gamma(1 + \delta + \cdots + \delta^t))^*\varepsilon)^*$ and the last section of the cited paper (called "Concluding remarks") contains some further information about the form of the intersection formulated as a byproduct of the proof which spans about fifteen pages (without Preliminaries). The proof is in many places based on insights that are certainly correct but not always easy to verify directly from the text. In particular, the proof often crucially relies on "the way" certain words are "built up" from words x and y, and/or u and v. This is exactly the kind of argument that is much easier to make if x and y (u and v) are seen as images of a binary morphism.

2 Preliminaries

We shall denote the longest common prefix (suffix resp.) of u and v by $u \wedge v$ ($u \wedge_s v$ resp.). Two words are *prefix-comparable* (*suffix-comparable* resp.) if one of them is a prefix (suffix resp.) of the other. By $u \leq_s v$ we denote that u is a suffix of v. If we want to say that u is a suffix of some sufficiently large power of v, we say that u is a suffix of v^*. Concepts of concatenation, prefix and suffix are extended to pairs in the obvious way.

We shall use the standard notation of regular expressions to describe certain sets of words. Note that $\{u,v\}^*$ is an alternative notation for $(u+v)^*$.

© Springer Nature Switzerland AG 2019
R. Mercaş and D. Reidenbach (Eds.): WORDS 2019, LNCS 11682, pp. 217–225, 2019.
https://doi.org/10.1007/978-3-030-28796-2_17

A pair of noncommuting words is also called a *binary code*. We need the following properties of binary codes (see [1, Lemma 3.1]). If u and v do not commute, then the word $z = uv \wedge vu$ is prefix-comparable with all words in $\{u, v\}^*$. Moreover, there are distinct letters a_u and a_v such that za_u is prefix-comparable with each word in $u\{u, v\}^*$ and za_v is prefix comparable with each word in $v\{u, v\}^*$. We shall use these facts for suffixes analogously. In particular, the above facts imply a weak version of the Periodicity lemma in the following form:

Lemma 1. *If w is a common suffix of u^* and v^* and $|w| \geq |u| + |v|$, then u and v commute.*

A binary morphism $g : \{a, b\}^* \rightarrow \Sigma^*$ is called *marked* if $\text{pref}_1(g(a)) \neq \text{pref}_1(g(b))$, where $\text{pref}_1(u)$ denotes the first letter of u. The *marked version* of a morphism g' is the morphism g defined by $g(u) = \alpha^{-1}g'(u)\alpha$ where $\alpha = g'(ab) \wedge g'(ba)$. It is easy to see that the definition of g is correct, and that g is marked.

We shall also use the following basic facts about submonoids of a free monoid (see [7, Chap. 1, Sect. 1.2]). Let $M \subseteq \Sigma^*$ be a monoid, where Σ is an alphabet. Then there is a unique minimal (w.r.t. inclusion) generating set B of M. The cardinality (possibly infinite) of B is the *rank* of M. If elements of B satisfy no non-trivial relation, then M is *free*. A monoid with the minimal generating set $\{u, v\}$ is free if and only if u and v do not commute. If M is not free, then there exists a unique minimal (w.r.t inclusion) monoid $F \subseteq \Sigma^*$ containing M, called the *free hull* of M. It is convenient to assume that the free hull of M is Σ^* itself. This assumption often does not harm generality, since we can consider Γ as the set of letters instead of Σ, where Γ is the minimal generating set of the free hull of M, and to use the unique embedding $M \hookrightarrow \Gamma^*$. The important advantage is the fact that if Σ^* is the free hull of M, then $\Sigma = \{\text{pref}_1(b) \mid b \in B\}$, where B is the minimal generating set of M.

The minimal generating set B of M is a *prefix code* if $u, uv \in B$ implies $v \in B$.

3   The Result

We are given four words x, y, u, $v \in \Sigma^*$ such that $xy \neq yx$ and $uv \neq vu$, and we are interested in the intersection $I = \{x, y\}^* \cap \{u, v\}^*$. As we pointed out in Preliminaries, we can assume that

$$\Sigma = \{\text{pref}_1(x), \text{pref}_1(y), \text{pref}_1(u), \text{pref}_1(v)\}.$$

This implies that if $\{x, y\}^* \cap \{u, v\}^*$ contains a nonempty word, then either $\text{pref}_1(x) \neq \text{pref}_1(y)$ or $\text{pref}_1(u) \neq \text{pref}_1(v)$. Without loss of generality, let $\text{pref}_1(u) \neq \text{pref}_1(v)$, and let $\alpha = xy \wedge yx$.

We set $A = \{a, b\}$, and define morphisms $g', h : A^* \rightarrow \Sigma^*$ by $g'(a) = x$, $g'(b) = y$, $h(a) = u$, $h(b) = v$. Our aim now is to investigate the set

$$\mathbf{C}(g', h) = \{(r, s) \in A^* \times A^* \mid g'(r) = h(s)\}.$$

We shall call pairs $(r, s) \in \mathbf{C}(g', h)$ *solutions*. A description of solutions is slightly stronger result than description of the intersection I since there is a one to one correspondence between I and $\mathbf{C}(g', h)$ given by $w \leftrightarrow (r, s)$, where $w = g'(r) = h(s)$. This also means that $\mathbf{C}(g', h)$ is a free subsemigroup of $A^* \times A^*$, and elements of its minimal generating set shall be called *minimal solutions*.

Note that h is marked, and let g be the marked version of g', that is, $g(u) = \alpha^{-1} g'(u) \alpha$, where $\alpha = g'(ab) \wedge g'(ba)$. The crucial tool for investigation of the structure of $\mathbf{C}(g', h)$ is its description in terms of an analogous set

$$\mathbf{C}(g, h) = \{(r, s) \in A^* \times A^* \mid g(r) = h(s)\},$$

which is easy to describe. Since both g and h are marked, the construction of a candidate pair $(e, f) \in \mathbf{C}(g, h)$ is deterministic as soon as the first letter of e (or f) is chosen. This observation immediately implies that $\mathbf{C}(g, h)$ is generated by at most two minimal elements.

We have that (r, s) is a solution if and only if

$$\alpha g(r) = h(s)\alpha. \tag{1}$$

Therefore, also the construction of (r, s) is deterministic, unless there is a pair (p, q) such that

$$\alpha g(p) = h(q). \tag{2}$$

Clearly, there is at most one such pair (p, q), and if it does not exist, then there is at most one minimal solution.

From now on, we shall assume that (p, q) exists.

Lemma 2. *If (r, s) is a solution, then $(p^{-1}rp, q^{-1}sq) \in \mathbf{C}(g, h)$.*

Proof. From (1) and (2), we have $\alpha g(rp) = h(sq)$, which implies that p is a prefix of rp, q is a prefix of sq, and

$$g(p^{-1}rp) = h(q^{-1}sq).$$

□

Therefore there is a mapping $\pi : \mathbf{C}(g', h) \to \mathbf{C}(g, h)$ defined by

$$\pi : (r, s) \mapsto (p^{-1}rp, q^{-1}sq).$$

One is tempted to conclude that π is an isomorphism, which would complete the characterization. However, π is not an isomorphism in general. The problem is that the inverse $\pi^{-1} : (e, f) \mapsto (pep^{-1}, qfq^{-1})$ may be ill-defined. Indeed, (p, q) need not be a suffix of (pe, qf). Instead we have the following characterization:

Lemma 3.

$$\mathbf{C}(g', h) = \{(pep^{-1}, qfq^{-1}) \mid (e, f) \in \mathbf{C}(g, h) \text{ and } (p, q) \leq_s (pe, qf)\}.$$

Proof. The inclusion $\subseteq$ is Lemma 2. The inclusion $\supseteq$ is a straightforward verification of

$$\alpha g(pep^{-1}) = h(qfq^{-1})\alpha.$$

$\square$

Lemma 4. *If (p,q) is a suffix of (pe^i, qf^i) with $i > 1$, then (p,q) is also a suffix of (pe, qf).*

Proof. If p is a suffix of pe^i, then p is a suffix of e^* which implies that p is a suffix of pe. Similarly for q. $\square$

The previous lemma shows that if $\mathbf{C}(g,h)$ has only one generator, then there is at most one minimal solution. It remains to suppose that there are two generators of $\mathbf{C}(g,h)$, namely

$$g(e_0) = h(f_0),$$
$$g(e_1) = h(f_1).$$

From now on, we shall use the notation $e_i = e(i)$ and $f_i = f(i)$ and see e and f as morphisms $\{0,1\}^* \to A^*$. Note that e and f are marked. Morphisms e and f play an important role in the literature on the binary Post correspondence problem and on binary equality sets. They were introduced in [2], where they were called "equality collectors". Later the name "successor morphisms" prevailed (see [3,4] and [5]).

We say that τ is the *block decomposition* of a solution (r,s) if $p^{-1}rp = e(\tau)$ and $q^{-1}sq = f(\tau)$. Lemma 2 says that each solution has a block decomposition. Let

$$T = \left\{ \tau \mid (pe(\tau)p^{-1}, qf(\tau)q^{-1}) \in \mathbf{C}(g', h) \right\}.$$

be the set of block decompositions of solutions. Due to

$$\mathbf{C}(g', h) = \left\{ (pe(\tau)p^{-1}, qf(\tau)q^{-1}) \mid \tau \in T \right\},$$

it is enough to characterize the set T. First, observe the following two facts.

Lemma 5. $\tau \in T$ *if and only if (p,q) is a suffix of $(pe(\tau), qf(\tau))$.*

Proof. This is a reformulation of Lemma 3. $\square$

Lemma 6. *The minimal generating set of T is a prefix code.*

Proof. Let τ_1 and $\tau_1\tau_2$ be two elements of T. Since p is a suffix of $pe(\tau_1)$, we have that $pe(\tau_2)$ is a suffix of $pe(\tau_1)e(\tau_2)$. Since p is also a suffix of $pe(\tau_1)e(\tau_2)$, we deduce that p is a suffix of $pe(\tau_2)$. Similarly, we obtain that q is a suffix of $qf(\tau_2)$. Hence $\tau_2 \in T$, and the claim follows. $\square$

The following is the key technical result.

Lemma 7. *If $\tau c \in T$ for some $c \in \{0,1\}$ and $\tau \in \{0,1\}^*$, then also $c \in T$.*

Proof. Without loss of generality, let $c = 0$. The claim follows from Lemma 4 if $\tau \in 0^*$. Let therefore $\tau = \tau'10^i$, and assume

$$(p, q) \leq_s \left(pe(\tau')e_1 e_0^i, qe(\tau')f_1 f_0^i \right) .$$

We want to show that (p, q) is a suffix of (pe_0, qf_0). This is equivalent to showing that (p, q) is a suffix of (e_0^*, f_0^*). Assume the contrary.

Minimality of (p, q) implies that p is a proper suffix of e_0. The equality $\alpha g(p) = h(q)$ implies $g(e_0 p^{-1})$ is a suffix of α. Assume, without loss of generality, that a is the first letter of e_0. Since $|\alpha| < |g(ab)|$, we have that $e_0 p^{-1}$ is a^m for some $m \geq 1$.

Let $z_f = f_1^* \wedge_s f_0^*$, and let c_0 and c_1 be distinct letters such that $c_1 z_f \leq_s f_1^*$ and $c_0 z_f \leq_s f_0^*$. Let moreover $z_h = h(c_0)^* \wedge_s h(c_1)^*$. Then $z_h h(z_f)$ is the longest common suffix of $h(f_0^*)$ and $h(f_1^*)$. Since α is a suffix of both $g(e_0^*)$ and $g(e_1^*)$, we deduce that α is a suffix of $z_h h(z_f)$ and hence

$$|\alpha| \leq |h(z_f)| + |z_h| .$$

Since q is a suffix of $qf(\tau')f_1 f_0^i$ and not a suffix of f_0^*, we obtain that $c_1 z_f f_0^i$ is a suffix of q. From $\alpha g(p) = h(q)$ and $e_0 = a^m p$, we now have $h(c_1 z_f f_0^{i-1}) \leq_s \alpha g(a^m)^{-1}$, which yields

$$|h(c_1 z_f)| + |g(a)| \leq |\alpha| .$$

The two inequalities above imply that $|h(c_1)| + |g(a)| \leq |\alpha|$ and $|h(c_1)| + |g(a)| \leq |z_h|$. Since z_h is a suffix of $h(c_1)^*$, α is a suffix of $h(a)^*$ and z_h and α are suffix comparable, the Periodicity lemma implies that $g(a)$ and $h(c_1)$ commute (see Lemma 1). Since both g and h are marked, we obtain that $f_0 \in c_1^*$ which contradicts $c_0 z_f \leq_s f_0^*$. $\quad\square$

We now have the following characterization of T.

Lemma 8. *If T contains a nonempty word, then either $T = \{0, 1\}^*$ or T is generated (up to the exchange of letters 0 and 1) by the set*

$$0 + \left(1 + 10 + \cdots 10^{(t-1)} \right)^* 10^t,$$

where $t > 0$ is the least integer such that $q \leq_s qf_1 f_0^t$.

Proof. Suppose that T contains a nonempty word.

If (p, q) is a suffix of both (pe_0, qf_0) and (pe_1, qf_1), then $T = \{0, 1\}^*$ by Lemma 5.

Otherwise, by Lemma 7, we can suppose (up to the exchange of 0 and 1) that $(p, q) \leq_s (pe_0, qf_0)$ and (p, q) is not a suffix of (pe_1, qf_1). Then (p, q) is a suffix of (e_0^*, f_0^*), hence there exists a least integer t such that $(p, q) \leq_s (pe_1 e_0^t, qf_1 f_0^t)$, that is, such that $10^t \in T$.

Let $\tau 10^i$ be an element of T. The minimality of (p, q) implies that p is a suffix of e_0. Since q is a suffix of both f_0^* and $qf(\tau 10^i)$, we deduce that q is a suffix of

$z_f f_0^i$ where $z_f = f_1^* \wedge_s f_0^*$. Therefore q is also a suffix of $q f_1 f_0^i$. Hence $i \geq t$, and $\tau 10^t$ is in T for any τ.

The characterization of the minimal generating set of T is now completed by Lemma 6. Namely, a word $\tau 10^t$ is a minimal generating element of T if and only if it does not start with 0 and does not contain a factor 10^t. $\square$

Altogether, we have the following characterization of $\mathbf{C}(g', h)$.

Theorem 1. *The set $\mathbf{C}(g', h)$ is generated by at most one minimal solution if either there is no pair (p, q) satisfying $\alpha g(p) = h(q)$, or if there are not two minimal pairs (e_0, f_0) and (e_1, f_1) such that $g(e_i) = h(f_i)$, $i = 0, 1$.*

If there are at least two minimal solutions, then

$$\mathbf{C}(g', h) = \{(pe(\tau)p^{-1}, qf(\tau)q^{-1}) \mid \tau \in T\},$$

where (up to the exchange of 0 and 1)

$$T = \left(0 + \left(1 + 10 + \cdots + 10^{t-1}\right)^* 10^t\right)^*$$

where t is the least integer such that $(p, q) \leq_s (pe_1 e_0^t, qf_1 f_0^t)$.

Note that the first claim is not "if and only if". The opposite implication does not hold as we show below in Example 4.

4   Comparison

We now map our proof to the structure and notation of the original paper. Theorem [6, Theorem 2] gives two options:

$$I = \{\beta, \gamma\}^* \tag{$*$}$$

$$I = \left(\beta_0 + \beta(\gamma(1 + \tau + \cdots + \tau^t))^* \epsilon\right)^* \tag{$**$}$$

The option $(*)$ corresponds to the unique minimal solution (if γ is empty) or to $T = \{0, 1\}^+$. Then we have

$$\beta = h\left(q f_0 q^{-1}\right), \qquad\qquad \gamma = h\left(q f_1 q^{-1}\right).$$

The option $(**)$ is further specified by [6, Theorem 3], and we return to it later.

The case analysis of [6] corresponds to the present paper as follows:

- **Case I**: (p, q) does not exist.
- **Case II**: (p, q) exists.
  - **Case II A**: No (e, f) exists.
  - **Case II B**: $\mathbf{C}(g, h)$ has a unique generator.
  - **Case II C**: Both (e_0, f_0) and (e_1, f_1) exist.

The **Case II C** is now divided into two main cases. In our terminology, the classification is based on smallest elements of T whose image is longer than q.

Subcase (i) corresponds to the situation when T contains elements $0\tau_0$ and $1\tau_1$ where q is both longer than $f(\tau_0)$ and than $f(\tau_1)$. First, it is shown that $\tau_0, \tau_1 \in 0^* \cup 1^*$. The case is then divided into the following situations:

- **Subcase (i) a:** $\tau_0 \in 0^*$ and $\tau_1 \in 1^*$ (then $T = \{0,1\}^+$),
- **Subcase (i) b:** $\tau_0 \in 0^*$ and $\tau_1 \in 0^+$ (T is infinitely generated),
- **Subcase (i) c:** $\tau_0 \in 1^+, \tau_1 \in 0^+$ (strictly speaking this cannot happen, but in [6], this case is reduced to a)).

Subcase (ii) corresponds to the situation when T contains $0\tau_0$ with the above properties but not $1\tau_1$. Again, there are subcases:

- **Subcase (ii) a:** $\tau_0 \in 0^*$ (T is infinitely generated),
- **Subcase (ii) b:** $\tau_0 \in 1^+$ (cannot happen).

The last mentioned subcase, namely **Case II C, Subcase (ii) b**, leads to a contradiction. This is the crucial, in some sense the only technically complicated case, and it corresponds to our Lemma 7. In [6], its proof is again divided into two cases and spans pages 198–202.

The two infinitely generated cases cover [6, Theorem 1 (**)] above, and are formulated by [6, Theorem 3] as two further options for the generating set of I:

$$\beta\gamma + \beta(\gamma\beta)^{\mathsf{t}}\left(\delta(1 + \gamma\beta + \cdots + (\gamma\beta)^{\mathsf{t}})^*\right)^* \delta\gamma \qquad \text{(Thm 3 *)}$$

$$\beta\gamma + \beta(\gamma\beta)^{\mathsf{t}}\left(\delta(1 + \gamma\beta + \cdots + (\gamma\beta)^{\mathsf{q}-1})^*\right)^* \delta(\beta(\gamma\beta)^{\mathsf{t}-\mathsf{q}})^{-1}. \qquad \text{(Thm 3 **)}$$

Here, we have $\gamma\beta = h(f_0)$, $\delta = h(f_1)$, and $q = \beta(\gamma\beta)^{\mathsf{t}}$. In the case (Thm 3 *), we have $t = \mathsf{t} + 1$ (where t is from our Lemma 8). In the case (Thm 3 **), we have $t = \mathsf{q}$ and $\delta = \delta'\beta(\gamma\beta)^{\mathsf{t}-\mathsf{q}}$.

5 Examples

We conclude by several examples. The first two are from [6].

Example 1.

$$g' : a \mapsto a \qquad\qquad b \mapsto a^m b,$$
$$h : a \mapsto a \qquad\qquad b \mapsto ba^m.$$

Then $\alpha = a^m$, $g = h$, hence $(e_0, f_0) = (a, a)$ and $(e_1, f_1) = (b, b)$. We have $(p, q) = (\epsilon, a^m)$, where ϵ denotes the empty word. Moreover, $t = m$,

$$T = \left(0 + (1 + 10 + \cdots + 10^{m-1})^* 10^m\right)^*.$$

Therefore the intersection is generated by the set

$$g'(e(0 + (1 + 10 + \cdots + 10^{m-1})^* 10^m))$$
$$= g'(a + (b + ba + \cdots + ba^{m-1})^* ba^m)$$
$$= \mathsf{a} + (\mathsf{a}^m\mathsf{b} + \mathsf{a}^m\mathsf{ba} + \cdots + \mathsf{a}^m\mathsf{ba}^{m-1})^* \mathsf{a}^m\mathsf{ba}^m.$$

Example 2.

$$g' : a \mapsto \mathsf{aab} \qquad\qquad b \mapsto \mathsf{aba},$$
$$h : a \mapsto \mathsf{a} \qquad\qquad b \mapsto \mathsf{baaba}.$$

Then $\alpha = \mathsf{a}$,

$$g : a \mapsto \mathsf{aba} \qquad\qquad b \mapsto \mathsf{baa},$$

$(e_0, f_0) = (aa, ab)$ and $(e_1, f_1) = (bb, ba)$. We have $(p, q) = (\epsilon, a)$. For T the "up to the exchange of letters 0 and 1" applies with $t = 1$, and

$$T = (1 + 0^+1)^*.$$

Therefore the intersection is

$$I = (g'(e(1 + 0^+1)))^* = (g'(bb + (aa)^+bb))^* = (\mathsf{abaaba} + (\mathsf{aabaab})^+\mathsf{abaaba})^*.$$

This can be also written in the form given in [6] as $I = (\mathsf{a}(\mathsf{abaaba})^*\mathsf{baaba})^*$.

The following example shows the possibility of (Thm 3 ∗∗).

Example 3.

$$g' : a \mapsto \mathsf{aa} \qquad\qquad b \mapsto \mathsf{a}^6\mathsf{b},$$
$$h : a \mapsto \mathsf{a} \qquad\qquad b \mapsto \mathsf{ba}^4.$$

Then $\alpha = \mathsf{a}^6$,

$$g : a \mapsto \mathsf{aa} \qquad\qquad b \mapsto \mathsf{ba}^6,$$

$(e_0, f_0) = (a, aa)$ and $(e_1, f_1) = (b, baa)$. We have $(p, q) = (\epsilon, a^6)$, $t = 2$, and

$$T = (0 + (1 + 10)^*100)^*.$$

The intersection is

$$I = (g'(e(0 + (1 + 10)^*100)))^* = (g'(a + (b + ba)^*baa))^*$$
$$= (\mathsf{aa} + (\mathsf{a}^6\mathsf{b} + \mathsf{a}^6\mathsf{baa})^*\mathsf{a}^6\mathsf{baaaa})^*.$$

The remarkable feature of this example is that f_0 is a suffix of f_1.

Example 4. Finally, Table 1 lists various situations in which the intersection is generated by at most one word. The notation is such that $\mathrm{pref}_1(g(a)) = \mathrm{pref}_1(h(a)) = \mathsf{a}$, $\mathrm{pref}_1(g(b)) = \mathrm{pref}_1(h(b)) = \mathsf{b}$ and $\mathrm{pref}_1(e_0) = \mathrm{pref}_1(h_0) = a$ (whenever applicable).

Particularly important is the last line where all three necessary conditions for infinitely generated intersection are met, yet the intersection contains the empty word only. Note that (p, q) is not a suffix of any $(pe(\tau), qf(\tau))$ in that case. This explains the formulation of the main Theorem 1.

Table 1. Intersections generated by at most one word.

$g'(a)$	$g'(b)$	$h(a)$	$h(b)$	α	(p,q)	(e_0, f_0)	(e_1, f_1)	I
aabb	ab	aba	bab	a	(b,a)	$\times$	(bbb, ba)	ababab*
aa	ab	aba	ba	a	(b,a)	$\times$	(b,b)	$\{\varepsilon\}$
aabb	ab	aba	babb	a	(b,a)	$\times$	$\times$	$\{\varepsilon\}$
aab	aba	aba	baa	a	$\times$	(a,a)	(b,b)	aba*
aab	abb	aba	bba	a	$\times$	(a,a)	(b,b)	$\{\varepsilon\}$
aabb	ab	abaa	bb	a	$\times$	$\times$	$\times$	abaabb*
aab	abb	aa	bb	a	$\times$	$\times$	$\times$	$\{\varepsilon\}$
aab	abb	aab	bba	a	$\times$	$\times$	(b,b)	aab*
aab	abb	aba	bab	a	$\times$	(a,a)	$\times$	$\{\varepsilon\}$
abaab	ababab	a	ba	aba	(ε, ab)	(a, abb)	(b, bbb)	$\{\varepsilon\}$

References

1. Choffrut, C., Karhumäki, J.: Combinatorics of words. In: Rozenberg, G., Salomaa, A. (eds.) Handbook of Formal Languages, pp. 329–438. Springer, Heidelberg (1997). https://doi.org/10.1007/978-3-642-59136-5_6
2. Ehrenfeucht, A., Karhumäki, J., Rozenberg, G.: The (generalized) Post correspondence problem with lists consisting of two words is decidable. Theoret. Comput. Sci. **21**(2), 119–144 (1982). https://doi.org/10.1016/0304-3975(89)90080-7
3. Halava, V., Harju, T., Hirvensalo, M.: Binary (generalized) Post correspondence problem. Theoret. Comput. Sci. **276**(1–2), 183–204 (2002). https://doi.org/10.1016/S0304-3975(01)00157-8
4. Halava, V., Holub, Š.: Reduction tree of the binary generalized Post correspondence problem. Int. J. Found. Comput. Sci. **22**(2), 473–490 (2011). https://doi.org/10.1142/S0129054111008143
5. Holub, Š.: Binary equality sets are generated by two words. J. Algebra **259**(1), 1–42 (2003). https://doi.org/10.1016/S0021-8693(02)00534-3
6. Karhumäki, J.: A note on intersections of free submonoids of a free monoid. Semigroup Forum **29**(1), 183–205 (1984). https://doi.org/10.1007/BF02573324
7. Lothaire, M.: Combinatorics on words. Cambridge Mathematical Library, Cambridge University Press, Cambridge (1997). https://doi.org/10.1017/CBO9780511566097

On Substitutions Closed Under Derivation: Examples

Václav Košík[1] and Štěpán Starosta[2]([✉]) [ID]

[1] Department of Mathematics, FNSPE, Czech Technical University in Prague,
Trojanova 13, 120 00 Praha 2, Czech Republic
[2] Department of Applied Mathematics, FIT, Czech Technical University in Prague,
Thákurova 9, 160 00 Praha 6, Czech Republic
stepan.starosta@fit.cvut.cz

Abstract. We study infinite words fixed by a morphism and their derived words. A derived word is a coding of return words to a factor. We exhibit two examples of sets of morphisms which are closed under derivation—any derived word with respect to any factor of the fixed point is again fixed by a morphism from this set. The first example involves standard episturmian morphisms, and the second concerns the period doubling morphism.

Keywords: Return word · Derived word · Fixed point of substitution · Arnoux–Rauzy word · Episturmian word · Period doubling morphism

1 Introduction

In 1998 Fabien Durand characterized primitive substitutive sequences, i.e., morphic images of fixed points of primitive substitutions. A crucial role in his characterization is played by the notion "derived word". Any primitive substitutive sequence $\mathbf{u}$ is uniformly recurrent, i.e., for each factor w, the distances between consecutive occurrences of w in $\mathbf{u}$ are bounded. Or equivalently, there are only finitely many gaps between neighbouring occurrences of w. An infinite word coding ordering of these gaps (seen as finite words) is called the derived word to w in $\mathbf{u}$ and is denoted by $\mathbf{d_u}(w)$.

The mentioned main result of [2] says that a uniformly recurrent word is primitive substitutive if and only if the set of derived words to all prefixes of $\mathbf{u}$ is finite. If moreover, $\mathbf{u}$ is fixed by a primitive substitution, then the derived word to a prefix w of $\mathbf{u}$ is fixed by a primitive substitution as well. In other words, given any primitive substitution φ, there exists a finite list $L = \{\varphi_1, \varphi_2, \ldots, \varphi_k\}$ of primitive substitutions such that for each prefix w of $\mathbf{u}$, the fixed point of φ, the derived word $\mathbf{d_u}(w)$ is fixed by a substitution φ_i from L. An algorithm which to a given Sturmian substitution creates such list L is described in [7].

On the other hand, if w is a non-prefix factor of $\mathbf{u}$, then it seems that $\mathbf{d_u}(w)$ is fixed by a substitution only exceptionally. In [5], this phenomenon is studied for fixed points of Sturmian substitutions. For this purpose, the following new notion has been introduced.

© Springer Nature Switzerland AG 2019
R. Mercaş and D. Reidenbach (Eds.): WORDS 2019, LNCS 11682, pp. 226–237, 2019.
https://doi.org/10.1007/978-3-030-28796-2_18

Definition 1. *A finite non-empty set M of primitive substitutions is said to be* closed under derivation *if the derived word $\mathbf{d}_{\mathbf{u}(\mathbf{w})}$ to any factor w of any fixed point $\mathbf{u}$ of $\varphi \in M$ is fixed (after a suitable renaming of letters) by a substitution $\psi \in M$. A primitive substitution ξ is said to be* closeable under derivation *if it belongs to a set M closed under derivation.*

Sturmian substitutions closeable under derivation are characterized in [5]. The aim of this contribution is to provide two new examples of sets M closed under derivation.

In our first example, in Sect. 4, the set M is a finite subset of the monoid of episturmian morphisms. In this case, all substitutions in M act on the same alphabet. In our second example, in Sect. 5, the substitutions in M act on alphabets with distinct cardinality. An inspiration for the second example comes from a recent result by Huang and Wen in [4], where a curious property of the period doubling substitution $\psi(a) = ab$ and $\psi(b) = aa$ was observed.

The article is organized as follows. The next section introduces necessary notions and definition. Section 3 deals with the set of derived words to factors of an infinite word in general. The last two sections contain the two examples of substitutions closeable under derivation.

2   Preliminaries

Let $\mathcal{A}$ denote an *alphabet*—a finite set of symbols. A *word* over $\mathcal{A}$ is a finite sequence $u = u_1 u_2 \cdots u_n$ where $u_i \in \mathcal{A}$ for all $i = 1, 2, \ldots, n$. The *length* of the word u is denoted by $|u|$ and is equal to n. The set of all words over $\mathcal{A}$ together with the operation concatenation forms a free monoid $\mathcal{A}^*$, its neutral element is the *empty word* ε. If $u = pws \in \mathcal{A}^*$, then w is a *factor* of u, p is a *prefix* of u, and s is a *suffix* of u. For $w = uv$, we write $u = wv^{-1}$ and $v = u^{-1}w$.

Let $\mathcal{B}$ be an alphabet. A *morphism* φ is a mapping $\varphi : \mathcal{A}^* \to \mathcal{B}^*$ such that $\varphi(uv) = \varphi(u)\varphi(v)$ for all $u, v \in \mathcal{A}^*$. A morphism $\varphi : \mathcal{A}^* \to \mathcal{A}^*$ is called *primitive* if there exists an iteration $k \in \mathbb{N}$ such that for any pair a, b of letters from $\mathcal{A}$, the letter a occurs in $\varphi^k(b)$. In accordance with Durand's terminology, a morphism φ is a *substitution* if there exist $a \in \mathcal{A}$ and $w \in \mathcal{A}^*, w \neq \varepsilon$ such that $\varphi(a) = aw$ and $|\varphi^n(a)|$ tends to infinity with growing n.

An *infinite word* over $\mathcal{A}$ is an infinite sequence $\mathbf{u} = u_0 u_1 u_2 \ldots$ from $\mathcal{A}^{\mathbb{N}}$. A finite word w of length n is a factor of $\mathbf{u}$ if there exists an index $i \in \mathbb{N}$, such that $w = u_i u_{i+1} \cdots u_{i+n-1}$. The index i is called an *occurrence* of w in $\mathbf{u}$. The set of all factors of $\mathbf{u}$ is denoted by $\mathcal{L}(\mathbf{u})$. If each factor w of $\mathbf{u}$ has infinitely many occurrences, then $\mathbf{u}$ is *recurrent*. A *return word* to w in $\mathbf{u}$ is a factor $r = u_i u_{i+1} \cdots u_{j-1}$, where $i < j$ are two consecutive occurrences of w in $\mathbf{u}$. The word rw is called a *complete return word* to w in $\mathbf{u}$ and obviously, rw is a factor of $\mathbf{u}$. The set of all return words to w in $\mathbf{u}$ is denoted by $\mathcal{R}_{\mathbf{u}}(w)$. If the set $\mathcal{R}_{\mathbf{u}}(w)$ is finite, say $\mathcal{R}_{\mathbf{u}}(w) = \{r_0, r_1, \ldots, r_{k-1}\}$, then $\mathbf{u}$ can be written as a concatenation $\mathbf{u} = p r_{i_0} r_{i_1} r_{i_2} \cdots$, where p is the prefix of $\mathbf{u}$ such that the factor w occurs in pw exactly once and $i_j \in \{0, 1, 2, \ldots, k-1\}$. The infinite word

$i_0 i_1 i_2 \cdots$ over the alphabet $\{0, 1, 2, \ldots, k-1\}$ is the *derived word to w in $\mathbf{u}$* and is denoted by $\mathbf{d_u}(w)$. A recurrent infinite word $\mathbf{u}$ is *uniformly recurrent* if the set $\mathcal{R_u}(w)$ is finite for all $w \in \mathcal{L}(\mathbf{u})$.

An infinite word $\mathbf{u} = u_0 u_1 u_2 \cdots \in \mathcal{A}^{\mathbb{N}}$ is *eventually periodic* if there exist integers k and n such that $k \geq 0, n > 0$ and for all i with $i \geq k$, $u_i = u_{i+n}$. An infinite word that is not eventually periodic is *aperiodic*.

The domain of a morphism $\varphi : \mathcal{A}^* \to \mathcal{B}^*$ is naturally extended to $\mathcal{A}^{\mathbb{N}}$ by putting $\varphi(\mathbf{u}) = \varphi(u_0 u_1 u_2 \cdots) = \varphi(u_0)\varphi(u_1)\varphi(u_2) \cdots$. We say that a morphism $\varphi : \mathcal{A}^* \to \mathcal{A}^*$ is a *substitution* if there exists a letter $a \in \mathcal{A}$ such that $\varphi(a) = aw$ for a non-empty word $w \in \mathcal{A}^*$ and $\varphi(b) \neq \varepsilon$ for all letters b. In other words, a substitution is a morphism having a fixed point which can be obtained as the limit $\lim_{k \to +\infty} \varphi^k(a)$. A word $\mathbf{u}$ is *purely substitutive* if there exists a substitution φ over $\mathcal{A}$ such that $\mathbf{u} = \varphi(\mathbf{u})$, i.e., $\mathbf{u}$ is a fixed point of φ. A word $\mathbf{v}$ over $\mathcal{B}$ is *substitutive* if $\mathbf{v} = \psi(\mathbf{u})$, where $\psi : \mathcal{A}^* \to \mathcal{B}^*$ is a morphism and $\mathbf{u}$ is a purely substitutive word. If $\mathbf{u}$ is fixed by a primitive substitution, then $\mathbf{v}$ is *primitive substitutive*. A well-known fact is that a primitive substitutive word is uniformly recurrent (c.f. [2]).

3   The Set of Derived Words to Factors of an Infinite Word

In this section we list several simple properties of the set

$$\mathrm{Der}_f(\mathbf{u}) = \{\mathbf{d_u}(w) \colon w \in \mathcal{L}(\mathbf{u})\}.$$

Since the derived words are often consider only with respect to a prefix, and we consider here the derived words with respect to all the factors, we add a lower index f in the notation Der_f to emphasize this fact.

First, we show that only some special factors need to be examined to describe $\mathrm{Der}_f(\mathbf{u})$. A letter $a \in \mathcal{A}$ is a *right extension* of $w \in \mathcal{L}(\mathbf{u})$ if $wa \in \mathcal{L}(\mathbf{u})$. Note that any factor of $\mathbf{u}$ has at least one right extension. A factor $w \in \mathcal{L}(\mathbf{u})$ is *right special* if it has at least two distinct right extensions. Analogously, we define *left special* factors.

A factor which is simultaneously right and left special is *bispecial*.

Proposition 2. *Let $\mathbf{u}$ be an infinite uniformly recurrent word over $\mathcal{A}$ and $w \in \mathcal{L}(\mathbf{u})$.*

(1) If w is not left special, then $\mathcal{R_u}(aw) = a\mathcal{R_u}(w)a^{-1}$, where $a \in \mathcal{A}$ is the unique left extension of w. Moreover, if w is not a prefix of $\mathbf{u}$, then $\mathbf{d_u}(aw) = \mathbf{d_u}(w)$.

(2) If w is not right special, then $\mathcal{R_u}(wa) = \mathcal{R_u}(w)$ and $\mathbf{d_u}(wa) = \mathbf{d_u}(w)$, where $a \in \mathcal{A}$ is the unique right extension of w.

Proof. Item (1): First assume that w is not left special and w is not a prefix of $\mathbf{u}$. The integer i is an occurrence of w in $\mathbf{u}$ if and only if $i - 1$ is an occurrence of aw in $\mathbf{u}$. Consequently, $r \in \mathcal{R}_{\mathbf{u}}(w)$ if and only if $ara^{-1} \in \mathcal{R}_{\mathbf{u}}(aw)$ and the ordering of the return words to w in $\mathbf{u}$ and the ordering the return words to aw in $\mathbf{u}$ coincide.

Let 0 be an occurrence of w, i.e., w is a prefix of $\mathbf{u}$. Then a return word r to w and rw have an occurrence 0. We have to show that even for such r the word ara^{-1} belongs to $\mathcal{R}_{\mathbf{u}}(aw)$. Indeed, the word $\mathbf{u}$ is recurrent and thus rw has an occurrence $j > 0$. As w is always preceded by the letter a and a is a suffix of r we can conclude that ara^{-1} is a return word to aw in $\mathbf{u}$.

Item (2): The proof is analogous. $\qquad \square$

We formulate a straightforward corollary of Proposition 2.

Proposition 3. *Let* $\mathbf{u}$ *be an aperiodic infinite uniformly recurrent word over* $\mathcal{A}$. *We have*
$$\mathrm{Der}_f(\mathbf{u}) = \{\mathbf{d}_{\mathbf{u}}(w) \colon w \text{ is a right special prefix of } \mathbf{u}\}$$
$$\cup \{\mathbf{d}_{\mathbf{u}}(w) \colon w \text{ is a bispecial factor of } \mathbf{u}\}.$$

Proof. Let w be a factor of $\mathbf{u}$. Let ws be the shortest right special factor of $\mathbf{u}$ having w as a prefix (its existence is guaranteed by the aperiodicity of $\mathbf{u}$). Repeatedly applying Item (2) of Proposition 2 for the factors ws' with s' being a prefix of s, we obtain $\mathbf{d}_{\mathbf{u}}(w) = \mathbf{d}_{\mathbf{u}}(ws)$.

Similarly, let pws be the shortest left special factor of $\mathbf{u}$ having ws as a suffix. If for all suffices p' of p, the word $p'ws$ is not a prefix of $\mathbf{u}$, we may repeatedly apply Item (1) of Proposition 2 to obtain $\mathbf{d}_{\mathbf{u}}(ws) = \mathbf{d}_{\mathbf{u}}(pws)$. Moreover, by the construction of pws, for a given w there is a unique bispecial factor pws, thus, in this case, we need to consider only $\mathbf{d}_{\mathbf{u}}(pws)$.

If for some suffix p' of p, the word $p'ws$ is a prefix of $\mathbf{u}$, we may repeatedly apply Item (1) of Proposition 2 to obtain $\mathbf{d}_{\mathbf{u}}(ws) = \mathbf{d}_{\mathbf{u}}(p'ws)$. Again, by the construction of $p'ws$, there is a unique such prefix $p'ws$ for a given w. Moreover, it is right special. Therefore, in this case, we need to take into account the word $\mathbf{d}_{\mathbf{u}}(p'ws)$.

Thus we obtain
$$\mathrm{Der}_f(\mathbf{u}) \subseteq \{\mathbf{d}_{\mathbf{u}}(w) \colon w \text{ is a right special prefix of } \mathbf{u}\}$$
$$\cup \{\mathbf{d}_{\mathbf{u}}(w) \colon w \text{ is a bispecial factor of } \mathbf{u}\},$$

and since the other inclusion is trivial, the proof is concluded. $\qquad \square$

The following claim is taken from Durand's article. His proof is constructive and provides an algorithm for finding a suitable morphism.

Proposition 4 ([2]). *Let* $\mathbf{u} \in \mathcal{A}^{\mathbb{N}}$ *be a fixed point of a primitive morphism* φ *and* w *be a prefix of* $\mathbf{u}$. *The word* $\mathbf{d}_{\mathbf{u}}(w)$ *is fixed by a primitive morphism as well.*

Proof (Sketch of the proof). We do not repeat the whole proof, we only describe the construction of a primitive morphism fixing $\mathbf{d}_{\mathbf{u}}(w)$.

As φ is primitive, its fixed point $\mathbf{u}$ is uniformly recurrent. Let $r_0, r_1, \ldots, r_{k-1}$ be the return words to w. Since $\mathbf{u}$ is fixed by φ, the image $\varphi(w)$ has a prefix

w and thus $\varphi(r_i w)$ has a prefix $\varphi(r_i)w$. As w is a prefix and a suffix of $\varphi(r_i)w$, the factor $\varphi(r_i)$ is a concatenation of several return words to w, i.e., we can find unique indices $s_1, s_2, \ldots, s_{\ell_i} \in \{0, 1, \ldots, k-1\}$ such that $\varphi(r_i) = r_{s_1} r_{s_2} \cdots r_{s_{\ell_i}}$. It is easy to check that the morphism given by

$$\delta : \quad i \mapsto s_1 s_2 \cdots s_{\ell_i} \quad \text{for each } i \in \{0, 1, \ldots, k-1\}$$

is primitive and fixes $\mathbf{d_u}(w)$. All details can be found in [2].

Proposition 5. *Let $\mathbf{u} \in \mathcal{A}^{\mathbb{N}}$ be a fixed point of a primitive morphism φ and $w \in \mathcal{L}(\mathbf{u})$. The word $\mathbf{d_u}(w)$ is primitive substitutive.*

Proof. Let pw be the shortest prefix of $\mathbf{u}$ containing the factor w. Let $r_0, r_1, \ldots, r_{k-1}$ denote the return words to pw and let $\tilde{r}_0, \tilde{r}_1, \ldots, \tilde{r}_{j-1}$ denote the return words to w. As w is a prefix and a suffix of the factor $p^{-1} r_i pw$, the word $p^{-1} r_i p$ can be written as concatenation of the return words to w, i.e., $p^{-1} r_i p = \tilde{r}_{s_1} \tilde{r}_{s_2} \cdots \tilde{r}_{s_{\ell_i}}$ for some unique indices $s_1, s_2, \ldots, s_{\ell_i} \in \{0, 1, \ldots, j-1\}$. Define the morphism $\psi : \{0, 1, \ldots, k-1\}^* \mapsto \{0, 1, \ldots, j-1\}^*$ by

$$\psi : \quad i \mapsto s_1 s_2 \cdots s_{\ell_i} \quad \text{for each } i \in \{0, 1, \ldots, k-1\}.$$

It follows that $\mathbf{d_u}(w) = \psi(\mathbf{d_u}(pw))$. By Proposition 4, $\mathbf{d_u}(pw)$ is fixed by a primitive substitution. $\qquad\square$

We finish this section by an example.

Example 6. *Recall the period doubling substitution*

$$\psi(a) = ab \quad \text{and} \quad \psi(b) = aa,$$

and its fixed point

$$\mathbf{z} = abaaababababaaabaaabaa \ldots.$$

- *Any occurrence of the letter b is preceded and followed by the letter a, therefore b is neither right nor left special. By Proposition 2,*

$$\mathbf{d_z}(b) = \mathbf{d_z}(ab) = \mathbf{d_z}(aba).$$

- *There are two return words to a in $\mathbf{z}$, namely $r_0 = ab$ and $r_1 = a$. We can write*

$$\mathbf{z} = r_0 r_1 r_1 r_0 r_0 r_0 r_1 r_1 r_0 r_1 r_1 r_0 r_1 \ldots \quad \text{and thus} \quad \mathbf{d_z}(a) = 0110001101101 \ldots.$$

The word $\mathbf{d_z}(a)$ is fixed by a substitution. To find it, we compute

$$\psi(r_0) = \psi(ab) = abaa = r_0 r_1 r_1 \quad \text{and} \quad \psi(r_1) = \psi(a) = ab = r_0.$$

It follows from the proof of Proposition 4 that $\mathbf{d_z}(a)$ is fixed by the substitution ξ determined by

$$\xi(0) = 011 \quad \text{and} \quad \xi(1) = 0.$$

4   Example 1: Standard Episturmian Morphisms

Let us recall the definition of standard Arnoux–Rauzy words and known results on morphisms fixing these words. All mentioned facts and further results can be found in the survey [3].

Definition 7. *An infinite word* $\mathbf{u} \in \mathcal{A}^{\mathbb{N}}$ *is* Arnoux–Rauzy *if*

1. $\mathbf{u}$ *has exactly one right special factor of each length;*
2. $wa \in \mathcal{L}(\mathbf{u})$ *for every right special factor* w *of* $\mathbf{u}$ *and every letter* $a \in \mathcal{A}$*;*
3. $\mathcal{L}(\mathbf{u})$ *is closed under reversal, i.e.,* $v_1 v_2 \cdots v_n \in \mathcal{L}(\mathbf{u})$ *implies* $v_n v_{n-1} \cdots v_1 \in \mathcal{L}(\mathbf{u})$.

An Arnoux–Rauzy word $\mathbf{u}$ *is* standard *if each of its prefixes is a left special factor of* $\mathbf{u}$.

The Arnoux–Rauzy words represent a generalization of Sturmian words to multiliteral alphabets and share many properties with Sturmian words. A property which is important for a description of their derived words is that Arnoux–Rauzy words are aperiodic and by [1] they are also uniformly recurrent. Let $\mathcal{M}_{\mathcal{A}}$ denote the monoid generated by *standard episturmian morphisms* L_a defined for every $a \in \mathcal{A}$ as follows:

$$L_a : \begin{cases} a \mapsto a, \\ b \mapsto ab & \text{for all } b \neq a. \end{cases}$$

To abbreviate the notation of the elements of the monoid $\mathcal{M}_{\mathcal{A}}$, we put

$$L_z = L_{z_1} \circ L_{z_2} \circ \cdots \circ L_{z_n} \quad \text{for} \quad z = z_1 z_2 z \cdots z_n \in \mathcal{A}^*.$$

A morphism $L_z \in \mathcal{M}_{\mathcal{A}}$ is primitive if and only if each letter from $\mathcal{A}$ occurs in z. Any primitive morphism in $\mathcal{M}_{\mathcal{A}}$ has only one fixed point and this fixed point is a standard Arnoux–Rauzy word. On the other hand, if a standard Arnoux–Rauzy word is fixed by a primitive substitution, then it is fixed by a primitive morphism from the monoid $\mathcal{M}_{\mathcal{A}}$.

Example 8. *Let us consider the Tribonacci word* $\mathbf{u}_\tau = abacabaabacababac\, abaa \ldots$ —*the fixed point of the morphism* $\tau : a \mapsto ab, b \mapsto ac, c \mapsto a$. *The word* $\mathbf{u}_\tau$ *is a standard Arnoux—Rauzy word over* $\{a, b, c\}$ *and it is fixed also by the primitive morphism* τ^3. *It is easy to check that* $\tau^3 = L_{abc}$ *and thus the Tribonacci word is fixed by a substitution from* $\mathcal{M}_{\mathcal{A}}$.

Medková in [8] studies derived words of Arnoux–Rauzy words. She considers all Arnoux–Rauzy (not only standard) words, but she describes derived words only to prefixes of infinite words. To quote a consequence of one of her results we need to recall the cyclic shift operation on $\mathcal{A}^*$:

$$\mathrm{cyc}(z_1 z_2 \cdots z_n) = z_n z_1 \cdots z_{n-1}.$$

Proposition 9 (Theorem 24 in [8]). *Let $L_z \in \mathcal{M}_\mathcal{A}, z \in \mathcal{A}^*$, be a primitive morphism and $\mathbf{u}$ be its fixed point. If w is a prefix of $\mathbf{u}$, then there exists $k \in \{1, 2, \ldots, |z|\}$ such that $\mathbf{d_u}(w)$ is fixed (up to a permutation of letters) by $L_{\mathrm{cyc}^k(z)}$. In particular, the word $\mathbf{d_u}(w)$ is a standard Arnoux–Rauzy word.*

Theorem 10. *Let z be a word in $\mathcal{A}^*$ such that each letter $a \in \mathcal{A}$ occurs in z at least once. The set*

$$M = \{ L_{\mathrm{cyc}^k(z)} \colon k \in \{1, 2, \ldots, |z|\} \}$$

is closed under derivation.

Proof. Let $\mathbf{v}$ be a fixed point of L_v with $v = \mathrm{cyc}^k(z)$ for some $k \in \{1, 2, \ldots, |z|\}$. Since z contains each letter from $\mathcal{A}$, the word v contains all letters form $\mathcal{A}$ as well and thus L_v is primitive.

As $\mathbf{v}$ is a standard Arnoux–Rauzy word, it follows from Definition 7 that each of its bispecial factor is a prefix of $\mathbf{v}$. By Proposition 3, only derived words to prefixes have to be considered. By Proposition 9, each such derived word is fixed (up to a permutation of letters) by a morphism $L_{\mathrm{cyc}^i(v)}$ for some $i \in \{1, 2, \ldots, |v|\}$. Obviously, this morphism belongs to M.

Example 11. *If we apply the previous theorem to the ternary word abc, we obtain that the set $M = \{L_{abc}, L_{bca}, L_{cab}\}$ is closed under derivation. Nevertheless, all the 3 morphisms in M fix the same word (up to a permutation of letters), namely the Tribonacci word. This word is fixed by the substitution τ given in Example 8. Therefore, the set $\{\tau\}$ is closed under derivation as well.*

5   Example 2: The Period Doubling Morphism

The aim of this section is to show that the period doubling substitution ψ determined by $\psi(a) = ab$ and $\psi(b) = aa$ is closeable under derivation. For this purpose, we first define the two following substitutions:

$$\nu : \begin{cases} 0 \mapsto 01, \\ 1 \mapsto 02020101, \\ 2 \mapsto 0202, \end{cases} \quad \text{and} \quad \xi : \begin{cases} 0 \mapsto 011, \\ 1 \mapsto 0. \end{cases} \tag{1}$$

Next, we deduce several auxiliary statements which help us to prove the following main theorem.

Theorem 12. *The sets $\{\psi, \xi, \nu\}$ and $\{\xi, \nu\}$ are closed under derivation.*

First, we focus on the derived words of the fixed point $\mathbf{z} = abaaababababaaabaaabaa\ldots$ of the substitution ψ. The following properties are immediate:

(I) $bb \notin \mathcal{L}(\mathbf{z})$. If $a^i \in \mathcal{L}(\mathbf{z})$, then $i \leq 3$.
(II) a and aa are bispecial factors of $\mathbf{z}$.

(III) Any bispecial factor of length more than 2 has a prefix ab and a suffix ba.

Proof. Recall that $\psi : a \mapsto ab, b \mapsto aa$. It follows from Item (I) that b and aaa can be neither a prefix nor a suffix of a bispecial factor. As $bb \notin \mathcal{L}(\mathbf{z})$, the factor baa is followed only by $\psi(a) = ab$. Therefore, baa is not a suffix of a bispecial factor. Similarly, the factor aab is preceded only by a. Thus, the factor aab is not a prefix of a bispecial factor. It follows that a bispecial factor must have the prefix ab and the suffix ba.

(IV) The longest common prefix of $\psi(a)$ and $\psi(b)$ is the letter a; the longest common suffix of $\psi(a)$ and $\psi(b)$ is the empty word. It implies that $\Phi(v) := \psi(v)a$ is bispecial whenever v is bispecial.

The converse of the very last property also holds (if $\Phi(v)$ is not too short).

Proposition 13. *Let w be a non-empty bispecial factor of $\mathbf{z}$ such that $w \neq a$ and $w \neq aa$. There exists a bispecial factor v such that $\Phi(v) = w$.*

Proof. As mentioned before, the bispecial factor w has a suffix ba and a prefix ab. Hence, there exists a factor v such that $\Phi(v) = \psi(v)a = w$ and a is both a prefix and a suffix of v. It remains to show that v is bispecial. If it is not right special, then v is followed only by a or b. But then w is followed only by b or a, respectively, since $\psi(va) = \psi(v)ab$ and $\psi(vb) = \psi(v)aa$. Thus, v is right special. Similarly, v is left special, and therefore bispecial.

Corollary 14. *Let w be a non-empty bispecial factor of $\mathbf{z}$. There exists $i \in \mathbb{N}$ such that*
$$w = \Phi^i(v) \quad \text{with } v \in \{a, aa\}.$$

Remark 15. Proposition 13 and its corollary are in fact a special case of a more general construction of bispecial factors given in [6]. From the point of view of this general construction, the bispecial factors a and aa are called initial, and Φ is the mapping generating all other bispecial factors from the initial bispecial factors in the spirit of Proposition 13.

As the fixed point $\mathbf{z}$ has a bispecial factor aa which is not a prefix of $\mathbf{z}$, the description of derived words to non-prefix factors is more complicated than in the case of a fixed point of a standard episturmian morphism. The following notion will be very useful for this purpose.

Definition 16. *Let w be a non-empty factor of a fixed point $\mathbf{x}$ of a substitution φ. Suppose there exist words y, y' and $u = u_1 u_2 \cdots u_n$ such that $ywy' = \varphi(u)$, $|y| < |\varphi(u_1)|$, $|y'| < |\varphi(u_n)|$, and $u \in \mathcal{L}(\mathbf{x})$. If there is exactly one occurrence of w in $\varphi(u)$, then we call u an ancestor of w. The set of all ancestors of w is denoted by $A(w)$. If there are more occurrences of w in $\varphi(u)$, then we say w allows an ambiguous ancestor.*

Example 17. *Given the fixed point* $\mathbf{z} = abaaababababaaabaaabaa\ldots$ *of the period doubling substitution* ψ, *the set of all ancestors of the factor* aa *is* $A(aa) = \{b\}$ *because* $\psi(b) = aa$ *and* $y = \varepsilon, y' = \varepsilon$. *Since* $\psi(ba) = aaab$, $y = a$, $y' = b$ *and there are two occurrences of* aa *in* $\psi(ba)$, *the factor* aa *allows an ambiguous ancestor. The prefix* aba *has two ancestors* aa *and* ab *and it does not allow an ambiguous ancestor.*

Proposition 18. *Let* $\mathbf{x}$ *be a fixed point of an injective substitution* φ *and* w *be a factor of* $\mathbf{x}$ *with a unique ancestor* u. *Assume* w *does not allow an ambiguous ancestor. We have* $\mathbf{d}_{\mathbf{x}}(w) = \mathbf{d}_{\mathbf{x}}(u)$.

Proof. The infinite word $\mathbf{x}$ can be written as $\mathbf{x} = z r_{i_0} r_{i_1} r_{i_2} \ldots$, where $r_{i_j} \in \mathcal{R}_{\mathbf{x}}(u)$ for all $j \in \mathbb{N}$. If u is a prefix, then $z = \varepsilon$. By the definition of a return word, u is a prefix of the word $r_{i_k} u \ldots$ for all $k \in \mathbb{N}$. Since u is the unique ancestor of w and w does not allow an ambiguous ancestor, there are exactly two occurrences of w in $\varphi(r_{i_k})\varphi(u)$. Let $\varphi(u) = ywy'$.

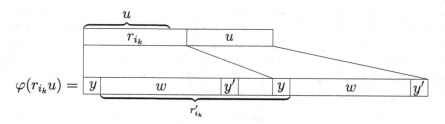

Fig. 1. An illustration of $r_{i_k} u$ and $\varphi(r_{i_k} u)$ in the proof of Proposition 18. For clarity of the figure, the depiction is a special case of non-overlapping factors w.

If we define $r'_{i_k} := y^{-1} \varphi(r_{i_k}) y$ as in Fig. 1, then $r'_{i_k} \in \mathcal{R}_{\mathbf{x}}(w)$ for all $k \in \mathbb{N}$ and we have

$$\mathbf{x} = \varphi(\mathbf{x}) = \varphi(z)\varphi(r_{i_0})\varphi(r_{i_1})\varphi(r_{i_2}) \cdots =$$
$$= \underbrace{\varphi(z)y}_{:=z'} \underbrace{y^{-1}\varphi(r_{i_0})y}_{r'_{i_0}} \underbrace{y^{-1}\varphi(r_{i_1})y}_{r'_{i_1}} \underbrace{y^{-1}\varphi(r_{i_2})y}_{r'_{i_2}} y^{-1} \cdots = z' r'_{i_0} r'_{i_1} r'_{i_2} \ldots.$$

The derived words of u and w are both $i_0 i_1 i_2 \ldots$.

Lemma 19. *Let* v *be a non-empty bispecial factor of the fixed point* $\mathbf{z}$ *of the period doubling substitution* ψ. *We have* $\mathbf{d}_{\mathbf{z}}(\Phi(v)) = \mathbf{d}_{\mathbf{z}}(v)$.

Proof. Since v is bispecial, by properties (II) and (III), the word a is a suffix of v and thus $\psi(v)$ has a suffix b. It implies that $\psi(v)$ is not right special. Therefore $\mathbf{d}_{\mathbf{z}}(\psi(v)) = \mathbf{d}_{\mathbf{z}}(\psi(v)a) = \mathbf{d}_{\mathbf{z}}(\Phi(v))$.

The word v is surely an ancestor of $\psi(v)$. We show that it is the only ancestor. Suppose there is another ancestor t with $t \neq v$. Thus, there exists y, y' such that

$y\psi(v)y' = \psi(t)$. By Definition 16 and the form of ψ, we conclude that the length of y and y' is at most 1. If $|y| = 0$, then the form of ψ implies $v = t$ and $y' = \varepsilon$, which is a contradiction with $t \neq v$. Similarly, y' cannot be empty. Therefore, y and y' are both letters. Thus, the last letter of $\psi(v)$ is the first letter of $\psi(a)$ or $\psi(b)$ which is in both cases the letter a—a contradiction following from the first paragraph of this proof. Therefore $A(\psi(v)) = \{v\}$ and it is not difficult to verify that $\psi(v)$ does not allow an ambiguous ancestor when it contains at least one letter b. By Proposition 18 we have $\mathbf{d_z}(v) = \mathbf{d_z}(\psi(v)) = \mathbf{d_z}(\Phi(v))$.

Proposition 20. *If w is a non-empty factor of $\mathbf{z}$, then $\mathbf{d_z}(w) = \mathbf{d_z}(a)$ or $\mathbf{d_z}(w) = \mathbf{d_z}(aa)$. If w is a non-empty prefix of $\mathbf{z}$, then $\mathbf{d_z}(w) = \mathbf{d_z}(a)$.*

Proof. By Proposition 3 we have to describe the derived words to right special prefixes and to bispecial factors only. First assume that w is a non-empty bispecial factor of $\mathbf{z}$. By Corollary 14, the factor w can be obtained by iteration of the mapping $\Phi(v) = \psi(v)a$ starting from the two bispecial factors a and aa. Thus, by repeated application of Lemma 19, we obtain that $\mathbf{d_z}(w)$ equals $\mathbf{d_z}(a)$ or $\mathbf{d_z}(aa)$.

Now assume that w is a right special prefix of $\mathbf{z}$. As the bispecial factor a is a prefix of $\mathbf{z}$ and image by Ψ of every letter starts with a, the bispecial factor $\Phi^k(a)$ is a prefix of $\mathbf{z}$ for each $k \in \mathbb{N}$. Therefore, any right special prefix w of $\mathbf{z}$ is left special as well. Since $\Phi^\ell(aa)$ is not a prefix of $\mathbf{z}$ for all positive ℓ and by Corollary 14, any right special prefix of $\mathbf{z}$ equals $\Phi^k(a)$ for some $k \in \mathbb{N}$. By Lemma 19, $\mathbf{d_z}(w) = \mathbf{d_z}(\Phi^k(a)) = \mathbf{d_z}(a)$.

Now we show that both derived words to a factor of $\mathbf{z}$ are fixed by primitive substitutions. We exploit the following simple tool.

Observation 21. *Let $\mathbf{v}$ be a fixed point of a morphism γ and let $\mathbf{u} = \alpha(\mathbf{v})$ where α is a morphism. If there exists a morphism β such that $\alpha\gamma = \beta\alpha$, then $\mathbf{u}$ is fixed by β.*

Proof. $\beta(\mathbf{u}) = \beta\alpha(\mathbf{v}) = \alpha\gamma(\mathbf{v}) = \alpha(\mathbf{v}) = \mathbf{u}$.

Proposition 22. *The derived word $\mathbf{d_z}(a)$ is fixed by ξ and the derived word $\mathbf{d_z}(aa)$ is fixed by ν (where ξ and ν are defined in (1)).*

Proof. In Example 6 above, we show, using Proposition 4, that the derived word $\mathbf{d_z}(a)$ is fixed by the substitution ξ.

It remains to consider $\mathbf{d_z}(aa)$. As $abaa$ is the shortest prefix of $\mathbf{z}$ containing the bispecial factor aa, we can use the construction from the proof of Definition 5 to find a morphism α such that $\mathbf{d_z}(aa) = \alpha(\mathbf{d_z}(abaa))$. In our case $p = ab$ and $w = aa$. According to Proposition 20, the derived word $\mathbf{d_z}(abaa)$ is fixed by ξ since $\mathbf{d_z}(a)$ is fixed by ξ. Thus, $\mathbf{d_z}(abaa)$ is over a binary alphabet, and so the prefix $abaa$ has exactly two return words, say r_0 and r_1. These two return words can be found in the prefix of $\mathbf{z}$ of length 16. They are

$$r_0 = abaaabab \quad \text{and} \quad r_1 = abaa.$$

It follows from the proof of Proposition 5 that $(ab)^{-1}r_0ab$ and $(ab)^{-1}r_1ab$ can be written as a concatenation of return words to aa. Specifically, $r_0' = a, r_1' = aababab, r_2' = aab$ are return words of aa and $(ab)^{-1}r_0ab = r_0'r_1'$ and $(ab)^{-1}r_1ab = r_0'r_2'$. Hence, according to this claim we have

$$\alpha(0) = 01,$$
$$\alpha(1) = 02.$$

Note that since $\mathbf{d_z}(abaa)$ is fixed by ξ, it is also fixed by ξ^2. By Observation 21, if the substitution ν satisfies $\alpha\xi^2 = \nu\alpha$, the proof is finished. This is very easy to verify:

$$\alpha\xi^2(0) = \alpha(01100) = 0102020101$$
$$\nu\alpha(0) = \nu(01) = 0102020101$$
$$\alpha\xi^2(1) = \alpha(011) = 010202$$
$$\nu\alpha(1) = \nu(02) = 010202.$$

Remark 23. The derived word $\mathbf{d_z}(aa)$ is also fixed by the morphism

$$\eta(0) = \varepsilon$$
$$\eta(1) = 010202$$
$$\eta(2) = 01.$$

The proof is the same as the proof of Proposition 22, but at the end we have to verify the equality $\alpha\xi = \eta\alpha$. The reason why we prefer ν to η is that η is an erasing non-primitive morphism.

Corollary 24. *If w is a non-empty factor of $\mathbf{z}$, then $\mathbf{d_z}(w)$ is fixed by ξ or ν.*

Proof. The corollary follows from Propositions 20 and 22.

We conclude this section by the proof of our main result. For this purpose we need one more ingredient. It is a modification of Item 5 from [2, Proposition 6]. Its proof is almost identical to the proof of the original statement and thus we omit it.

Lemma 25. *Let $\mathbf{u}$ be a uniformly recurrent word and let w be its factor. Set $\mathbf{v} = \mathbf{d_u}(w)$. For a factor x of $\mathbf{v}$, there exists a factor y of $\mathbf{u}$ such that $\mathbf{d_v}(x) = \mathbf{d_u}(y)$.*

Proof (Proof of Theorem 12). Let $\mathbf{v}$ be a fixed point of the primitive substitution ξ and x be a factor of $\mathbf{v}$. By Proposition 22, we have $\mathbf{v} = \mathbf{d_z}(a)$. By Lemma 25, there exists a factor y in $\mathbf{z}$ such that $\mathbf{d_v}(x) = \mathbf{d_z}(y)$. Proposition 20 implies that $\mathbf{d_v}(x)$ equals $\mathbf{d_z}(a)$ or $\mathbf{d_z}(aa)$. Therefore, $\mathbf{d_v}(x)$ is fixed by ξ or ν.

The same reasoning gives that the derived word to any factor of the fixed point of ν is fixed by ξ or by ν. By Definition 1, the set $\{\nu, \xi\}$ is closed under derivation.

As $\mathbf{d_z}(\varepsilon) = \mathbf{z}$ and the derived word to any non-empty factor of $\mathbf{z}$ is fixed by ξ or by ν, the set $\{\nu, \xi, \psi\}$ is also closed under derivation.

Acknowledgments. This work was supported by the Ministry of Education, Youth and Sports of the Czech Republic, project no. CZ.02.1.01/0.0/0.0/16_019/0000778. We also acknowledge financial support of the Grant Agency of the Czech Technical University in Prague, grant No. SGS17/193/OHK4/3T/14. We thank Michel Dekking for attracting our attention to the article [4].

References

1. Droubay, X., Justin, J., Pirillo, G.: Episturmian words and some constructions of de Luca and Rauzy. Theoret. Comput. Sci. **255**(1–2), 539–553 (2001). https://doi.org/10.1016/S0304-3975(99)00320-5
2. Durand, F.: A characterization of substitutive sequences using return words. Discrete Math. **179**(1—3), 89–101 (1998)
3. Glen, A., Justin, J.: Episturmian words: a survey. Theor. Inf. Appl. **43**(3), 403–442 (2009)
4. Huang, Y.K., Wen, Z.Y.: Envelope words and the reflexivity of the return word sequences in the period-doubling sequence. Preprint available at https://arxiv.org/abs/1703.07157 (2017)
5. Klouda, K., Pelantová, E., Starosta, Š.: Sturmian substitutions closed under derivation (2019, in preparation)
6. Klouda, K.: Bispecial factors in circular non-pushy D0L languages. Theoret. Comput. Sci. **445**, 63–74 (2012). https://doi.org/10.1016/j.tcs.2012.05.007
7. Klouda, K., Medková, K., Pelantová, E., Starosta, Š.: Fixed points of Sturmian morphisms and their derivated words. Theoret. Comput. Sci. **743**, 23–37 (2018). https://doi.org/10.1016/j.tcs.2018.06.037
8. Medková, K.: Derived sequences of Arnoux-Rauzy sequences. Submitted to WORDS 2019 (2019)

Templates for the k-Binomial Complexity of the Tribonacci Word

Marie Lejeune$^{(\boxtimes)}$, Michel Rigo, and Matthieu Rosenfeld

Department of Mathematics, University of Liège,
Allée de la Découverte 12 (B37), 4000 Liège, Belgium
{M.Lejeune,M.Rigo,M.Rosenfeld}@uliege.be

Abstract. Consider k-binomial equivalence: two finite words are equivalent if they share the same subwords of length at most k with the same multiplicities. With this relation, the k-binomial complexity of an infinite word **x** maps the integer n to the number of pairwise non-equivalent factors of length n occurring in **x**. In this paper based on the notion of template introduced by Currie et al., we show that, for all $k \geq 2$, the k-binomial complexity of the Tribonacci word coincides with its usual factor complexity $p(n) = 2n + 1$. A similar result was already known for Sturmian words, but the proof relies on completely different techniques that seemingly could not be applied for Tribonacci.

1 Introduction

Abelian equivalence of words has been investigated for quite a long time; e.g., in the sixties Erdős raised the question whether abelian squares can be avoided by an infinite word over an alphabet of size 4 [6,7,16]. Let Σ be a finite alphabet. We let Σ^* denote the set of all finite words over Σ. Two words u and v in Σ^* are *abelian equivalent* if one word is obtained by permuting the letters of the other word. More formally, u and v are abelian equivalent if $|u|_a = |v|_a$, for all $a \in \Sigma$, where we let $|u|_a$ denote the number of occurrences of the letter a in u.

Definition 1. *Let u be a word over an ordered alphabet $\{0, \ldots, k-1\}$. The* abelianization *or* Parikh *vector of u, denoted by $\Psi(u)$, is the column vector in $\mathbb{N}^k$*

$$(|u|_0, \ldots, |u|_{k-1})^{\mathsf{T}}.$$

With this notation, two words are abelian equivalent if and only if $\Psi(u) = \Psi(v)$. A possible generalization of abelian equivalence is the k-*binomial equivalence* based on binomial coefficient of words. An independent generalization is k-abelian equivalence where one counts factors of length at most k [9,10]. For a survey, see, for instance, [19]. We let the binomial coefficient $\binom{u}{v}$ denote the number of times v appears as a (not necessarily contiguous) subsequence of u. Let $k \geq 1$ be an integer. Two words u and v are k-*binomially equivalent*, denoted $u \sim_k v$, if $\binom{u}{x} = \binom{v}{x}$ for all words x of length at most k.

The first author is supported by a FNRS fellowship.

R. Mercaş and D. Reidenbach (Eds.): WORDS 2019, LNCS 11682, pp. 238–250, 2019.
https://doi.org/10.1007/978-3-030-28796-2_19

Definition 2. *Let* **x** *be an infinite word. The* k-*binomial complexity function of* **x** *is defined as*

$$b_{\mathbf{x},k} : \mathbb{N} \to \mathbb{N}, \ n \mapsto \# \left(\mathrm{Fac}_n(\mathbf{x}) / \sim_k \right)$$

where $\mathrm{Fac}_n(\mathbf{x})$ *is the set of factors of length n occurring in* **x**. *For $k = 1$, this measure of complexity is exactly the abelian complexity.*

The celebrated theorem of Morse–Hedlund [3,4] characterizes ultimately periodic words in terms of a bounded factor complexity function. Hence, aperiodic words with the lowest factor complexity are exactly the Sturmian words characterized by $p_{\mathbf{x}}(n) = n + 1$. It is also a well-known result of Cobham that the factor complexity of any k-automatic sequence is in $\mathcal{O}(n)$. The Tribonacci word has a factor complexity $2n + 1$.

We collect the known facts about the k-binomial complexity. For all $k \geq 2$, Sturmian words have a k-binomial complexity which is the same as their factor complexity, i.e., $b_{\mathbf{x},k}(n) = n + 1$ for all n. Since $b_{\mathbf{x},k}(n) \leq b_{\mathbf{x},k+1}(n)$, the proof consists in showing that any two distinct factors of length n occurring in a given Sturmian word are never 2-binomially equivalent [17, Thm. 7]. However, the Thue–Morse word has a bounded k-binomial complexity [17, Thm. 13]. So we have a striking difference with the most usual complexity measures. Naturally, the bound on the k-binomial complexity of the Thue–Morse word depends on the parameter k because when k tends to infinity, the k-binomial equivalence gets closer to equality of factors, i.e. $b_{\mathbf{x},k}(n) = p_{\mathbf{x},k}(n)$ for all $n \leq k$, and the Thue–Morse word has a factor complexity in $\Theta(n)$. The precise results are recalled below.

Theorem 3 [17]. *Let $k \geq 1$. There exists $C_k > 0$ such that the k-binomial complexity of the Thue–Morse word* **t** *satisfies $b_{\mathbf{t},k}(n) \leq C_k$ for all $n \geq 0$.*

Theorem 4 [11]. *We let* **t** *denote the Thue–Morse word over a 2-letter alphabet. Let k be a positive integer. For all $n \leq 2^k - 1$, we have $b_{\mathbf{t},k}(n) = p_{\mathbf{t}}(n)$. For all $n \geq 2^k$, we have*

$$b_{\mathbf{t},k}(n) = \begin{cases} 3 \cdot 2^k - 3, & \text{if } n \equiv 0 \pmod{2^k}; \\ 3 \cdot 2^k - 4, & \text{otherwise.} \end{cases}$$

When investigating this new k-binomial complexity measure, it is naturally interesting to consider various well-known words or families of words. A natural choice is therefore to try computing the k-binomial complexity of the Tribonacci word $\mathcal{T} = 010201001020101\cdots$, fixed point of the morphism $\tau : 0 \mapsto 01, 1 \mapsto 02, 2 \mapsto 0$. From computer experiments, the second author made the conjecture in 2014 that, for all $k \geq 2$, its k-binomial complexity is the same as the usual factor complexity $2n + 1$ [18]. As in the Sturmian case, it is enough to show that given any two distinct factors of length n occurring in the Tribonacci word, these two factors are not 2-binomially equivalent. Surprisingly, classical combinatorial techniques seemed to be unsuccessful. We make an extensive use of the concepts of *templates* and their ancestors similar to what can be found in [1,2,8] where avoidance of abelian repetitions is considered. Closely related, let us also mention

[5] where a morphic word avoiding three consecutive factors of the same size and same sum is given. Recently Liétard proposed algorithmic proofs for morphic words avoiding additive powers [13].

This paper is organized as follows. In Sect. 2, we recall the notion of Parikh matrix and extend it to our k-binomial context. In particular, this extended matrix is built from the classical one using the Kronecker product: binomial coefficients can be nicely represented in terms of this product. In Sect. 3 we define and adapt the notions of templates and ancestors to our purpose. To solve our problem, we need to show the finiteness of some set of realizable ancestors. To that end, in Sect. 4, we first get several bounds related to Parikh vectors of factors of the Tribonacci word. Consequently, we deduce bounds on the realizable ancestors. We put together the results of these last two sections to establish the main theorem in Sect. 5. Similarly to [1,2,8,13,15], our proof is a computer-assisted one.

2   Basics

Let $\Sigma = \{0, \ldots, s-1\}$ be an ordered alphabet of size s. As mentioned in the introduction, it is enough to consider 2-binomial equivalence but everything in this section generalizes well to k-binomial equivalence.

Definition 5. *Let w be a finite word over Σ. We will make an extensive use of its extended Parikh vector denoted by $\Phi(w)$ and defined as follows. We set*

$$\Phi(w) := \left(|w|_0, \ldots, |w|_{s-1}, \binom{w}{00}, \binom{w}{01}, \ldots, \binom{w}{(s-1)(s-1)} \right)^{\mathsf{T}}.$$

It is a column vector of size $s(s+1)$ and we assume that the s^2 subwords of length 2 are lexicographically ordered.

Take the word $u = 10010201010$ which is a factor of length 11 occurring in the Tribonacci word. Its extended Parikh vector is given by

$$\Phi(u) = \left(6, 4, 1, 15, 11, 3, 13, 6, 2, 3, 2, 0 \right)^{\mathsf{T}}.$$

With this notation, $\Phi(u) = \Phi(v)$ if and only if $u \sim_2 v$.

For a vector $\mathbf{d} \in \mathbb{Z}^n$, $n \geq s$, we let $\mathbf{d}|_{0,\ldots,s-1}$ denote the vector in $\mathbb{Z}^s$ made of the first s coordinates of $\mathbf{d}$. In particular over an alphabet of size s, $\Phi(w)|_{0,\ldots,s-1} = \Psi(w)$.

We let $A \otimes B$ denote the usual Kronecker product of two matrices $A \in \mathbb{Z}^{m \times n}$ and $B \in \mathbb{Z}^{p \times q}$. It is a block-matrix in $\mathbb{Z}^{mp \times nq}$ defined by

$$A \otimes B := \begin{pmatrix} a_{11}B & \cdots & a_{1n}B \\ \vdots & \ddots & \vdots \\ a_{m1}B & \cdots & a_{mn}B \end{pmatrix}.$$

Let m be an integer. We let $P_m \in \mathbb{Z}^{(m(m+1)) \times m^2}$ denote the matrix such that for all, i, j,

$$[P_m]_{i,j} = \begin{cases} 1, & \text{if } i = j + m; \\ 0, & \text{otherwise.} \end{cases}$$

This matrix adds m zeros at the beginning of a column vector of size m^2. We start with two straightforward lemmas and the introduction of an extended Parikh matrix.

Lemma 6. *Let u and v be two words over an alphabet of size s. We have*

$$\Phi(uv) = \Phi(u) + \Phi(v) + P_s\big(\Psi(u) \otimes \Psi(v)\big).$$

Proof. The first two terms in the statement take into account the separate contributions of u and v to the different coefficients. Nevertheless, subwords of length 2 can also be obtained by taking their first letter in u and their second one in v. This is exactly the contribution of the third term. Observe that $\Psi(u) \otimes \Psi(v)$ is a column vector of size s^2. Applying P_s will add s zeros on top because the contribution of individual letters has already been taken into account in the first two terms. $\square$

The classical *Parikh matrix* M'_h associated with a morphism h is a useful tool in combinatorics on words (not to be confused with the notion of Parikh matrix of a word introduced in 2000 by Mateescu et al.). Over an ordered s-letter alphabet, it is defined from its columns as a $s \times s$ matrix

$$M'_h = \big(\Psi(h(0)) \cdots \Psi(h(s-1))\big)$$

and it readily satisfies

$$\Psi(h(u)) = M'_h \Psi(u), \quad \forall u \in \Sigma^*.$$

For the Tribonacci morphism, it is given by

$$M'_\tau = \begin{pmatrix} 1 & 1 & 1 \\ 1 & 0 & 0 \\ 0 & 1 & 0 \end{pmatrix}. \tag{1}$$

Definition 7. *Mimicking the Parikh matrix and its use, one can define an extended Parikh matrix M_h associated with a morphism h defined over an ordered s-letter alphabet. It is a $s(s+1) \times s(s+1)$ matrix satisfying*

$$\Phi(h(u)) = M_h \Phi(u), \quad \forall u \in \Sigma^*. \tag{2}$$

The existence of the extended Parikh matrix satisfying (2) is ensured by the next result.

Lemma 8. *Let M_h' be the Parikh matrix associated with some morphism h. The extended Parikh matrix of h has the following form:*

$$M_h = \begin{pmatrix} M_h' & 0 & \cdots & 0 \\ \star & & & \\ \star & & M_h' \otimes M_h' & \\ \star & & & \end{pmatrix}.$$

In particular, $\det(M_h) = \det(M_h')^{2s+1}$. Moreover, if the alphabet is of size s, then $M_h P_s = P_s(M_h' \otimes M_h')$. If M_h' is non-singular, then M_h is non-singular and M_h^{-1} is a block-triangular matrix of the same form as M_h with diagonal blocks $M_h'^{-1}$ and $M_h'^{-1} \otimes M_h'^{-1}$.

Proof. Since the first s components of $\Phi(u)$ give the usual Parikh vector, M_h' is the upper-left corner of M_h. For the last s^2 components of $\Phi(u)$ dealing with binomial coefficients of subwords of length 2, it is shown in [11] that, for all $a, b \in \Sigma$,

$$\binom{h(u)}{ab} = \sum_{c \in \Sigma} \binom{h(c)}{ab} |u|_c + \sum_{x_1 x_2 \in \Sigma^2} \binom{h(x_1)}{a}\binom{h(x_2)}{b}\binom{u}{x_1 x_2}.$$

In this expression, the first sum corresponds to the $s \times s$ submatrices marked as $\star$ and the second sum exactly corresponds to the Kronecker product $M_h' \otimes M_h'$. Indeed, if we index M_h' on Σ and $M_h' \otimes M_h'$ on Σ^2, we have

$$\binom{h(x_1)}{a}\binom{h(x_2)}{b} = [M_h']_{a,x_1}[M_h']_{b,x_2} = [M_h' \otimes M_h']_{ab,x_1 x_2}. \qquad \square$$

This extended Parikh matrix was also used in [14] (for avoidance problems).

3   Templates and Ancestors

For this section, let $h : \Sigma^* \to \Sigma^*$ be any primitive (prolongable) morphism. Let M_h' be its Parikh matrix and M_h be its extended Parikh matrix. We let $s := \#\Sigma$. Recall that a prefix (resp., a suffix) of a word w is *proper* if it is different from w (and thus, possibly empty).

Definition 9. *The* language *of h, denoted by $\mathcal{L}(h)$, is the set of factors of any of its non-empty fixed points (if h is primitive, they all have the same language). The set $\mathrm{Pref}(h)$ (resp., $\mathrm{Suff}(h)$) is the set of proper prefixes (resp., proper suffixes) of the words in $\{h(a) \mid a \in \Sigma\}$, e.g., for the Tribonacci morphism, $\mathrm{Pref}(\tau) = \{\varepsilon, 0\}$ and $\mathrm{Suff}(\tau) = \{\varepsilon, 1, 2\}$. Such a notation can be extended to h^n. If $u \in \mathcal{L}(h)$, there exist a shortest $p_u \in \mathrm{Pref}(h)$, a shortest $s_u \in \mathrm{Suff}(h)$ and $u' \in \mathcal{L}(h)$ such that $h(u') = p_u u s_u$.*

In the following definition, the index **b** (resp., **e**) stands for beginning (resp., end).

Definition 10. *A template is a 5-tuple of the form* $t = [\mathbf{d}, \mathbf{D_b}, \mathbf{D_e}, a_1, a_2]$ *where* $a_1, a_2 \in \Sigma$, $\mathbf{d} \in \mathbb{Z}^{s(s+1)}$ *and* $\mathbf{D_b}, \mathbf{D_e} \in \mathbb{Z}^s$. *A pair of words* (u, v) *is a* realization *of (or* realizes*) the template* t *if:*

- $\Phi(u) - \Phi(v) = \mathbf{d} + P_s\big(\mathbf{D_b} \otimes \Psi(u) + \Psi(u) \otimes \mathbf{D_e}\big)$,
- *there exist* u' *and* v' *such that* $u = u'a_1$ *and* $v = v'a_2$.

A template t *is* realizable *by* h *if there is a pair of words in* $\mathcal{L}(h)$ *that realize* t.

Given two factors u and v, the template of the form $[\Phi(u) - \Phi(v), \mathbf{0}, \mathbf{0}, a_1, a_2]$ is obviously realizable by h, where a_1 (resp., a_2) is the last letter of u (resp., v).

Due to the presence of P_s in the above definition, note that if a template is realizable by a pair (u, v), then the corresponding vector $\mathbf{d}$ is such that $\mathbf{d}|_{0,\ldots,s-1} = \Psi(u) - \Psi(v)$.

Remark 11. There exist an infinite number of realizable templates. Actually, for any choice of words u, v and vectors $\mathbf{D_b}, \mathbf{D_e}$ in $\mathbb{Z}^s$, there exists a convenient $\mathbf{d} \in \mathbb{Z}^{s(s+1)}$.

Lemma 12. *Let* h *be a primitive morphism. Let* $T := \{[\mathbf{0}, \mathbf{0}, \mathbf{0}, a_1, a_2] : a_1 \neq a_2\}$. *The factorial complexity and the 2-binomial complexity of any fixed point of* h *are equal if and only if all templates from* T *are non-realizable by* h.

Proof. The factorial complexity is not the same as the 2-binomial complexity if and only if there exists a pair of factors (u, v) such that $u \neq v$ and $\Phi(u) = \Phi(v)$.

The two words of any realization of an element in T are 2-binomially equivalent and are different since they do not have the same last letter. Thus, if there is a realization of an element of T then the factorial complexity and the 2-binomial complexity are not equal.

Now, for the other direction, suppose that the two complexity functions are not equal: we have a pair of words (u, v) such that $u \neq v$ and $\Phi(u) = \Phi(v)$. Since $u \neq v$ and $|u| = |v|$, there exist $u', v', s \in \Sigma^*$ and $a, b \in \Sigma$ with $a \neq b$ such that $u = u'as$ and $v = v'bs$ (observe that s is the longest common suffix of u and v). Then $\Phi(u'a) = \Phi(v'b)$ so the pair $(u'a, v'b)$ realizes $[\mathbf{0}, \mathbf{0}, \mathbf{0}, a, b]$, which belongs to T. □

The idea in the next definition is that any long factor of a fixed point of a morphism must be the image of a shorter factor, up to (short) prefix and suffix. So the relation corresponds to the various relationships among the binomial coefficients that must hold if this is to be the case. For more details, the reader is invited to read the proof of [12, Lemma 15].

Definition 13. *Let* $t' = [\mathbf{d'}, \mathbf{D'_b}, \mathbf{D'_e}, a'_1, a'_2]$ *and* $t = [\mathbf{d}, \mathbf{D_b}, \mathbf{D_e}, a_1, a_2]$ *be two templates and* h *be a morphism. We say that* t' *is a* parent *by* h *of* t *if there exist* $p_u, p_v \in \mathrm{Pref}(h)$ *and* $s_u, s_v \in \mathrm{Suff}(h)$ *such that:*

- $\mathbf{d'}$ *is given by*

$$M_h \mathbf{d'} = \mathbf{d} + \Phi(p_u s_u) - \Phi(p_v s_v) + P_s\big(\Psi(p_v) \otimes \mathbf{d}|_{0,\ldots,s-1} + \mathbf{d}|_{0,\ldots,s-1} \otimes \Psi(s_v)\big)$$
$$- P_s\big((\mathbf{D_b} + \Psi(p_u) - \Psi(p_v)) \otimes \Psi(p_u s_u) + \Psi(p_u s_u) \otimes (\mathbf{D_e} + \Psi(s_u) - \Psi(s_v))\big);$$

- *the value of $\mathbf{D_b'}$ is given by $M_h' \mathbf{D_b'} = \mathbf{D_b} + \Psi(p_u) - \Psi(p_v)$;*
- *the value of $\mathbf{D_e'}$ is given by $M_h' \mathbf{D_e'} = \mathbf{D_e} + \Psi(s_u) - \Psi(s_v)$;*
- $a_1 s_u$ *is a suffix of $h(a_1')$;*
- $a_2 s_u$ *is a suffix of $h(a_2')$.*

We let $\mathrm{Par}_h(t)$ denote the set of parents by h of t.

Remark 14. Observe that for any given template t, $\mathrm{Par}_h(t)$ is finite and easy to compute as long as M_h and M_h' are non-singular. Indeed, the sets $\mathrm{Pref}(h)$ and $\mathrm{Suff}(h)$ are finite. For the Tribonacci word, $\#\mathrm{Pref}(\tau) = 2$, $\#\mathrm{Suff}(\tau) = 3$ and thus $\#\mathrm{Par}_\tau(t) \le 36$. At this stage, it is not required for a parent to be realizable.

More interestingly there is a link between preimages of the realization by h of a template and realization by h of the parents of the template. We make that link explicit in the following Lemma.

Lemma 15. *Let h be a morphism. Assume that $\det(M_h') = \pm 1$. Let t be a template, $u, v, v', u' \in \mathcal{L}(h)$, $p_u, p_v \in \mathrm{Pref}(h)$ and $s_u, s_v \in \mathrm{Suff}(h)$ such that:*

- $h(u') = p_u u s_u$ *and $h(v') = p_v v s_v$;*
- s_u *is a proper suffix of the image of the last letter of u';*
- s_v *is a proper suffix of the image of the last letter of v';*
- (u, v) *realizes t.*

Then there exists a parent t' of t such that (u', v') realizes t'.

This motivates the following definitions.

Definition 16. *A template t' is an ancestor by h (resp., realizable ancestor) of a template t if there exists a sequence of $n \ge 1$ templates (resp., realizable templates) $t = t_1, t_2, \ldots, t_n = t'$ such that for all $i \in \{1, \ldots, n-1\}$, t_{i+1} is a parent by h of t_i. For a template t, we denote by $\mathrm{RAnc}_h(t)$ the set of all the realizable ancestors by h of t. We may omit "by h" when the morphism is clear from the context.*

Let $|h| = \max_{a \in \Sigma} |h(a)|$, usually called the *width* of h. To prove that two different factors of Tribonacci are never 2-binomially equivalent, we will make use of Lemma 12. The next result will help us to show that no template of the set T defined in Lemma 12 is realizable by τ.

Proposition 17. *Let L be a positive integer. Let h be a primitive morphism and t_0 be a template. If there exists a pair of words in $\mathcal{L}(h)$ that is a realization of t_0, then*

- *either t_0 has a realization $(u, v) \in \mathcal{L}(h) \times \mathcal{L}(h)$ such that $\min(|u|, |v|) \le L$ or,*
- *there exists a realization $(u, v) \in \mathcal{L}(h) \times \mathcal{L}(h)$ of a template t of $\mathrm{RAnc}_h(t_0)$ with $L \le \min(|u|, |v|) \le |h| L$.*

Proof. Let (u, v) be a pair of factors of $\mathcal{L}(h)$ realizing t_0. If $\min(|u|, |v|) \leq L$, there is nothing left to prove. Assume therefore that $\min(|u|, |v|) > L$.

Since v is a factor of $\mathcal{L}(h)$, there are sequences of words $v = v_1, v_2, \ldots, v_n \in \Sigma^*$, $p_1, \ldots, p_{n-1} \in \mathrm{Pref}(h)$ and $s_1, \ldots, s_{n-1} \in \mathrm{Suff}(h)$ such that, for all $i < n$, $h(v_{i+1}) = p_i v_i s_i$ and $L \leq |v_n| \leq |h| L$. Moreover we may force that, for all $i < n$, s_i is a proper suffix of the image of the last letter of v_{i+1}.

Similarly, since u is a factor of $\mathcal{L}(h)$, there are sequences of words $u = u_1, u_2, \ldots, u_\ell \in \Sigma^*$ and $p'_1, \ldots, p'_{\ell-1} \in \mathrm{Pref}(h)$ and $s'_1, \ldots, s'_{\ell-1} \in \mathrm{Suff}(h)$ such that, for all $i < \ell$, $h(u_{i+1}) = p'_i u_i s'_i$ and $L \leq |u_\ell| \leq L|h|$.

Let $m = \min(n, \ell)$. We can simply apply Lemma 15 inductively m times. We obtain a template t' which is an ancestor of t_0 and is realized by (u_m, v_m). Since $m = \min(n, \ell)$, $L \leq \min(|u_m|, |v_m|) \leq |h| L$. This concludes the proof. $\square$

4   Bounding Realizable Templates for the Tribonacci Word

Recall that τ denote the Tribonacci morphism and that $\mathcal{T}$ is the Tribonacci word. The matrix M'_τ was given in (1). Since it is primitive, we may use Perron's theorem. Densities of letters $0, 1, 2$ exist and are denoted respectively by α_0, α_1 and α_2. Let $\theta \approx 1.839$ be the Perron eigenvalue of τ. Recall that $\boldsymbol{\alpha} = \begin{pmatrix} \alpha_0 & \alpha_1 & \alpha_2 \end{pmatrix}^{\mathsf{T}}$ is an eigenvector of τ associated with θ. Let

$$\Delta = \{(\delta_0, \delta_1) : -1.5 \leq \delta_0, \delta_1, \delta_0 + \delta_1 \leq 1.5\}.$$

4.1   Bounds on Extended Parikh Vectors

We can obtain two different kinds of bounds on extended Parikh vectors of factors of the Tribonacci word. First we essentially take care of the large eigenvalues.

Proposition 18. *Let $\mathbf{r}$ be a left eigenvector of M_τ having λ as associated eigenvalue. If $|\lambda| < \theta$, then there exists a constant $C(\mathbf{r})$ such that, for all factors w of $\mathcal{T}$,*

$$\frac{|\mathbf{r} \cdot \Phi(w)|}{|w|} \leq C(\mathbf{r}).$$

One can see [12] for the details. If we fix $n, \ell \in \mathbb{N}$, the bound given in the proof is the following one:

$$\max_{\substack{u \in \mathcal{L}(\tau) \\ |u| \leq \ell}} \left\{ \frac{|\mathbf{r} \cdot \Phi(u)|}{|u|}, \frac{|\lambda|^n}{\iota(\ell, n)\theta^n} \max_{\substack{u \in \mathcal{L}(\tau) \\ |u| \leq \ell}} \frac{|\mathbf{r} \cdot \Phi(u)|}{|u|} + c_3(\mathbf{r}) \frac{\iota(\ell, n)\theta^n}{\iota(\ell, n)\theta^n - |\lambda|^n} \right\},$$

where

$$\iota(\ell, n) = \frac{\ell}{\ell + \theta^n \left(2 + \frac{1.5}{\theta - 1}\right)}$$

and

$$c_3(\mathbf{r}) = \max_{\substack{p \in \mathrm{Pref}(\tau^n) \\ s \in \mathrm{Suff}(\tau^n)}} \left\{ |\mathbf{r} \cdot P_3 \left(\Psi(p) \otimes \boldsymbol{\alpha} + \boldsymbol{\alpha} \otimes \Psi(s) \right)| \right.$$

$$\left. + \frac{1}{\ell} \max_{\boldsymbol{\delta} \in [-1.5, 1.5]^3} |\mathbf{r} \cdot \left(\Phi(ps) + P_3 \left(\Psi(p) \otimes \boldsymbol{\delta} + \boldsymbol{\delta} \otimes \Psi(s) \right) \right)| \right\}.$$

Moreover, integers n and ℓ have to verify

$$\frac{|\lambda|^n}{\iota(\ell, n)\theta^n} < 1.$$

The bound given by Proposition 18 will only be useful for the eigenvalues λ such that $|\lambda| \geq 1$. When $|\lambda| < 1$, the following result is stronger.

Proposition 19. *Let $\mathbf{r}$ be an eigenvector of M_τ and λ be the associated eigenvalue. If $|\lambda| < 1$, then there exists a constant $C(\mathbf{r})$ such that for all factors w of $\mathcal{T}$,*

$$|\mathbf{r} \cdot \Phi(w)| \leq C(\mathbf{r}).$$

For every $n \in \mathbb{N}$, the constant

$$C(\mathbf{r}) = \frac{1}{1 - |\lambda|^n} \max_{\substack{p \in \mathrm{Pref}(\tau^n) \\ s \in \mathrm{Suff}(\tau^n)}} \max_{\boldsymbol{\delta} \in [-1.5, 1.5]^3} |\mathbf{r} \cdot \left(\Phi(ps) + P_3 \left(\Psi(p) \otimes \boldsymbol{\delta} + \boldsymbol{\delta} \otimes \Psi(s) \right) \right)|$$

is convenient. See [12] for details.

4.2   Bounds on Templates

This subsection contains several lemmas giving necessary conditions on templates to be realizable by τ.

Lemma 20. *Let λ be an eigenvalue of M_τ such that $|\lambda| < 1$. For every left eigenvector $\mathbf{r}$ of M_τ associated with λ and for every realizable template $t = [\mathbf{d}, \mathbf{D_b}, \mathbf{D_e}, a_1, a_2]$,*

$$\min_{(\delta_0, \delta_1) \in \Delta} \left| \mathbf{r} \cdot \left(\mathbf{d} + P_3 \left(\mathbf{D_b} \otimes \begin{pmatrix} \delta_0 \\ \delta_1 \\ -\delta_0 - \delta_1 \end{pmatrix} + \begin{pmatrix} \delta_0 \\ \delta_1 \\ -\delta_0 - \delta_1 \end{pmatrix} \otimes \mathbf{D_e} \right) \right) \right| \leq 2C(\mathbf{r})$$

where $C(\mathbf{r})$ is the constant from Proposition 19.

This bound is not so easy to use because of the complicated minimum. It can be computed using tools from optimization. However, we can simply use this bound as follows.

For the sake of notation, let

$$f(\delta_0, \delta_1) = \mathbf{r} \cdot \left(\mathbf{d} + P_3 \left(\mathbf{D_b} \otimes \begin{pmatrix} \delta_0 \\ \delta_1 \\ -\delta_0 - \delta_1 \end{pmatrix} + \begin{pmatrix} \delta_0 \\ \delta_1 \\ -\delta_0 - \delta_1 \end{pmatrix} \otimes \mathbf{D_e} \right) \right).$$

Then

$$\min_{(\delta_0, \delta_1) \in \Delta} |f(\delta_0, \delta_1)| \geq \sqrt{ \min_{(\delta_0, \delta_1) \in \Delta} \mathrm{Re}\left(f(\delta_0, \delta_1)\right)^2 + \min_{(\delta_0, \delta_1) \in \Delta} \mathrm{Im}\left(f(\delta_0, \delta_1)\right)^2 }.$$

Let I_{Re} and I_{Im} be intervals such that

$$I_{Re} = \left[\min_{(\delta_0, \delta_1) \in \Delta} \mathrm{Re}\left(f(\delta_0, \delta_1)\right), \max_{(\delta_0, \delta_1) \in \Delta} \mathrm{Re}\left(f(\delta_0, \delta_1)\right) \right]$$

and

$$I_{Im} = \left[\min_{(\delta_0, \delta_1) \in \Delta} \mathrm{Im}\left(f(\delta_0, \delta_1)\right), \max_{(\delta_0, \delta_1) \in \Delta} \mathrm{Im}\left(f(\delta_0, \delta_1)\right) \right].$$

Then

$$\min_{(\delta_0, \delta_1) \in \Delta} |f(\delta_0, \delta_1)| \geq \sqrt{ \min_{y \in I_{Re}} y^2 + \min_{y \in I_{Im}} y^2 }.$$

Thus any template for which this last quantity is greater than $2C(\mathbf{r})$ is not realizable.

Observe that each of the four interval bounds is reached for a vertex of the polytope, that is $\begin{pmatrix} \delta_0 \\ \delta_1 \end{pmatrix} \in \left\{ \begin{pmatrix} 1.5 \\ -1.5 \end{pmatrix}, \begin{pmatrix} 1.5 \\ 0 \end{pmatrix}, \begin{pmatrix} 0 \\ 1.5 \end{pmatrix}, \begin{pmatrix} -1.5 \\ 1.5 \end{pmatrix}, \begin{pmatrix} -1.5 \\ 0 \end{pmatrix}, \begin{pmatrix} 0 \\ -1.5 \end{pmatrix} \right\}$. This is due to the fact that f is linear (and thus convex) over the convex set Δ.

This allows us to remove many templates from the set of templates, but this is not enough to obtain a finite set, so we need to somehow use the bounds on the other eigenvectors as well.

Lemma 21. *Let L be a positive integer. Let λ be an eigenvalue of M_τ such that $|\lambda| < \theta$. Then, for all eigenvectors $\mathbf{r}$ of M_τ associated with λ, there exists a constant $C(\mathbf{r})$ such that for any template $t = [\mathbf{d}, \mathbf{D_b}, \mathbf{D_e}, a_1, a_2]$ realized by a pair of factors of the Tribonacci word (u, v) with $|u| \geq L$, we have*

$$|\mathbf{r} \cdot P_3 \left(\mathbf{D_b} \otimes \boldsymbol{\alpha} + \boldsymbol{\alpha} \otimes \mathbf{D_e} \right)| \leq$$
$$\frac{2L - \sum_{i=1}^{3} \mathbf{d}_i}{L} C(\mathbf{r}) + \max_{(\delta_0, \delta_1) \in \Delta} \frac{|\mathbf{r} \cdot \left(\mathbf{d} + P_3 \left(\mathbf{D_b} \otimes \boldsymbol{\delta} + \boldsymbol{\delta} \otimes \mathbf{D_e} \right) \right)|}{L}.$$

The quantity of the l.h.s. and the first term on the r.h.s. are straightforward to compute. For the last term, it is not difficult to show that the maximum is in fact necessarily reached on a vertex of the polytope, that is

$$
\max_{(\delta_0,\delta_1)\in\varDelta} \frac{|\mathbf{r}\cdot(\mathbf{d}+P_3\,(\mathbf{D_b}\otimes\boldsymbol{\delta}+\boldsymbol{\delta}\otimes\mathbf{D_e}))|}{L} \leq
$$

$$
\max_{\binom{\delta_0}{\delta_1}\in\left\{\binom{1.5}{0},\binom{1.5}{-1.5},\binom{0}{1.5},\binom{-1.5}{0},\binom{-1.5}{1.5},\binom{0}{-1.5}\right\}} \frac{|\mathbf{r}\cdot(\mathbf{d}+P_3\,(\mathbf{D_b}\otimes\boldsymbol{\delta}+\boldsymbol{\delta}\otimes\mathbf{D_e}))|}{L}.
$$

5   Proof of the Main Result

With all these lemmas, we are ready to show our main result.

Theorem 22. *Two factors of the Tribonacci word are 2-binomially equivalent if and only if they are equal.*

Proof. Let $T = \{[\mathbf{0},\mathbf{0},\mathbf{0},a_1,a_2] : a_1 \neq a_2\}$. Let us show that no template from T is realizable. Let $L = 15$. We can easily check with a computer that no pair of factors of $\mathcal{T}$ with $\min(|u|,|v|) \leq L$ realizes a template t from the set T. Indeed, since for all $t \in T$, $\mathbf{d} = \mathbf{0}$, $\mathbf{D_b} = \mathbf{0}$ and $\mathbf{D_e} = \mathbf{0}$, we know that a pair of words (u,v) realizes t if and only if $\Phi(u) - \Phi(v) = 0$. It just suffices to check that for all $n \leq L$, $b_{\mathcal{T},2}(n) = p_{\mathcal{T}}(n)$.

Now, from Proposition 17, if $t \in T$ is realized then one of its ancestors is realized by a pair (u,v) with $L \leq \min(|u|,|v|) \leq 2L$.

Lemmas 20 and 21 give us two sets of inequalities that any template realized by a pair (u,v) of factors of Tribonacci with $|u| \geq L$ must respect. Let X be the set of templates that respect the bounds. Let $A_0 = T$ and, for all i, let $A_{i+1} = \{\mathrm{Par}_\tau(t) \cap X : t \in A_i\}$. Then clearly $\mathrm{RAnc}_\tau(t) \subseteq \bigcup_{i\in\mathbb{N}} A_i$. Each A_i can be easily computed and it can be checked by a computer program that the set $\bigcup_{i\in\mathbb{N}} A_i$ is finite.

We can finally check with a computer that there is no pair (u,v) of factors of $\mathcal{T}$ with $L \leq \min(|u|,|v|) \leq 2L$ that realizes any element of $\bigcup_{i\in\mathbb{N}} A_i$. Thus no template of T is realizable. By Lemma 12, we can conclude that the 2-binomial complexity of the Tribonacci word is equal to its factorial complexity.   □

Accompanying this paper is an implementation in Mathematica of all the computations described in this theorem and in the previous lemmas and propositions. We also have a C++ implementation that is much faster, but uses machine floating point arithmetic whose accuracy cannot be guaranteed (in this case, however, we obtain exactly the same set of templates). Diagonalizing the matrix of Tribonacci gives 4 eigenvectors to which Lemma 20 can be applied. Since there are two pairs of conjugate complex vectors, it is useless to keep more than one of each pair. However, by taking a linear combination of these two, we get another eigenvector to which we can apply Lemma 20 (in practice we only do that once, but we could take as many vectors as we want from this 2-dimensional space). For this conjugation reason, we also only keep 4 of the 6 eigenvectors that correspond to an eigenvalue of norm less than 1. For each of these 7 eigenvectors, we choose[1] $\ell = 600$ and the best $1 \leq n \leq 6$ when applying Lemma 20 or Lemma 21. The rest is done as described in the article. We obtain a set of 241544 templates.

6   Conclusion

We used an algorithm to show that the 2-binomial complexity of the Tribonacci word is equal to its factorial complexity. It seems that our method can be turned into an algorithm that can decide under some mild conditions whether the factorial complexity of a given morphic word is equal to its k-binomial complexity. In fact, by keeping track of the first letter of each word in templates, the "if" in Proposition 17 can be replaced by an "if and only if" (some technicalities could allow us to apply it even if the matrix is singular). Moreover, with arguments similar to the ideas from [15], one could show that we also have bounds on the eigenvectors that correspond to larger eigenvalues and that the number of templates that respect these bounds is always finite (one might need no eigenvalue has norm 1).

Observe that the notion of template was first introduced in the context of avoidability of abelian powers [8] and, as one could expect, it seems that our technique also gives a decision algorithm for the avoidability of k-binomial powers in morphic words (and even avoidability of patterns in the k-binomial sense).

References

1. Aberkane, A., Currie, J.: A cyclic binary morphism avoiding abelian fourth powers. Theoret. Comput. Sci. **410**, 44–52 (2009)
2. Aberkane, A., Currie, J., Rampersad, N.: The number of ternary words avoiding abelian cubes grows exponentially. J. Integer Seq. **7** (2004). Article 04.2.7
3. Allouche, J.-P., Shallit, J.: Automatic Sequences. Theory, Applications, Generalizations. Cambridge University Press, Cambridge (2003)
4. Berthé, V., Rigo, M. (eds.): Combinatorics, Automata and Number Theory. Encyclopedia Mathematics Applications, vol. 135. Cambridge University Press, Cambridge (2010)
5. Cassaigne, J., Currie, J., Schaeffer, L., Shallit, J.: Avoiding three consecutive blocks of the same size and same sum. J. ACM **61**(2) (2014). Art. 10
6. Cassaigne, J., Richomme, G., Saari, K., Zamboni, L.Q.: Avoiding Abelian powers in binary words with bounded Abelian complexity. Int. J. Found. Comput. Sci. **22**, 905–920 (2011)
7. Currie, J., Rampersad, N.: Recurrent words with constant Abelian complexity. Adv. Appl. Math. **47**, 116–124 (2011)
8. Currie, J., Rampersad, N.: Fixed points avoiding Abelian k-powers. J. Combin. Theory Ser. A **119**, 942–948 (2012)

---

[1] Remember that we work on τ^n and that increasing n and ℓ tend to give us better bounds but increases the computation time.

9. Karhumäki, J., Saarela, A., Zamboni, L.Q.: On a generalization of Abelian equivalence and complexity of infinite words. J. Combin. Theory Ser. A **120**, 2189–2206 (2013)
10. Karhumäki, J., Saarela, A., Zamboni, L.Q.: Variations of the Morse-Hedlund theorem for k-Abelian equivalence. In: Shur, A.M., Volkov, M.V. (eds.) DLT 2014. LNCS, vol. 8633, pp. 203–214. Springer, Cham (2014). https://doi.org/10.1007/978-3-319-09698-8_18
11. Lejeune, M., Leroy, J., Rigo, M.: Computing the k-binomial complexity of the Thue-Morse word, 34 p. arXiv:1812.07330
12. Lejeune, M., Rigo, M., Rosenfeld, M.: Templates for the k-binomial complexity of the Tribonacci word (long version), 23 p. http://hdl.handle.net/2268/234215
13. Liétard, F.: Avoiding additive powers, talk at Mons TCS days, Bordeaux, 10–14 September 2018
14. Rao, M., Rigo, M., Salimov, P.: Avoiding 2-binomial squares and cubes. Theoret. Comput. Sci. **572**, 83–91 (2015)
15. Rao, M., Rosenfeld, M.: Avoiding two consecutive blocks of same size and same sum over $\mathbb{Z}^2$. SIAM J. Disc. Math. **32**(4), 2381–2397 (2018)
16. Richomme, G., Saari, K., Zamboni, L.Q.: Balance and Abelian complexity of the Tribonacci word. Adv. Appl. Math. **45**, 212–231 (2010)
17. Rigo, M., Salimov, P.: Another generalization of abelian equivalence: binomial complexity of infinite words. Theoret. Comput. Sci. **601**, 47–57 (2015)
18. Rigo, M.: Invited talk. Streams II, Lorentz Center, Leiden, January 2014
19. Rigo, M.: Relations on words. Indag. Math. (N.S.) **28**, 183–204 (2017)

Derived Sequences of Arnoux–Rauzy Sequences

Kateřina Medková[✉]

Department of Mathematics, Faculty of Nuclear Science and Physical Engineering,
Czech Technical University in Prague, Prague, Czech Republic
medkokat@fjfi.cvut.cz

Abstract. For an Arnoux–Rauzy sequence $\mathbf{u}$ we describe the set $\mathrm{Der}(\mathbf{u})$ of derived sequences corresponding to all nonempty prefixes of $\mathbf{u}$ using the normalized directive sequence of $\mathbf{u}$. As a corollary, we show that all derived sequences of $\mathbf{u}$ are also Arnoux–Rauzy sequences. Moreover, if $\mathbf{u}$ is primitive substitutive, we precisely determine the cardinality of the set $\mathrm{Der}(\mathbf{u})$.

Keywords: Arnoux–Rauzy sequence · Derived sequence · Return word

1 Introduction

Derived sequences were introduced by Durand [4] to characterize the primitive substitutive sequences, i.e., the sequences which are morphic images of fixed points of primitive morphisms.

Let $\mathbf{u} = u_0 u_1 u_2 \cdots$ be a recurrent sequence. An *occurrence* of the factor w in $\mathbf{u}$ is the index i such that w is a prefix of the sequence $u_i u_{i+1} u_{i+2} \cdots$. Let $i < j$ be two consecutive occurrences of w in $\mathbf{u}$. Then the word $u_i u_{i+1} \cdots u_{j-1}$ is a *return word* to w in $\mathbf{u}$. We take into consideration only the sequence $\mathbf{u}$ for which each factor w has finitely many return words, and we denote these return words by $r_0, r_1, \ldots, r_{k-1}$. Such a sequence is called *uniformly recurrent*. In addition, if w is a prefix of $\mathbf{u}$, then the sequence $\mathbf{u}$ can be written as the unique concatenation of the return words to w: $\mathbf{u} = r_{d_0} r_{d_1} r_{d_2} \cdots$ with all $d_i \in \{0, 1, \ldots, k-1\}$. The ordering of the return words in this concatenation is coded by the sequence $\mathbf{d}_{\mathbf{u}}(w) = d_0 d_1 d_2 \cdots$ which is called the *derived sequence* of $\mathbf{u}$ with respect to w.

Return words and derived sequences were especially studied in the case of *Sturmian sequences*, which are the aperiodic binary sequences having the least factor complexity possible. Every Sturmian sequence $\mathbf{u}$ has exactly one left and one right special factor per length. The factor w is left (right, respectively) special if the words aw, bw (wa, wb, respectively) are factors of $\mathbf{u}$ for two different letters

This work was supported by the project CZ.02.1.01/0.0/0.0/16_019/0000778 from European Regional Development Fund and by the grant No. SGS17/193/OHK4/3T/14 from the Grant Agency of the CTU in Prague.

© Springer Nature Switzerland AG 2019
R. Mercaş and D. Reidenbach (Eds.): WORDS 2019, LNCS 11682, pp. 251–263, 2019.
https://doi.org/10.1007/978-3-030-28796-2_20

a, b. Moreover, a Sturmian sequence is *standard* if all its prefixes are left special factors.

Vuillon [14] showed that a binary sequence is Sturmian if and only if each of its factors has exactly two return words. This property implies that the derived sequence with respect to each prefix of a Sturmian sequence is Sturmian as well. The derived sequences of standard Sturmian sequences were precisely described in [1], where the one-to-one correspondence between standard Sturmian sequences and continued fractions of irrational numbers from the interval $(0, 1)$ is used. Clearly, this approach does not work in the non-standard case, but using a special representation of Sturmian sequences by Sturmian morphisms, we can deal with it, too. This technique is basically used in [12] to study the derived sequences of fixed points of primitive Sturmian morphisms.

As is well known, Sturmian sequences have various generalizations for multiletter alphabets. The first one was introduced by Arnoux and Rauzy [2]: a uniformly recurrent sequence **u** over $\mathcal{A}$ is called *Arnoux–Rauzy* if it has exactly one left and one right special factor per length and all left (right, respectively) special factors appear in **u** immediately preceded (followed, respectively) by all letters from $\mathcal{A}$.

Many properties of the Arnoux–Rauzy sequences are known (see for example the survey [8]). For our considerations the work [9] is especially important since its authors showed that each factor of an Arnoux–Rauzy sequence over $\mathcal{A}$ has exactly $\#\mathcal{A}$ return words. It means that the derived sequences of Arnoux–Rauzy sequences over $\mathcal{A}$ can be considered over the same alphabet $\mathcal{A}$. Nevertheless, such a property does not characterize Arnoux–Rauzy sequences if $\#\mathcal{A} > 2$. For example, by [6] the sequences coding interval exchange transformations can have this property, too. More generally, the sequences over $\mathcal{A}$ each of whose factors has exactly $\#\mathcal{A}$ return words were studied in [3].

The aim of this paper is to study the derived sequences of Arnoux–Rauzy sequences. Let us emphasize that the description of derived sequences of standard Arnoux–Rauzy sequences can be easily deduced from the work of Justin and Vuillon [9], while here we cover also the more complicated case of non-standard Arnoux–Rauzy sequences. As in [12], our main tool is a special representation of Arnoux–Rauzy sequences, namely the directive sequences containing pure episturmian morphisms (see Sect. 2.3). Since these directive sequences need not be unique, in [7] the authors introduce so-called *normalized directive sequences* and show that these representations are unique. Moreover, they have also other useful properties which allow us to use them for a construction of derived sequences (see Sects. 2.3 and 3).

For every Arnoux–Rauzy sequence **u** we describe the set Der(**u**) of derived sequences with respect to all nonempty prefixes of **u** (see Theorem 24). As a corollary, we show that every derived sequence of an Arnoux–Rauzy sequence is an Arnoux–Rauzy sequence as well. By Durand's fundamental result [4] the sequence **u** is primitive substitutive if and only if the set Der(**u**) is finite. Here we precisely determine the cardinality of Der(**u**) for all primitive substitutive Arnoux–Rauzy sequences (see Corollary 29). It generalizes the results from [12], where the cardinality of Der(**u**) is bounded for the fixed points of primitive Sturmian morphisms.

2   Preliminaries

2.1   Words, Sequences and Morphisms

An *alphabet* $\mathcal{A}$ is a finite set of symbols called *letters*. A *word* of length n over $\mathcal{A}$ is a string $u = u_0 u_1 \cdots u_{n-1}$, where all $u_i \in \mathcal{A}$. The *length* of u is denoted by $|u| = n$. The unique word ε of length 0 is called the *empty word*. The symbol $\mathcal{A}^*$ denotes the set of all finite words over $\mathcal{A}$ and $\mathcal{A}^+ = \mathcal{A}^* \setminus \{\varepsilon\}$. By $|u|_a$ we denote the number of copies of the letter a used in u. The *reversal* of a word $u = u_0 u_1 \cdots u_{n-1}$ is the word $u_{n-1} \cdots u_1 u_0$.

A *sequence* over $\mathcal{A}$ is a right infinite string $\mathbf{u} = u_0 u_1 u_2 \cdots \in \mathcal{A}^{\mathbb{N}}$ with letters $u_i \in \mathcal{A}$ for all $i \in \mathbb{N} = \{0, 1, 2, \ldots\}$. A sequence $\mathbf{u}$ is *eventually periodic* if $\mathbf{u} = wvvv \cdots = wv^{\omega}$ for some $v, w \in \mathcal{A}^*$; otherwise it is *aperiodic*.

A word w of length n is a *factor* of $\mathbf{u} = u_0 u_1 u_2 \cdots$ if there is an index i such that $w = u_i u_{i+1} u_{i+2} \cdots u_{i+n-1}$. The index i is called an *occurrence* of w in $\mathbf{u}$. Further, if $i = 0$, then w is a *prefix* of $\mathbf{u}$. We will also use the abbreviated notation $u_i u_{i+1} \cdots u_{j-1} = \mathbf{u}_{[i,j)}$ and $u_i u_{i+1} \cdots = \mathbf{u}_{[i,\infty)}$ for all integers $0 \le i < j$.

The *language* $\mathcal{F}(\mathbf{u})$ of a sequence $\mathbf{u}$ is the set of all its factors. A factor w of $\mathbf{u}$ is *right special* (*left special*, resp.) if there exist at least two letters $a, b \in \mathcal{A}$ such that $wa, wb \in \mathcal{F}(\mathbf{u})$ ($aw, bw \in \mathcal{F}(\mathbf{u})$, resp.).

If each factor w of $\mathbf{u}$ has infinitely many occurrences in $\mathbf{u}$, the sequence $\mathbf{u}$ is *recurrent*. Moreover, if the distances between two consecutive occurrences of w are bounded, then $\mathbf{u}$ is *uniformly recurrent*.

A *morphism* over $\mathcal{A}^*$ is a mapping $\psi : \mathcal{A}^* \mapsto \mathcal{A}^*$ such that $\psi(vw) = \psi(v)\psi(w)$ for all $v, w \in \mathcal{A}^*$. We consider only non-erasing morphisms for which $\psi(a) \neq \varepsilon$ for every $a \in \mathcal{A}$. Then the domain of the morphism ψ can be naturally extended to $\mathcal{A}^{\mathbb{N}}$ by $\psi(u_0 u_1 \cdots) = \psi(u_0)\psi(u_1) \cdots$. A morphism ψ is *primitive* if there is $k \in \mathbb{N}$ such that for every $a, b \in \mathcal{A}$ the letter a occurs in $\psi^k(b)$.

A *fixed point* of a morphism ψ is a sequence $\mathbf{u}$ such that $\psi(\mathbf{u}) = \mathbf{u}$. A sequence $\mathbf{v}$ is *primitive substitutive* if $\mathbf{v} = \theta(\mathbf{u})$, where θ is a morphism and $\mathbf{u}$ is a fixed point of a primitive morphism.

A *permutation* P on $\mathcal{A}$ is a morphism over $\mathcal{A}^*$ such that $\{P(a) : a \in \mathcal{A}\} = \mathcal{A}$. The *order* of the permutation P is the smallest integer $n > 0$ such that $P^n = \text{Id}$.

2.2   Return Words and Derived Sequences

Let $i < j$ be two consecutive occurrences of a factor w in a recurrent sequence $\mathbf{u}$. Then the word $\mathbf{u}_{[i,j)}$ is a *return word* to w in $\mathbf{u}$. The set of all return words to w in $\mathbf{u}$ is denoted $\mathcal{R}_{\mathbf{u}}(w)$. If the sequence $\mathbf{u}$ is uniformly recurrent, then every factor w of $\mathbf{u}$ has a finite number of return words, we denote them $\mathcal{R}_{\mathbf{u}}(w) = \{r_0, r_1, \ldots, r_{k-1}\}$. In addition, if w is a prefix of $\mathbf{u}$, the sequence $\mathbf{u}$ can be written as the unique concatenation of these return words: $\mathbf{u} = r_{d_0} r_{d_1} r_{d_2} \cdots$ and the *derived sequence* of $\mathbf{u}$ to the prefix w is the sequence $\mathbf{d}_{\mathbf{u}}(w) = d_0 d_1 d_2 \cdots$ over the alphabet of cardinality $\#\mathcal{R}_{\mathbf{u}}(w) = k$. Originally, Durand [4] fixed this alphabet to the set $\{0, 1, \ldots, k-1\}$ and required that for $i < j$ the first occurrence of r_i in $\mathbf{u}$ is less than the first occurrence of r_j in $\mathbf{u}$. In particular, this means that his derived sequences always start with the letter 0. In this article, we do not need to fix the alphabet of derived sequences: two derived sequences which differ only by a permutation of letters are identified with one another.

We consider only aperiodic and uniformly recurrent sequences $\mathbf{u}$. Our aim is to describe the set

$$\mathrm{Der}(\mathbf{u}) = \{\mathbf{d_u}(w)\colon w \text{ is a nonempty prefix of } \mathbf{u}\}.$$

Let us emphasize that we study only derived sequences with respect to nonempty prefixes since the derived sequence with respect to the empty word is trivial.

Clearly, if a nonempty prefix w of $\mathbf{u}$ is not right special, then there exists a unique letter a such that $wa \in \mathcal{F}(\mathbf{u})$. Thus the occurrences of w and wa coincide, and so $\mathcal{R}_{\mathbf{u}}(w) = \mathcal{R}_{\mathbf{u}}(wa)$ and $\mathbf{d_u}(w) = \mathbf{d_u}(wa)$. Since $\mathbf{u}$ is aperiodic, w is a prefix of some right special prefix of $\mathbf{u}$. Therefore, it suffices to take into consideration only right special prefixes of $\mathbf{u}$, i.e.,

$$\mathrm{Der}(\mathbf{u}) = \{\mathbf{d_u}(w)\colon w \text{ is a nonempty right special prefix of } \mathbf{u}\}.$$

2.3 Episturmian and Arnoux–Rauzy Sequences

Definition 1. *A sequence* $\mathbf{u} \in \mathcal{A}^{\mathbb{N}}$ *is* episturmian *if its language is closed under reversal and* $\mathbf{u}$ *has at most one right special factor of each length.*

An episturmian sequence $\mathbf{u} \in \mathcal{A}^{\mathbb{N}}$ *is an* Arnoux–Rauzy sequence *if* $\mathbf{u}$ *has exactly one right special factor of each length and* $wa \in \mathcal{F}(\mathbf{u})$ *for every right special factor* w *of* $\mathbf{u}$ *and every letter* $a \in \mathcal{A}$. *An Arnoux–Rauzy sequence* $\mathbf{u}$ *is* standard *if each of its prefixes is a left special factor of* $\mathbf{u}$.

The Arnoux–Rauzy sequences over $\mathcal{A}$ are sometimes called $\#\mathcal{A}$-strict episturmian sequences, since there are also epistumian sequences which are not Arnoux–Rauzy (e.g., see [8]). In the binary case, the set of all Arnoux–Rauzy sequences coincides with the set of all Sturmian sequences. Clearly, all Arnoux–Rauzy sequences are aperiodic and by [5] they are also uniformly recurrent.

Example 2. The Tribonacci sequence $\mathbf{u}_\tau = abacabaabacababacabaa\cdots$ which is the fixed point of the morphism $\tau : a \to ab,\ b \to ac,\ c \to a$ is a standard Arnoux–Rauzy sequence over $\{a, b, c\}$.

In the sequel, we will use the description of episturmian sequences in terms of sequences of pure episturmian morphisms. We follow the notation from [7].

Definition 3. *For every* $a \in \mathcal{A}$ *we define* elementary episturmian morphisms:

$$L_a : \begin{cases} a \to a \\ b \to ab & \text{for all } b \neq a \end{cases} \quad \text{and} \quad R_a : \begin{cases} a \to a \\ b \to ba & \text{for all } b \neq a. \end{cases}$$

These $2\#\mathcal{A}$ *morphisms generate the monoid* $\mathcal{M}_{\mathcal{A}} = \langle L_a, R_a : a \in \mathcal{A} \rangle$ *of* pure episturmian morphisms.

Let us remark that *episturmian morphisms* are the morphisms obtained by composition of pure episturmian morphisms and permutations (e.g., see [8,10]). All episturmian morphisms are injective.

Definition 4. *For a given alphabet $\mathcal{A}$ we define a new alphabet $\bar{\mathcal{A}} = \{\bar{a} : a \in \mathcal{A}\}$ and we consider words and sequences over the alphabet $\mathcal{A} \cup \bar{\mathcal{A}}$ called* spinned. *We put $\varphi_a = L_a$ and $\varphi_{\bar{a}} = R_a$ for every letter $a \in \mathcal{A}$. Then for every spinned word $z = z_0 z_1 \cdots z_{n-1} \in (\mathcal{A} \cup \bar{\mathcal{A}})^*$ we write*

$$\varphi_z = \varphi_{z_0} \varphi_{z_1} \cdots \varphi_{z_{n-1}} \in \mathcal{M}_{\mathcal{A}}$$

and we say that z is a directive word *of the morphism φ_z. A spinned word is L-spinned (R-spinned, respectively) if all its letters are from $\mathcal{A}$ ($\bar{\mathcal{A}}$, respectively). The* opposite word $\bar{z}$ *of a spinned word z is obtained from z by switching spins of all its letters.*

Example 5. The words $\bar{a}a\bar{b}\bar{c}a$, abc, $\bar{b}\bar{b}$ are spinned words over $\{a, b, c, \bar{a}, \bar{b}, \bar{c}\}$. The word $z = \bar{a}a\bar{b}\bar{c}a$ directs the morphism $\psi = \varphi_z = \varphi_{\bar{a}a\bar{b}\bar{c}a} = R_a L_a R_b R_c L_a$. The word abc is L-spinned, while $\bar{b}\bar{b}$ is R-spinned. The opposite word of z is $a\bar{a}bc\bar{a}$.

Pure episturmian morphisms can have more than one directive word, i.e., the monoid $\mathcal{M}_{\mathcal{A}}$ is not free. Nevertheless, the presentation of the monoid $\mathcal{M}_{\mathcal{A}}$ is known. Here we state it in the notion of directive words using the so-called *block-transformation* from [11], but it also follows from more general presentation of the whole episturmian monoid as stated in [13].

Definition 6. *A* block-transformation *in the word z is the replacement of the factor $av\bar{a}$ of z, where $a \in \mathcal{A}$ and $v \in (\mathcal{A} \setminus \{a\})^*$, by the opposite word $\bar{a}\bar{v}a$ or vice-versa.*

Proposition 7 ([11]). *Let z, z' be two spinned words over $\mathcal{A} \cup \bar{\mathcal{A}}$. Then $\varphi_z = \varphi_{z'}$ if and only if we can pass from z to z' by a chain of block-transformations.*

Example 8. Using the block-transformations from Definition 6 we may rewrite $\bar{a}a\bar{b}\bar{c}a \longleftrightarrow a\bar{a}\bar{b}\bar{c}a \longleftrightarrow aab\bar{c}\bar{a}$, and so by Proposition 7 all these words direct the same morphism, i.e., $\varphi_{\bar{a}a\bar{b}\bar{c}a} = \varphi_{a\bar{a}\bar{b}\bar{c}a} = \varphi_{aab\bar{c}\bar{a}}$.

The following theorem extends the notion of directive words to infinite episturmian sequences.

Theorem 9 ([10]). *A sequence $\mathbf{u}$ is episturmian if and only if there exists a spinned sequence $\mathbf{z} = z_0 z_1 z_2 \cdots \in (\mathcal{A} \cup \bar{\mathcal{A}})^{\mathbb{N}}$ and an infinite sequence $(\mathbf{u}^{(i)})_{i \geq 0}$ of recurrent sequences such that $\mathbf{u}^{(0)} = \mathbf{u}$ and*

$$\mathbf{u}^{(i)} = \varphi_{z_i}(\mathbf{u}^{(i+1)}).$$

This sequence $\mathbf{z}$ is called a directive sequence *of $\mathbf{u}$.*

Let us notice that the directive sequence from Theorem 9 is the same object as the directive sequence from the construction of episturmian sequences using *palindromic closures* (e.g., see Sect. 3 in [8]).

Proposition 10 ([10]).

(i) A spinned sequence $\mathbf{z} \in (\mathcal{A} \cup \bar{\mathcal{A}})^{\mathbb{N}}$ which has infinitely many L-spinned letters directs the unique episturmian sequence $\mathbf{u}$. Moreover, the sequence $\mathbf{u}$ starts with the left-most L-spinned letter in $\mathbf{z}$.

(ii) A spinned sequence $\mathbf{z} \in (\mathcal{A} \cup \bar{\mathcal{A}})^{\mathbb{N}}$ which contains finitely many L-spinned letters directs one episturmian sequence for each $\bar{a} \in \bar{\mathcal{A}}$ which occurs in $\mathbf{z}$ infinitely many times.

Proposition 10 implies that some directive sequences direct more than one episturmian sequence. In addition, an episturmian sequence can have more than one directive sequence. However, in [7] the authors describe all directive sequences which direct the same episturmian sequence. Here we state this result only for the case of aperiodic episturmian sequences.

Theorem 11 (Theorem 4.1 in [7]). Two spinned sequences $\mathbf{z}^{(1)}$ and $\mathbf{z}^{(2)}$ direct the same aperiodic episturmian sequence if and only if one of the following cases holds for some i, j such that $\{i, j\} = \{1, 2\}$:

(i) $\mathbf{z}^{(i)} = \prod_{n \geq 1} u^{(n)}$, $\mathbf{z}^{(j)} = \prod_{n \geq 1} v^{(n)}$, where $u^{(n)}$, $v^{(n)}$ are spinned words such that $\varphi_{u^{(n)}} = \varphi_{v^{(n)}}$ for all $n \geq 1$;

(ii) $\mathbf{z}^{(i)} = wa \prod_{n \geq 1} u^{(n)} x^{(n)}$, $\mathbf{z}^{(j)} = w'\bar{a} \prod_{n \geq 1} \bar{u}^{(n)} y^{(n)}$, where w, w' are spinned words such that $\varphi_w = \varphi_{w'}$, a is an L-spinned letter and for all $n \geq 1$, $u^{(n)}$ is a nonempty a-free L-spinned word, $\bar{u}^{(n)}$ is the opposite word of $u^{(n)}$ and $x^{(n)}$, $y^{(n)}$ are nonempty spinned words over $\{a, \bar{a}\}$ such that $|x^{(n)}| = |y^{(n)}|$ and $|x^{(n)}|_a = |y^{(n)}|_a$.

Item (i) is based on block-transformations of the directive words of episturmian morphisms, while Item (ii) brings new relations. Now we define the normalized directive sequences which are unique for all aperiodic episturmian sequences.

Definition 12. A spinned sequence $\mathbf{z} \in (\mathcal{A} \cup \bar{\mathcal{A}})^{\mathbb{N}}$ is normalized if it contains infinitely many L-spinned letters, but no factor from the set $\{\bar{a}\bar{\mathcal{A}}^* a : a \in \mathcal{A}\}$.

Theorem 13 (Theorem 5.2 in [7]). Any aperiodic episturmian sequence $\mathbf{u}$ has a unique normalized directive sequence.

Every normalized spinned sequence directs exactly one episturmian sequence, see Proposition 10. Moreover, the normalized directive sequences can be constructed using Theorem 13. If a directive sequence does not contain infinitely many L-spinned letters, then we use Item (ii) to find another one with infinitely many L-spinned letters. If a directive sequence contains infinitely many L-spinned letters, then it can be normalized from left to right by repeated applications of Item (i) (see [7] for more details).

The Arnoux–Rauzy sequences can be easily recognised by their directive sequences (e.g., see Sect. 2.3 in [8]).

Proposition 14. *An episturmian sequence* $\mathbf{u} \in \mathcal{A}^{\mathbb{N}}$ *with the directive sequence* $\mathbf{z}$ *is an Arnoux–Rauzy sequence over* $\mathcal{A}$ *if and only if for every* $a \in \mathcal{A}$ *the letter* a *or* $\bar{a}$ *occurs infinitely many times in* $\mathbf{z}$.

Remark 15. Theorem 9 and Proposition 14 immediately imply that for an Arnoux–Rauzy sequence $\mathbf{u}$ each sequence $\mathbf{u}^{(i)}$ from Theorem 9 is an Arnoux–Rauzy sequence with a directive sequence $\mathbf{z}_{[i,\infty)} = z_i z_{i+1} \cdots$.

Example 16. By Propositions 10 and 14, the spinned sequence $\mathbf{y} = a(a\bar{b}\bar{c}\bar{a})^{\omega}$ directs the unique Arnoux–Rauzy sequence $\mathbf{u}$ over $\{a, b, c\}$. Obviously, $\mathbf{y}$ is not normalized. We can normalize it using Item (i) of Theorem 11. First we set $u^{(1)} = aa\bar{b}\bar{c}$, $u^{(2k)} = \bar{a}a$ and $u^{(2k+1)} = \bar{b}\bar{c}$ for all $k > 0$ and make the block-transformations in all even blocks. We get $\mathbf{y}' = aa\bar{b}\bar{c}(a\bar{a}\bar{b}\bar{c})^{\omega}$. Then we set $u^{(1)} = aa\bar{b}\bar{c}a$ and $u^{(k)} = \bar{a}\bar{b}\bar{c}a$ for all $k > 1$. After the relevant block-transformations we get $\mathbf{y}'' = aa\bar{b}\bar{c}a(abc\bar{a})^{\omega}$. Finally we set $u^{(1)} = aa\bar{b}\bar{c}aa$, $u^{(2k)} = bc$ and $u^{(2k+1)} = \bar{a}a$ for all $k > 0$, which leads us to the normalized sequence $\mathbf{y}''' = aa\bar{b}\bar{c}aa(bca\bar{a})^{\omega}$.

By Proposition 10, the spinned sequence $\mathbf{z} = (\bar{a}\bar{b}\bar{c})^{\omega}$ directs three Arnoux–Rauzy sequences $\mathbf{u}^{(a)}$, $\mathbf{u}^{(b)}$, $\mathbf{u}^{(c)}$ starting with the letters a, b, c, respectively. Using Item (ii) of Theorem 11 we find their normalized directive sequences $\mathbf{z}^{(a)} = a(bc\bar{a})^{\omega}$, $\mathbf{z}^{(b)} = \bar{a}b(ca\bar{b})^{\omega}$ and $\mathbf{z}^{(c)} = \bar{a}\bar{b}c(ab\bar{c})^{\omega}$, respectively.

Justin and Vuillon [9] completely describe the return words to any factor of an episturmian sequence. In particular, an Arnoux–Rauzy sequence has the same number of return words to each of its factors.

Proposition 17 ([9]). *Let* $\mathbf{u}$ *be an Arnoux–Rauzy sequence over* $\mathcal{A}$. *Then every factor* w *of* $\mathbf{u}$ *has exactly* $\#\mathcal{A}$ *different return words.*

3   Derived Sequences of Episturmian Preimages

In this section we study the relations between the derived sequences of a given Arnoux–Rauzy sequence and the derived sequences of its preimage under the morphisms L_a or R_a. In the binary case, these relations are completely analogous to those described in Section 3 of [12]. Proposition 19 can be also deduced from the results in [9].

For simplicity, we now define the return words and the derived sequence with respect to the empty prefix ε of a sequence $\mathbf{u}$ over $\mathcal{A}$ as $\mathcal{R}_{\mathbf{u}}(\varepsilon) = \mathcal{A}$ and $\mathbf{d}_{\mathbf{u}}(\varepsilon) = \mathbf{u}$. We start with an auxiliary lemma which follows directly from the form of the morphism L_a.

Lemma 18. *Let* $\mathbf{u}, \mathbf{u}'$ *be Arnoux–Rauzy sequences over* $\mathcal{A}$ *such that* $\mathbf{u} = L_a(\mathbf{u}')$ *for some* $a \in \mathcal{A}$. *For each factor* $pa \in \mathcal{F}(\mathbf{u})$ *with the prefix* a *there is exactly one word* $p' \in \mathcal{F}(\mathbf{u}')$ *such that* $pa = L_a(p')a$.

Proposition 19. *Let* $\mathbf{u}$ *and* $\mathbf{u}'$ *be Arnoux–Rauzy sequences over* $\mathcal{A}$ *such that* $\mathbf{u} = L_a(\mathbf{u}')$ *for some* $a \in \mathcal{A}$.

(i) If w is a nonempty right special prefix of $\mathbf{u}$, then there exists a right special prefix w' of $\mathbf{u}'$ such that $w = L_a(w')a$ and $\mathbf{d_u}(w) = \mathbf{d_{u'}}(w')$.

(ii) If w' is a right special prefix of $\mathbf{u}'$, then $w = L_a(w')a$ is a right special prefix of $\mathbf{u}$ and $\mathbf{d_{u'}}(w') = \mathbf{d_u}(w)$.

Proof. We start with Item (i). For a nonempty right special prefix w of $\mathbf{u}$ we denote its return words $\mathcal{R_u}(w) = \{r_c : c \in \mathcal{A}\}$ and its derived sequence $\mathbf{d_u}(w) = d_0 d_1 \cdots$. Thus $\mathbf{u} = r_{d_0} r_{d_1} \cdots$. By the form of the morphism L_a, the sequence $\mathbf{u}$ starts with the letter a and a is also separating in $\mathbf{u}$, i.e., every factor of $\mathbf{u}$ of length two contains the letter a. Since w is nonempty right special prefix, it both starts and ends with the letter a and by Lemma 18 there is a unique prefix w' of $\mathbf{u}'$ such that $w = L_a(w')a$. Since w is a right special factor of the Arnoux–Rauzy sequence $\mathbf{u}$, the word $wc = L_a(w')ac \in \mathcal{F}(\mathbf{u})$ for every $c \in \mathcal{A}$. Thus $w'c \in \mathcal{F}(\mathbf{u}')$ for every $c \in \mathcal{A}$ and so w' is a right special factor of $\mathbf{u}'$. In addition, all return words r_c start with the letter a and so by Lemma 18 there are uniquely given words r'_c such that $r_c = L_a(r'_c)$ for all $c \in \mathcal{A}$. Since L_a is injective, we have $\mathbf{u}' = r'_{d_0} r'_{d_1} \cdots$.

Now it suffices to prove that the set $\{|r'_{d_0} \cdots r'_{d_j}| : j \in \mathbb{N}\} \cup \{0\}$ is the set of all occurrences of w' in $\mathbf{u}'$. Then the words r'_c, $c \in \mathcal{A}$, are return words to w' in $\mathbf{u}'$ and $\mathbf{d_u}(w) = \mathbf{d_{u'}}(w')$. Let $i > 0$ be an occurrence of w' in $\mathbf{u}'$. It means that $\mathbf{u}'_{[0,i)} w' c$ is a prefix of $\mathbf{u}'$ for some $c \in \mathcal{A}$. Then $L_a(\mathbf{u}'_{[0,i)} w' c)$ is a prefix of $\mathbf{u}$, the word $L_a(w'c)$ has a prefix $L_a(w')a = w$ and $|L_a(\mathbf{u}'_{[0,i)})|$ is an occurrence of w in $\mathbf{u}$. Thus $L_a(\mathbf{u}'_{[0,i)}) = r_{d_0} \cdots r_{d_j}$ for some $j \in \mathbb{N}$ and by injectivity of L_a, it follows that $\mathbf{u}'_{[0,i)} = r'_{d_0} \cdots r'_{d_j}$ and so $i = |r'_{d_0} \cdots r'_{d_j}|$ for some $j \in \mathbb{N}$.

Conversely, we suppose that $i = |r'_{d_0} \cdots r'_{d_j}|$ for some $j \in \mathbb{N}$. If we denote $p = r_{d_0} \cdots r_{d_j}$, then pw is a prefix of $\mathbf{u}$ and by Lemma 18 there is a unique prefix p' of $\mathbf{u}'$ such that $p = L_a(p')$. Clearly, $p'w'$ is also a prefix of $\mathbf{u}'$ and by injectivity of L_a we can conclude that $p' = r'_{d_0} \cdots r'_{d_j}$. Thus i is an occurrence of w' in $\mathbf{u}'$.

To prove Item (ii) we suppose that w' is a right special prefix of $\mathbf{u}'$. We denote its return words $\mathcal{R_{u'}}(w') = \{r'_c : c \in \mathcal{A}\}$ and its derived sequence $\mathbf{d_{u'}}(w') = d_0 d_1 \cdots$. Thus $\mathbf{u}' = r'_{d_0} r'_{d_1} \cdots$. If we set $w = L_a(w')a$ and $r_c = L_a(r'_c)$ for all $c \in \mathcal{A}$, we get $\mathbf{u} = r_{d_0} r_{d_1} \cdots$. Now it remains to prove that w is a right special prefix of $\mathbf{u}$ and the set $\{|r_{d_0} \cdots r_{d_j}| : j \in \mathbb{N}\} \cup \{0\}$ is the set of all occurrences of w in $\mathbf{u}$. We skip these proofs since the arguments are completely analogous to those used in the proof of Item (i). $\qed$

Proposition 20. *Let $\mathbf{u}$ and $\mathbf{u}'$ be Arnoux–Rauzy sequences over $\mathcal{A}$ such that $\mathbf{u} = R_a(\mathbf{u}')$ for some $a \in \mathcal{A}$ and let $\mathbf{u}$ start with the letter $b \in \mathcal{A}, b \neq a$.*

(i) *If w is a nonempty right special prefix of $\mathbf{u}$, then there exists a nonempty right special prefix w' of $\mathbf{u}'$ such that $w = R_a(w')$ and $\mathbf{d_u}(w) = \mathbf{d_{u'}}(w')$.*

(ii) *If w' is a nonempty right special prefix of $\mathbf{u}'$, then $w = R_a(w')$ is a nonempty right special prefix of $\mathbf{u}$ and $\mathbf{d_{u'}}(w') = \mathbf{d_u}(w)$.*

Proof. The morphisms L_a and R_a are conjugate, i.e., $aR_a(x) = L_a(x)a$ for every word $x \in \mathcal{A}^*$. Thus for the Arnoux–Rauzy sequence $\mathbf{v} = a\mathbf{u}$ we get $\mathbf{v} = aR_a(\mathbf{u}') = L_a(\mathbf{u}')$, since the conjugacy holds for every prefix of $\mathbf{u}'$.

Let w be a nonempty right special prefix of $\mathbf{u}$ and let (i_n) be the increasing sequence of the occurrences of w in $\mathbf{u}$. By the form of the morphism R_a, each letter $b \neq a$ (excluding the first letter of $\mathbf{u}$) is preceded by the letter a. Thus the sequence (i_n) is also the sequence of the occurrences of the word aw in $\mathbf{v}$ and $\mathbf{d_u}(w) = \mathbf{d_v}(aw)$. Moreover, aw is a right special prefix of $\mathbf{v}$ and so we can apply Proposition 19 and find the right special prefix w' of $\mathbf{u}'$ such that $aw = L_a(w')a = aR_a(w')$ and $\mathbf{d_u}(w) = \mathbf{d_v}(aw) = \mathbf{d_{u'}}(w')$. The proof of Item (ii) is similar and so we skip it.

Propositions 19 and 20 can be also restated as follows.

Corollary 21. *Let $\mathbf{u}, \mathbf{u}'$ be Arnoux–Rauzy sequences over $\mathcal{A}$ and $a \in \mathcal{A}$.*

(i) *If $\mathbf{u} = L_a(\mathbf{u}')$, then $\mathrm{Der}(\mathbf{u}) = \mathrm{Der}(\mathbf{u}') \cup \{\mathbf{u}'\}$.*
(ii) *If $\mathbf{u} = R_a(\mathbf{u}')$ and $\mathbf{u}$ starts with a letter $b \in \mathcal{A}, b \neq a$, then $\mathrm{Der}(\mathbf{u}) = \mathrm{Der}(\mathbf{u}')$.*

4   Derived Sequences of Arnoux–Rauzy Sequences

First, we introduce a transformation Δ on the set of normalized directive sequences. Subsequently, we use this transformation to describe the set $\mathrm{Der}(\mathbf{u})$ of derived sequences of an Arnoux–Rauzy sequence $\mathbf{u}$.

Definition 22. *Let $\mathbf{z} = z_0 z_1 z_2 \cdots$ be a normalized spinned sequence and let k be the unique index such that z_k is an L-spinned letter and $z_0 z_1 \cdots z_{k-1}$ is an R-spinned word (or is empty). Then $\Delta(\mathbf{z}) = \mathbf{z}_{[k+1,\infty)} = z_{k+1} z_{k+2} z_{k+3} \cdots$.*

Clearly, if $\mathbf{z}$ is the normalized directive sequence of an Arnoux–Rauzy sequence $\mathbf{u}$, then $\Delta(\mathbf{z})$ is the normalized directive sequence of an Arnoux–Rauzy sequence as well. For every integer $m \geq 1$ we let $\mathbf{d}_m$ denote the Arnoux–Rauzy sequence directed by $\Delta^m(\mathbf{z})$ and we also set $\mathbf{d}_0 = \mathbf{u}$.

Example 23. For the normalized spinned sequence $\mathbf{z} = \bar{c}ba(\bar{c}b\bar{a}b)^\omega$ we get $\Delta(\mathbf{z}) = a(\bar{c}b\bar{a}b)^\omega$, $\Delta^2(\mathbf{z}) = (\bar{c}b\bar{a}b)^\omega$, $\Delta^3(\mathbf{z}) = (\bar{a}b\bar{c}b)^\omega$ and $\Delta^4(\mathbf{z}) = (\bar{c}b\bar{a}b)^\omega = \Delta^2(\mathbf{z})$.

Theorem 24. *Let $\mathbf{u}$ be an Arnoux–Rauzy sequence over $\mathcal{A}$ with the normalized directive sequence $\mathbf{z}$. Then $\mathbf{d}$ is the derived sequence with respect to a nonempty prefix of $\mathbf{u}$ if and only if $\mathbf{d} = \mathbf{d}_m$ for some $m \geq 1$, i.e., $\mathbf{d}$ is an Arnoux–Rauzy sequence directed by $\Delta^m(\mathbf{z})$ for some $m \geq 1$.*

Proof. ($\Rightarrow$) We consider a nonempty right special prefix w of $\mathbf{u}$ and prove that $\mathbf{d_u}(w) = \mathbf{d}_m$ for some $m \geq 1$. In fact, we prove that for every $i \in \mathbb{N}$ and a right special prefix v of $\mathbf{d}_i$ there is a right special prefix v' of $\mathbf{d}_{i+1}$ such that $|v'| < |v|$ and $\mathbf{d_{d_i}}(v) = \mathbf{d_{d_{i+1}}}(v')$. Then starting with a nonempty right special prefix w of $\mathbf{u}$ we eventually find the index $m \geq 1$ and the prefix w'' of $\mathbf{d}_m$ such that $w'' = \varepsilon$ and so $\mathbf{d_u}(w) = \mathbf{d_{d_m}}(\varepsilon) = \mathbf{d}_m$.

Since $\mathbf{z}$ is normalized, $\Delta^i(\mathbf{z}) = \mathbf{y}$ is also normalized and so it has a prefix $\bar{x}a$ for some $a \in \mathcal{A}$ and $\bar{x} \in (\bar{\mathcal{A}} \setminus \{\bar{a}\})^*$. If $\bar{x} = \varepsilon$, then $\Delta^{i+1}(\mathbf{z}) = \mathbf{y}_{[1,\infty)}$ and so

$\mathbf{d}_i = L_a(\mathbf{d}_{i+1})$. By Proposition 19 there is a right special prefix v' of $\mathbf{d}_{i+1}$ such that $v = L_a(v')a$ and $\mathbf{d}_{\mathbf{d}_i}(v) = \mathbf{d}_{\mathbf{d}_{i+1}}(v')$. If $\bar{x}$ is nonempty, we denote $|\bar{x}| = n$. Then $\Delta^{i+1}(\mathbf{z}) = \mathbf{y}_{[n+1,\infty)}$. Let us denote $\mathbf{u}^{(\ell)}$ the sequence directed by $\mathbf{y}_{[\ell,\infty)}$ for all $\ell \in \mathbb{N}$. In particular, $\mathbf{u}^{(0)} = \mathbf{d}_i$, $\mathbf{u}^{(n+1)} = \mathbf{d}_{i+1}$ and $\mathbf{u}^{(\ell)} = \varphi_{y_\ell}(\mathbf{u}^{(\ell+1)})$ for all $\ell \in \mathbb{N}$. By Proposition 10 all sequences $\mathbf{u}^{(0)}, \ldots, \mathbf{u}^{(n)}$ starts with the letter a and so by Proposition 20 there are nonempty right special prefixes $v^{(\ell)}$ of $\mathbf{u}^{(\ell)}$ for all $\ell = 0, \ldots, n$ such that $v^{(0)} = v$, $v^{(\ell)} = \varphi_{y_\ell}(v^{(\ell+1)})$ for all $\ell = 0, \ldots, n-1$ and

$$\mathbf{d}_{\mathbf{d}_i}(v) = \mathbf{d}_{\mathbf{u}^{(1)}}(v^{(1)}) = \cdots = \mathbf{d}_{\mathbf{u}^{(n)}}(v^{(n)}).$$

By Proposition 19 there is a right special prefix v' of $\mathbf{u}^{(n+1)} = \mathbf{d}_{i+1}$ such that $v^{(n)} = L_a(v')a$ and $\mathbf{d}_{\mathbf{u}^{(n)}}(v^{(n)}) = \mathbf{d}_{\mathbf{d}_{i+1}}(v')$. Since we also have

$$|v| > |v^{(1)}| > |v^{(2)}| > \cdots > |v^{(n)}| > |v'|,$$

v' is the desired right special prefix of $\mathbf{d}_{i+1}$.

($\Leftarrow$) For arbitrary $m \geq 1$ we find a nonempty right special prefix w of $\mathbf{u}$ such that $\mathbf{d}_{\mathbf{u}}(w) = \mathbf{d}_m$. We set $\mathbf{z} = z_0 z_1 \cdots z_i \Delta^m(\mathbf{z})$ for some $i \in \mathbb{N}$ and we let $\mathbf{u}^{(\ell)}$ denote the sequence directed by $\mathbf{z}_{[\ell,\infty)}$ for all $\ell \in \mathbb{N}$. In particular, $\mathbf{u}^{(0)} = \mathbf{u}$ and $\mathbf{u}^{(i+1)} = \mathbf{d}_m$. Now we take the right special prefix ε of $\mathbf{d}_m$ and using Propositions 19 and 20 we successively find right special prefixes $w^{(\ell)}$ of $\mathbf{u}^{(\ell)}$ for all $\ell = i, \ldots, 0$. Since z_i is L-spinned, the inequalities $0 < |w^{(i)}| < |w^{(i-1)}| < \cdots < |w^{(0)}|$ hold and

$$\mathbf{d}_m = \mathbf{d}_{\mathbf{d}_m}(\varepsilon) = \mathbf{d}_{\mathbf{u}^{(i)}}(w^{(i)}) = \cdots = \mathbf{d}_{\mathbf{u}^{(1)}}(w^{(1)}) = \mathbf{d}_{\mathbf{u}}(w^{(0)}).$$

Then $w^{(0)}$ is the desired prefix w of $\mathbf{u}$.

Corollary 25. *All derived sequences with respect to nonempty prefixes of a given Arnoux–Rauzy sequence over $\mathcal{A}$ are Arnoux–Rauzy sequences over $\mathcal{A}$ as well.*

Proof. This follows directly from Theorems 24 and 9.

By Durand's result [4] the set $\mathrm{Der}(\mathbf{u})$ is finite if and only if $\mathbf{u}$ is a primitive substitutive sequence. An Arnoux–Rauzy sequence is primitive substitutive if and only if its normalized directive sequence is eventually periodic. Indeed, a pure episturmian morphism is primitive if and only if its directive word contains at least one letter a or $\bar{a}$ for every $a \in \mathcal{A}$ and the normalization of an eventually periodic directive sequence always produces an eventually periodic normalized directive sequence (see [7] for more details).

Now we specify the cardinality of $\mathrm{Der}(\mathbf{u})$ according to the normalized directive sequence of an Arnoux–Rauzy sequence $\mathbf{u}$. Let us recall that two derived sequences $\mathbf{d}^{(1)}, \mathbf{d}^{(2)}$ such that $\mathbf{d}^{(1)} = P(\mathbf{d}^{(2)})$ for some permutation P are considered as equal since their structure is the same. Let us emphasize that a permutation P on $\mathcal{A}$ can be naturally extended to the alphabet $\mathcal{A} \cup \bar{\mathcal{A}}$: P acts on the letters from $\mathcal{A}$ without any changes and for every letter $\bar{a} \in \bar{\mathcal{A}}$ we put $P(\bar{a}) = \bar{b}$ if $P(a) = b$.

Observation 26. *Let* $\mathbf{u}^{(1)}$, $\mathbf{u}^{(2)}$ *be Arnoux–Rauzy sequences with the normalized directive sequences* $\mathbf{z}^{(1)}$, $\mathbf{z}^{(2)}$, *respectively, and let* P *be a permutation. Then* $\mathbf{u}^{(1)} = P(\mathbf{u}^{(2)})$ *if and only if* $\mathbf{z}^{(1)} = P(\mathbf{z}^{(2)})$.

Lemma 27. *Let* $\mathbf{z}$ *be the normalized directive sequence and let* $k < \ell$ *be the minimal indices such that there is a permutation* P *satisfying* $\mathbf{z}_{[\ell,\infty)} = P(\mathbf{z}_{[k,\infty)})$. *We denote* $x = \mathbf{z}_{[0,k)}$, $y = \mathbf{z}_{[k,\ell)}$ *and* n *the order of the permutation* P. *Then* $\mathbf{z}$ *is eventually periodic:* $\mathbf{z} = x \left(yP(y) \cdots P^{n-1}(y) \right)^{\omega}$. *Moreover, every sequence* $\mathbf{z}_{[i,\infty)}$ *with* $i \geq \ell$ *is equal (up to permutation of letters) to the sequence* $\mathbf{z}_{[j,\infty)}$ *for some* $j \in \{k, \ldots, \ell - 1\}$ *and if* $\mathbf{z}_{[i,\infty)} = Q(\mathbf{z}_{[j,\infty)})$ *for some* $j < i$ *and a permutation* Q, *then* $i \geq \ell$.

Proof. In the notation from the statement we can write

$$\mathbf{z} = xy\mathbf{z}_{[\ell,\infty)} = xyP(\mathbf{z}_{[k,\infty)}) = xyP(y\mathbf{z}_{[\ell,\infty)}) = xyP(y)P^2(\mathbf{z}_{[k,\infty)}) = \cdots$$
$$= xyP(y) \cdots P^{n-1}(y)P^n(y)P^{n+1}(y) \cdots = x \left(yP(y) \cdots P^{n-1}(y) \right)^{\omega}.$$

Moreover, for every $i \geq l$ we can write $\mathbf{z}_{[i,\infty)} = P(\mathbf{z}_{[i-\ell+k,\infty)})$. Thus eventually we get $\mathbf{z}_{[i,\infty)} = P^m(\mathbf{z}_{[i',\infty)})$ for some positive integer m and an index i' such that $k \leq i' < \ell$. The last part of the statement clearly holds since otherwise it leads us to the contrary with the minimality of the indices k, ℓ. ∎

Corollary 28. *Let* $\mathbf{u}$ *be an Arnoux–Rauzy sequence over* $\mathcal{A}$ *with the aperiodic normalized directive sequence and let* v, w *be two distinct nonempty right special prefixes of* $\mathbf{u}$. *Then the derived sequences with respect to* v *and* w *are distinct, i.e.,* $\mathbf{d}_\mathbf{u}(v) \neq P(\mathbf{d}_\mathbf{u}(w))$ *for any permutation* P.

Proof. We argue by contradiction. By Theorem 24 all derived sequences with respect to nonempty prefixes of $\mathbf{u}$ are the elements of the sequence $(\mathbf{d}_m)_{m \geq 1}$. Thus we can suppose that $\mathbf{d}_m = P(\mathbf{d}_\ell)$ for some positive integers m, ℓ and a permutation P. Since v, w are distinct right special prefixes, we get $m \neq \ell$. By Observation 26, it means that $\Delta^m(\mathbf{z}) = P(\Delta^\ell(\mathbf{z}))$ and so by Lemma 27 $\mathbf{z}$ is eventually periodic, which is the contradiction. ∎

Corollary 29. *Let* $\mathbf{u}$ *be an Arnoux–Rauzy sequence over* $\mathcal{A}$ *with the eventually periodic normalized directive sequence* $\mathbf{z} = x \left(yP(y) \cdots P^{n-1}(y) \right)^{\omega}$, *where the words* $x \in (\mathcal{A} \cup \bar{\mathcal{A}})^*$, $y \in (\mathcal{A} \cup \bar{\mathcal{A}})^+$ *are the shortest possible and* P *is a permutation with the order* n. *We denote* $|x|_L$, $|xy|_L$ *the numbers of L-spinned letters in the words* x, xy, *respectively.*

(i) *If the last letters of both* x, y *are L-spinned, then* $\#\mathrm{Der}(\mathbf{u}) = |xy|_L - 1$. *More precisely, there are* $|x|_L - 1$ *derived sequences belonging to exactly one nonempty right special prefix of* $\mathbf{u}$ *and* $|y|_L$ *derived sequences belonging to infinitely many right special prefixes of* $\mathbf{u}$.

(ii) *If the last letter of* x *or* y *is R-spinned or* $x = \varepsilon$, *then* $\#\mathrm{Der}(\mathbf{u}) = |xy|_L$. *More precisely, there are* $|x|_L$ *derived sequences belonging to exactly one nonempty right special prefix of* $\mathbf{u}$ *and* $|y|_L$ *derived sequences belonging to infinitely many right special prefixes of* $\mathbf{u}$.

Proof. By Theorem 24 all elements of $\mathrm{Der}(\mathbf{u})$ are the elements of the sequence $(\mathbf{d}_m)_{m\geq 1}$. To prove Item (i), we have to show that the sequence $(\mathbf{d}_m)_{m\geq 1}$ has pre-period $|x|_L - 1$ and period $|y|_L$ (up to permutation of letters). However, the sequence $(\Delta^m(\mathbf{z}))_{m\geq 1}$ has the same pre-period and period, see Observation 26. Now it suffices to apply Lemma 27 with $k = |x|$ and $\ell = |xy|$. Since both x and y end with L-spinned letters, the sequences that occur once in $(\Delta^m(\mathbf{z}))_{m\geq 1}$ are exactly the elements of the set $\{\mathbf{z}_{[i,\infty)} : 0 < i < |x|\}$ for which z_{i-1} is L-spinned. Thus they are the sequences $\Delta(\mathbf{z}), \ldots, \Delta^{|x|_L - 1}(\mathbf{z})$. Similarly, the sequences that occur (up to permutation of letters) infinitely many times in $(\Delta^m(\mathbf{z}))_{m\geq 1}$ are exactly the elements of the set $\{\mathbf{z}_{[i,\infty)} : |x| \leq i < |xy|\}$ for which z_{i-1} is L-spinned, so they are the sequences $\Delta^{|x|_L}(\mathbf{z}), \ldots, \Delta^{|xy|_L - 1}(\mathbf{z})$.

We prove Item (ii) analogously. It suffices to realize that if x or y ends with an R-spinned letter or x is the empty word, then the periodic part of $(\Delta^m(\mathbf{z}))_{m\geq 1}$ starts with the element $\Delta^{|x|_L + 1}(\mathbf{z})$. Thus the sequence $(\Delta^m(\mathbf{z}))_{m\geq 1}$ has pre-period $|x|_L$ and period $|y|_L$ (up to permutation of letters).

Example 30. The Arnoux–Rauzy sequence $\mathbf{u}$ is directed by the normalized directive sequence $\mathbf{z} = \bar{c}ba(\bar{c}b\bar{a}b)^\omega = \bar{c}ba(\bar{c}bP(\bar{c}b))^\omega$ for the permutation $P : a \to c, b \to b, c \to a$ with the order 2. By Item (i) of Corollary 29, the sequence $\mathbf{u}$ has two derived sequences: $\mathbf{d}_1$ directed by $\Delta(\mathbf{z}) = a(\bar{c}b\bar{a}b)^\omega$ belonging to the shortest nonempty right special prefix of $\mathbf{u}$ and $\mathbf{d}_2$ directed by $\Delta^2(\mathbf{z}) = (\bar{c}b\bar{a}b)^\omega$ belonging to all the others right special prefixes of $\mathbf{u}$.

The Tribonacci sequence $\mathbf{u}_\tau$ from Example 2 is directed by the normalized directive sequence $\mathbf{z} = (abc)^\omega = (aP(a)P^2(a))^\omega$ for the permutation $P : a \to b, b \to c, c \to a$ with the order 3. Then by Item (ii) of Corollary 29, $\mathbf{u}_\tau$ has one derived sequence $\mathbf{d}$ directed by $(abc)^\omega$. In other words, the derived sequence with respect to any prefix of $\mathbf{u}_\tau$ is the sequence $\mathbf{u}_\tau$ itself.

The Arnoux–Rauzy sequence $\mathbf{u}^{(a)}$ directed by the normalized directive sequence $\mathbf{z}^{(a)} = a(bc\bar{a})^\omega$ (see Example 16) has by Item (ii) of Corollary 29 three derived sequences $\mathbf{d}_1, \mathbf{d}_2, \mathbf{d}_3$ directed by $\Delta(\mathbf{z}^{(a)}) = (bc\bar{a})^\omega$, $\Delta^2(\mathbf{z}^{(a)}) = (c\bar{a}b)^\omega$, $\Delta^3(\mathbf{z}^{(a)}) = (\bar{a}bc)^\omega$, respectively. The sequence $\mathbf{d}_1$ is the derived sequence with respect to the shortest nonempty right special prefix of $\mathbf{u}^{(a)}$, while both $\mathbf{d}_2, \mathbf{d}_3$ belong to infinitely many right special prefixes of $\mathbf{u}^{(a)}$.

References

1. Araújo, I.M., Bruyère, V.: Words derivated from Sturmian words. Theoret. Comput. Sci. **340**, 204–219 (2005)
2. Arnoux, P., Rauzy, G.: Représentation géométrique de suites de complexité $2n+1$. Bull. de la Société Mathématique de France **119**, 199–215 (1991)
3. Balková, L., Pelantová, E., Steiner, W.: Sequences with constant number of return words. Monatsh. Math. **155**, 251–263 (2008)
4. Durand, F.: A characterization of substitutive sequences using return words. Discrete Math. **179**, 89–101 (1998)
5. Droubay, X., Justin, J., Pirillo, G.: Episturmian words and some constructions of de Luca and Rauzy. Theoret. Comput. Sci. **225**(1–2), 539–553 (2001)

6. Ferenczi, S., Holton, C., Zamboni, L.Q.: Structure of three interval exchange transformations I. An arithmetic study. Ann. Inst. Fourier (Grenobles) **51**, 861–901 (2001)
7. Glen, A., Levé, F., Richomme, G.: Directive words of episturmian words: equivalences and normalization. RAIRO-Theoret. Inf. Appl. **43**, 299–319 (2009)
8. Glen, A., Justin, J.: Episturmian words: a survey. RAIRO-Theoret. Inf. Appl. **43**, 403–442 (2009)
9. Justin, J., Vuillon, L.: Return words in Sturmian and episturmian words. RAIRO-Theoret. Inf. Appl. **34**, 343–356 (2000)
10. Justin, J., Pirillo, G.: Episturmian words and episturmian morphisms. Theoret. Comput. Sci. **276**(1–2), 281–313 (2002)
11. Justin, J., Pirillo, G.: Episturmian words: shifts, morphisms and numeration systems. Int. J. Found. Comput. Sci. **15**(2), 329–348 (2004)
12. Klouda, K., Medková, K., Pelantová, E., Starosta, Š.: Fixed points of Sturmian morphisms and their derivated words. Theoret. Comput. Sci. **743**, 23–37 (2018)
13. Richomme, G.: Lyndon morphisms. Bull. Belg. Math. Soc. Simon Stevin **10**, 761–785 (2003)
14. Vuillon, L.: A characterization of Sturmian words by return words. Eur. J. Combin. **22**, 263–275 (2001)

New Results on Pseudosquare Avoidance

Tim Ng[1], Pascal Ochem[2], Narad Rampersad[3], and Jeffrey Shallit[1(✉)]

[1] School of Computer Science, University of Waterloo,
Waterloo, ON N2L 3G1, Canada
{tim.ng,shallit}@uwaterloo.ca
[2] LIRMM, CNRS, Université de Montpellier, Montpellier, France
ochem@lirmm.fr
[3] Department of Mathematics/Statistics, University of Winnipeg,
515 Portage Ave., Winnipeg, MB R3B 2E9, Canada
narad.rampersad@gmail.com

Abstract. We start by considering binary words containing the minimum possible numbers of squares and antisquares (where an antisquare is a word of the form $x\overline{x}$), and we completely classify which possibilities can occur. We consider avoiding $xp(x)$, where p is any permutation of the underlying alphabet, and $xt(x)$, where t is any transformation of the underlying alphabet. Finally, we prove the existence of an infinite binary word simultaneously avoiding all occurrences of $xh(x)$ for *every* nonerasing morphism h and all sufficiently large words x.

1 Introduction

Let x, v be words. We say that v is a *factor* of x if there exist words u, w such that $x = uvw$. For example, or is a factor of word.

By a *square* we mean a nonempty word of the form xx, like the French word couscous. The *order* of a square xx is $|x|$, the length of x. It is easy to see that every binary word of length at least 4 contains a square factor. However, in a classic paper from combinatorics on words, Entringer, Jackson, and Schatz [7] constructed an infinite binary word containing, as factors, only 5 distinct squares: 0^2, 1^2, $(01)^2$, $(10)^2$, and $(11)^2$. This bound of 5 squares was improved to 3 by Fraenkel and Simpson [9]; it is optimal. For some other constructions also achieving the bound 3, see [2,10,14,15].

Instead of considering squares, one could consider *antisquares*: these are binary words of the form $x\overline{x}$, where $\overline{x}$ is a coding that maps $0 \to 1$ and $1 \to 0$. For example, 01101001 is an antisquare. (They should not be confused with the different notion of antipower recently introduced by Fici, Restivo, Silva, and Zamboni [8].) Clearly it is possible to construct an infinite binary word that avoids all antisquares, but only in a trivial way: the only such words are $0^\omega = 000\cdots$ and $1^\omega = 111\cdots$. Similarly, the only infinite binary words with exactly one antisquare are 01^ω and 10^ω. However, it is easy to see that every word in $\{1000, 10000\}^\omega$ has exactly two antisquares—namely 01 and 10—and hence there are infinitely many such words that are aperiodic.

R. Mercaş and D. Reidenbach (Eds.): WORDS 2019, LNCS 11682, pp. 264–274, 2019.
https://doi.org/10.1007/978-3-030-28796-2_21

Several writers have considered variations on these results. For example, Blanchet-Sadri, Choi, and Mercaş [3] considered avoiding large squares in partial words. Chiniforooshan, Kari, and Zhu [4] studied avoiding words of the form $x\theta(x)$, where θ is an antimorphic involution. Their results implicitly suggest the general problem of simultaneously avoiding what we might call *pseudosquares*: patterns of the form xx', where x' belongs to some (possibly infinite) class of modifications of x.

This paper has two goals. First, for all integers $a, b \geq 0$ we determine whether there is an infinite binary word having at most a squares and b antisquares. If this is not possible, we determine the length of the longest finite binary word with this property.

Second, we apply our results to discuss the simultaneous avoidance of xx', where x' belongs to some class of modifications of x. We consider three cases:

(a) where $x' = p(x)$ for a permutation p of the underlying alphabet;
(b) where $x' = t(x)$ for a transformation t of the underlying alphabet; and
(c) where $x' = h(x)$ for an arbitrary nonerasing morphism.

In particular, we prove the existence of an infinite binary word that avoids $xh(x)$ simultaneously for all nonerasing morphisms h and all sufficiently long words x.

2   Simultaneous Avoidance of Squares and Antisquares

We are interested in binary words where the number of distinct factors that are squares and antisquares is bounded. More specifically, we completely solve this problem determining in every case the length of the longest word having at most a distinct squares and at most b distinct antisquares. Our results are summarized in the following table. If (one-sided) infinite words are possible, this is denoted by writing ∞ for the length.

The results in the first two columns and first three rows (that is, for $a \leq 2$ and $b \leq 1$) are very easy. We first explain the first two columns:

Proposition 1.

(a) For $a \geq 0$, the longest binary word with a squares and 0 antisquares has length $2a + 1$.

(b) For $a \geq 0$, the longest binary word with a squares and 1 antisquare has length $2a + 2$.

Proof.

(a) If a binary word has no antisquares, then in particular it has no occurrences of either 01 or 10. Thus it must contain only one type of letter. If it has length $2a + 2$, then it has $a + 1$ squares, of order $1, 2, \ldots, a + 1$. If it has length $2a + 1$, it has a squares. So $2a + 1$ is optimal.

(b) If a length-n binary word w has only one antisquare, this antisquare must be either 01 or 10; without loss of generality, assume it is 01. Then w is either of the form $0^{n-1}1$ or 01^{n-1}. Such a word clearly has $\lfloor (n - 1)/2 \rfloor$ squares.

a\\b	0	1	2	3	4	5	6	7	8	9	10	11	12	13	$\cdots$
0	1	2	3	3	3	3	3	3	3	3	3	3	3	3	$\cdots$
1	3	4	7	7	7	7	7	7	7	7	7	7	7	7	$\cdots$
2	5	6	11	11	11	11	12	12	12	13	15	18	18	18	$\cdots$
3	7	8	15	15	15	20	20	20	24	29	34	53	98	∞	$\cdots$
4	9	10	19	19	27	31	45	56	233	∞	∞	∞	∞		$\cdots$
5	11	12	27	27	40	∞	∞	∞	∞						$\cdots$
6	13	14	35	38	313	∞	$\cdots$								
7	15	16	45	∞	∞	$\cdots$									
8	17	18	147	∞	$\cdots$										
9	19	20	∞	$\cdots$											
10	21	22	∞	$\cdots$											
$\vdots$															

Fig. 1. Length of longest binary word having at most a squares and b antisquares

We next explain the first three rows: if a binary word has no squares, its length is clearly bounded by 3, as we remarked earlier. If it has one square, a simple argument shows it has length at most 7. Finally, if it has two squares, already Entringer, Jackson, and Schatz [7, Thm. 2] observed that it has length at most 18.

For all the remaining finite entries, we obtained the result through the usual backtrack search method, and we omit the details.

In what follows, we provide the lexicographically least binary words achieving the "important" bounds in Fig. 1.

$(3, 12)$: 0010001100101110001011001110001100101110001011000
11100101110001011001110001100101110001011001110011011

$(4, 8)$: 00001000001100001011000001100010110000010111000101100001011
100000101100000110001011000001011100010110000011000101100001011100000
10111000001011000001100010110000010111000101100001011100000
1011000001100010110000101110001011000001100010110000010000

$(5, 4)$: 000100000101000000010101000000101000001

$(6, 4)$: 000010000000101000000001101000000010100001101000000010100000011010
0000010100001101000000010100001101000000010100001101000000010100
0000110100000010100001101000000010100001101000000010100000011010
0000010100001101000000010100001101000000010100001101000000010100
0000110100000010100001101000000010100001101000000011010111010

$(7, 2)$: 00000100000001010000001000010100000010000101

(8, 2) : 001000010100010001000101000001000100010000101000

001000100101000001000100000101000100010100000010001

000100000101000100010001010000010001000100000101010

It now remains to prove the results labeled ∞. First, we introduce some morphisms. Let the morphisms $h_{3,13}$, $h_{4,9}$, $h_{5,5}$, $h_{7,3}$, $h_{9,2}$ be defined as follows:

(a) $h_{3,13}$: $0 \to 0010110011100011$

$\qquad 1 \to 001011000111$

$\qquad 2 \to 00101110$

(b) $h_{4,9}$: $0 \to 0000101110000011000010110000011000101100001011100010110$

$\qquad 1 \to 0000101110000011000010110000011000101100000101110001011$

$\qquad 2 \to 0000101110000011000010110000010111000101100000110001011$

This is a 55-uniform morphism.

(c) $h_{5,5}$: $0 \to 101000001011000010100001101011000001$

$\qquad 1 \to 101000001011000011010110000101000001$

$\qquad 2 \to 101000001010001100001010000011000$

This is a 36-uniform morphism.

(d) $h_{7,3}$: $0 \to 0100100100001010000$

$\qquad 1 \to 01001001000001$

$\qquad 2 \to 0100100101000$

(e) $h_{9,2}$: $0 \to 0001000100000001000101$

$\qquad 1 \to 0000010001000100000101$

$\qquad 2 \to 0000001000100000010100$

This is a 22-uniform morphism.

Theorem 1. *Let* **w** *be an infinite squarefree sequence over the alphabet* $\{0, 1, 2\}$. *Then* $h_{a,b}(\mathbf{w})$ *contains exactly a squares and b antisquares. More precisely*

(a) $h_{3,13}(\mathbf{w})$ *contains the squares* 0^2, 1^2, *and* $(01)^2$ *and the antisquares* 01, 10, 0011, 0110, 1001, 1100, 000111, 001110, 011100, 100011, 110001, 111000, *and* 10010110.

(b) $h_{4,9}(\mathbf{w})$ *contains the squares* 0^2, 1^2, $(00)^2$, *and* $(01)^2$ *and the antisquares* 01, 10, 0011, 0110, 1100, 011100, 110001, 111000, *and* 1110000011.

(c) $h_{5,5}(\mathbf{w})$ *contains the squares* 0^2, 1^2, $(00)^2$, $(01)^2$, *and* $(10)^2$ *and the antisquares* 01, 10, 0011, 0110, *and* 1100.

(d) $h_{7,3}(\mathbf{w})$ *contains the squares* 0^2, $(00)^2$, $(01)^2$, $(10)^2$, $(001)^2$, $(010)^2$, *and* $(100)^2$ *and the antisquares* 01, 10, *and* 1001.

(e) $h_{9,2}(\mathbf{w})$ *contains the squares* 0^2, $(00)^2$, $(01)^2$, $(10)^2$, $(000)^2$, $(0001)^2$, $(0010)^2$, $(0100)^2$, *and* $(1000)^2$ *and the antisquares* 01 *and* 10.

Proof. Let h be any of the morphisms above. We first show that large squares are avoided. The h-images of the letters have been ordered such that $|h(0)| \geq |h(1)| \geq |h(2)|$. A computer check shows that for every letter i and every ternary word $\mathbf{w}$, the factor $h(i)$ appears in $h(\mathbf{w})$ only as the h-image of i. Another computer check shows that for every ternary squarefree word $\mathbf{w}$, the only squares uu with $|u| \leq 2|h(0)| - 2$ that appear in $h(w)$ are the ones we claim. If $h(w)$ contains a square uu with $|u| \geq 2|h(0)| - 1$, then u contains the full h-image of some letter. Thus, uu is a factor of $h(avbvc)$ with a, b, c single letters and v a nonempty word. Moreover, $a \neq b$ and $b \neq c$, since otherwise $avbvc$ would contain a square. It follows that $u = ph(v)s$, so that p is a suffix of $h(a)$, $h(b) = sp$, and s is a prefix of $h(c)$. Thus, $h(abc)$ contains the square $psps$ with period $|ps| = |h(b)|$. Since $5 < |h(2)| \leq |h(b)| \leq |h(0)| < 2|h(0)| - 2$, this contradicts our computer check, which rules out squares with period at least 5 and at most $2|h(0)| - 2$.

To show that large antisquares are avoided, it suffices to exhibit a factor f such that f is uniformly recurrent in $h(\mathbf{w})$ and $\overline{f}$ is not a factor of $h(\mathbf{w})$. We use $f = 0101$ for $h_{3,13}$ and $f = 0^4$ for the other morphisms.

Remark 1. The uniform morphisms were found as follows: for increasing values of q, our program looks (by backtracking) for a binary word of length $3q$ corresponding to the image $h(012)$ of 012 by a suitable q-uniform morphism h. Given a candidate h, we check that $h(w)$ has at most a squares and b antisquares for every squarefree word $\mathbf{w}$ up to some length. Standard optimizations are applied to the backtracking. Squares and antisquares are counted naively (recomputed from scratch at every step), which is sufficient since the morphisms found are not too large.

Remark 2. The morphisms $h_{3,13}$ and $h_{7,3}$ are not uniform. However, we can construct uniform morphisms with the same properties as follows. Let m be the 18-uniform squarefree morphism given by

$$0 \rightarrow 021012102012021201$$
$$1 \rightarrow 021012102120210201$$
$$2 \rightarrow 021012102120102012.$$

Notice that $m(0)$, $m(1)$, and (2) contain 6 occurrences of each letter. So the 216-uniform morphism $h'_{3,13} = h_{3,13} \circ m$ is such that $h'_{3,13}(\mathbf{w})$ and $h_{3,13}(\mathbf{w})$ contain the same squares and antisquares. Similarly, for binary words with the same squares and antisquares as $h_{7,3}(w)$, we can use the 276-uniform morphism $h_{7,3} \circ m$. However in this case, we have found the following smaller morphism, which is 29-uniform.

$$0 \rightarrow 00101000010010010100000101001$$
$$1 \rightarrow 00101000010010010000101001000$$
$$2 \rightarrow 00101000010010010000101000001.$$

Corollary 1. *There exists an infinite binary word having at most ten distinct squares and antisquares as factors, but the longest binary word having nine or fewer distinct squares and antisquares is of length 45.*

Remark 3. A word of length 45 with a total of nine distinct squares and anti-squares is

$$0000010000000101000000100001010000000010000101.$$

Corollary 2. *Every infinite word having at most ten distinct squares and anti-squares has critical exponent at least 5, and there is such a word having 5-powers but no powers of higher exponent.*

Proof. By the usual backtracking approach, we can easily verify that the longest finite word having at most ten distinct antisquares, and critical exponent <5 is of length 57. One such example is

$$010001010000100100100001010010010100001001001000010100010.$$

On the other hand, if $\mathbf{w}$ is any squarefree ternary infinite word, then from above we know that the only possible squares that can occur in $h_{5,5}(\mathbf{w})$ are of the form x^2 for $x \in \{0, 1, 00, 01, 10\}$. It is now easy to verify that the largest power of 0 that occurs in $h_{5,5}(\mathbf{w})$ is 0^5; the largest power of 1 that occurs is 1^2; the largest power of 01 that occurs is $(01)^{5/2}$; and the largest power of 10 that occurs is $(10)^{5/2}$.

Proposition 2. *Every infinite cubefree binary word has a total of at least 23 distinct squares and antisquares.*

Proof. By the usual backtracking method.

Remark 4. In the final version of this paper, we plan to provide the optimal bound.

3 Pseudosquare Avoidance

In this section we discuss avoiding xx' where x' belongs to some large class of modifications of x'. This is in the spirit of previous results [5,12,16], where one is interested in avoiding factors of low Kolmogorov complexity. The problems we study are not quite so general, but our results are effective, and we obtain explicit bounds.

3.1 Avoiding Pseudosquares for Permutations

Here we are interested in avoiding patterns of the form $xp(x)$, for *all* codings p that are permutations of the underlying alphabet. Of course, this is impossible for words of length ≥ 2 strictly as stated, since every word of length 2 is of the form $ap(a)$ where p is the permutation sending the letter a to $p(a)$. Thus it is reasonable to ask about avoiding $xp(x)$ for all words x of length $\geq n$. Our first result shows this is impossible for $n = 2$.

Theorem 2. *For all finite alphabets Σ, and for all words w of length ≥ 10 over Σ, there exists a permutation p of Σ and a factor of w of the form xx', where $x' = p(x)$, and $|x| \geq 2$.*

Proof. Using the usual tree-traversal technique, where we extend the alphabet size at each length extension.

We now turn to the case of larger n. For $n \geq 3$, we can avoid all factors of the form $xp(x)$ over the binary alphabet. Of course, this case is particularly simple, since there are only two permutations of the alphabet: the identity permutation that leaves letters invariant, and the map $x \rightarrow \bar{x}$, which changes 0 to 1 and vice versa.

Theorem 3. *There exists an infinite word $\mathbf{w}$ over the binary alphabet $\Sigma_2 = \{0, 1\}$ that avoids xx and $x\bar{x}$ for all x with $|x| \geq 3$.*

Proof. We can use the morphism $h_{5,5}$ in Theorem 1(c). Alternatively, a simpler proof comes from the fixed point of the morphism

$$
\begin{array}{ll}
0 \rightarrow 01 & 1 \rightarrow 23 \\
2 \rightarrow 24 & 3 \rightarrow 51 \\
4 \rightarrow 06 & 5 \rightarrow 01 \\
6 \rightarrow 74 & 7 \rightarrow 24
\end{array}
$$

followed by the coding $n \rightarrow n \bmod 2$. We can now use `Walnut` [13] to verify that the resulting 2-automatic word has the desired property. This word has exactly 5 distinct squares:

$$0^2, 1^2, (00)^2, (01)^2, (10)^2,$$

and exactly 6 distinct antisquares:

$$01, 10, 0011, 0110, 1001, 1100.$$

3.2   Avoiding Pseudosquares for Transformations

In the previous subsection we considered permutations of the alphabet. We now generalize this to *transformations* of the alphabet, or, in other words, to arbitrary codings (letter-to-letter morphisms).

Theorem 4.

(a) *For all finite alphabets Σ, and all words w of length ≥ 31 over Σ, there exists a transformation $t : \Sigma^* \rightarrow \Sigma^*$ such that w contains a factor of the form $xt(x)$ for $|x| \geq 3$.*

(b) *For all finite alphabets Σ, and all words w of length ≥ 16 over Σ, there exists a transformation t of Σ such that w contains a factor of the form xx', where $x' = t(x)$ or $x = t(x')$ and $|x| \geq 3$.*

Proof. Using the usual tree-traversal technique, where we extend the alphabet size at each length extension.

We now specialize to the binary alphabet. This case is particularly simple, since in addition to the two permutations of the alphabet, the only other transformations are the ones sending both $0, 1$ to a single letter (either 0 or 1).

Theorem 5. *There exists an infinite word* **w** *over the binary alphabet* $\Sigma_2 = \{0,1\}$ *avoiding* 0^4, 1^4, *and* xx *and* $x\bar{x}$ *for every* x *with* $|x| \geq 4$. *In other words,* **w** *avoids both* $xt(x)$ *and* $t(x)x$ *for* $|x| \geq 4$ *and all transformations* t.

Proof. Use the fixed point of the morphism

$$
\begin{array}{ll}
0 \to 01 & \qquad 1 \to 23 \\
2 \to 45 & \qquad 3 \to 21 \\
4 \to 23 & \qquad 5 \to 42
\end{array}
$$

followed by the coding $n \to \lfloor n/3 \rfloor$. The result can now easily be verified with Walnut.

3.3   Avoiding Pseudosquares with Morphic Images

In this subsection we consider simultaneously avoiding all patterns of the form $xh(x)$, for all morphisms h defined over $\Sigma_k = \{0, 1, \ldots, k-1\}$. Clearly this is impossible if h is allowed to be erasing (that is, some images are allowed to be empty), or if x consists of a single letter. So once again we consider the question for sufficiently long x.

For this version of the problem, it is particularly hard to obtain experimental data, because the problem of determining, given x and y, whether there is a morphism h such that $y = h(x)$, is NP-complete [1, 6].

Theorem 6. *No infinite word over a finite alphabet avoids all factors of the form* $xh(x)$, *for all nonerasing morphisms* h, *with* $|x| \geq 4$.

Proof. Let **w** be a potential counter-example to Theorem 6. Without loss of generality, we can assume that **w** is uniformly recurrent (see, e.g., [11, Lemma 2.4]). We use a, b, and c to denote distinct letters and u and v to denote non-empty finite words.

We call a word *basic* if it is of the form au such that $|u| = 3$ and u does not contain the letter a. Suppose, to get a contradiction, that **w** contains a basic factor. Since u is recurrent, the factor au extends to $auvu$, which is a forbidden occurrence of $xh(x)$.

Suppose, to get a contradiction, that **w** contains a factor aaa. Since $\mathbf{w} \neq a^\omega$, the word **w** contains $baaa$, which is a basic factor. So **w** avoids aaa for every letter a.

Suppose, to get a contradiction, that **w** contains three consecutive distinct letters abc. To avoid a basic factor, abc must extend to $abca$. Then $abca$ must extend to $abcab$, and so on. Thus **w** must contain $(abc)^{8/3} = abcabcab$, which is a forbidden occurrence of $xh(x)$. So **w** avoids abc.

Since **w** avoids *aaa*, *abc*, and the basic factor *abbc*, it must be that **w** is a binary word.

Suppose, to get a contradiction, that **w** contains both 0100 and 1011. The factor 0100 extends to 01001. Since **w** is uniformly recurrent and contains 11, the word **w** contains 01001u11, which is a forbidden occurrence of $xh(x)$. So **w** does not contain both 0100 and 1011, and we assume without loss of generality that **w** avoids 0100.

Using the usual tree-traversal technique, we check that no infinite binary word avoids 000, 111, 0100, and every square xx with $|x| \geq 4$. Thus, **w** does not exist.

Theorem 7. *There exists an infinite binary word that avoids all factors of the form $xh(x)$ and $h(x)x$, for all nonerasing binary morphisms h, with $|x| \geq 5$.*

Proof. Let **w** $= m(\mathbf{t})$, where **t** is any ternary squarefree word **t** and m is the 57-uniform morphism given below.

$$0 \rightarrow 101000110010100110001011001010110001010100011001011000110$$

$$1 \rightarrow 101000110010100110001010110001101001100010101000110101001$$

$$2 \rightarrow 101000110010100110001010100011010011000101011000110101001$$

We use u and v to denote non-empty words and a to denote a letter. We will need the following properties of **w**.

(a) The only squares occurring in **w** are 00, 11, 0101, and 1010.
(b) **w** does not contain any factor uvu with $|u| \geq 15$ and $|v| \leq 6$.
(c) **w** does not contain any of the following factors: 111, 110011, 101011001, 100110101, 0100010, 00100.
(d) Every factor of **w** of length 13 contains 000, 001100, 010100110, or 011001010.
(e) Every factor of **w** of length 12 contains 0101 or 1010.
(f) Every factor of **w** of length at least 5, except 00010 and 01000, contains 0011, 1100, 0101, 1010, 0110, 1001, or 10001.

The proofs of (a) and (b) are similar to the proof of Theorem 1. The other properties can be checked by inspecting factors of **w** with bounded length. The following cases show that **w** contains no factor of the form $xh(x)$ or $h(x)x$ with $|x| \geq 5$.

– We rule out $h(0) = h(1)$, as $h(x)$ contains $h(0)^5$, which contradicts (a).

– We rule out $h(a) = a$, as $xh(x) = h(x)x = xx$ is a square with period at least 5, which contradicts (a).

– We rule out $|x| \geq 13$: By (e), x contains 0101 or 1010. By (a), **w** contains no square with period at least 3, which forces $|h(0)| = |h(1)| = 1$. By the previous cases, the only remaining possibility is $h(a) = \overline{a}$. By (d), x contains a factor $v \in \{000, 001100, 010100110, 011001010\}$. Thus, $h(x)$ contains the factor $\overline{v} \in \{111, 110011, 101011001, 100110101\}$, which contradicts (c).

- We rule out that x contains $a\bar{a}a\bar{a}$ or $aa\overline{aa}$: Notice that both letters 0 and 1 are contained in a square. By (a), $|h(0)| \le 2$ and $|h(1)| \le 2$. A computer check shows that **w** contains no factor of the form $xh(x)$ or $h(x)x$ such that $5 \le |x| \le 12$, $|h(0)| \le 2$, and $|h(1)| \le 2$.

- We rule out that x contains $a\overline{aa}a$ or 10001: in every case $h(x)$ contains a factor uvu such that v is square or a cube. By (a), $\min\{|h(0)|, |h(1)|\} \le 2$ and $|v| \le 6$. By (b), this means that $\max\{|h(0)|, |h(1)|\} \le 14$. Again, we only have to consider the case $5 \le |x| \le 12$, so that $xh(x)$ and $h(x)x$ have bounded length and can be ruled out by computer check.

- By (f), the only remaining possibilities are $x = 00010$ and $x = 01000$. Notice that if $h(000)$ is in **w**, then $h(0) \in \{0, 01, 10\}$. Suppose that $x = 00010$. An occurrence of $h(x)x$ contains 0000 if $h(0) \in \{0, 10\}$ and contains 0100010 if $h(0) = 01$. This contradicts (a) and (c), respectively. An occurrence of $xh(x)$ contains 00100 if $h(0) \in \{0, 01\}$ and contains 10101010 if $h(0) = 10$. This contradicts (c) and (a), respectively. The case $x = 01000$ is symmetrical by reversal.

4 Future Work

In the future we could consider similar questions for abelian avoidability problems.

Acknowledgments. We thank the referees for many useful suggestions.

References

1. Angluin, D.: Finding patterns common to a set of strings. J. Comput. Syst. Sci. **21**, 46–62 (1980)
2. Badkobeh, G., Crochemore, M.: Fewest repetitions in infinite binary words. RAIRO Inform. Théor. App. **46**, 17–31 (2012)
3. Blanchet-Sadri, F., Choi, I., Mercaş, R.: Avoiding large squares in partial words. Theoret. Comput. Sci. **412**, 3752–3758 (2011)
4. Chiniforooshan, E., Kari, L., Xu, Z.: Pseudopower avoidance. Fund. Inform. **114**(1), 55–72 (2012)
5. Durand, B., Levin, L., Shen, A.: Complex tilings. J. Symbolic Logic **73**, 593–613 (2008)
6. Ehrenfeucht, A., Rozenberg, G.: Finding a homomorphism between two words is NP-complete. Inform. Process. Lett. **9**, 86–88 (1979)
7. Entringer, R.C., Jackson, D.E., Schatz, J.A.: On nonrepetitive sequences. J. Combin. Theory. Ser. A **16**, 159–164 (1974)
8. Fici, G., Restivo, A., Silva, M., Zamboni, L.Q.: Anti-powers in infinite words. J. Combin. Theory Ser. A **157**, 109–119 (2018)
9. Fraenkel, A.S., Simpson, J.: How many squares must a binary sequence contain? Electron. J. Combinatorics **2**, #R2 (1995)

10. Harju, T., Nowotka, D.: Binary words with few squares. Bull. Eur. Assoc. Theor. Comput. Sci. **89**, 164–166 (2006)
11. Luca, A.D., Varricchio, S.: Finiteness and iteration conditions for semigroups. Theoret. Comput. Sci. **87**, 315–327 (1991)
12. Miller, J.S.: Two notes on subshifts. Proc. Am. Math. Soc. **140**, 1617–1622 (2012)
13. Mousavi, H.: Automatic theorem proving in Walnut (2016). http://arxiv.org/abs/1603.06017
14. Ochem, P.: A generator of morphisms for infinite words. RAIRO Inform. Théor. Appl. **40**, 427–441 (2006)
15. Rampersad, N., Shallit, J., Wang, M.: Avoiding large squares in infinite binary words. Theoret. Comput. Sci. **339**, 19–34 (2005)
16. Rumyantsev, A.Y., Ushakov, M.A.: Forbidden substrings, Kolmogorov complexity and almost periodic sequences. In: Durand, B., Thomas, W. (eds.) STACS 2006. LNCS, vol. 3884, pp. 396–407. Springer, Heidelberg (2006). https://doi.org/10.1007/11672142_32

Every Nonnegative Real Number
Is an Abelian Critical Exponent

Jarkko Peltomäki[1,2,3](✉) [ID] and Markus A. Whiteland[3](✉)

[1] The Turku Collegium for Science and Medicine TCSM, University of Turku,
Turku, Finland
[2] Turku Centre for Computer Science TUCS, Turku, Finland
[3] Department of Mathematics and Statistics, University of Turku, Turku, Finland
{jspelt,mawhit}@utu.fi

Abstract. The abelian critical exponent of an infinite word w is defined
as the maximum ratio between the exponent and the period of an abelian
power occurring in w. It was shown by Fici et al. that the set of finite
abelian critical exponents of Sturmian words coincides with the Lagrange
spectrum. This spectrum contains every large enough positive real number. We construct words whose abelian critical exponents fill the remaining gaps, that is, we prove that for each nonnegative real number θ there
exists an infinite word having abelian critical exponent θ. We also extend
this result to the k-abelian setting.

Keywords: Abelian equivalence · k-abelian equivalence ·
Critical exponent · Sturmian word

1 Introduction

The study of powers and their avoidance has been one of the central themes in
combinatorics on words; see [2, Chap. 4]. The central notion here is that of the
critical exponent which measures the maximum exponent of a power occurring in
a given word. Recently it has been popular to generalize the notion of a power
using some equivalence relation in place of the usual equality of words. For
example, abelian equivalence (see the references of [6]), and its generalizations
k-abelian equivalence [3,9] and binomial equivalence [16,19] have been popular
options.

Two words u and v are *abelian equivalent*, written $u \sim v$, if they are permutations of each other. An *abelian power* of exponent e and period m is a word
of the form $u_0 \cdots u_{e-1}$ such that $m = |u_0|$ and $u_0, \ldots, u_{e-1}$ are nonempty and
abelian equivalent. For example, $01 \cdot 10$ (a square) and $abc \cdot bca \cdot cab$ (a cube)
are abelian powers. Now it is possible to define the abelian critical exponent
of an infinite word as the maximum exponent of an abelian power occurring in
it. However, this does not give any interesting information on abelian powers
occurring in Sturmian words or, more generally, in words with bounded abelian

© Springer Nature Switzerland AG 2019
R. Mercaş and D. Reidenbach (Eds.): WORDS 2019, LNCS 11682, pp. 275–285, 2019.
https://doi.org/10.1007/978-3-030-28796-2_22

complexity because such words contain abelian powers of arbitrarily high exponent [18]. In order to capture more information on abelian powers of an infinite word to a single quantity, it was proposed in [6] to define the *abelian critical exponent* $\mathcal{A}c(\mathbf{w})$ of an infinite word $\mathbf{w}$ as the quantity

$$\limsup_{m \to \infty} \frac{\mathcal{A}e_{\mathbf{w}}(m)}{m},$$

where $\mathcal{A}e_{\mathbf{w}}(m)$ is the supremum of exponents of abelian powers of period m occurring in $\mathbf{w}$. This notion turns out to be much more interesting. For example, $\mathcal{A}c(\mathbf{f}) = \sqrt{5}$ for the Fibonacci word $\mathbf{f}$, the fixed point of the substitution $0 \mapsto 01$, $1 \mapsto 0$ [6, Theorem 5.14]. Furthermore $\sqrt{5}$ is the minimum abelian critical exponent among all Sturmian words [6, Theorem 5.14]. It follows that for each Sturmian word $\mathbf{s}$ and each $\delta > 0$, there exists an increasing sequence (m_i) of integers such that $\mathbf{s}$ contains an abelian power of period m_i and total length greater than $(\sqrt{5} - \delta)m_i^2$. Notice that if $\mathbf{w}$ does not contain abelian powers with arbitrarily large exponent, then $\mathcal{A}c(\mathbf{w}) = 0$. Many examples of such words are known; see, e.g., [2, Chap. 4.6]. It is also possible that $\mathcal{A}c(\mathbf{w}) = \infty$. Take for example the Thue-Morse word $\mathbf{t}$, the fixed point of the substitution $0 \mapsto 01$, $1 \mapsto 10$. Indeed, it is straightforward to see that $\mathbf{t}$ can be factored as a product of abelian equivalent words of length $2n$ for all $n \geq 0$. This shows that $\mathcal{A}c(\mathbf{t}) = \infty$.

Further study in [6] showed the surprising fact that the set of finite abelian critical exponents of Sturmian words equals the Lagrange spectrum $\mathcal{L}$. The Lagrange constant of an irrational α is the infimum of the real numbers λ such that for every $c > \lambda$ the inequality $|\alpha - n/m| < 1/cm^2$ has only finitely many rational solutions n/m. The Lagrange spectrum is the set of finite Lagrange constants of irrational numbers. The Lagrange spectrum has been extensively studied in number theory since the works of Markov [12,13] in the 19th century. The famous theorems of Markov show that the initial part of $\mathcal{L}$ inside the interval $[\sqrt{5}, 3)$ is discrete. Later in 1947 Hall proved that $\mathcal{L}$ contains a half-line [8]. After a series of improvements by multiple authors, it was finally determined by Freiman in 1975 [7] that the largest half-line contained in the Lagrange spectrum is $[c_F, \infty)$, where

$$c_F = \frac{2221564096 + 283748\sqrt{462}}{491993569} = 4.5278295661\ldots$$

Good sources for information on the Lagrange spectrum are the monograph of Cusick and Flahive [4] and Aigner's book [1]. See also the recent book [17] of Reutenauer for a more word-combinatorial flavor.

The connection between the Lagrange spectrum and abelian critical exponents of Sturmian words shows that each real number larger than c_F is the abelian critical exponent of some infinite word. This raises the obvious question of whether this can be extended to hold for all nonnegative numbers. In this paper, we answer the question in the positive. The main result of this paper is the following theorem.

Theorem 1. *Let θ be a nonnegative real number. Then there exists an infinite word $\mathbf{w}$ such that $\mathcal{A}c(\mathbf{w}) = \theta$. The word $\mathbf{w}$ can be taken over an alphabet of at most three letters.*

This result should be compared with a result of Krieger and Shallit stating that every real number $\theta > 1$ is a critical exponent (in the usual sense) of some infinite word [10]. Notice that here the number of letters required tends to infinity when θ tends to 1 [10], but in our setting we need at most three letters.

We prove an analogue of Theorem 1 for k-abelian critical exponents; see Sect. 3 for the extension and the necessary definitions.

Our proof method is to exploit the properties of the Lagrange spectrum, that is, the fact that Theorem 1 is already known to be true for all reals greater than c_F. The idea is to find a suitable N-uniform substitution σ such that each abelian power in $\sigma(\mathbf{w})$ can be decoded to an abelian power in $\mathbf{w}$ with the same exponent. This means, in essence, that the abelian powers in $\sigma(\mathbf{w})$ are the abelian powers of $\mathbf{w}$ blown up by a factor of N. Roughly speaking, the ratio of exponents and periods corresponding to $\mathcal{A}c(\mathbf{w})$ gets divided by N, that is, $\mathcal{A}c(\sigma(\mathbf{w})) = \mathcal{A}c(\mathbf{w})/N$. The conclusion is that Theorem 1 is true for each real in the interval $[c_F/N, \infty)$, where $[c_F, \infty)$ is the largest half-line contained in the Lagrange spectrum. We may choose N to be arbitrarily large, so Theorem 1 follows. The extension of Theorem 1 to the k-abelian setting is proved using the same ideas.

We use the usual notions and notation from combinatorics on words. If the reader encounters anything undefined, we refer him or her to [11]. Even though we mention Sturmian words several times in this paper, we do not need any properties of these binary words. For their definition, we refer the reader to [11, Chap. 2] and [14, Chap. 4].

2   Proof of Theorem 1

Let θ be a nonnegative real number. If $\theta = 0$, then θ is the abelian critical exponent of any infinite word that avoids abelian powers with large enough exponent. Such words exist by [5] (abelian fourth powers are avoidable over two letters); see also [2, Chap. 4.6].

Assume then that $\theta > 0$, and let N be an integer such that $N\theta \in [c_F, \infty)$. Let $\mathbf{w}$ be an infinite binary word. Our aim is to find an N-uniform substitution f defined on a two-letter alphabet with the following properties:

(i) If an abelian power $u_0 \cdots u_{e-1}$ occurs in $\mathbf{w}$, then $f(u_0) \cdots f(u_{e-1})$ is an abelian power occurring in $f(\mathbf{w})$.

(ii) If an abelian power $u_0 \cdots u_{e-1}$, $e \geq N$, occurs in $f(\mathbf{w})$, then $\mathbf{w}$ contains an abelian power $v_0 \cdots v_{e-1}$ with $|v_0| = |u_0|/N$.

Let us show how to prove Theorem 1 under the assumption that such f exists.

Let $\mathbf{s}$ be a Sturmian word having $\mathcal{A}c(\mathbf{s}) = N\theta$. In fact, any binary word $\mathbf{s}$ with $\mathcal{A}c(\mathbf{s}) = N\theta$ will do, we just know that such a Sturmian word exists by the results of [6]. We claim that $\mathcal{A}c(f(\mathbf{s})) = \theta$. This proves Theorem 1 when $\theta > 0$ (assuming that $f(\mathbf{w})$ has at most three letters).

By Property (i), we have $\mathcal{A}e_{f(\mathbf{s})}(tN) \geq \mathcal{A}e_{\mathbf{s}}(t)$ for all positive integers t. Since $\mathcal{A}c(\mathbf{s}) > 0$, the word $\mathbf{s}$ contains abelian powers of arbitrarily high exponent, and thus by Property (i) the word $f(\mathbf{s})$ contains abelian powers of arbitrarily high exponent and period divisible by N. If $\mathcal{A}e_{f(\mathbf{s})}(tN) \geq N$, then $\mathcal{A}e_{f(\mathbf{s})}(tN) \leq \mathcal{A}e_{\mathbf{s}}(t)$ by Property (ii). Therefore there exists a sequence (t_i) such that $\mathcal{A}e_{\mathbf{s}}(t_i) = \mathcal{A}e_{f(\mathbf{s})}(t_i N)$ for all i. Hence

$$\limsup_{i\to\infty} \frac{\mathcal{A}e_{f(\mathbf{s})}(t_i N)}{t_i N} = \limsup_{i\to\infty} \frac{\mathcal{A}e_{\mathbf{s}}(t_i)}{t_i N} = \frac{1}{N} \limsup_{i\to\infty} \frac{\mathcal{A}e_{\mathbf{s}}(t_i)}{t_i} = \frac{1}{N}\mathcal{A}c(\mathbf{s}) = \theta,$$

so $\mathcal{A}c(f(\mathbf{s})) \geq \theta$. If $\mathcal{A}c(f(\mathbf{s})) > \theta$, then there exists an increasing sequence (ℓ_i) such that

$$\frac{\mathcal{A}e_{f(\mathbf{s})}(\ell_i)}{\ell_i} > \theta > 0$$

for all i. By the preceding, only finitely many of the numbers in the sequence (ℓ_i) are divisible by N. By Property (ii), we thus have $\mathcal{A}e_{f(\mathbf{s})}(\ell_i) \geq N$ only for finitely many i meaning that

$$\frac{\mathcal{A}e_{f(\mathbf{s})}(\ell_i)}{\ell_i} < \frac{N}{\ell_i}$$

for i large enough. This is impossible as $N/\ell_i \to 0$ as $i \to \infty$. The conclusion is that $\mathcal{A}c(f(\mathbf{s})) = \theta$. This concludes the proof of Theorem 1.

Let us then show how to choose a suitable substitution f. Let N be a fixed positive and *even* integer, and define the N-uniform substitution $\sigma\colon \{0,1\}^* \to \{0,1,\#\}^*$ by

$$0 \mapsto \#0^{N-1},$$
$$1 \mapsto \#1^{N-1}.$$

Lemma 2. *The substitution σ satisfies Property (i).*

Proof. Property (i) trivially holds for any nonerasing substitution.   □

Before showing that the substitution σ satisfies Property (ii), we show that the period of an abelian power with large enough exponent is divisible by N, the length of the substitution σ.

Lemma 3. *Let $\mathbf{w}$ be an infinite binary word. If an abelian power $u_0 \cdots u_{e-1}$, with $e \geq N$, occurs in $\sigma(\mathbf{w})$, then N divides $|u_0|$.*

Proof. Let $m = |u_0|$, and write $m = tN + r$ for some $t \geq 0$ and $0 \leq r < N$. The claim is thus that $r = 0$. Assume, for a contradiction, that $r > 0$. Observe that for $\sigma(\mathbf{w}) = a_0 a_1 \cdots$, where $a_n \in \{0,1,\#\}$ for each $n \geq 0$, we have $a_n = \#$ if and only if $n \equiv 0 \pmod{N}$. Let us denote the position of the occurrence of u_j in $\sigma(\mathbf{w})$ by i_j, that is,

$$u_j = a_{i_j} a_{i_j+1} \cdots a_{i_j+m-1}.$$

Observe that $i_j = i_0 + jm$, and $i_j \equiv i_0 + jr \pmod{N}$ for each $j = 0, \ldots, e - 1$. Notice also that the number of occurrences of the letter $\#$ in u_j equals the number of indices k in the set $\{i_j, i_j + 1, \ldots, i_j + m - 1\}$ for which $k \equiv 0 \pmod{N}$. Let $n_j = i_j \bmod N$. If $n_j = 0$, then we may compute the value $|u_j|_\#$ as follows:

$$|u_j|_\# = \left\lceil \frac{m}{N} \right\rceil = \left\lceil \frac{tN + r}{N} \right\rceil = t + \left\lceil \frac{r}{N} \right\rceil = t + 1$$

since $0 < r < N$ by assumption. If $n_j > 0$, then none of the first $N - n_j$ letters of u_j equals $\#$. The value $|u_j|_\#$ is thus computed as follows:

$$|u_j|_\# = \left\lceil \frac{m - (N - n_j)}{N} \right\rceil = \left\lceil \frac{tN + r - (N - n_j)}{N} \right\rceil = t - 1 + \left\lceil \frac{r + n_j}{N} \right\rceil.$$

We conclude that $|u_j|_\# = t + 1$ if and only if $n_j = 0$ or $n_j > N - r$, and otherwise $|u_j|_\# = t$.

We exhibit two words u_{j_1} and u_{j_2} from the abelian power for which the number of occurrences of the letter $\#$ differ. This contradiction proves our claim. Since $e \geq N$, we see that the numbers n_j, $n_j \equiv n_0 + jr \pmod{N}$, $j = 0, \ldots, e-1$, form the coset $n_0 + \langle r \rangle$ of the subgroup $\langle r \rangle$ of $\mathbb{Z}/N\mathbb{Z}$. Let now $d = \gcd(r, N)$, so that $\langle r \rangle = \{0, d, 2d, \ldots (N/d - 1)d\}$. For example, if $\gcd(r, N) = 1$, then $\langle r \rangle = \mathbb{Z}/N\mathbb{Z}$. There thus exists an index j_1 such that the letter $\#$ occurs among the first d letters of u_{j_1}. This means that either $n_{j_1} = 0$ or

$$n_{j_1} > N - d \geq N - r.$$

Thus $|u_{j_1}|_\# = t + 1$ as was concluded previously. Similarly, there exists an index j_2 such that the letter $\#$ occurs among the d letters immediately preceding u_{j_2}. This means that

$$0 < n_{j_2} \leq d.$$

In this case

$$n_{j_2} + r \leq d + r \leq d + N - d = N$$

since $r \leq N - \gcd(r, N) = N - d$. We thus have $n_{j_2} \leq N - r$ implying that $|u_{j_2}|_\# = t$ as was concluded previously. This concludes the proof. $\square$

Remark 4. The above result may be slightly generalized. Indeed, notice that the only structural properties of σ used in the above proof are that σ is uniform, the images of the letters begin with $\#$, and the images of the letters contain no other occurrences of $\#$. In fact, the property that both images of letters begin with $\#$ is not important, it is only required that $\#$ occurs at the same position in both $\sigma(0)$ and $\sigma(1)$. We are thus led to the following generalization of Lemma 3. Let $\varphi \colon \{0,1\}^* \to \{0,1,\#\}^*$ be a uniform substitution defined by $\varphi(0) = u\#v$, $\varphi(1) = u'\#v'$, where $u, u', v, v' \in \{0,1\}^*$, $|u| = |u'|$, and $|v| = |v'|$. Let $\mathbf{w}$ be a binary word. If an abelian power $u_0 \cdots u_{e-1}$, $e \geq |u\#v|$, occurs in $\varphi(\mathbf{w})$, then $|u\#v|$ divides $|u_0|$. We shall need this generalization later in Sect. 3.

Lemma 5. *The substitution σ satisfies Property (ii).*

Proof. Let $u_0 \cdots u_{e-1}$, $e \geq N$, be an abelian power occurring in $\sigma(\mathbf{w})$. It follows by Lemma 3 that N divides the length of u_0. Our aim is to show that the abelian power $u_0 \cdots u_{e-1}$ can be shifted (to the left or the right) to obtain another abelian power $u_0' \cdots u_{e-1}'$ with $|u_0'| = |u_0|$ such that each u_i' begins with the letter #. Before doing so, let us show how the main claim follows from this. Because σ is injective, as is readily verified, there exist unique factors $v_0, \ldots, v_{e-1}$ of $\mathbf{w}$ of length $|u_0|/N$ such that $\sigma(v_i) = u_i'$ for $i = 0, \ldots, e-1$. Notice that $v_0 \cdots v_{e-1}$ is a factor of $\mathbf{w}$. Clearly the words v_i are abelian equivalent as $|v_i|_0 = |u_i'|_0/(N-1)$ and $|u_i'|_0 = |u_j'|_0$ for all j. We conclude that the word $v_0 \cdots v_{e-1}$ is an abelian power in $\mathbf{w}$.

Let us again write $\sigma(\mathbf{w}) = a_0 a_1 \cdots$ with $a_n \in \{0, 1, \#\}$ for each $n \geq 0$. Let u_0 have the position i in $\sigma(\mathbf{w})$, and let $n = i \mod N$. If $n = 0$ then we are done since we may choose $u_i' = u_i$ in the above (recall that N divides $|u_0|$). Also, if $n = 1$, each word u_j, $j = 0, \ldots, e-1$, is immediately preceded by # in $\sigma(\mathbf{w})$ and, moreover, each of the words ends with #. By setting $u_j' = \#u_j\#^{-1}$, we see that $\#u_0 \cdots u_{e-1} = u_0' \cdots u_{e-1}'\#$ occurs in $\sigma(\mathbf{w})$, and clearly $u_j' \sim u_0$ for each $j = 0, \ldots, e-1$. Thus $u_0' \cdots u_{e-1}'$ is an abelian power of the claimed form. Assume now that $n > 1$. Without loss of generality, we assume that u_0 begins with 0 so, in fact, u_0 begins with $0^{N-n}\#$. By the form of the substitution, u_0 is preceded by $\#0^{n-1}$ in $\sigma(\mathbf{w})$. We claim that each of the words u_j, $j = 0, \ldots, e-1$, begins with $0^{N-n}\#$ and ends with $\#0^{n-1}$. Let us first show that u_1 begins with $0^{N-n}\#$ (and thus that u_0 ends with $\#0^{n-1}$). Assume for a contradiction that u_1 begins with $1^{N-n}\#$ (whence u_0 ends with $\#1^{n-1}$), and say that u_1 ends with $\#c^{n-1}$ where $c \in \{0, 1\}$. Now the word $\#0^{n-1}u_0(\#1^{n-1})^{-1}$ is the image of a factor x of $\mathbf{w}$. Similarly, the word $\#1^{n-1}u_1(\#c^{n-1})^{-1}$ is the image of a factor y of $\mathbf{w}$ with $|x| = |y|$. We may write

$$|u_0|_1 = |x|_1(N-1) + n - 1$$

and

$$|u_1|_1 = |y|_1(N-1) - (n-1) + \delta_{c=1} \cdot (n-1),$$

where $\delta_{c=1} = 1$ if $c = 1$, and otherwise $\delta_{c=1} = 0$. Since $u_0 \sim u_1$, by rearranging the terms, we obtain

$$(|y|_1 - |x|_1)(N-1) = (2 - \delta_{c=1})(n-1).$$

Notice here that $1 \leq 2 - \delta_{c=1} \leq 2$ and that $n > 1$. The right side of the inequality is positive, so $|y|_1 - |x|_1 \geq 1$. Since $N > n$, it must be that $|y|_1 - |x|_1 < 2 - \delta_{c=1} \leq 2$. We conclude that $|y|_1 - |x|_1 = 1$ and, furthermore, $\delta_{c=1} = 0$. We now have

$$N - 1 = 2(n-1),$$

which is impossible since N was chosen to be even. This contradiction shows that u_1 begins with $0^{N-n}\#$ as well. A symmetric argument shows that u_1 ends with $\#0^{n-1}$. We may repeat the above argument to show that each of the words u_j, $j = 0, \ldots, e-1$, begins with $0^{N-n}\#$ and ends with $\#0^{n-1}$.

To finish off the proof, we choose $u'_j = \#0^{n-1}u_j(\#0^{n-1})^{-1}$ for each $j = 0, \ldots, e-1$. Observe that $\#0^{n-1}u_0 \cdots u_{e-1} = u'_0 \cdots u'_{e-1}\#0^{n-1}$ and that $u'_0 \sim u'_j$ for each $j = 0, \ldots, e-1$. We have thus exhibited an abelian power of the claimed form thus concluding the proof. $\square$

Since the substitution σ satisfies Properties (i)–(ii) and $\sigma(\mathbf{w})$ has at most three letters, Theorem 1 is proved.

3   Extension to the k-abelian Setting

In this section, we consider a generalization of abelian equivalence. Let k be a positive integer. Two words u and v are k-abelian equivalent, written $u \sim_k v$, if $|u|_w = |v|_w$ for all nonempty words w of length at most k [9]. For words of length at least $k - 1$, we can equivalently say that $u \sim_k v$ if and only if u and v share a common prefix and a common suffix of length $k - 1$ and $|u|_w = |v|_w$ for each word w of length k [9, Lemma 2.4]. The k-abelian equivalence relation is a congruence relation. Notice that 1-abelian equivalence is simply abelian equivalence. Moreover, if $u \sim_{k+1} v$, then $u \sim_k v$.

A nonempty word $u_0 \cdots u_{e-1}$ is a k-abelian power of exponent e and period m if $|u_0| = m$ and $u_0 \sim_k \cdots \sim_k u_{e-1}$. It was proved in [9, Theorem 5.4] using Szemerédi's theorem that every infinite word having bounded k-abelian complexity contains k-abelian powers of arbitrarily high exponent. Sturmian words are particular examples of such words, so each Sturmian word contains k-abelian powers of arbitrarily high exponent; an alternative proof of this fact is given in [15, Lemma 3.10]

Let $\mathbf{w}$ be an infinite word. Then we set $\mathcal{A}e_{k,\mathbf{w}}(m)$ to be the supremum of the exponents of k-abelian powers of period m occurring in $\mathbf{w}$. We define the k-abelian critical exponent of $\mathbf{w}$ to be the quantity

$$\limsup_{m \to \infty} \frac{\mathcal{A}e_{k,\mathbf{w}}(m)}{m},$$

and we denote it by $\mathcal{A}c_k(\mathbf{w})$. This generalization of the abelian critical exponent is considered in the preprint [15], where the authors of this paper study the set of finite k-abelian critical exponents of Sturmian words. This set, dubbed as the k-Lagrange spectrum, is similarly complicated as the Lagrange spectrum. When $k > 1$, the least accumulation point of the k-Lagrange spectrum is $\sqrt{5}/(2k - 1)$, and the spectrum is dense in the interval $(\sqrt{5}/(2k - 1), \infty)$.

Next we prove the following analogue of Theorem 1.

Theorem 6. *Let θ be a nonnegative real number. Then there exists an infinite word $\mathbf{w}$ such that $\mathcal{A}c_k(\mathbf{w}) = \theta$. The word $\mathbf{w}$ can be taken over an alphabet of at most three letters.*

Similar to Sect. 2, we wish to find a substitution f defined on a two-letter alphabet with the following properties:

(i') If an abelian power $u_0 \cdots u_{e-1}$ occurs in $\mathbf{w}$, then $f(u_0) \cdots f(u_{e-1})$ is a k-abelian power occurring in $f(\mathbf{w})$.

(ii') If a k-abelian power $u_0 \cdots u_{e-1}$, $e \geq N$, occurs in $f(\mathbf{w})$, then $\mathbf{w}$ contains an abelian power $v_0 \cdots v_{e-1}$ with $|v_0| = |u_0|/N$.

Given such a substitution f, Theorem 6 is proved exactly as Theorem 1 was proved in Sect. 2. The case $k = 1$ is handled by Theorem 1, so we may assume that $k > 1$.

Let $N \geq 2k - 1$ be a fixed integer, and define the N-uniform substitution $\tau \colon \{0,1\}^* \to \{0,1,\#\}^*$ by

$$0 \mapsto \#0^{k-2}0^{N-2k+2}0^{k-1},$$
$$1 \mapsto \#0^{k-2}1^{N-2k+2}0^{k-1}.$$

Let u and v be two words of length greater than $2k - 2$. Suppose that $\mathrm{pref}_{k-1}(u) = \mathrm{pref}_{k-1}(v)$ and $\mathrm{suff}_{k-1}(u) = \mathrm{suff}_{k-1}(v)$, that is, assume that they share a common prefix of length $k - 1$ and a common suffix of length $k - 1$. One easily checks that then $uv \sim_k vu$. Remark then that it follows that $xuvy \sim_k xvuy$ for all words x and y because $\sim_k$ is a congruence.

Lemma 7. *The substitution τ satisfies Property (i').*

Proof. By the form of the substitution τ, we have $\mathrm{pref}_{k-1}(\tau(0)) = \mathrm{pref}_{k-1}(\tau(1))$ and $\mathrm{suff}_{k-1}(\tau(0)) = \mathrm{suff}_{k-1}(\tau(1))$. Therefore $\tau(0)\tau(1) \sim_k \tau(1)\tau(0)$, and hence $\tau(u_i) \sim_k \tau(0)^{|u_i|_0}\tau(1)^{|u_i|_1}$ for $i = 0, \ldots, e-1$. Let $u_0 \cdots u_{e-1}$ be an abelian power in $\mathbf{w}$. Since the words $u_0, \ldots, u_{e-1}$ are abelian equivalent, we have

$$\tau(u_0), \ldots, \tau(u_{e-1}) \sim_k \tau(0)^{|u_0|_0}\tau(1)^{|u_0|_1},$$

so $\tau(u_0) \sim_k \cdots \sim_k \tau(u_{e-1})$. □

Lemma 8. *If $u_0 \cdots u_{e-1}$, $e \geq N$, is a k-abelian power occurring in $\tau(\mathbf{w})$, then N divides $|u_0|$.*

Proof. Since $u_0 \cdots u_{e-1}$ is a k-abelian power, it is an abelian power. Observe now that the substitution τ is as in Remark 4. Thus N divides $|u_0|$. □

Lemma 9. *The substitution τ satisfies Property (ii').*

Proof. Suppose that a k-abelian power $u_0 \cdots u_{e-1}$ with $e \geq N$ occurs in $\tau(\mathbf{s})$. By Lemma 8, N divides $|u_0|$. Similar to the proof of Lemma 5, we want to show that the k-abelian power $u_0 \cdots u_{e-1}$ can be shifted (to the left or the right) to obtain another k-abelian power $u_0' \cdots u_{e-1}'$, $|u_0'| = |u_0|$, such that each u_i' begins with $\#$. Then a slight modification of the argument presented in the first paragraph of the proof of Lemma 5 proves the claim. Indeed, given the preimages $v_0, \ldots, v_{e-1}$ of $u_0', \ldots, u_{e-1}'$, we see that $|v_i|_0 = |u_i'|_{\#0^{k-1}}$ for all i. Since $u_0' \sim_k \cdots \sim_k u_{e-1}'$, we have $|u_i'|_{\#0^{k-1}} = |u_j'|_{\#0^{k-1}}$ for all i and j, and it follows that $v_0 \sim \cdots \sim v_{e-1}$.

Let p be the common prefix of length $k - 1$ of the words $u_0, \ldots, u_{e-1}$ and similarly q be the common suffix of length $k - 1$ of these words. Suppose first that $\#$ occurs in p, that is, $p = 0^r \# 0^s$ with $r + s = k - 2$. As each occurrence of $\#$ is preceded by 0^{k-1} and N divides $|u_i|$, the word u_{e-1} is followed by 0^r. Thus we may set $u_i' = (0^r)^{-1} u_i 0^r$ for $i = 0, \ldots, e - 1$. The same r factors $0^r \# 0^{s+1}$, $0^{r-1} \# 0^{s+2}, \ldots, 0 \# 0^{s+r}$ of length k were removed from each u_i and the same r factors of length k were added to each u_i' (the final $k - 1$ factors of $q0^r$ of length k) during the shift. Thus $u_0' \sim_k \cdots \sim_k u_{e-1}'$. If $\#$ occurs in q, that is, say $q = 0^r \# 0^s$ with $r + s = k - 2$ then, like above, we may set $u_i' = \# 0^s u_i (\# 0^s)^{-1}$ for $i = 0, \ldots, e - 1$. Suppose then that some word u_i begins with $0^{k-1} \#$. It is straightforward to see that then all of the words $u_0, \ldots, u_{e-1}$ begin with $0^{k-1} \#$ and, furthermore, that u_{e-1} is followed by $0^{k-1} \#$. Setting $u_i' = (0^{k-1})^{-1} u_i 0^{k-1}$ for $i = 0, \ldots, e - 1$ gives the claim as above.

By the preceding paragraph, we may assume that the occurrence of p as the prefix of u_i is a proper factor of $\tau(c_i)$ for a letter c_i, that is, we may write $\tau(c_i) = x_i p y_i$, with x_i and y_i nonempty, for this occurrence of p. Moreover, the preceding paragraph tells that we may assume that q is a proper suffix of x_i (otherwise $\#$ occurs in q). Since p has length $k - 1$, it is clear from the form of the substitution τ that the letter c_i is uniquely determined by p. Since N divides $|u_i|$, it follows that $c_0 = \ldots = c_{e-1}$. This means that each u_i is preceded by x_0. We still need to know that u_{e-1} ends with x_0; the words $u_0, \ldots, u_{e-2}$ must end with x_0. Since N divides $|u_i|$, the suffix q of u_{e-1} occurs in $\tau(d)$, $d \in \{0, 1\}$, in the same position as the occurrence of q preceding $p y_0$ in $\tau(c_0)$. Now q has length $k - 1$, so its occurrence preceding $p y_i$ in $\tau(c_i)$ uniquely determines c_i, and hence its occurrence in $\tau(d)$ in the same position uniquely determines d. Therefore $d = c_0$ and u_{e-1} ends with x_0. We may now set $u_i' = x_0 u_i x_0^{-1}$ for $i = 0, \ldots, e - 1$. The suffix x_0 of u_i is preceded by 0^{k-1} and u_i has prefix p of length $k - 1$, so exactly the same factors of length $k - 1$ are added and removed when shifting each u_i to u_i'. Thus $u_0' \sim_k \cdots \sim_k u_{e-1}'$. $\square$

Since τ satisfies Properties (i') and (ii'), Theorem 6 follows.

4   Concluding Remarks

Theorem 1 raises the following question.

Question 10. Given a nonnegative real number θ, does there exist an infinite binary word having k-abelian critical exponent θ?

We conjecture that the question has a positive answer. To use the presented method, the marker letter $\#$ needs to be replaced by a suitable binary word ensuring that Properties (ii) and (ii') hold. There seems to be no obvious choice, at least no obvious choice leading to reasonable proofs. Perhaps another method is required. It would certainly be very interesting if the answer to the above question turned out to be negative. Nevertheless, we leave the question open.

The k-abelian equivalence is a refinement of abelian equivalence that "tends" to the usual equality of words as $k \to \infty$. As mentioned in the introduction, it is typical to consider the maximum exponent $\sup_{m \geq 1} \exp(m)$ for the equality relation, not the superior limit of the ratio between the maximum exponent $\exp(m)$ and period m as is done here for abelian equivalence and k-abelian equivalence. What then happens if we consider the unorthodox notion? Does an analogue to Theorem 1 hold? The answer is yes. The following result is proved by the authors in the preprint [15].

Proposition 11 [15, Proposition 3.17]. *Given an infinite word* **w**, *let* $E(\mathbf{w})$ *be the quantity*

$$\limsup_{m \to \infty} \frac{\exp(m)}{m},$$

where $\exp(m)$ *is the supremum of (integral) exponents of powers of period* m *occurring in* **w**. *For each nonnegative* θ, *there exists a Sturmian word* **s** *such that* $E(\mathbf{s}) = \theta$.

References

1. Aigner, M.: Markov's Theorem and 100 Years of the Uniqueness Conjecture. Springer, Cham (2013). https://doi.org/10.1007/978-3-319-00888-2
2. Berthé, V., Rigo, M. (eds.): Combinatorics, Words and Symbolic Dynamics. Encyclopedia of Mathematics and Its Applications, vol. 159. Cambridge University Press, Cambridge (2016)
3. Cassaigne, J., Karhumäki, J., Saarela, A.: On growth and fluctuation of k-abelian complexity. Eur. J. Comb. **65**, 92–105 (2017). https://doi.org/10.1016/j.ejc.2017.05.006
4. Cusick, T.W., Flahive, M.E.: The Markoff and Lagrange Spectra. Mathematical Surveys and Monographs, vol. 30. American Mathematical Society, Providence (1989)
5. Dekking, F.M.: Strongly non-repetitive sequences and progression-free sets. J. Comb. Theory Ser. A **27**(2), 181–185 (1979). https://doi.org/10.1016/0097-3165(79)90044-X
6. Fici, G., et al.: Abelian powers and repetitions in Sturmian words. Theoret. Comput. Sci. **635**, 16–34 (2016). https://doi.org/10.1016/j.tcs.2016.04.039
7. Freiman, G.A.: Diophantine approximation and geometry of numbers (Markov's problem). Kalininskii Gosudarstvennyi Universitet, Kalinin (1975). (Russian)
8. Hall Jr., M.: On the sum and products of continued fractions. Ann. of Math. **48**(4), 966–993 (1947). https://doi.org/10.2307/1969389
9. Karhumäki, J., Saarela, A., Zamboni, L.Q.: On a generalization of Abelian equivalence and complexity of infinite words. J. Combin. Theory Ser. A **120**, 2189–2206 (2013). https://doi.org/10.1016/j.jcta.2013.08.008
10. Krieger, D., Shallit, J.: Every real number greater than 1 is a critical exponent. Theoret. Comput. Sci. **381**, 177–182 (2007). https://doi.org/10.1016/j.tcs.2007.04.037
11. Lothaire, M.: Algebraic Combinatorics on Words. Encyclopedia of Mathematics and Its Applications, vol. 90. Cambridge University Press, Cambridge (2002)
12. Markov, A.A.: Sur les formes quadratiques binaires indéfinies. Math. Ann. **15**(3–4), 381–406 (1879). https://doi.org/10.1007/BF02086269
13. Markov, A.A.: Sur les formes quadratiques binaires indéfinies ii. Math. Ann. **17**(3), 379–399 (1880). https://doi.org/10.1007/BF01446234

14. Peltomäki, J.: Privileged words and Sturmian words. Ph.D. dissertation, Turku Centre for Computer Science, University of Turku, Turku (2016). http://urn.fi/URN:ISBN:978-952-12-3422-4

15. Peltomäki, J., Whiteland, M.A.: On k-abelian equivalence and generalized Lagrange spectra (2018). arXiv:1809.09047 (under review)

16. Rao, M., Rigo, M., Salimov, P.: Avoiding 2-binomial squares and cubes. Theoret. Comput. Sci. **572**, 83–91 (2015). https://doi.org/10.1016/j.tcs.2015.01.029

17. Reutenauer, C.: From Christoffel Words to Markoff Numbers. Oxford University Press, Oxford (2019)

18. Richomme, G., Saari, K., Zamboni, L.Q.: Abelian complexity of minimal subshifts. J. Lond. Math. Soc. **83**(1), 79–95 (2011). https://doi.org/10.1112/jlms/jdq063

19. Rigo, M., Salimov, P.: Another generalization of abelian equivalence: binomial complexity of infinite words. Theoret. Comput. Sci. **601**, 47–57 (2015). https://doi.org/10.1016/j.tcs.2015.07.025

Rich Words Containing
Two Given Factors

Josef Rukavicka[✉]

Department of Mathematics,
Faculty of Nuclear Sciences and Physical Engineering,
Czech Technical University in Prague,
Trojanova 13, 120 01 Prague 2, Czech Republic
josef.rukavicka@seznam.cz

Abstract. A finite word w with $|w| = n$ contains at most $n+1$ distinct palindromic factors. If the bound $n+1$ is attained, the word w is called *rich*. Let $\mathrm{F}(w)$ be the set of factors of the word w. It is known that there are pairs of rich words that cannot be factors of a same rich word. However it is an open question how to decide for a given pair of rich words u, v if there is a rich word w such that $\{u, v\} \subseteq \mathrm{F}(w)$. We present a response to this open question:

If w_1, w_2, w are rich words, $m = \max\{|w_1|, |w_2|\}$, and $\{w_1, w_2\} \subseteq \mathrm{F}(w)$ then there exists also a rich word $\bar{w}$ such that $\{w_1, w_2\} \subseteq \mathrm{F}(\bar{w})$ and $|\bar{w}| \leq m2^{k(m)+2}$, where $k(m) = (q+1)m^2(4q^{10}m)^{\log_2 m}$ and q is the size of the alphabet. Hence it is enough to check all rich words of length equal or lower to $m2^{k(m)+2}$ in order to decide if there is a rich word containing factors w_1, w_2.

1   Introduction

In the last years there have appeared several articles dealing with rich words; see, for instance, [1–3,5]. Recall that a palindrome is a word that reads the same forwards and backwards, for example "noon" and "level". If a word w of length n contains $n + 1$ distinct palindromic factors then the word w is called *rich*. It is known that a word of length n can contain at most $n + 1$ palindromic factors including the empty word. The notion of a rich word has been extended also to infinite words. An infinite word is called rich if its every finite factor is rich [3,4].

Let $\mathrm{lps}(w)$ and $\mathrm{lpp}(w)$ denote the longest palindromic suffix and the longest palindromic prefix of a word w, respectively. The authors of [1] showed the following property of rich words:

Proposition 1. *If r, t are two factors of a rich word w such that $\mathrm{lps}(r) = \mathrm{lps}(t)$ and $\mathrm{lpp}(r) = \mathrm{lpp}(t)$, then $r = t$.*

Two related open questions can be found:

- In [5]: Is the condition in Proposition 1 sufficient for two rich words u and v to be factors of the same rich word?

© Springer Nature Switzerland AG 2019
R. Mercaş and D. Reidenbach (Eds.): WORDS 2019, LNCS 11682, pp. 286–298, 2019.
https://doi.org/10.1007/978-3-030-28796-2_23

– In [3]: We do not know how to decide whether two rich words u and v are factors of a same rich word w.

In the current article we present a response to the question from [3] in the following form: We prove that if w_1, w_2, w are rich words, $m = \max\{|w_1|, |w_2|\}$, and $\{w_1, w_2\} \subseteq F(w)$ then there exists a rich word $\bar{w}$ such that $\{w_1, w_2\} \subseteq F(\bar{w})$ and $|\bar{w}| \leq m2^{k(m)+2}$, where $k(m) = (q+1)m^2(4q^{10}m)^{\log_2 m}$ and q is the size of the alphabet. Thus it is enough to check all rich words of length equal or lower to $m2^{k(m)+2}$ in order to decide if there is a rich word containing factors w_1, w_2. However it is a rather theoretic way how to check the existence of such a word, since the number of words needed to be checked grows "pretty rapidly" with the length of the factors in question.

We describe the basic ideas of the proof. If w is a rich word, then let a be a letter such that $\text{lps}(wa) = a \, \text{lpps}(w)a$, where lpps denotes the longest proper palindromic suffix. It is known and easy to show that wa is a rich word [5, Proof of Theorem 2.1]. Thus every rich word w can be richly extended to a word wa. We will call wa a *standard extension* of w. If there is a letter b such that $a \neq b$ and wb is also a rich word, then we call the longest palindromic suffix of wb a *flexed palindrome*; the explication of the terminology is that wb is not a standard extension of w, hence wb is "flexed" from the standard extension. We define a set Γ of pairs of rich words (w, r), where r is a flexed palindrome of w, the longest palindromic prefix of w does not contain the factor r, and $|r| \geq |\bar{r}|$ for each flexed palindrome $\bar{r}$ of w. If $(w, r) \in \Gamma$, w_1 is the prefix of w with $|w_1| = |r| - 1$ and w_2 is the suffix of w with $|w_2| = |r| - 1$ then we construct a rich word $\bar{w}$ possessing the following properties:

– The word w_1 is a prefix of $\bar{w}$ and the word w_2 is a suffix of $\bar{w}$.
– The number of occurrences of r in $\bar{w}$ is strictly smaller than the number of occurrences of r in w.
– The set of flexed palindromes of $\bar{w}$ is a subset of the set of flexed palindromes of w.

Iterative applying of this construction will allow us for a given rich word w with a prefix w_1 and a suffix w_2 to construct a rich word t containing factors w_1, w_2 and having no flexed palindrome longer than m, where $m = \max\{|w_1|, |w_2|\}$.

Another important, but simple, observation is that if w is a rich word with prefix u such that the number of flexed palindromes in w is less than k and u has exactly one occurrence in w then there is an upper bound for the length of w. We show this upper bound as a function of k and consequently we derive an upper bound for the length of t.

2 Preliminaries

Let A be a finite alphabet with $q = |A|$. The elements of A will be called letters. Let ϵ denote the empty word.
Let A^* be the set of all finite words over A including the empty word and let $A^n \subset A^*$ be the set of all words of length n.

Let $R \subset A^*$ denote the set of all rich words.

Let $F(w) \subset A^*$ denote the set of all factors of $w \in A^*$; we state explicitly that $\epsilon, w \in F(w)$. Let $F(S) = \bigcup_{v \in S} F(v)$, where $S \subseteq A^*$.

Let $F_p(w) \subseteq F(w)$ be set of all palindromic factors of $w \in A^*$.

Let $\mathrm{Prf}(w)$ and $\mathrm{Suf}(w)$ be the set of all prefixes and all suffixes of $w \in A^*$ respectively; we define that $\{\epsilon, w\} \subseteq \mathrm{Prf}(w) \cap \mathrm{Suf}(w)$.

Let w^R denote the reversal of $w \in A^*$; formally if $w = w_1 w_2 \ldots w_k$ then $w^R = w_k \ldots w_2 w_1$, where $w_i \in A$ and $i \in \{1, 2, \ldots, k\}$. In addition we define that $\epsilon^R = \epsilon$.

Let $\mathrm{lps}(w)$ and $\mathrm{lpp}(w)$ denote the longest palindromic suffix and the longest palindromic prefix of $w \in A^*$ respectively. We define that $\mathrm{lps}(\epsilon) = \mathrm{lpp}(\epsilon) = \epsilon$.

Let $\mathrm{lpps}(w)$ and $\mathrm{lppp}(w)$ denote the longest proper palindromic suffix and the longest proper palindromic prefix of $w \in A^*$ respectively, where $|w| \geq 2$.

Let $\mathrm{trim}(w) = v$, where $v, w \in A^*$, $x, y \in A$, $w = xvy$, and $|w| \geq 2$.

Let $\mathrm{rtrim}(w) = v$, where $v, w \in A^*$, $y \in A$, $w = vy$, and $|w| \geq 1$.

Let $\mathrm{ltrim}(w) = v$, where $v, w \in A^*$, $x \in A$, $w = xv$, and $|w| \geq 1$.

Example 2. If $A = \{1, 2, 3, 4, 5\}$ and $w = 124135$, then $\mathrm{trim}(w) = 2413$, $\mathrm{ltrim}(w) = 24135$, and $\mathrm{rtrim}(w) = 12413$.

Let $\mathrm{pc}(w)$ be the palindromic closure of $w \in A^*$; formally $\mathrm{pc}(w) = uvu^R$, where $w = uv$ and $v = \mathrm{lps}(w)$. Note that $\mathrm{pc}(w)$ is a palindrome.

Let $\mathrm{MinLenWord}(U)$ and $\mathrm{MaxLenWord}(U)$ be the shortest and the longest word from the set U respectively, where either $U \subseteq \mathrm{Prf}(w)$ or $U \subseteq \mathrm{Suf}(w)$ for some $w \in A^*$. If $U = \emptyset$ then we define $\mathrm{MinLenWord}(U) = \epsilon$ and $\mathrm{MaxLenWord}(U) = \epsilon$.

Let $\mathrm{lcp}(w_1, w_2)$ be the longest common prefix of words $w_1, w_2 \in A^*$; formally $\mathrm{lcp}(w_1, w_2) = \mathrm{MaxLenWord}(\mathrm{Prf}(w_1) \cap \mathrm{Prf}(w_2))$.

Let $\mathrm{lcs}(w_1, w_2)$ be the longest common suffix of words $w_1, w_2 \in A^*$; formally $\mathrm{lcs}(w_1, w_2) = \mathrm{MaxLenWord}(\mathrm{Suf}(w_1) \cap \mathrm{Suf}(w_2))$.

Let $\mathrm{occur}(u, v)$ be the number of occurrences of v in u, where $u, v \in A^*$ and $|v| > 0$; formally $\mathrm{occur}(u, v) = |\{w \mid w \in \mathrm{Suf}(u) \text{ and } v \in \mathrm{Prf}(w)\}|$. We call a factor v *unioccurrent* in u if $\mathrm{occur}(u, v) = 1$.

Recall the notion of a *complete return* [2]: Given a word w and factors $r, u \in F(w)$, we call the factor r a complete return to u in w if r contains exactly two occurrences of u, one as a prefix and one as a suffix.

We list some known properties of rich words that we use in our article. All of them can be found, for instance, in [2].

Proposition 3. *If $w, u \in R$, $|w| \geq 1$, $|u| \geq 1$, and $u \in F_p(w)$ then all complete returns to u in w are palindromes.*

Proposition 4. *If $w \in R$ and $p \in F(w)$ then $p, p^R \in R$.*

Proposition 5. *A word w is rich if and only if every prefix $p \in \mathrm{Prf}(w)$ has a unioccurrent palindromic suffix.*

3   Standard Extensions and Flexed Palindromes

We start with a formal definition of a standard extension and a flexed palindrome introduced at the beginning of the article.

Definition 6. *Let $j \geq 0$ be a nonnegative integer, $w \in R$, and $|w| \geq 2$. We define* $\mathrm{StdExt}(w, j)$ *as follows:*

- $\mathrm{StdExt}(w, 0) = w$.
- $\mathrm{StdExt}(w, 1) = wa$ *such that* $\mathrm{lps}(wa) = a\,\mathrm{lpps}(w)a$ *and* $a \in A$.
- $\mathrm{StdExt}(w, j) = \mathrm{StdExt}(\mathrm{StdExt}(w, j-1), 1)$, *where* $j > 1$.

Let $\mathrm{StdExt}(w) = \{\mathrm{StdExt}(w, j) \mid j \geq 0\}$. *If* $p \in \mathrm{StdExt}(w)$ *then we call* p *a standard extension of* w.

Let $\mathrm{T}(w) = \{\mathrm{lps}(ub) \mid ub \in \mathrm{Prf}(w)$ *and* $b \in A$ *and* $ub \neq \mathrm{StdExt}(u, 1)\}$. *If* $r \in \mathrm{T}(w)$ *then we call* r *a flexed palindrome of* w.

For a given rich word $w \in R$ having a flexed palindrome r we define a *standard palindromic replacement* of r to be the longest palindromic suffix of a standard extension of a prefix p of w such that $\mathrm{lps}(px) = r$, where px is a prefix of w and $x \in A$. The idea is that we can "replace" r with the standard palindromic replacement.

Definition 7. *Let* $\mathrm{stdPalRep}(w, r) = \mathrm{lps}(\mathrm{StdExt}(p, 1))$, *where* $w, r \in R$, $r \in \mathrm{T}(w)$, $px \in \mathrm{Prf}(w)$, $x \in A$, *and* $\mathrm{lps}(px) = r$.
We call $\mathrm{stdPalRep}(w, r)$ *a standard palindromic replacement of* r *in* w.

Example 8. If $A = \{0, 1\}$ and $w = 110101100110011$ then $001100 \in \mathrm{T}(w)$, $\mathrm{lps}(1101011001100) = 001100$, $\mathrm{StdExt}(110101100110, 1) = 1101011001101$, and $\mathrm{stdPalRep}(w, 001100) = \mathrm{lps}(1101011001101) = 1011001101$.

We show that the length of a flexed palindrome r is less than the length of the standard palindromic replacement $\mathrm{stdPalRep}(w, r)$.

Lemma 9. *If* $ux, uy \in R$, $x, y \in A$, $x \neq y$, *and* $ux = \mathrm{StdExt}(u, 1)$ *then* $|\mathrm{lps}(ux)| > |\mathrm{lps}(uy)|$.

Proof. Let $yty = \mathrm{lps}(uy)$. From the definition of a standard extension we have $\mathrm{lps}(ux) = xvx$, where $v = \mathrm{lpps}(u)$ and hence $t \in \mathrm{Suf}(v)$. Since $y \neq x$ we have also $yt \in \mathrm{Suf}(v)$. The lemma follows.

An obvious corollary is that a flexed palindrome of w is not a prefix of w.

Corollary 10. *If* $w, r \in R$ *and* $r \in \mathrm{T}(w)$ *then* $r \notin \mathrm{Prf}(w)$.

In [5] the standard extension has been used to prove that each rich word w can be extended "richly"; this means that there is $a \in A$ such that wa is rich.

Lemma 11. *If* $w \in R$ *and* $|w| \geq 2$ *then* $\mathrm{StdExt}(w) \subset R$.

Proof. Obviously it is enough to prove that $\mathrm{StdExt}(w,1) \in \mathrm{R}$, since for every $t \in \mathrm{StdExt}(w) \setminus \{w\}$ there is a rich word $\bar{t}$ such that $t = \mathrm{StdExt}(\bar{t},1)$.

Let $xpx = \mathrm{lps}(\mathrm{StdExt}(w,1))$, where $x \in \mathrm{A}$. Proposition 5 implies that we need to prove that xpx is unioccurrent in $\mathrm{StdExt}(w,1)$. Realize that p is unioccurrent in w, hence xpx is unioccurrent in $\mathrm{StdExt}(w,1)$.

To simplify the proofs of the paper we introduce a function $\mathrm{MaxStdExt}(u,v)$ to be the longest prefix z of u such that z is also a standard extension of v:

Definition 12. *Let* $\mathrm{MaxStdExt}(u,v) = \mathrm{MaxLenWord}(\{\mathrm{StdExt}(v) \cap \mathrm{Prf}(u)\})$, *where* $u \in \mathrm{R}$ *and* $v \in \mathrm{Prf}(u)$. *We call* $\mathrm{MaxStdExt}(u,v)$ *a maximal standard extension of* v *in* u.

The next lemma shows that if a rich word contains factors ypx and ypy, where p is a palindrome, p is not a prefix of w, x,y are distinct letters, and ypx "occurs" before ypy in w then ypy is a flexed palindrome.

Lemma 13. *If* $w,v,p \in \mathrm{R}$, $v \in \mathrm{Prf}(w)$, $p \notin \mathrm{Prf}(w)$, $x,y \in \mathrm{A}$, $x \neq y$, $ypx \in \mathrm{Suf}(v)$, $ypy \notin \mathrm{F}(v)$, *and* $ypy \in \mathrm{F}(w)$ *then* $ypy \in \mathrm{T}(w)$.

Proof. Let $\bar{v}$ be such that $\bar{v}y \in \mathrm{Prf}(w)$, $ypy \in \mathrm{Suf}(\bar{v}y)$, and $\mathrm{occur}(\bar{v}y, ypy) = 1$. Let $u = \mathrm{lps}(\bar{v})$. Because $p \notin \mathrm{Prf}(w)$ it follows that $u = \mathrm{lpps}(\bar{v}) = \mathrm{lps}(\bar{v})$ and thus there is $z \in \mathrm{A}$ such that $zu \in \mathrm{Suf}(\bar{v})$. Obviously $v \in \mathrm{Prf}(\bar{v})$ and hence $\mathrm{occur}(\bar{v},p) > 1$. Then Proposition 5 implies that $\mathrm{occur}(u,p) > 1$. It follows that $yp \in \mathrm{Suf}(u) \cap \mathrm{Prf}(u)$, $z \neq y$, and Lemma 9 implies that $ypy \in \mathrm{T}(w)$. The word w with is its factors is depicted on Fig. 1. This completes the proof.

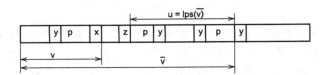

Fig. 1. Structure of the word w for Lemma 13.

4   Removing Flexed Points

We define formally the set Γ mentioned in the introduction. An element (w,r) of the set Γ represents a rich word w for which we are able to construct a new rich word $\bar{w}$ such that $\bar{w}$ does not contain the flexed palindrome r, but $\bar{w}$ have certain common prefixes and suffixes with w. We require that r is one of the longest flexed palindromes of w and that r is not a factor of the longest palindromic prefix of w. In addition we require that $|r| > 2$ so that the standard extension of $\mathrm{rtrim}(r)$ would be defined.

Definition 14. *Let Γ be a set defined as follows: $(w, r) \in \Gamma$ if*

1. $w, r \in \mathrm{R}$ and $|r| > 2$ and $r \in \mathrm{T}(w)$ and
2. $r \notin \mathrm{F}(\mathrm{lpp}(w))$ and
3. $|r| \geq |\bar{r}|$ for each $\bar{r} \in \mathrm{T}(w)$.

Given $(w, r) \in \Gamma$, we need to express w as a concatenation of its factors having some special properties. For this reason we define a function $\mathrm{parse}(w, r)$:

Definition 15. *If $(w, r) \in \Gamma$ then let $\mathrm{parse}(w, r) = (v, z, t)$, where*

- *$v, z, t \in \mathrm{R}$ and $vzt = w$ and*
- *$r \in \mathrm{Suf}(v)$ and $\mathrm{occur}(w, r) = \mathrm{occur}(v, r)$ and*
- *$vz = \mathrm{MaxStdExt}(vzt, v)$.*

Remark 16. The prefix v is the shortest prefix of w that contains all occurrences of r. The prefix vz is the maximal standard extension of v in w, and t is such that $vzt = w$. It is easy to see that v, z, t exist and are uniquely determined for $(w, r) \in \Gamma$.

The next simple lemma is necessary for the following definition of a *reduced prefix*.

Lemma 17. *Let $(w, r) \in \Gamma$, let $(v, z, t) = \mathrm{parse}(w, r)$, and let $\bar{v}$ be such that $v = \bar{v}\,\mathrm{lps}(v)$.*

- *If $\mathrm{occur}(\bar{v}r, r) > 1$ then there is a word $\bar{g}$ such that $\bar{g}rz \in \mathrm{Prf}(v)$ and $\mathrm{occur}(\bar{g}rz, r) < \mathrm{occur}(v, r)$*
- *If $\mathrm{occur}(\bar{v}r, r) = 1$ then $U \neq \emptyset$ and $r \notin \mathrm{F}(U)$, where $U = \{u \mid u \in \mathrm{Prf}(\mathrm{pc}(\bar{v}\,\mathrm{rtrim}(r))) \text{ and } \mathrm{ltrim}(r)z \in \mathrm{Suf}(u)\}$.*

Proof. It follows from Property 2 of Definition 14 that there is $h \in \mathrm{Prf}(w)$ such that $w = hz^R\,\mathrm{lps}(v)zt$. Note that $\mathrm{lps}(v) \neq v$ since $r \in \mathrm{T}(w)$ and thus $r \notin \mathrm{Prf}(w)$, see Corollary 10. It is clear that $r \in \mathrm{Prf}(\mathrm{lps}(v)) \cap \mathrm{Suf}(\mathrm{lps}(v))$. This implies that $hz^Rr \in \mathrm{Prf}(w)$. Note that $\bar{v} = hz^R$. We distinguish two cases as stated in the Lemma:

- $\mathrm{occur}(\bar{v}r, r) > 1$: Let g be the complete return to r in v such that $g \in \mathrm{Suf}(hz^Rr)$. Clearly $rz \in \mathrm{Prf}(g)$ and $z^Rr \in \mathrm{Suf}(g)$, since $r \notin \mathrm{F}(\mathrm{ltrim}(r)z)$; recall $r \in \mathrm{Suf}(v)$ and $\mathrm{occur}(v, r) = \mathrm{occur}(vzt, r)$. Let $\bar{g}$ be such that $\bar{g}g = hz^Rr$.
- If $\mathrm{occur}(\bar{v}r, r) = 1$: Let $\bar{u} = \mathrm{stdPalRep}(hz^Rr, r)$. Clearly $\mathrm{lps}(hz^Rr) = r$ and $\bar{u} \neq r$. Because $z^R\,\mathrm{rtrim}(r) \in \mathrm{Suf}(hz^R\,\mathrm{rtrim}(r))$, then obviously $U \neq \emptyset$ and $r \notin \mathrm{F}(U)$.

The word w with is its factors is depicted on Fig. 2. This completes the proof.

For an element $(w, r) \in \Gamma$ we define a function $\mathrm{rdcPrf}(w, r)$ (the reduced prefix), which is a prefix of the palindromic closure of some prefix of w. In Theorem 28 we show that the concatenation of $\mathrm{rdcPrf}(w, r)$ and t is a rich word having a strictly smaller number of occurrences of r than in w, where $(v, z, t) = \mathrm{parse}(w, r)$. This reducing of occurrences of r is the key for removing all "long" flexed palindromes as explained in the introduction.

Definition 18. *If $w, r \in \Gamma$ and $(v, z, t) = \mathrm{parse}(w, r)$ then let $\mathrm{rdcPrf}(w, r)$ be defined as follows. Following the notation and the proof of Lemma 17 we distinguish two cases:*

- *$\mathrm{occur}(\bar{v}r, r) > 1$: We define $\mathrm{rdcPrf}(w, r) = \bar{g}rz$.*
- *$\mathrm{occur}(\bar{v}r, r) = 1$: We define $\mathrm{rdcPrf}(w, r) = \mathrm{MinLenWord}(U)$.*

We call $\mathrm{rdcPrf}(w, r)$ the reduced prefix of w by r.

Figure 2 depicts the factors of the word w used for construction of the reduced prefix of w.

Remark 19. Note in Definition 18 in the second case, where $\mathrm{occur}(\bar{v}r, r) = 1$, it may happen that the reduced prefix $\mathrm{rdcPrf}(w, r)$ is not a prefix of w. However it is a prefix of a palindromic closure of $hz^R \mathrm{rtrim}(r)$, hence the number of flexed palindromes remains the same; formally $| \mathrm{T}(hz^R \mathrm{rtrim}(r))) | = | \mathrm{T}(\mathrm{rdcPrf}(w, r)) |$. Realize that $\mathrm{pc}(t) \in \mathrm{StdExt}(t)$ for each $t \in \mathrm{R}$ and $|t| \geq 2$.

In the first case, where $\mathrm{occur}(\bar{v}r, r) > 1$, the reduced prefix $\mathrm{rdcPrf}(w, r)$ is always a prefix of w.

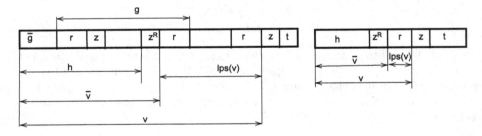

Fig. 2. Construction of the reduced prefix. Case 1 and 2.

To clarify the definition of the reduced prefix $\mathrm{rdcPrf}(w, r)$ we present below two examples representing those two cases in the definition. For both examples we consider that $\mathrm{A} = \{1, 2, 3, 4, 5, 6, 7, 8, 9\}$.

Example 20. If $w = 12399932239993244239993225522 3993$ and $r = 999$ then $v = 1239993223999324423999$, $z = 322$, $t = 55223993$, $\mathrm{lps}(v) = 999324423999$, $h = 1239993$, $w = hz^R \mathrm{lps}(v)zt$, $g = 9993223999 \in \mathrm{Suf}(hz^Rr) = \mathrm{Suf}(1239993223999)$, $\bar{g} = 123$, and $\mathrm{rdcPrf}(w, r) = 123999322$.

Example 21. If $w = 123999599932239949$ and $r = 999$ then $v = 1239995999$, $z = 32$, $t = 239949$, $\mathrm{lps}(v) = 9995999$, $h = 1$, $w = hz^R \mathrm{lps}(v)zt$, $\mathrm{StdExt}(hz^R \mathrm{rtrim}(r), 1) = \mathrm{StdExt}(12399, 1) = 123993$, $\bar{u} = \mathrm{stdPalRep}(123999, 999) = 3993$, $\mathrm{pc}(12399) = 12399321$, $U = \{1239932\}$, and $\mathrm{rdcPrf}(w, r) = 1239932$.

Using the reduced prefix we can now define the word $\mathrm{rdcWrd}(w, r)$ (a *reduced word*):

Definition 22. *Let* $\mathrm{rdcWrd}(w, r) = \mathrm{rdcPrf}(w, r)t$, *where* $(v, z, t) = \mathrm{parse}(w, r)$ *and* $(w, r) \in \Gamma$. *We call* $\mathrm{rdcWrd}(w, r)$ *the reduced word of* w *by* r.

We show that the reduced word $\mathrm{rdcWrd}(w, r)$ and w have the same prefix and suffix of length $|r| - 1$.

Lemma 23. *If* $(w, r) \in \Gamma$ *and* $u = \mathrm{rdcWrd}(w, r)$ *then* $|\mathrm{lcp}(u, w)| \geq |r| - 1$ *and* $|\mathrm{lcs}(u, w)| \geq |r| - 1$.

Proof. From the construction of the reduce prefix and the reduced word, it is easy to see that $\mathrm{rtrim}(r) \in F(\mathrm{lcp}(u, v))$ and $\mathrm{ltrim}(r) \in F(\mathrm{lcs}(u, v))$. The lemma follows.

As already mentioned the reduced prefix $\mathrm{rdcPrf}(w, r)$ is not necessarily a prefix of w. In such a case $\mathrm{rdcPrf}(w, r) \in \mathrm{Prf}(\mathrm{pc}(\bar{v}\,\mathrm{rtrim}(r)))$, see Definition 18. We show that every palindrome from the set $F(\mathrm{rdcPrf}(w, r)) \backslash F(\bar{v}\,\mathrm{rtrim}(r)))$ contains as a factor the standard palindromic replacement $\bar{u}$ of r in w and we show that $\bar{u}$ is not a factor of w. This will be important when proving richness of the word $\mathrm{rdcWrd}(w, r)$.

Let $F(w, r) = \{u \mid u \in F(w) \text{ and } r \notin F(u)\} \subseteq F(w)$, where $w, r \in A^*$. The set $F(w, r)$ contains factors of w that do not contain the factor r. Let $F_p(w, r) = F_p(w) \cap F(w, r)$.

Proposition 24. *If* $(w, r) \in \Gamma$, $(v, z, t) = \mathrm{parse}(w, r)$, $u = \mathrm{rdcPrf}(w, r)$, $\bar{u} = \mathrm{stdPalRep}(w, r)$, *and* $\bar{v}$ *is such that* $v = \bar{v}\,\mathrm{lps}(v)$ *then* $F_p(u, \bar{u}) \subseteq F_p(\bar{v}\,\mathrm{rtrim}(r))$ *and* $\bar{u} \notin F_p(w)$.

Proof. From the properties of the palindromic closure it is easy to see that $F_p(\mathrm{pc}(f), \mathrm{lps}(f)) \subseteq F_p(f)$ for each $f \in R$. It means that every palindromic factor of $\mathrm{pc}(f)$ that is not a factor of f contains the factor $\mathrm{lps}(f)$. It follows that $F_p(u, \bar{u}) \subseteq F_p(\mathrm{rtrim}(v))$.

We show that $\mathrm{occur}(w, \bar{u}) = 0$. Let $\bar{u} = xtx$ and $r = ypy$, where $x, y \in A$. Obviously $x \neq y$. Lemma 9 implies that $|\bar{u}| > |r|$. It follows that $py \in \mathrm{Prf}(t)$, and $yp \in \mathrm{Suf}(t)$. Thus $xty \in F(w)$. Lemma 13 implies that $\bar{u} \in F_p(w)$ if and only if $\bar{u} \in T(w)$. Since $|\bar{u}| > |r|$, this would be a contradiction to Property 3 of Definition 14. Hence $\bar{u} \notin F_p(w)$. This completes the proof.

We define a set Mergeable which contains 3-tuples (d, g, t) of rich words such that, among other properties, dg and gt are rich. Later we prove that the "merge" dgt of dg and gt is also rich. Let $\mathrm{flt}(p) = A \cap \mathrm{Prf}(p)$ be the first letter of a word $p \in A^*$ with $|p| \geq 1$.

Definition 25. *We define a set* Mergeable *as follows:* $(d, g, t) \in$ Mergeable *if*

1. $d, g, t, dg, gt, dg\,\mathrm{flt}(t) \in \mathrm{R}$ *and*
2. $\mathrm{lps}(dg\,\mathrm{flt}(t)) \in \mathrm{T}(dg\,\mathrm{flt}(t))$ *and*
3. $\mathrm{lps}(gp) \notin \mathrm{F}(dg)$ *for each* $p \in \mathrm{Prf}(t)$ *with* $|p| \geq 1$.

Let $(d, g, t) \in$ Mergeable. The following proposition shows that dgt is a rich word. This will allow us from a rich word of the form $dgwgt$ to construct a rich word dgt. In other words this will allow us to remove the factor w from a rich word, and thus to reduce the number of occurrences of flexed palindromes.

Proposition 26. *If* $(d, g, t) \in$ Mergeable *then*

- $dgt \in \mathrm{R}$ *and*
- $\mathrm{lps}(dgp) = \mathrm{lps}(gp)$ *for each* $p \in \mathrm{Prf}(t)$ *with* $|p| \geq 1$.

Proof. From Definition 25 it follows immediately that the Proposition holds for $(d, g, \mathrm{flt}(t))$.

Suppose that the Proposition holds for $(d, g, \bar{p})$, where $\bar{p} \in \mathrm{Prf}(t)$ with $1 \leq |\bar{p}| < |t|$. We show that the Proposition holds for (d, g, p) and (h, g, p), where $p \in \mathrm{Prf}(t)$ with $|p| = |\bar{p}| + 1$. From the property that a finite rich word w of length n has $n+1$ palindromic factors it follows that $|\mathrm{F}_p(w)| = |\mathrm{F}_p(\mathrm{rtrim}(w))| + 1$. This and Property 3 of Definition 25 imply that $\mathrm{lps}(gp) \notin \mathrm{F}(\mathrm{lps}(dg\bar{p}))$. Consequently $\mathrm{lps}(gp) = \mathrm{lps}(dgp)$ and $dgp \in \mathrm{R}$, see Proposition 5. This completes the proof.

We prove that the set of flexed palindromes of the word dgt that are not factors of prefix dg, where $(d, g, t) \in$ Mergeable, does not depend on the prefix d.

Proposition 27. *If* $(d, g, t), (h, g, t) \in$ Mergeable, $|d| \geq 1$, *and* $|h| \geq 1$ *then* $\mathrm{T}(dgt) \setminus \mathrm{T}(dg) = \mathrm{T}(hgt) \setminus \mathrm{T}(hg)$.

Proof. To get a contradiction, suppose that there is $p \in \mathrm{Prf}(t)$ with $|p| \geq 1$ such that $\mathrm{lps}(dgp) \in \mathrm{T}(dgp)$ and $\mathrm{lps}(hgp) \notin \mathrm{T}(hgp)$. If $|p| > 1$ then $|\mathrm{lps}(dgp)| \leq |\mathrm{lps}(dg\,\mathrm{rtrim}(p))|$ and $\mathrm{trim}(\mathrm{lps}(hgp)) = \mathrm{lps}(hg\,\mathrm{rtrim}(p))$, which is a contradiction, because $\mathrm{lps}(dg\,\mathrm{rtrim}(p)) = \mathrm{lps}(hg\,\mathrm{rtrim}(p)) = \mathrm{lps}(g\,\mathrm{rtrim}(p))$, see Proposition 26. If $|p| = 1$ the proposition holds because of Property 2 of Definition 25. This completes the proof.

The main theorem of the paper states that the reduced word $\mathrm{rdcWrd}(w, r)$ is rich, where $(w, r) \in \Gamma$. In addition the theorem asserts that the set of flexed palindromes of $\mathrm{rdcWrd}(w, r)$ is a subset of the set of flexed palindromes of the word w, the number of occurrences of r is strictly smaller in $\mathrm{rdcWrd}(w, r)$ than in w, and the longest common prefix and suffix of $\mathrm{rdcWrd}(w, r)$ and w are longer than $|r| - 1$.

Theorem 28. *If* $(w, r) \in \Gamma$ *then*

- $\mathrm{rdcWrd}(w, r) \in \mathrm{R}$ *and* $\mathrm{T}(\mathrm{rdcWrd}(w, r)) \subseteq \mathrm{T}(w)$ *and*
- $\mathrm{occur}(\mathrm{rdcWrd}(w, r), r) < \mathrm{occur}(w, r)$ *and*
- $|\mathrm{lcp}(\mathrm{rdcWrd}(w, r), w)| \geq |r| - 1$ *and* $|\mathrm{lcs}(\mathrm{rdcWrd}(w, r), w)| \geq |r| - 1$.

Proof. Recall that $\mathrm{rdcWrd}(w, r) = ut$, where $(v, z, t) = \mathrm{parse}(w, r)$ and $u = \mathrm{rdcPrf}(w, r)$. If $|t| = 0$ then $\mathrm{rdcWrd}(w, r) \in \mathrm{R}$ and $\mathrm{T}(\mathrm{rdcWrd}(w, r)) \subseteq \mathrm{T}(w)$.

Let d be such that $\mathrm{rdcPrf}(w, r) = d \, \mathrm{ltrim}(r) z$. If $|t| > 0$ then we are going to show that $(d, \mathrm{ltrim}(r) z, t) \in \mathrm{Mergeable}$. Obviously $d \, \mathrm{ltrim}(r) z, \mathrm{ltrim}(r) z t \in \mathrm{R}$; recall that $\mathrm{ltrim}(r) z t \in \mathrm{Suf}(w)$. We need to show that Property 3 of Definition 25 is satisfied: Because $vz = \mathrm{MaxStdExt}(vzt, v)$ it follows that $\mathrm{lps}(vz \, \mathrm{flt}(t)) \in \mathrm{T}(w)$. This and $\mathrm{occur}(\mathrm{ltrim}(r) z t, r) = 0$ imply that $|\mathrm{lps}(vzp)| \leq |\mathrm{ltrim}(r) z p|$ for each $p \in \mathrm{Prf}(t)$ with $|p| \geq 1$. In consequence $\mathrm{lps}(\mathrm{ltrim}(r) z p) = \mathrm{lps}(vzp)$. Proposition 24 and $\mathrm{occur}(vzp, \mathrm{lps}(vzp)) = 1$ imply that $\mathrm{lps}(vzp) \notin \mathrm{F}(d \, \mathrm{ltrim}(r) z)$. The other properties of Definition 25 are clearly also fulfilled. Hence $(d, \mathrm{ltrim}(r) z, t) \in \mathrm{Mergeable}$. Thus from Proposition 26 we get that $d \, \mathrm{ltrim}(r) z t \in \mathrm{R}$.

Let $\bar{w}$ be such that $w = \bar{w} \, \mathrm{ltrim}(r) z t$. Obviously $(\bar{w}, \mathrm{ltrim}(r) z, t) \in \mathrm{Mergeable}$. Then Proposition 27 asserts that $\mathrm{T}(\mathrm{rdcWrd}(w, r)) \subseteq \mathrm{T}(w)$.

The fact that $\mathrm{occur}(ut, r) < \mathrm{occur}(w, r)$ follows Lemma 17 and Definition 18. Note that $\mathrm{occur}(\mathrm{rdcPrf}(w, r), r) < \mathrm{occur}(w, r)$.

The properties $|\mathrm{lcp}(\mathrm{rdcWrd}(w, r), w)| \geq |r| - 1$ and $|\mathrm{lcs}(\mathrm{rdcWrd}(w, r), w)| \geq |r| - 1$ follow from Lemma 23.

This completes the proof.

Two more examples will illuminate the construction of $\mathrm{rdcWrd}(w, r)$. The examples are again based on the two cases of Definition 18. For both example we consider that $\mathrm{A} = \{1, 2, 3, 4, 5, 6, 7, 8\}$.

Example 29. If $w = 12145656547745656545656547874$ and $r = 656$ then $v = 12145656547745656545656$, $z = 547$, $t = 874$, $\mathrm{lps}(v) = 656545656$, $u = \mathrm{rdcPrf}(w, r) = 12145656547$, and $\mathrm{rdcWrd}(w, r) = ut = 12145656547874$.

Example 30. If $w = 12145656547874$ and $r = 656$ then $v = 12145656$, $z = 54$, $t = 7874$, $\mathrm{lps}(v) = 656$, $u = \mathrm{rdcPrf}(w, r) = 12145654$, and $\mathrm{rdcWrd}(w, r) = ut = 121456547874$.

For a finite set S, we can consider that the set S is well-ordered. No matter how, we just need a function that selects one element from S. Let the function $\mathrm{selectFirst}(S)$ returns the first element of S. If S is an empty set, then we define $\mathrm{selectFirst}(S) = \epsilon$.

If a rich word w has a factor u, then the palindromic closure of w is rich and contains the factor u^R. Hence for us when constructing a rich word containing given factors, it does not matter if w contains u or u^R. We introduce the notion of a *reverse-unioccurrent* factor. Moreover we define a function $\mathrm{ruo}(w, u, v)$ (a *reverse-unioccurrence* of u, v in w) which returns a factor of w such that u, v are reverse-unioccurrent in this factor; in addition we require u or u^R to be a prefix and v or v^R to be a suffix of $\mathrm{ruo}(w, u, v)$.

Definition 31. *If $|\{u, u^R\} \cap \mathrm{F}(w)| = 1$ then we say that a word u is reverse-unioccurrent in w, where $w, u \in \mathrm{R}$.*

If $w_1, w_2, w \in \mathrm{R}$, $w_1, w_2 \in \mathrm{F}(w)$, and there is $t \in \mathrm{Prf}(w)$ such that $w_1 \in \mathrm{F}(t)$ and $\{w_2, w_2^R\} \cap \mathrm{F}(t) = \emptyset$ then let $\mathrm{M}(w, w_1, w_2) \subset \mathrm{F}(w)$ such that $t \in \mathrm{M}(w, w_1, w_2)$ if:

– $t \in \mathrm{F}(w)$ *and* w_1, w_2 *are reverse-unioccurrent in* t *and*
– $\{w_1, w_1^R\} \cap \mathrm{Prf}(t) \neq \emptyset$ *and* $\{w_2, w_2^R\} \cap \mathrm{Suf}(t) \neq \emptyset$.

Let $\mathrm{ruo}(w, w_1, w_2) = \mathrm{selectFirst}(\mathrm{M}(w, w_1, w_2))$.

Remark 32. It is not difficult to see that the function $\mathrm{ruo}(r, w_1, w_2)$ is well defined and the set $\mathrm{M}(w, w_1, w_2)$ is nonempty.

We define the function $\mathrm{elmWrd}(w, w_1, w_2)$ (*eliminated word*) that constructs a rich word from w by "eliminating all" flexed palindromes longer than $m = \max\{|w_1|, |w_2|\}$ and keeping the prefix w_1 and the suffix w_2 of w.

Definition 33. *Let* $\mathrm{maxFlxPal}(w) = \{r \mid (w, r) \in \Gamma\}$. *If* $w, w_1, w_2 \in \mathrm{R}$, $m = \max\{|w_1|, |w_2|\}$, $w_1 \in \mathrm{Prf}(w)$, *and* $w_2 \in \mathrm{Suf}(w)$, *then let* $\mathrm{elmWrd}(w, w_1, w_2)$ *be the result of the following procedure:*

```
01 INPUT: w,m,w_1,w_2;
02 res: = ruo(w,w_1,w_2);
03 r := selectFirst(maxFlxPal(res));
04 WHILE r is longer than m
05 DO
06   res := rdcWrd(res,r);
07   res := ruo(res,w_1,w_2);
08   r := selectFirst(maxFlxPal(res));
09 END-DO;
10 RETURN res;
```

The calls of the function ruo on the lines 02 and 07 guarantee that w_1, w_2 are reverse-unioccurrent in the word res and that $\{w_1, w_1^R\} \cap \mathrm{Prf}(res) \neq \emptyset$ and $\{w_2, w_2^R\} \cap \mathrm{Suf}(res) \neq \emptyset$. Realize that it is not guaranteed that w_1, w_2 are reverse-unioccurrent in $\mathrm{rdcWrd}(res, r)$, even if w_1, w_2 are reverse-unioccurrent in res.

Clearly, the facts that $\bar{t}$ is reverse-unioccurrent in a rich word t and $\bar{t} \in \mathrm{Prf}(t)$ imply that $\mathrm{lpp}(t) \in \mathrm{Prf}(\bar{t})$; realize that if $d \in \mathrm{F}(\mathrm{lpp}(\bar{t}))$ then $d^R \in \mathrm{F}(\mathrm{lpp}(t))$ also, since palindromes are closed under reversal. Thus if r is a flexed palindrome of t longer than the prefix $\bar{t}$, then r is not a factor of $\mathrm{lpp}(t)$ and hence r satisfies Property 2 of Definition 14.

Let $r = \mathrm{selectFirst}(\mathrm{maxFlxPal}(w))$. The call of the function $\mathrm{rdcWrd}(res, r)$ on the line 06 contains valid parameters, since if $r \neq \epsilon$ and $|r| > m$ then $(w, r) \in \Gamma$.

In addition, because $|r| > \max\{|w_1|, |w_2|\}$, Theorem 28 asserts that $\{w_1, w_1^R\} \cap \mathrm{Prf}(\mathrm{rdcWrd}(res, r)) \neq \emptyset$ and $\{w_2, w_2^R\} \cap \mathrm{Suf}(\mathrm{rdcWrd}(res, r)) \neq \emptyset$; consequently $\{w_1, w_1^R\} \cap \mathrm{Prf}(res) \neq \emptyset$ and $\{w_2, w_2^R\} \cap \mathrm{Suf}(res) \neq \emptyset$ on the line 06.

Moreover Theorem 28 implies that the procedure finishes after a finite number of iterations, because $\mathrm{occur}(\mathrm{rdcWrd}(w, r), r) < \mathrm{occur}(w, r)$ and $\mathrm{T}(\mathrm{rdcWrd}(w, r)) \subseteq \mathrm{T}(w)$. The number of iterations is bounded by the number $\sum_{r \in \mathrm{T}(w)} \mathrm{occur}(w, r)$. Note that several occurrences of r may be "eliminated" in one iteration. Hence we proved the following lemma:

Lemma 34. *If $w \in R$, $w_1 \in Prf(w)$, $w_2 \in Suf(w)$, $m = \max\{|w_1|, |w_2|\}$, and $t = elmWrd(w, w_1, w_2)$ then*

- $t \in R$ *and for each $r \in T(t)$ we have $|r| \leq m$ and*
- $\{w_1, w_1^R\} \cap Prf(t) \neq \emptyset$ *and $\{w_2, w_2^R\} \cap Suf(t) \neq \emptyset$.*

5 Words with Limited Number of Flexed Points

What is the maximal length of a word u such that w is reverse-unioccurrent in u, w is a prefix of u, and u has a given maximal number of flexed palindromes? The proposition below answers this question.

Proposition 35. *If $u, w \in R$, $|u| \geq 1$, $|v| \geq 1$, $w \in Prf(u)$, $|T(u) \setminus T(w)| \leq k$, $|w| \leq m$, and w is reverse-unioccurrent in u then $|u| \leq m2^{k+1}$.*

Proof. Obviously $|pc(u)| < 2|u|$, $pc(u) \in StdExt(u)$, and w is not reverse-unioccurrent in $pc(u)$, since $w^R \in Suf(pc(u))$. It follows that if $v_1, v_2 \in Prf(\bar{u})$ such that v_1 is reverse-unioccurrent in $\bar{u}$, $v_1 \in Prf(v_2)$, $|T(v_2) \setminus T(v_1)| = 1$, and $lps(v_2) \in T(v_2)$ then $|rtrim(v_2)| < 2|v_1|$, since $rtrim(v_2) \in StdExt(v_1)$ and $pc(v_1) \in StdExt(v_1)$ also. This implies that $|v_2| \leq 2|v_1|$. The proposition follows.

Remark 36. The proof asserts that if v_1, v_2 are two prefixes of a word u such that the longest palindromic suffix of v_2 is the only flexed palindrome in v_2 which is not a factor of v_1, then v_2 is at most twice longer than v_1 on condition that v_1 is reverse-unioccurrent in $ltrim(v_2)$. Less formally it means that the length of a word can grow at most twice before the next flexed palindrome appears. Note that for $k = 1$ we have $|u| \leq 2m$, which makes sense, since the palindromic closure of a word u is at most twice longer than u.

In [4] the author showed an upper bound for the number of palindromic factors of given length in a rich word. Recall that $q = |A|$.

Proposition 37 ([4], **Corollary 2.23**)**.** *If $w \in R$ and $n > 0$ then*

$$|F_p(w) \cap A^n| \leq (q+1)n(4q^{10}n)^{\log_2 n}.$$

Proposition 37 implies an upper bound for the number of flexed palindromes:

Lemma 38. *If $w \in R$, $n > 0$, and $A^{\leq n} = \bigcup_{j=0}^{n} A^j$ then*

$$|T(w) \cap A^{\leq n}| \leq (q+1)n^2(4q^{10}n)^{\log_2 n}.$$

Proof. Just realize that $\sum_{j=1}^{n}(q+1)j(4q^{10}j)^{\log_2 j} \leq (q+1)n^2(4q^{10}n)^{\log_2 n}$.

From Lemmas 34, 38 and Proposition 35 we obtain the result of the article:

Corollary 39. *If w, w_1, w_2 are rich words, $w_1, w_2 \in F(w)$, $m = \max\{|w_1|, |w_2|\}$ then there exists also a rich word $\bar{w}$ such that $w_1, w_2 \in F(\bar{w})$ and $|\bar{w}| \leq m2^{k(m)+2}$, where $k(m) = (q+1)m^2(4q^{10}m)^{\log_2 m}$.*

Proof. Without loss of generality, suppose that there is $\bar{t} \in \mathrm{Prf}(w)$ such that $w_1 \in \mathrm{Prf}(\bar{t})$ and $\{w_2, w_2^R\} \cap \mathrm{F}(\bar{t}) = \emptyset$. Then the function $\mathrm{ruo}(w, w_1, w_2)$ is well-defined. Let $t \in \mathrm{ruo}(w, w_1, w_2)$. Consider the word $g = \mathrm{elmWrd}(t, w_1, w_2)$. Let $k(m) = (q+1)m^2(4q^{10}m)^{\log_2 m}$. Lemma 38 and Proposition 35 imply that $|g| \geq m2^{k(m)+1}$. Lemma 34 implies that $g \in \mathrm{R}$, $\{w_1, w_1^R\} \cap \mathrm{F}(g) \neq \emptyset$, and $\{w_2, w_2^R\} \cap \mathrm{F}(g) \neq \emptyset$. Let $\bar{w} = \mathrm{pc}(g)$. It follows that $w_1, w_2 \in \mathrm{F}(\bar{w})$. Because $|\mathrm{pc}(g)| \leq 2|g|$, the corollary follows.

Acknowledgments. The author wishes to thank to Štěpán Starosta for his useful comments. The author acknowledges support by the Czech Science Foundation grant GAČR 13-03538S and by the Grant Agency of the Czech Technical University in Prague, grant No. SGS14/205/OHK4/3T/14.

References

1. Bucci, M., De Luca, A., Glen, A., Zamboni, L.Q.: A new characteristic property of rich words. Theor. Comput. Sci. **410**, 2860–2863 (2009)
2. Glen, A., Justin, J., Widmer, S., Zamboni, L.Q.: Palindromic richness. Eur. J. Comb. **30**, 510–531 (2009)
3. Pelantová, E., Starosta, Š.: On words with the zero palindromic defect. In: Brlek, S., Dolce, F., Reutenauer, C., Vandomme, É. (eds.) WORDS 2017. LNCS, vol. 10432, pp. 59–71. Springer, Cham (2017). https://doi.org/10.1007/978-3-319-66396-8_7
4. Rukavicka, J.: An upper bound for palindromic and factor complexity of rich words. (2018, submitted for publication). https://arxiv.org/abs/1810.03573
5. Vesti, J.: Extensions of rich words. Theor. Comput. Sci. **548**, 14–24 (2014)

Mortality and Synchronization of Unambiguous Finite Automata

Andrew Ryzhikov[✉]

LIGM, Université Paris-Est, Marne-la-Vallée, France
ryzhikov.andrew@gmail.com

Abstract. We study mortal words and words of minimum non-zero rank (in particular, synchronizing words) in strongly connected unambiguous automata. We show that every n-state strongly connected unambiguous automaton admits a word of minimum non-zero rank of length at most n^5, and this word can be found in polynomial time. We show that for words of minimum rank this upper bound can be lowered to $O(n^3(\log n)^4)$ for prefix automata of finite codes and to $O(n^3 \log n)$ for prefix automata of complete finite codes. We also provide quadratic lower bounds on the length of shortest mortal words for several classes of deterministic automata.

Keywords: Unambiguous automaton · Synchronizing word · Mortal word

1 Introduction

The questions of mortality and synchronization are the most important reachability problems for finite automata. Both these problems can be considered as finding a word of some particular (either minimum or minimum non-zero) rank in a given automaton $\mathcal{A}$. In this paper we concentrate on extremal problems regarding the length of a shortest mortal or synchronizing word. The length of a shortest mortal word shows the minimum time for a system to become uncontrollable, while the length of a shortest synchronizing word indicates the minimum time to get total control over the current state of the system.

Let $\mathcal{A}$ be an n-state strongly connected unambiguous automaton. Carpi proved an upper bound of $\frac{1}{2}rn^3$ on the length of a shortest word of minimum rank if $\mathcal{A}$ is complete [6], thus providing an upper bound to a problem which can be considered as a broad generalization of the Černý conjecture (see [4,16]). Kiefer and Mascle extended his result and proved an upper bound of n^5 on the length of a shortest mortal word if $\mathcal{A}$ is non-complete [10]. However, no polynomial bound on the length of a shortest word of minimum non-zero rank was known in the case where $\mathcal{A}$ is non-complete. In this paper, we provide such a bound by extending the proof of Kiefer and Mascle, thus partially answering a question stated in the list of research problems of [4]. Our proof also implies that

© Springer Nature Switzerland AG 2019
R. Mercaş and D. Reidenbach (Eds.): WORDS 2019, LNCS 11682, pp. 299–311, 2019.
https://doi.org/10.1007/978-3-030-28796-2_24

it can be checked in polynomial time whether a strongly connected unambiguous automaton is synchronizing.

The paper is organized as follows. In Sect. 2 we present the definitions and preliminary results used in the paper. In Sect. 3 we present the idea of the proof of Kiefer and Mascle on bounding the lengths of mortal words in strongly connected unambiguous automata. In Sect. 4 we show how to extend this proof to obtain an n^5 upper bound on the length of shortest words of minimum non-zero rank. In Sect. 5 we refine the bounds on the length of shortest words of minimum rank for prefix automata of finite codes. In Sect. 6 we provide lower bounds on the length of shortest mortal words in several classes of deterministic automata. In Sect. 7 we prove a sufficient condition for recognizable codes to be synchronizing.

2   Main Definitions

A *nondeterministic finite automaton* (which we simply call an *automaton* in this paper) is a triple $\mathcal{A} = (Q, \Sigma, \Delta)$, where Q is a finite set of states, Σ is a finite alphabet, and Δ is a transition relation $Q \times \Sigma \to 2^Q$, where 2^Q is the family of all subsets of Q. Note that our definition of an automaton does not include any initial or accepting states. We say that a word w *kills* a state q if $\Delta(q, w)$ is empty, otherwise we say that q *survives* w. The transition relation is naturally extended to $2^Q \times \Sigma \to 2^Q$ by taking $\Delta(S, a) = \cup_{q \in S} \Delta(q, a)$ for $S \subseteq Q$ and $a \in \Sigma$, and then to $Q \times \Sigma^* \to 2^Q$ by taking $\Delta(q, wa) = \Delta(\Delta(q, w), a)$ for $w \in \Sigma^*, a \in \Sigma$ and $q \in Q$. When the transition relation is clear from the context, we denote $\Delta(q, w)$ by $q.w$. We also denote by $w.q$ the set $\{p \mid q \in p.w\}$. More generally, we denote by $S.w$ and $w.S$ the sets $\{q.w \mid q \in S\}$ and $\{w.q \mid q \in S\}$ for $S \subseteq Q$, $w \in \Sigma^*$. An automaton is *unambiguous* if for any two states $p, q \in Q$ and any word $w \in \Sigma^*$ there is at most one path from p to q labeled by w. An automaton is *deterministic* if for every state $q \in Q$ and every letter $a \in \Sigma$ the set $\Delta(q, w)$ has cardinality at most one. Clearly, every deterministic automaton is unambiguous. An automaton is called *strongly connected* if for every state p and every state q there is a word w such that $q \in p.w$. Everywhere in the paper we assume the automata to be strongly connected. By ϵ we denote the empty word.

Provided a word $w \in \Sigma^*$ for an automaton $\mathcal{A} = (Q, \Sigma, \Delta)$, one can naturally assign a matrix $M(w)$ to it as follows. Fix some ordering $q_1, \ldots, q_n$ of the set Q of states. The entry $M(w)_{i,j}$ of the matrix $M(w)$ is the number of paths from q_i to q_j labeled by w. As follows from the definition, each entry of such a matrix for an unambiguous automaton belongs to the set $\{0, 1\}$. For the mapping $M : \Sigma^* \to \{0, 1\}^{n \times n}$ for any words $w_1, w_2 \in \Sigma^*$ we have $M(w_1 w_2) = M(w_1) M(w_2)$. We denote $\mathcal{M}(\mathcal{A}) = \{M(w) \mid w \in \Sigma^*\}$.

Using this morphism, we define the rank of a word w in an unambiguous automaton $\mathcal{A}$ as the Boolean rank of the matrix $M(w)$ [8]. By Boolean rank we mean the rank over the Boolean semiring $\{0, 1\}$ (having $1 + 1 = 1$), that is, the smallest r such that $M(w)$ can be represented as a product of an $n \times r$ and $r \times n$ zero-one matrices. Given an automaton $\mathcal{A}$, we call a word *synchronizing* if it has rank 1, and *mortal* if it has rank 0. An automaton is called *non-complete* if it admits a mortal word, otherwise it is called *complete*. An automaton admitting a synchronizing word is called *synchronizing*. The paper [2] surveys some questions on the structure of unambiguous automata and words of particular rank for them.

A *code* is a set of words (called *codewords*) such that no word can be represented as a concatenation of codewords in two different ways. A word w is called a *factor* of w' if there exist words u, v such that $w' = uwv$. A word $w \in \Sigma^*$ is called *non-mortal* for a code X over Σ if it is a factor of a word in X^*, otherwise it is called *mortal*. A code is called *complete* if it does not admit a mortal word. Thus, complete codes are exactly the codes such that every word is a factor of a decodable message. A word w is called *synchronizing* for a code X if for every pair of words u, v such that $uwv \in X^*$ we have $uw, wv \in X^*$. If the word w is synchronizing, then seeing a word ww in a decodable message $uwwv$ allows to separate the decoding process into two independent parts for uw and wv. A presence of a synchronizing word in a message allows to restart the decoding process after transmission errors or when a part of a message is skipped.

Unambiguous automata serve as a very powerful tool for studying codes. Let $\mathcal{A}$ be a strongly connected unambiguous automaton, and q be some its state. The set of *first return words* of q is then defined as the set of all words labeling paths from q to q and not containing q as an intermediate state. For a strongly connected unambiguous automaton $\mathcal{A}$ and a state q in it, the set of first return words is always a recognizable (by some finite automaton) code [4]. The converse is also true: for every recognizable code X there is a strongly connected unambiguous automaton $\mathcal{A}$ such that X is the set of first return words of some its state. Moreover, a word is mortal (synchronizing) for $\mathcal{A}$ if and only if it is mortal (synchronizing) for X [4]. More generally, the minimum non-zero rank of words for $\mathcal{A}$ is related to the minimum number of interpretations of words with respect to X (called the degree of X, see Sect. 9.6 of [4]).

3   The Result of Kiefer and Mascle

Let $\mathcal{A} = (Q, \Sigma, \Delta)$ be an n-state strongly connected unambiguous automaton. Fix an ordering $q_1, \ldots, q_n$ of the set Q. For a subset S of states denote by $\mathrm{vect}(S)$ the n-dimensional *characteristic (row) vector* of S, that is, a vector $v \in \{0,1\}^n$ with ith component equal to 1 if $q_i \in S$ and equal to 0 otherwise.

A set $C \subseteq Q$ is called *mergeable* if there exists a word w and a state q' such that for every state $q \in R$ we have $q' \in q.w$. A set $R \subseteq Q$ is called *coreachable* if there exists a word w and a state q' such that for every state $q \in C$ we have $q \in q'.w$. Characteristic vectors of mergeable and coreachable sets are exactly non-zero columns and rows of matrices in the monoid $\mathcal{M}(\mathcal{A})$.

Let y, z be a pair of words. We denote by $\mathcal{P}(y, z)$ the set of all states q which are reached by y and survive z, that is such that $y.q$ and $q.z$ are non-empty. Studying $\mathcal{P}(y, z)$ is the central idea of the discussed result. Assume without loss of generality that $\mathcal{P}(y, z) = \{q_1, \ldots, q_k\}$. Then there exist two families of non-empty sets: a family $\mathcal{C} = \{C_1, \ldots, C_k\}$ of mergeable and a family $\mathcal{R} = \{R_1, \ldots, R_k\}$ of coreachable sets such that $C_i = y.q_i$, $R_i = q_i.z$ for $1 \leq i \leq k$.

The rank of the word yz is at most k. Indeed, we can represent the matrix $M(yz)$ as a product of the matrix with columns $\mathrm{vect}(C_1)^T, \dots, \mathrm{vect}(C_k)^T$ and the matrix with rows $\mathrm{vect}(R_1), \dots, \mathrm{vect}(R_k)$ (by v^T we denote the column which is the transposition of the row v):

$$M(yz) = \mathrm{vect}(C_1)^T \cdot \mathrm{vect}(R_1) + \dots + \mathrm{vect}(C_k)^T \cdot \mathrm{vect}(R_k).$$

Without additional restrictions on the words y and z, the sets from $\mathcal{C}$ and $\mathcal{R}$ can intersect. However, the relations between the sets C_i and R_j are not arbitrary. First, for a mergeable set C and a coreachable set R the intersection $C \cap R$ has size at most one. Second, if C_i intersects C_j then R_i and R_j are disjoint and vice versa. Some further results in this direction are obtained in [13].

In [10] Kiefer and Mascle show how to construct a pair y, z of words of length $O(n^4)$ such that no pair of states in $\mathcal{P}(y, z)$ is coreachable or mergeable. In particular this means that no pair of sets in $\mathcal{C}$ and no pair of sets in $\mathcal{R}$ intersects. The algorithm decreases the number of states in $\mathcal{P}(y', z')$ by iteratively constructing the words y', z'. This is done by concatenating words z_q for different states q with the property that no state coreachable with q survives z_q. Thus we obtain the word z forcing that no pair of states in $\mathcal{P}(\epsilon, z)$ is coreachable. The same is done symmetrically for mergeable states by the word y. Such a pair y, z allows to manipulate unambiguous automata much easier. In particular, one can kill a state in $\mathcal{P}(y, z)$ by a word of linear length (Lemma 14 of [10]). The final mortal word is then obtained by subsequently killing all the remaining states in $\mathcal{P}(y, z)$. The rank of yz for such y, z is exactly the cardinality of $\mathcal{P}(y, z)$.

The proof of Carpi [6] (with some small modifications) can be seen as a similar algorithm. The main difference is that the structure of complete automata (provided in particular by Proposition 3.4 of [6]) makes it possible to get a better upper bound. In particular, it is enough to perform only r steps of the algorithm producing the pair y, z, and there is no need then to kill the states in $\mathcal{P}(y, z)$ afterwards.

4   Words of Minimum Non-zero Rank

In this section we extend the result of Kiefer and Mascle described in Sect. 3 to words of minimum non-zero rank. First we observe that the word yz constructed in Lemma 11 of [10] is not mortal. Indeed, every word w_q produced in Lemma 10 of [10] does not kill the state q. Then the algorithm in Lemma 11 consists of taking a state $p.w$ such that p survives w, and applying $w_{p.w}$ to it, thus p is not killed by $ww_{p.w}$. After getting rid of coreachable states the symmetric version for mergeable states which are not yet killed is performed, and by the same reason it is still maintained that some state survives.

Let now w be a word of minimum non-zero rank r such that at least one state $q_i \in \mathcal{P}(y, z)$ is not killed by zwy. Such a word always exists, since we can take $w = w_1 w' w_2$ where w' is a word of minimum non-zero rank, and the words w_1 and w_2 are such that w_1 maps a state $q_i.z$ to a state which is not killed by w', and w_2 maps a state in $q_i.zw_1w'$ to a state surviving y. The pair $yzwy, z$

has then the same properties as y, z but the cardinality of the set $\mathcal{P}(y, zwyz)$ is exactly r. This means there exist exactly r states in $\mathcal{P}(y, z)$ which are mapped by zw to $y.\,\mathcal{P}(y, z)$, and all the other states are mapped to the complement of this set. Let q_i, q_j be a pair of states such that zw sends exactly one of them to $y.\,\mathcal{P}(y, z)$, and the other one to the complement of this set. The next lemma is an analogue of Lemma 14 from [10] for the case of words of minimum non-zero rank instead of mortal words.

Lemma 1. *One can compute in polynomial time a word $x \in \Sigma^*$ with $|x| \le n$ such that exactly one of the sets $q_i.zxyz$ and $q_j.zxyz$ is empty.*

Proof. It is enough to compute a word x such that exactly one of the sets $q_i.zx$ and $q_j.zx$ intersects $y.\,\mathcal{P}(y, z)$. Define e_i, e_j and f to be the characteristic vectors of $q_i.z$, $q_j.z$ and $y.\,\mathcal{P}(y, z)$ respectively. As shown in the proof of Lemma 14 of [10], we have that $e_i M(x) f^T \le 1$ and $e_j M(x) f^T \le 1$. Since all the entries are non-negative and integer, the only two possible values are thus 0 and 1.

Let V be the subspace of $\mathbb{R}^n$ spanned by the vectors $(e_i - e_j)M(x)$ for $x \in \Sigma^*$. This subspace can be seen as the smallest subspace containing the vector $e_i - e_j$ and closed under multiplying by $M(a)$ for all $a \in \Sigma$. Thus we can compute a basis B of V by subsequently finding for an already added word u a letter $a \in \Sigma$ such that $(e_i - e_j)M(ua)$ is not in the already generated subspace and adding this vector to the basis. Since the dimension of V is at most n, every vector in this basis corresponds to a word of length at most n.

As shown before, the possible values of $(e_i - e_j)M(x)f^T$ belong to the set $\{-1, 0, 1\}$. Since there exists a word w (defined above) with $(e_i-e_j)M(w)f^T \ne 0$, we have that V is not orthogonal to f. Thus there exists a word x corresponding to a vector in B such that $(e_i - e_j)M(x)f^T \in \{-1, 1\}$, and $|x| \le n$. The word x has then the required properties. $\square$

Now we can follow the proof of [10] up to using Lemma 14, and then use Lemma 1 of this paper instead. This lemma allows to kill one state of the set $\mathcal{P}(y, z)$ while maintaining that the constructed word is not mortal. Thus we can proceed to Lemma 15 of [10] and construct iteratively a word of minimum non-zero rank having the same upper bound of n^5 on its length. We summarize these results in the following theorem.

Theorem 2. *Every strongly connected unambiguous automaton $\mathcal{A}$ with n states has a word of minimum non-zero rank of length at most n^5. Such a word can be found in polynomial time.*

In particular, this means that it can be checked in polynomial time whether a finite code is synchronizing (for example, by checking that its prefix automaton is synchronizing, see Sect. 5 for the definition). It is natural to ask whether a faster algorithm to check synchronizability of a finite code (or a strongly connected unambiguous automaton in general) exists.

The main contribution of Theorem 2 is the first known polynomial upper bound on the length of shortest words of minimum non-zero rank for the non-complete case. Additionally it can be seen as a way to make the proof of Kiefer and Mascle (in the general case of words of minimum rank) more uniform, since it still works for the complete case and does not require the proof of Carpi for the special case of complete automata.

If a non-complete strongly connected deterministic automaton has a synchronizing word of length ℓ, then it has a mortal word of length at most $\ell + n$, where n is the number of states. To find this word it is enough to kill the only state in the image of the synchronizing word, which can be done by a word of linear length. A similar relation for strongly connected unambiguous automata is provided by the following result.

Proposition 3. *Let $\mathcal{A}$ be an n-state strongly connected non-complete synchronizing unambiguous automaton. If $\mathcal{A}$ has a synchronizing word of length ℓ, then it has a mortal word of length $3\ell + n$.*

Proof. Let w be a synchronizing word for $\mathcal{A} = (Q, \Sigma, \Delta)$. If ww is mortal, we are done. Otherwise there exists exactly one state q which survives w and is reached from some state by w. Indeed, by definition every word of rank 1 has sets C, R of states such that $Q.w = R$, $w.Q = C$ and w maps every state in C to every state in R. Thus, if there are two different states in $R \cap C$, the automaton is not unambiguous. Hence we can apply Lemma 14 of [10] for $y = w$ and $z = w$ and get a word x of length at most n such that the word $wxww$ of length at most $3\ell + n$ is mortal.   □

Obviously, no relation in the other direction exists in the general case: one can consider an arbitrary complete deterministic synchronizing automaton and add a nowhere defined new letter to it. Then the length of a shortest synchronizing word is not bounded by any non-trivial function of the length of a shortest mortal word which is equal to 1.

5   Prefix Automata

A natural automaton associated to a finite code is its prefix automaton. Let X be a finite code over an alphabet Σ. One can then construct the *prefix automaton* $\mathcal{A} = (Q, \Sigma, \Delta)$ of X as follows. The states of $\mathcal{A}$ are the proper prefixes of words in X. For a state $q \in \Sigma^*$ and a letter $a \in \Sigma$ the set $\Delta(q, a)$ contains qa if qa is a proper prefix of a word in X and contains ϵ if qa is a word in X (these situations can occur at the same time). It is easy to see that the prefix automaton of a finite code is always strongly connected and unambiguous. The number of its states provides a lower bound on the total length of all words in the code. The next lemma generalizes the log-log lemma (Lemma 16 of [3], see also [5]) from complete prefix codes to general complete codes.

Lemma 4 (Log-log lemma for general complete codes). *Let $\mathcal{A}$ be the prefix automaton of a finite code X over an alphabet of size m. Then there exists a word of length $\lceil \log_m n \rceil$ and rank at most $\lceil \log_m n \rceil$ for this automaton, where n is the number of its states.*

Proof. Denote $r = \lceil \log_m n \rceil$ and let i be the state corresponding to ϵ. Similar to the prefix case, we first show that there exists a word w of length r such that for every state q every path from q labeled by w contains the state i. Suppose that this is not true and hence for every word w of this length there is a state q_w having a path labeled by w which does not contain the state i. Denote by q'_w a state in $q_w.w$ such that the path from q_w to q'_w labeled by w does not contain i. Observe that for different words w_1, w_2 the states q'_{w_1}, q'_{w_2} are different, and none of these states is the state i. Thus the total number of states in $\mathcal{A}$ is at least $m^r + 1 = m^{\lceil \log_m n \rceil} + 1 > n$, which is a contradiction.

Now we will show that w has rank at most r. Denote by W the set of prefixes of w. Since for every state q all the paths going from q and labeled by w contain the state i, every row of $M(w)$ is a linear combination of vectors $\text{vect}(i.w')$ for $w' \in W$ with coefficients 0 and 1. Hence the rank of $M(w)$ is at most r. □

Provided a short word of small rank, we can perform the general algorithm of [10] described in Sect. 3 starting with the pair w, w and thus improve the upper bound. Let w be a word of rank k for a strongly connected unambiguous automaton $\mathcal{A}$. Then the set $\mathcal{P}(w, w)$ contains at most k^2 states. Indeed, by definition of rank the matrix $M(w)$ can be represented as $\sum_{i=1}^{k} \text{vect}(C_i)^T \cdot \text{vect}(R_i)$ for some families $C_1, \ldots, C_k$ and $R_1, \ldots, R_k$ of states, where w maps every state of C_i to every state of R_i. Every state in $\mathcal{P}(w, w)$ is then contained in some C_i and in some R_j. If there are two different states contained in the same pair C_i and R_j, the automaton $\mathcal{A}$ is not unambiguous. Hence the cardinality of $\mathcal{P}(w, w)$ is at most k^2. If the rank of ww is zero, we have constructed a mortal word for $\mathcal{A}$. Otherwise we can start with the pair w, w and apply the iterative algorithm described in Sect. 3 to get a pair y', z' of words such that no pair of states in $\mathcal{P}(y'w, wz')$ is coreachable or mergeable. Since $\mathcal{P}(w, w)$ contains at most k^2 states, it remains to perform at most k^2 steps to get rid of all mergeable and coreachable states by y' and z'. Then we can kill the states in $\mathcal{P}(y'w, wz')$ one by one as described in Lemma 15 of [10]. The length of a mortal word thus constructed is at most $k^2 n + (k^2 + 1)(\frac{1}{4}k^2(n+2)^2(n-1) + 2|w|)$, see the bound in Lemma 15 of [10].

By Lemma 4 we can find in polynomial time a word w of logarithmic length and logarithmic rank (by enumerating all words of that length). By starting with the pair w, w we get the following.

Theorem 5. *Let $\mathcal{A}$ be the prefix automaton of a finite non-complete code X over an alphabet of size m. Then there exists a mortal word of length $O(n^3(\log_m n)^4)$, where n is the number of states in $\mathcal{A}$.*

For the case of complete automata we can get a slightly better bound by using the main result of [6], since by Lemma 4 the minimum rank of a word in a complete prefix automaton of a finite code is at most logarithmic.

Corollary 6. *Let $\mathcal{A}$ be the prefix automaton of a finite complete code X over an alphabet of size m. Then the length of a word of minimum rank for $\mathcal{A}$ is at most $\frac{1}{2}n^3 \cdot \lceil \log_m n \rceil$, where n is the number of states in $\mathcal{A}$.*

Now we proceed to a much stronger bound for the special case of non-complete finite prefix codes. A code is called *prefix* if none of its codewords is a prefix of another codeword. If k is the length of the longest codeword of a finite non-complete prefix code, an upper bound of $2k^2$ on the length of a shortest mortal word can be easily obtained (see, e.g., [12]). We improve this bound.

Theorem 7. *Let X be a non-complete finite prefix code, and let k be the length of a longest word in X. Then there exists a mortal word of length at most $\frac{3}{2}(k^2 + k)$.*

Proof. Consider the prefix automaton $\mathcal{A} = (Q, \Sigma, \delta)$ of X (since X is a prefix code, this automaton is deterministic). Observe that $Q' = Q.a^k$ is a subset of Q of size at most k. For any word $w \in \Sigma^*$ the set $Q'.w$ does not contain two different states corresponding to prefixes of the same length.

Now we start with the set Q' of active states and consequently perform the following algorithm: while there is at least one active state, take the active state corresponding to the longest prefix and kill it by sending it to a state q such that $q.a$ is empty for some $a \in \Sigma$ and then applying a. Since the set of active states consists of states corresponding to prefixes of pairwise different lengths, it takes a word of length at most $(1 + k) + (2 + k) + \ldots + (k + k) = \frac{k(k+1)}{2} + k^2 = \frac{3k^2+k}{2}$ to map each state first to the state corresponding to the empty prefix and then to kill it. Together with the word a^k we get the required. $\square$

As proved in [14], the code $\Sigma^k \setminus \{u\}$ for any unbordered word u (that is, a word having no prefix equal to its suffix) has the length of a shortest mortal word equal to $k^2 + k - 1$. We conjecture that the order of this bound is tight, that is, the optimal upper bound is $k^2 + O(k)$.

6   Mortality Lower Bounds for Deterministic Automata

In this section we provide lower bounds on the length of shortest mortal words in several classes of deterministic automata. It is known that every n-state non-complete deterministic automaton admits a mortal word of length at most $\frac{n(n+1)}{2}$, and the bound is tight [15] (the lower bound is the best known even for unambiguous automata). For strongly connected binary automata the best known lower bound is of order $\frac{n^2}{4} + O(n)$, see, e.g., [11]. We concentrate on several subclasses of strongly connected automata and show similar lower bounds for them. We start with a short proof of a lower bound of Pribavkina for Huffman decoders [14], basing directly on the structure of the automaton. Then we use a similar technique to provide quadratic upper bounds for two other classes of automata. For all figures of this section dashed lines represent transitions for a, dotted lines for b, and solid lines for both a and b.

A strongly connected deterministic automaton is called a *Huffman decoder* if it has a state which is contained in every cycle. Huffman decoders are decoders of the star of finite prefix codes [5].

Proposition 8. *For every odd $n \geq 3$ there exists an n-state binary non-complete Huffman decoder $\mathcal{A}$ with the length of a shortest mortal word equal to $\frac{n^2+4n-1}{4}$.*

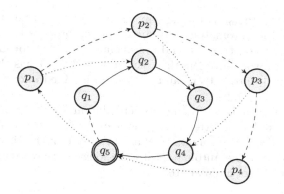

Fig. 1. A Huffman decoder with 9 states.

Proof. Consider the following construction of a non-complete deterministic automaton $A = (Q, \{a, b\}, \delta)$. The automaton has a cycle $q_1, \ldots, q_m \in Q$ such that it is possible to leave this cycle only at the state q_m. More formally, $\delta(q_i, x) = q_{i+1}$ for $1 \leq i \leq m-1$ and $x \in \{a, b\}$. The state q_m is mapped to q_1 by a.

Besides the cycle $q_1, \ldots, q_m$, $\mathcal{A}$ has a chain $p_1, \ldots, p_{m-1}$ of states leading to the only state with an undefined transition, which is p_{m-1}. More precisely, the letter a moves the states of the chain (except p_{m-1}) one step forward: $\delta(p_i, a) = p_{i+1}$ for $1 \leq i \leq m-2$. The letter b returns the states of the chain to the cycle (to the same position as like they did not leave it): $\delta(p_i, b) = q_{i+1}$ for $1 \leq i \leq m-1$. The only undefined transition is by the letter a at the state p_{m-1}. Finally, the state q_m is mapped to p_1 by b, which is the only way to reach the chain $p_1, \ldots, p_{m-1}$ from the cycle $q_1, \ldots, q_m$. See Fig. 1 for the example with $m = 5$.

In the automaton $\mathcal{A}$ thus constructed all the cycles pass through the state q_m, hence it is a Huffman decoder.

Observe that the word $ba^{m-1}(bba^{m-1})^{m-1}$ of length $m^2 + m - 1$ is mortal for $\mathcal{A}$. We are going to show that it is in fact a shortest mortal word.

Indeed, $\mathcal{A}$ has period m, which means that Q can be partitioned into m equivalence classes such that no pair of states from different classes can be mapped to the same state. More precisely, the classes are $\{q_i, p_i\}$ for $1 \leq i \leq m-1$ and $\{q_m\}$. The only way to decrease the number of active classes is to map a state of some active class to the state p_{m-1} and then apply a. We call a class active if at least one state from it is active. Let us start with the set $\{q_1, \ldots, q_m\}$ of active states, one from each class. To kill a state from this set (by sending it to p_{m-1} first) we have to apply the word ba^{m-1}, and during the application of this word no other active state leaves the set $\{q_1, \ldots, q_m\}$. Further, after any application of ba^{m-1} the state q_m cannot be active, thus we have to apply b to make q_m active so that we can kill another active state. We have to perform that for each of m classes, thus the length of a shortest mortal word is at least $m + (m+1)(m-1) = m^2 + m - 1$.   □

If we take the state q_m to be the only initial and accepting state, the prefix code decoded by the automaton constructed in the proof is the code $\{a, b\}^m \setminus \{ba^{m-1}\}$. The same construction and arguments show that for any alphabet Σ and an unbordered word u of length m the code $\Sigma^m \setminus \{u\}$ has the same length of a shortest mortal word [14] (recall that a word is called *unbordered* is none of its prefix equals to its suffix).

We proceed with a lower bound for circular automata. An automaton is called *circular* if it has a letter acting as a cyclic permutation on the whole set of its states. Circular automata play a crucial role in the theory of synchronizing automata since the automata with the longest known length of shortest synchronizing words are circular [16].

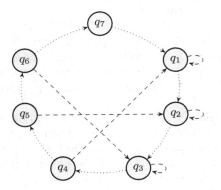

Fig. 2. A circular automaton with 7 states.

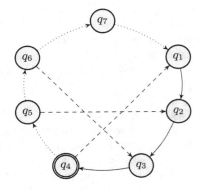

Fig. 3. A circular Huffman decoder with 7 states.

Proposition 9. *For every odd $n \geq 3$ there exists an n-state non-complete deterministic circular automaton $\mathcal{A}$ with the length of a shortest mortal word equal to $\frac{n^2 + 2n + 1}{4}$.*

Proof. Consider the following construction of a non-complete deterministic automaton $\mathcal{A} = (Q, \{a, b\}, \delta)$. The automaton has a cycle $q_1, \ldots, q_n \in Q$ which is induced by the letter b. This means that $\delta(q_i, b) = q_{i+1}$ for $1 \leq i \leq n - 1$, and $\delta(q_n, b) = q_1$. The letter a is undefined for q_n. Let $n = 2m + 1$. For q_i, $1 \leq i \leq m$, a acts as a self-loop. Finally, for q_i, $m + 1 \leq i \leq 2m$, a sends the state m steps back on the cycle: $\delta(q_i, a) = q_{i-m}$. See Fig. 2 for an example.

Observe that $a(b^{m+1}a)^m$ is a mortal word of length $m^2 + 2m + 1 = \frac{n^2 + 2n + 1}{4}$. We will show that this is in fact a shortest mortal word. Let us start with a set $\{q_n, q_1, \ldots, q_m\}$ of active states, and let w be a shortest word killing all these states. There are two options: either w first maps a pair of active states to the same state, or it first kills one state. In both cases after that the set of active states is exactly $\{q_1, \ldots, q_m\}$. No pair of active states can be mapped to the same state after that, so the only way to decrease the number of active states is to map a state to q_n and then to apply a. After each application of a the set of active states is a subset of $\{q_1, \ldots, q_m\}$, and the number of active states is decreased by at most 1. Thus, every mortal word has length at least $1 + (m+2)m$. □

Proposition 10. *For every odd* $n \geq 3$ *there exists an* n-*state non-complete circular Huffman decoder with the length of a shortest mortal word at least* $\frac{n^2}{8}$.

Proof. We modify the construction of Proposition 9. Consider the following construction of a non-complete deterministic automaton $\mathcal{A} = (Q, \{a, b\}, \delta)$. The automaton has a cycle $q_1, \ldots, q_n \in Q$ which is induced by the letter b. This means that $\delta(q_i, b) = q_{i+1}$ for $1 \leq i \leq n - 1$, and $\delta(q_n, b) = q_1$. The letter a is undefined for q_n. Let $n = 2m + 1$. For q_i, $1 \leq i \leq m$, a sends the state q_i one step forward on the cycle: $\delta(q_i, a) = q_{i+1}$. Finally, for q_i, $m + 1 \leq i \leq 2m$, a sends the state m steps back on the cycle: $\delta(q_i, a) = q_{i-m}$. See Fig. 3 for an example. The proof of the lower bound is similar to the previous two proofs, so we omit it because of the space constraints. □

It is an interesting question whether there is a connection of the length of shortest mortal words with properties of graphs and matrices, similar to the synchronization case [1].

7 Codes of Full Combinatorial Rank

In this section we use unambiguous automata to prove a result about synchronizing codes. A *bi-infinite word* over an alphabet Σ is a bi-infinite sequence of symbols from Σ, i.e. a mapping $w : \mathbb{Z} \to \Sigma$. A bi-infinite word w is *periodic* if there exists an integer k such that $w(i) = w(i + k)$ for every $i \in \mathbb{Z}$. A finite word which is not a power of a shorter word is called *primitive*.

Proposition 11. *Let* X *be a recognizable (by a finite automaton) code consisting of at least two words. If* X *is not synchronizing, then there exists an infinite number of bi-infinite periodic words having two different factorizations over* X.

Proof. Construct a strongly connected unambiguous automaton $\mathcal{A} = (Q, \Sigma, \Delta)$ such that X is the set of first return words of some of its state i (such an automaton always exists, see Sect. 4 of [4]). If $\mathcal{A}$ is not synchronizing, the minimum non-zero rank r of words in $\mathcal{A}$ is at least 2. Then there exists an idempotent $e \in \Sigma^*$ of rank r stabilizing i (Theorem 9.3.10 of [4]), that is, $i \in i.e$, and for every state $q \in Q$ we have $q.e^2 = q.e$. Consider the set W of all non-mortal words eue with $u \in \Sigma^*$. Let S be the set of fixed points of e, that is, states q with $q \in q.e$. The set W acts on S as a transitive permutation group of degree r, with $r \geq 2$ (Theorem 9.3.10 of [4]). Thus, there exists a word $w = eme \in W$

having $q \in i.w$, where $q \in S$, $q \neq i$. Let ℓ be the smallest positive number such that $i \in i.w^\ell$. The states from S in the sets $\{i\}, i.w, i.w^2, \ldots, i.w^{\ell-1}$ are pairwise different. Thus, the word $w^\ell \in X^*$ is not a power of a shorter word from X^*.

Let x_1, x_2 be two words in X. At least one of them, say x_1, does not commute with $w' = eem$, that is, $x_1 w' \neq w' x_1$. Then x_1 and w' are not the powers of the same word. We are going to show that the words $w' x_1^k$ are primitive for large enough k. Indeed, assume that $u^t = w' x_1^k$ for $t > 1$. If $|x_1^k| \geq |x_1| + |u|$, then by the periodicity lemma [7] we get that u and x_1 are powers of the same word, and thus w' is also a power of this word, which is not possible. If $|x_1^k| < |x_1| + |u|$, then $t = 2$ and $|w'| + |x_1| \geq |x_1^{k-1}|$, which is false for k large enough.

Denote $w_k = em x_1^k e$. Since the word $w' x_1^k = eem x_1^k$ is primitive for large enough k, so is w_k. Since the word x_1 stabilizes i, the word w_k is not mortal for every k since $i \in q'.w_k$, where $q' \in S$ is a state such that $i \in q'.w$. Moreover, w_k has rank r, and we have $i \in q'.w_k$, so $i \notin i.w_k$. Thus in the same way as for w, we get for every large enough k a word $(w_k)^{\ell_k} \in X^*$ which is not a power of a shorter word in X^*. To every word w_k we put in correspondence a bi-infinite word obtained by repeating w_k. Since the lengths of the words w_k increase and the words are primitive for large enough k, all these bi-infinite words are pairwise different. Each of them has two different factorizations over X.   □

The converse statement is not true. To see that, it is enough to consider a non-synchronizing code (e.g., $\{aaaa, aabb, bbaa, bbbb\}$) which is contained in a synchronizing code (e.g., $\{aaaa, aabb, bbaa, bbbb, abab, baba\}$).

The *combinatorial rank* of a set X of words over an alphabet Σ is the minimum cardinality of a set $Y \subseteq \Sigma^*$ such that $X \subseteq Y^*$. That is, it is the minimum cardinality of a set of words such that every word of X can be written as a concatenation of words from Y. For example, the combinatorial rank of every binary code is exactly two. By the result of [9] every finite code X with the combinatorial rank equal to the cardinality of X has only finite number of bi-infinite words with two different factorizations over X. By Proposition 11 this means that this code is synchronizing. Thus we get the following.

Theorem 12. *Every finite code such that its cardinality equals its combinatorial rank is synchronizing.*

Since every two-word code has combinatorial rank 2, we get in particular that every two-word code is synchronizing. Consider the following examples. For the two-word code $\{x, y\}$ with $x = abab$, $y = baba$ none of the codewords is synchronizing, but xy and yx are. For the two-word code $\{x, y\}$ with $x = b$, $y = abab$ none of the codewords is synchronizing, xy is not synchronizing, but yx is. For the two-word code $\{x, y\}$ with $x = ababa$, $y = bab$ none of the codewords is synchronizing, xy and yx are not synchronizing, but xx and yy are. We conjecture that for every two-word code X there always exists a synchronizing word in X^2.

Acknowledgments. I am grateful to Dominique Perrin for many helpful discussions and his constant interest to this work. I also thank Vladimir Gusev, Stefan Kiefer and

Elena Pribavkina for their useful comments on an early version of this manuscript, and anonymous reviewers for their suggestions on the presentation of the paper.

References

1. Ananichev, D.S., Volkov, M.V., Gusev, V.V.: Primitive digraphs with large exponents and slowly synchronizing automata. J. Math. Sci. **192**(3), 263–278 (2013). https://doi.org/10.1007/s10958-013-1392-8
2. Béal, M.P., Czeizler, E., Kari, J., Perrin, D.: Unambiguous automata. Math. Comput. Sci. **1**(4), 625–638 (2008). https://doi.org/10.1007/s11786-007-0027-1
3. Berlinkov, M.V., Szykuła, M.: Algebraic synchronization criterion and computing reset words. Inf. Sci. **369**, 718–730 (2016). https://doi.org/10.1016/j.ins.2016.07.049
4. Berstel, J., Perrin, D., Reutenauer, C.: Codes and Automata. Encyclopedia of Mathematics and its Applications, vol. 129. Cambridge University Press, Cambridge (2010)
5. Biskup, M.T., Plandowski, W.: Shortest synchronizing strings for huffman codes. Theoret. Comput. Sci. **410**(38), 3925–3941 (2009). https://doi.org/10.1016/j.tcs.2009.06.005
6. Carpi, A.: On synchronizing unambiguous automata. Theoret. Comput. Sci. **60**(3), 285–296 (1988). https://doi.org/10.1016/0304-3975(88)90114-4
7. Fine, N.J., Wilf, H.S.: Uniqueness theorems for periodic functions. Proc. Am. Math. Soc. **16**(1), 109–114 (1965)
8. Froidure, V.: Ranks of binary relations. Semigroup Forum **54**(1), 381–401 (1997). https://doi.org/10.1007/BF02676619
9. Karhumäki, J., Manuch, J., Plandowski, W.: A defect theorem for bi-infinite words. Theoret. Comput. Sci. **292**(1), 237–243 (2003). https://doi.org/10.1016/S0304-3975(01)00225-0
10. Kiefer, S., Mascle, C.: On finite monoids over nonnegative integer matrices and short killing words. In: Niedermeier, R., Paul, C. (eds.) 36th International Symposium on Theoretical Aspects of Computer Science (STACS 2019). Leibniz International Proceedings in Informatics (LIPIcs), vol. 126, pp. 43:1–43:13. Schloss Dagstuhl-Leibniz-Zentrum fuer Informatik, Dagstuhl (2019). https://doi.org/10.4230/LIPIcs.STACS.2019.43
11. Martugin, P.: A series of slowly synchronizing automata with a zero state over a small alphabet. Inf. Comput. **206**(9), 1197–1203 (2008). https://doi.org/10.1016/j.ic.2008.03.020
12. Néraud, J., Selmi, C.: On codes with a finite deciphering delay: constructing uncompletable words. Theoret. Comput. Sci. **255**(1), 151–162 (2001). https://doi.org/10.1016/S0304-3975(99)00160-7
13. Pouzet, M., Woodrow, R., Zaguia, N.: Generating boxes from ordered sets and graphs. Order **9**(2), 111–126 (1992). https://doi.org/10.1007/BF00814404
14. Pribavkina, E.V.: Slowly synchronizing automata with zero and noncomplete sets. Math. Notes **90**(3), 422–430 (2011). https://doi.org/10.1134/S0001434611090094
15. Rystsov, I.K.: Reset words for commutative and solvable automata. Theoret. Comput. Sci. **172**(1), 273–279 (1997). https://doi.org/10.1016/S0304-3975(96)00136-3
16. Volkov, M.V.: Synchronizing automata and the Černý conjecture. In: Martín-Vide, C., Otto, F., Fernau, H. (eds.) LATA 2008. LNCS, vol. 5196, pp. 11–27. Springer, Heidelberg (2008). https://doi.org/10.1007/978-3-540-88282-4_4

On Discrete Idempotent Paths

Luigi Santocanale[(✉)] [iD]

Laboratoire d'Informatique et des Systèmes, UMR 7020,
Aix-Marseille Université, CNRS, Marseille, France
luigi.santocanale@lis-lab.fr

Abstract. The set of discrete lattice paths from $(0,0)$ to (n,n) with North and East steps (i.e. words $w \in \{x,y\}^*$ such that $|w|_x = |w|_y = n$) has a canonical monoid structure inherited from the bijection with the set of join-continuous maps from the chain $\{0,1,\ldots,n\}$ to itself. We explicitly describe this monoid structure and, relying on a general characterization of idempotent join-continuous maps from a complete lattice to itself, we characterize idempotent paths as upper zigzag paths. We argue that these paths are counted by the odd Fibonacci numbers. Our method yields a geometric/combinatorial proof of counting results, due to Howie and to Laradji and Umar, for idempotents in monoids of monotone endomaps on finite chains.

Keywords: Discrete path · Idempotent · Join-continuous map

1 Introduction

Discrete lattice paths from $(0,0)$ to (n,m) with North and East steps have a standard representation as words $w \in \{x,y\}^*$ such that $|w|_x = n$ and $|w|_y = m$. The set $P(n,m)$ of these paths, with the dominance ordering, is a distributive lattice (and therefore of a Heyting algebra), see e.g. [2,8,9,18]. A simple proof that the dominance ordering is a lattice relies on the bijective correspondence between these paths and monotone maps from the chain $\{1,\ldots,n\}$ to the chain $\{0,1,\ldots,m\}$, see e.g. [2,3]. In turn, these maps bijectively correspond to join-continuous maps from $\{0,1,\ldots,n\}$ to $\{0,1,\ldots,m\}$ (those order preserving maps that sends 0 to 0). Join-continuous maps from a complete lattice to itself form, when given the pointwise ordering, a complete lattice in which composition distributes with joins. This kind of algebraic structure combining a monoid operation with a lattice structure is called a quantale [19] or (roughly speaking) a residuated lattice [10]. Therefore, the aforementioned bijection also witnesses a richer structure for $P(n,n)$, that of a quantale and of a residuated lattice. The set $P(n,n)$ is actually a *star-autonomous* quantale or, as a residuated lattice, *involutive*, see [12].

Partially supported by the "LIA LYSM AMU CNRS ECM INdAM" and by the "LIA LIRCO".

R. Mercaş and D. Reidenbach (Eds.): WORDS 2019, LNCS 11682, pp. 312–325, 2019.
https://doi.org/10.1007/978-3-030-28796-2_25

A main aim of this paper is to draw attention to the interplay between the algebraic/enumerative combinatorics of paths and these algebraic structures (lattices, Heyting algebras, quantales, residuated lattices) that, curiously, are all related to logic. We focus in this paper on the monoid structure that corresponds under the bijection to function composition—which, from a logical perspective, can be understood as a sort of non-commutative conjunction. In the literature, the monoid structure appears to be less known than the lattice structure. A notable exception is the work [17] where a different kind of lattice paths, related to Delannoy paths, are considered so to represent monoids of injective order-preserving partial transformations on chains.

We explicitly describe the monoid structure of $P(n, n)$ and characterize those paths that are idempotents. Our characterization relies on a general characterization of idempotent join-continuous maps from a complete lattice to itself. When the complete lattice is the chain $\{0, 1, \ldots, n\}$, this characterization yields a description of idempotent paths as those paths whose all North-East turns are above the line $y = x + \frac{1}{2}$ and whose all East-North turns are below this line. We call these paths *upper zigzag*. We use this characterization to provide a geometric/combinatorial proof that upper zigzag paths in $P(n, n)$ are counted by the odd Fibonacci numbers f_{2n+1}. Simple algebraic connections among the monoid structure on $P(n, n)$, the monoid $\mathcal{O}_n$ of order preserving maps from $\{1, \ldots, n\}$ to itself, and the submonoid $\mathcal{O}_n^n$ of $\mathcal{O}_n$ of maps fixing n, yield a geometric/combinatorial proof of counting results due to Howie [13] (the number of idempotents in $\mathcal{O}_n$ is the even Fibonacci numbers f_{2n}) and Laradji and Umar [16] (the number of idempotents in $\mathcal{O}_n^n$ is the odd Fibonacci numbers f_{2n-1}).

2   A Product on Paths

In the following, $P(n, m)$ shall denote the set of words $w \in \{x, y\}^*$ such that $|w|_x = n$ and $|w|_y = m$. We identify a word $w \in P(n, m)$ with a discrete path from $(0, 0)$ to (n, m) which uses only East and North steps of length 1. For example, the word $yxxxyxyyxy \in P(5, 5)$ is identified with the path in Fig. 1.

Let L_0, L_1 be complete lattices. A map $f : L_0 \rightarrow L_1$ is *join-continuous* if $f(\bigvee X) = \bigvee f(X)$, for each subset X of L_0. We use $Q_\vee(L_0, L_1)$ to denote the set of join-continuous maps from L_0 to L_1. If $L_0 = L_1 = L$, then we write $Q_\vee(L)$ for $Q_\vee(L, L)$.

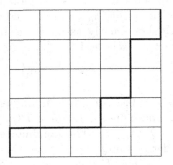

Fig. 1. The path $yxxxyxyyxy$.

The set $Q_\vee(L_0, L_1)$ can be ordered pointwise (i.e. $f \leq g$ if and only if $f(x) \leq g(x)$, for each $x \in L_0$); with this ordering it is a complete lattice. Function composition distributes over (possibly infinite) joins:

$$\left(\bigvee_{j \in J} g_j\right) \circ \left(\bigvee_{i \in I} f_i\right) = \bigvee_{j \in J, i \in I} (g_j \circ f_i), \tag{1}$$

whenever L_0, L_1, L_2 are complete lattices, $\{\, f_i \mid i \in I \,\} \subseteq Q_\vee(L_0, L_1)$ and $\{\, g_j \mid j \in J \,\} \subseteq Q_\vee(L_1, L_2)$. A *quantale* (see [19]) is a complete lattice endowed with a semigroup operation $\circ$ satisfying the distributive law (1). Thus, $Q_\vee(L)$ is a quantale, for each complete lattice $Q_\vee(L)$.

For $k \geq 0$, we shall use $\mathbb{I}_k$ to denote the chain $\{\, 0, 1, \ldots, k \,\}$. Notice that $f : \mathbb{I}_n \to \mathbb{I}_m$ is join-continuous if and only if it is monotone (or order-preserving) and $f(0) = 0$. For each $n, m \geq 0$, there is a well-known bijective correspondence between paths in $P(n, m)$ and join-continuous maps in $Q_\vee(\mathbb{I}_n, \mathbb{I}_m)$; next, we recall this bijection. If $w \in P(n, m)$, then the occurrences of y in w split w into $m + 1$ (possibly empty) blocks of contiguous xs, that we index by the numbers $0, \ldots, m$:

$$w = \mathbf{bl}_0^{w,x} \cdot y \cdot \mathbf{bl}_1^{w,x} \cdot y \ldots \mathbf{bl}_{m-1}^{w,x} \cdot y \cdot \mathbf{bl}_m^{w,x} \,.$$

We call the words $\mathbf{bl}_0^{w,x}, \mathbf{bl}_1^{w,x}, \ldots, \mathbf{bl}_m^{w,x} \in \{\, x \,\}^*$ the x-blocks of w. Given $i \in \{\, 1, \ldots, n \,\}$, the index of the block of the i-th occurrence of the letter x in w is denoted by $\mathbf{blno}_i^{w,x}$. We have therefore $\mathbf{blno}_i^{w,x} \in \{\, 0, \ldots, m \,\}$. Notice that $\mathbf{blno}_i^{w,x}$ equals the number of ys preceding the i-th occurrence of x in w so, in particular, $\mathbf{blno}_i^{w,x}$ can be interpreted as the height of the i-th occurrence of x when w is considered as a path. Similar definitions, $\mathbf{bl}_j^{w,y}$ and $\mathbf{blno}_j^{w,y}$, for $j = 1, \ldots, m$, are given for the blocks obtained by splitting w by means of the xs:

$$w = \mathbf{bl}_0^{w,y} \cdot x \cdot \mathbf{bl}_1^{w,y} \cdot x \ldots \mathbf{bl}_{n-1}^{w,y} \cdot x \cdot \mathbf{bl}_n^{w,y} \,.$$

The map $\mathbf{blno}^{w,x}$, sending $i \in \{\, 1, \ldots, n \,\}$ to $\mathbf{blno}_i^{w,x}$, is monotone from the chain $\{\, 1, \ldots, n \,\}$ to the chain $\{\, 0, 1, \ldots, m \,\}$. There is an obvious bijective correspondence from the set of monotone maps from $\{\, 1, \ldots, n \,\}$ to $\mathbb{I}_m = \{\, 0, 1, \ldots, m \,\}$ to the set $Q_\vee(\mathbb{I}_n, \mathbb{I}_m)$ obtained by extending a monotone f by setting $f(0) := 0$. We shall tacitly assume this bijection and, accordingly, we set $\mathbf{blno}_0^{w,x} := 0$. Next, by setting $\mathbf{blno}_{n+1}^{w,x} := m$, we notice that

$$|\mathbf{bl}_i^{w,y}| = \mathbf{blno}_{i+1}^{w,x} - \mathbf{blno}_i^{w,x} \,,$$

for $i = 0, \ldots, n$, so w is uniquely determined by the map $\mathbf{blno}^{w,x}$. Therefore, the mapping sending $w \in P(n, m)$ to $\mathbf{blno}^{w,x}$ is a bijection from $P(n, m)$ to the set $Q_\vee(\mathbb{I}_n, \mathbb{I}_m)$. The dominance ordering on $P(n, m)$ arises from the pointwise ordering on $Q_\vee(\mathbb{I}_n, \mathbb{I}_m)$ via the bijection.

For $w \in P(n, m)$ and $u \in P(m, k)$, the product $w \otimes u$ is defined by concatenating the x-blocks of w and the y-blocks of u:

Definition 1. *For $w \in P(n, m)$ and $u \in P(m, k)$, we let*

$$w \otimes u := \mathbf{bl}_0^{w,x} \cdot \mathbf{bl}_0^{u,y} \cdot \mathbf{bl}_1^{w,x} \cdot \mathbf{bl}_1^{u,y} \ldots \mathbf{bl}_m^{w,x} \cdot \mathbf{bl}_m^{u,y} \,.$$

Example 2. Let $w = yxxyxy$ and $u = xyxyyx$, so the x-blocks of w are $\epsilon, xx, x, \epsilon$ and the y-blocks of u are $\epsilon, y, yy, \epsilon$; we have $w \otimes u = xxyxyy$. We can trace the original blocks by inserting vertical bars in $w \otimes u$ so to separate $\mathbf{bl}_i^{w,x} \mathbf{bl}_i^{u,y}$

from $\mathbf{bl}_{i+1}^{w,x}\mathbf{bl}_{i+1}^{w,y}$, $i = 0,\ldots,m-1$. That is, we can write $w \otimes_{tr} u = |xxy|xyy|$, so $w \otimes u$ is obtained from $w \otimes_{tr} u$ by deleting vertical bars. Notice that also w and u can be recovered from $w \otimes_{tr} u$, for example w is obtained from $w \otimes_{tr} u$ by deleting the letter y and then renaming the vertical bars to the letter y. Figure 2 suggests that $\otimes$ is a form of synchronisation product, obtained by shuffling the x-blocks of w with the y-blocks of u so to give "priority" to all the xs (that is, the xs precede the ys in each block). It can be argued that there are other similar products, for example, the one where the ys precede the xs in each block, so $w \oplus u = yxxyyx$. It is easy to see that $w \oplus u = (u^{\star} \otimes w^{\star})^{\star}$, where $w^{\star}$ is the image of w along the morphism that exchanges the letters x and y.

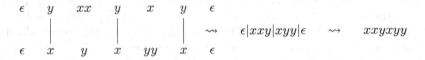

Fig. 2. Construction of the product $yxxyxy \otimes xyxyyx$.

Proposition 3. *The product $\otimes$ corresponds, under the bijection, to function composition. That is, we have*

$$\mathbf{blno}^{w \otimes u, x} = \mathbf{blno}^{u,x} \circ \mathbf{blno}^{w,x}.$$

Proof. In order to count the number of ys preceding the i-th occurrence of x in $w \otimes u$, it is enough to identify the block number j of this occurrence in w, and then count how many ys precede the j-th occurrence of x in u. That is, we have $\mathbf{blno}_i^{w \otimes u, x} = \mathbf{blno}_j^{u,x}$ with $j = \mathbf{blno}_i^{w,x}$.   □

Remark 4. Let us exemplify how the algebraic structure of $Q_{\vee}(\mathbb{I}_n, \mathbb{I}_m)$ yields combinatorial identities. The product is a function $\otimes : P(n,m) \times P(m,k) \to P(n,k)$, so we study how many preimages a word $w \in P(n,k)$ might have. By reverting the operational description of the product previously given, this amounts to inserting m vertical bars marking the beginning-end of blocks (so to guess a word of the form $u_0 \otimes_{tr} u_1$) under a constraint that we describe next. Each position can be barred more than once, so adding j bars can be done in $\binom{n+k+j}{j}$ ways. The only constraint we need to satisfy is the following. Recall that a position $\ell \in \{0,\ldots,n+k\}$ is a *North-East turn* (or a *descent*), see [15], if $\ell > 0$, $w_{\ell-1} = y$ and $w_\ell = x$. If a position is a North-East turn, then such a position is necessarily barred. Let us illustrate this with the word $xxyxyy$ which has just one descent, which is necessarily barred: $xxy|xyy$. Assuming $m = 3$, we need to add two more vertical bars. For example, for $x|xy|xy|y$ we obtain the following decomposition:

$$x|xy|xy|y \rightsquigarrow (x|x|x|, |y|y|y) \rightsquigarrow (xyxyxy, xyxyxy).$$

Therefore, if w has i descents, then these positions are barred, while the other $m-i$ barred positions can be chosen arbitrarily, and there are $\binom{n+k+m-i}{m-i}$ ways to do this. Recall that there are $\binom{n}{i}\binom{k}{i}$ words $w \in P(n,k)$ with i descents, since such a w is determined by the subsets of $\{1, \ldots, n\}$ and $\{1, \ldots, k\}$ of cardinality i, determining the descents. Summing up w.r.t. the number of descents, we obtain the following formulas:

$$\binom{n+m}{n}\binom{m+k}{k} = \sum_{i=0}^{m}\binom{n+m+k-i}{m-i}\binom{n}{i}\binom{k}{i}, \quad \binom{2n}{n}^2 = \sum_{i=0}^{n}\binom{3n-i}{n-i}\binom{n}{i}^2.$$

Similar kind of combinatorial transformations and identities appear in [5,6,11], yet it is not clear to us at the moment of writing whether these works relate in some way to the product of paths studied here.

Remark 5. The previous remark also shows that if $w \in P(n,k)$ has $m \geq 0$ descents, then there is a canonical factorization $w = w_0 \otimes w_1$ with $w_0 \in P(n,m)$ and $w_1 \in P(m,k)$. It is readily seen that, via the bijection, this is the standard epi-mono factorization in the category of join-semilattices. The word $xxyxyy$, barred at its unique descent as $xxy|xyy$, is decomposed into $xxyx$ and $yxyy$.

Remark 6. As in [17], many semigroup-theoretic properties of the monoid $Q_\vee(\mathbb{I}_n)$ can be read out of (and computed from) the bijection with $P(n,n)$. For example

$$\mathrm{card}(\{f \in Q_\vee(\mathbb{I}_n) \mid \mathrm{card}(Image(f)) = k+1\}) = \binom{n}{k}^2$$

since, as in the previous remark, a path with k North-East turns corresponds to a join-continuous map f such that $\mathrm{card}(Image(f)) = k+1$. Similarly

$$\mathrm{card}(\{f \in Q_\vee(\mathbb{I}_n) \mid \max(Image(f)) = k\}) = \binom{n+k-1}{k}$$

since a map $f \in Q_\vee(\mathbb{I}_n)$ such that $\max(Image(f)) = k$ (i.e. $f(n) = k$) corresponds to a path in $P(n,k)$ whose last step is an East step, thus to a path in $P(n-1,k)$. A similar argument can be used to count maps $f \in \mathcal{O}_n$ such that $f(n) = k$, cf. [16, Proposition 3.7].

Remark 7. Further properties of the monoid $Q_\vee(\mathbb{I}_n)$ can be easily verified, for example, this monoid is aperiodic. For the next observation, see also [16, Proposition 2.3] and [17, Theorem 3.4]. Recall that $f \in Q_\vee(\mathbb{I}_n)$ is *nilpotent* if, for some $\ell \geq 0$, f^ℓ is the bottom of the lattice, that is, it is the constant map with value 0. It is easily seen that f is nilpotent if and only if $f(x) < x$, for each $x = 1, \ldots, n$. Therefore, a path is nilpotent if and only it lies below the diagonal, that is, it is a Dyck path. Therefore, there are $\frac{1}{n+1}\binom{2n}{n}$ nilpotents in $Q_\vee(\mathbb{I}_n)$.

3   Idempotent Join-Continuous Maps as Emmentalers

We provide in this section a characterization of idempotent join-continuous maps from a complete lattice to itself. The characterization originates from the notion of EA-duet used to study some elementary subquotients in the category of lattices, see [20, Definition 9.1].

Definition 8. *An* emmentaler *of a complete lattice L is a collection $\mathcal{E} = \{\,[y_i, x_i] \mid i \in I\,\}$ of closed intervals of L such that*

- $[y_i, x_i] \cap [y_j, x_j] = \emptyset$, *for $i, j \in I$ with $i \neq j$,*
- $\{\, y_i \mid i \in I\,\}$ *is a subset of L closed under arbitrary joins,*
- $\{\, x_i \mid i \in I\,\}$ *is a subset of L closed under arbitrary meets.*

The main result of this section is the following statement.

Theorem 9. *For an arbitrary complete lattice L, there is a bijection between idempotent join-continuous maps from L to L and emmentalers of L.*

For an emmentaler $\mathcal{E} = \{\,[y_i, x_i] \mid i \in I\,\}$ of L, we let

$$J(\mathcal{E}) := \{\, y_i \mid i \in I\,\}, \qquad\qquad M(\mathcal{E}) := \{\, x_i \mid i \in I\,\}$$
$$\mathrm{int}_{\mathcal{E}}(z) := \bigvee\{\, y \in J(\mathcal{E}) \mid y \leq z\,\}, \qquad \mathrm{cl}_{\mathcal{E}}(z) := \bigwedge\{\, x \in M(\mathcal{E}) \mid z \leq x\,\}.$$

It is a standard fact that $\mathrm{cl}_{\mathcal{E}}$ is a closure operator on L (that is, it is a monotone inflating idempotent map from L to itself) and that $\mathrm{int}_{\mathcal{E}}$ is an interior operator on L (that is, a monotone, deflating, and idempotent endomap of L). In the following statements an emmentaler $\mathcal{E} = \{\,[y_i, x_i] \mid i \in I\,\}$ is fixed.

Lemma 10. *For each $i \in I$, $x_i = \mathrm{cl}_{\mathcal{E}}(y_i)$ and $\mathrm{int}_{\mathcal{E}}(x_i) = y_i$. Therefore $\mathrm{int}_{\mathcal{E}}$ restricts to an order isomorphism from $M(\mathcal{E})$ to $J(\mathcal{E})$ whose inverse is $\mathrm{cl}_{\mathcal{E}}$.*

Proof. Clearly, $\mathrm{cl}_{\mathcal{E}}(y_i) \leq x_i$. Let us suppose that $y_i \leq x_j$ yet $x_i \not\leq x_j$, then $y_i \leq x_j \wedge x_i < x_i$ and $x_j \wedge x_i = x_\ell$ for some $\ell \in I$ with $\ell \neq i$. But then $x_\ell \in [y_\ell, x_\ell] \cap [y_i, x_i]$, a contradiction. The equality $\mathrm{int}_{\mathcal{E}}(x_i) = y_i$ is proved similarly. $\qquad\square$

In view of the following lemma we think of $\mathcal{E}$ as a sublattice of L with prescribed holes/fillings, whence the naming "emmentaler".

Lemma 11. *If $\mathcal{E}$ is an emmentaler of L, then $\bigcup \mathcal{E}$ is a subset of L closed under arbitrary joins and meets. Moreover, the map sending $z \in [y_i, x_i]$ to y_i is a complete lattice homomorphism from $\bigcup \mathcal{E}$ to $J(\mathcal{E})$.*

Proof. Let $\{\, z_k \mid k \in K\,\}$ with $z_k \in [y_k, x_k]$ for each $k \in K$. Then, for some $j \in I$,

$$y_j = \bigvee_{k \in K} y_k \leq \bigvee_{k \in K} z_k \leq \bigvee_{k \in K} x_k \leq \mathrm{cl}_{\mathcal{E}}\left(\bigvee_{k \in K} x_k\right)$$
$$= \bigvee_{M(\mathcal{E})}\{\, x_k \mid k \in K\,\} = \bigvee_{M(\mathcal{E})}\{\, \mathrm{cl}_{\mathcal{E}}(y_k) \mid k \in K\,\} = \mathrm{cl}_{\mathcal{E}}\left(\bigvee_{k \in K} y_k\right) = \mathrm{cl}_{\mathcal{E}}(y_i) = x_j\,,$$

$$(2)$$

where in the second line we have used the fact that $\mathrm{cl}_{\mathcal{E}}(\bigvee_{k \in K} x_k)$ is the join in $M(\mathcal{E})$ of the family $\{\, x_k \mid k \in K\,\}$ and also the fact that $\mathrm{cl}_{\mathcal{E}}$ is an order isomorphism (so it is join-continuous) from $J(\mathcal{E})$ to $M(\mathcal{E})$. Therefore, $\bigvee_{k \in K} z_k \in \bigcup \mathcal{E}$ and, in a similar way, $\bigwedge_{k \in K} z_k \in \bigcup \mathcal{E}$.

Next, let $\pi : \bigcup \mathcal{E} \to J(\mathcal{E})$ be the map sending $z \in [y_i, x_i]$ to $y_i \in J(\mathcal{E})$. The computations in (2) show that π is join-continuous. With similar computations it is seen that $\bigwedge_{k \in K} z_k$ is sent to $\mathbf{int}_{\mathcal{E}}(\bigwedge_{k \in K} y_k)$ which is the meet of the family $\{ y_k \mid k \in K \}$ within $J(\mathcal{E})$. Therefore, π is meet-continuous as well. □

We recall next some facts on adjoint pairs of maps, see e.g. [4, §7]. Two monotone maps $f, g : L \to L$ form an *adjoint pair* if $f(x) \leq y$ if and only if $x \leq g(y)$, for each $x, y \in L$. More precisely, f is *left* (or *lower*) adjoint to g, and g is *right* (or *upper*) adjoint to f. Each map determines the other: that is, if f is left adjoint to g and g', then $g = g'$; if g is right adjoint to f and f', then $f = f'$. If L is a complete lattice, then a monotone $f : L \to L$ is a left adjoint (that is, there exists g for which f is left adjoint to g) if and only if it is join-continuous; under the same assumption, a monotone $g : L \to L$ is a right adjoint if and only if it is meet-continuous.

Proposition 12. *If $\mathcal{E}$ is an emmentaler of L, then the maps $f_{\mathcal{E}}$ and $g_{\mathcal{E}}$ defined by*

$$f_{\mathcal{E}}(z) := \mathbf{int}_{\mathcal{E}}(\mathbf{cl}_{\mathcal{E}}(z)), \qquad g_{\mathcal{E}}(z) := \mathbf{cl}_{\mathcal{E}}(\mathbf{int}_{\mathcal{E}}(z)),$$

are idempotent and adjoint to each other. In particular, $f_{\mathcal{E}}$ is join-continuous, so it belongs to $Q_\vee(L)$.

Proof. Clearly, $f_{\mathcal{E}}$ is idempotent:

$$\mathbf{int}_{\mathcal{E}}(\mathbf{cl}_{\mathcal{E}}(\mathbf{int}_{\mathcal{E}}(\mathbf{cl}_{\mathcal{E}}(z)))) = \mathbf{int}_{\mathcal{E}}(\mathbf{cl}_{\mathcal{E}}(z)),$$

since $\mathbf{cl}_{\mathcal{E}}(z) = x_i$ for some $i \in I$ and $\mathbf{cl}_{\mathcal{E}}(\mathbf{int}_{\mathcal{E}}(x_i)) = x_i$. In a similar way, $g_{\mathcal{E}}$ is idempotent. Let us argue that $f_{\mathcal{E}}$ and $g_{\mathcal{E}}$ are adjoint. If $z_0 \leq \mathbf{cl}_{\mathcal{E}}(\mathbf{int}_{\mathcal{E}}(z_1))$, then $\mathbf{cl}_{\mathcal{E}}(z_0) \leq \mathbf{cl}_{\mathcal{E}}(\mathbf{cl}_{\mathcal{E}}(\mathbf{int}_{\mathcal{E}}(z_1))) = \mathbf{cl}_{\mathcal{E}}(\mathbf{int}_{\mathcal{E}}(z_1))$ and $\mathbf{int}_{\mathcal{E}}(\mathbf{cl}_{\mathcal{E}}(z_0)) \leq \mathbf{int}_{\mathcal{E}}(\mathbf{cl}_{\mathcal{E}}(\mathbf{int}_{\mathcal{E}}(z_1))) = \mathbf{int}_{\mathcal{E}}(z_1) \leq z_1$. Similarly, if $\mathbf{int}_{\mathcal{E}}(\mathbf{cl}_{\mathcal{E}}(z_0)) \leq z_1$, then $z_0 \leq \mathbf{cl}_{\mathcal{E}}(\mathbf{int}_{\mathcal{E}}(z_1))$. □

Lemma 13. $J(\mathcal{E}) = Image(f_{\mathcal{E}})$ *and* $M(\mathcal{E}) = Image(g_{\mathcal{E}})$.

Proof. Clearly, if $y = \mathbf{int}_{\mathcal{E}}(\mathbf{cl}_{\mathcal{E}}(z))$ for some $z \in L$, then $y \in J(\mathcal{E})$. Conversely, if $y \in J(\mathcal{E})$, then $y = \mathbf{int}_{\mathcal{E}}(\mathbf{cl}_{\mathcal{E}}(y))$, so $y \in Image(f_{\mathcal{E}})$. The other equality is proved similarly. □

For the next proposition, recall that if f, g are adjoint, then $f \circ g \circ f = f$ and $g \circ f \circ g = g$.

Proposition 14. *Let $f \in Q_\vee(L)$ be idempotent and let g be its right adjoint. Then*

1. $y \leq g(y)$, *for each* $y \in Image(f)$,
2. *the collection of intervals* $\mathcal{E}_f := \{ [y, g(y)] \mid y \in Image(f) \}$ *is an emmentaler of L,*
3. $J(\mathcal{E}_f) = Image(f)$ *and* $M(\mathcal{E}_f) = Image(g)$.

Proof. If $y \in Image(f)$, then $y = f(y)$ and therefore the relation $y \leq g(y)$ follows from $f(y) \leq y$. The subset $Image(f)$ is closed under arbitrary joins since f is join-continuous. Similarly, $Image(g)$ is closed under arbitrary meets, since g is meet-continuous. Let us show that $\{ g(y) \mid y \in Image(f) \} = Image(g)$. To this end, observe that if $x = g(z)$ for some $z \in L$, then $x = g(z) = g(f(g(z)))$, so $x = g(y)$ with $y = f(g(z))$.

Finally, let $z \in [y_1, g(y_1)] \cap [y_2, g(y_2)]$. Then $y_i = f(y_i) \leq f(z) \leq f(g(y_i))$. We already observed that $f(g(y_i)) = y_i$, so $y_i = f(z)$, for $i = 1, 2$. We have therefore $y_1 = y_2$ and $g(y_1) = g(y_2)$.   □

Lemma 15. *If $f \in Q_\vee(L)$ is idempotent then, for each $x \in L$,*

1. $\mathbf{int}_{\mathcal{E}_f}(x) \leq f(x)$,
2. *if $f(x) \leq x$, then $f(x) = \mathbf{int}_{\mathcal{E}_f}(x)$,*
3. *if $x \in M(\mathcal{E}_f)$, then $f(x) \leq x$, and so $f(x) = \mathbf{int}_{\mathcal{E}_f}(x)$.*

Proof. 1. Recall that $\mathbf{int}_{\mathcal{E}_f}(x) \leq x$ and $\mathbf{int}_{\mathcal{E}_f}(x) \in J(\mathcal{E}_f) = Image(f)$, so $\mathbf{int}_{\mathcal{E}_f}(x)$ is a fixed point of f. Then, using monotonicity, $\mathbf{int}_{\mathcal{E}_f}(x) = f(\mathbf{int}_{\mathcal{E}_f}(x)) \leq f(x)$.
2. From $f(x) \leq x$ and recalling that $\mathbf{int}_{\mathcal{E}_f}(x)$ is the greatest element of $J(\mathcal{E}_f) = Image(f)$ below x, it immediately follows that $f(x) \leq \mathbf{int}_{\mathcal{E}_f}(x)$.
3. Recall that $M(\mathcal{E}_f) = Image(g)$, where g is right adjoint to f. Let z be such that $x = g(z)$, so we aim at proving that $f(g(z)) \leq g(z)$. This is follows from $f(f(g(z))) = f(g(z)) \leq z$ and adjointness.

     □

Proposition 16. *For each idempotent $f \in Q_\vee(L)$, we have $f = \mathbf{int}_{\mathcal{E}_f} \circ \mathbf{cl}_{\mathcal{E}_f} = f_{\mathcal{E}_f}$.*

Proof. Since $\mathbf{cl}_{\mathcal{E}_f}(z) \in M(\mathcal{E}_f)$, then $f(\mathbf{cl}_{\mathcal{E}_f}(z)) = \mathbf{int}_{\mathcal{E}_f}(\mathbf{cl}_{\mathcal{E}_f}(z))$, by the previous Lemma. Therefore we need to prove that $f(\mathbf{cl}_{\mathcal{E}_f}(z)) = f(z)$. This immediately follows from the relation $\mathbf{cl}_{\mathcal{E}_f} = g \circ f$ that we prove next.

We show that $g(f(z))$ is the least element of $Image(g)$ above z. We have $z \leq g(f(z)) \in Image(g)$ by adjointness. Suppose now that $x \in Image(g)$ and $z \leq x$. If $y \in L$ is such that $x = g(y)$, then $z \leq g(y)$ yields $f(z) \leq y$ and $g(f(z)) \leq g(y) = x$.   □

We can now give a proof of the main result of this section, Theorem 9.

Proof (Theorem 9). We argue that the mappings $\mathcal{E} \mapsto f_{\mathcal{E}}$ and $f \mapsto \mathcal{E}_f$ are inverse to each other.

We have seen (Proposition 16) that, for an idempotent $f \in Q_\vee(L)$, $f_{\mathcal{E}_f} = f$. Given an emmentaler $\mathcal{E}$, we have $J(\mathcal{E}) = Image(f_{\mathcal{E}})$ by Lemma 13, and $J(\mathcal{E}_{f_{\mathcal{E}}}) = Image(f_{\mathcal{E}})$, by Proposition 14. Therefore, $J(\mathcal{E}) = J(\mathcal{E}_{f_{\mathcal{E}}})$ and, similarly, $M(\mathcal{E}) = M(\mathcal{E}_{f_{\mathcal{E}}})$. Since the two sets $J(\mathcal{E})$ and $M(\mathcal{E})$ completely determine an emmentaler, we have $\mathcal{E} = \mathcal{E}_{f_{\mathcal{E}}}$.   □

4   Idempotent Discrete Paths

It is easily seen that an emmentaler of the chain $\mathbb{I}_n$ can be described by an alternating sequence of the form

$$0 = y_0 \leq x_0 < y_1 \leq x_1 < y_2 \leq \ldots < y_k \leq x_k = n\,,$$

so $J(\mathcal{E}) = \{0, y_1, \ldots, y_k\}$ and $M(\mathcal{E}) = \{x_1, x_2, \ldots, x_{k-1}, n\}$. Indeed, $J(\mathcal{E})$ is closed under arbitrary joins if and only if $0 \in J(\mathcal{E})$, while $M(\mathcal{E})$ is closed under arbitrary meets if and only if $n \in M(\mathcal{E})$.

The correspondences between idempotents of $Q_\vee(\mathbb{I}_n)$, their paths, and emmentalers can be made explicit as follows: for $y \in J(\mathcal{E})$ such that $y \neq 0$, the path corresponding to $f_{\mathcal{E}}$ touches the point (y, y) coming from the left of the diagonal; for $x \in M(\mathcal{E}) \setminus J(\mathcal{E})$, the path corresponding to $f_{\mathcal{E}}$ touches (x, x) coming from below the diagonal. For $\mathcal{E} = \{0 < 1 < 2 \leq 2 < 3 < 4\}$, with $J(\mathcal{E}) = \{0, 2, 3\}$ and $M(\mathcal{E}) = \{1, 2, 4\}$, the path corresponding to $f_{\mathcal{E}}$ is illustrated in Fig. 3. On the left of the figure, points of the form (x, x) with $x \in M(\mathcal{E})$ are squared, while points of the form (y, y) with $y \in J(\mathcal{E})$ are circled.

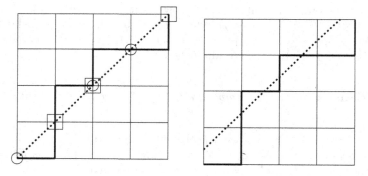

Fig. 3. Idempotent path corresponding to $\{0 < 1 < 2 \leq 2 < 3 < 4\}$.

Our next goal is to give a geometric characterization of idempotent paths using their North-East and East-North turns. To this end, observe that we can describe North-East turns of a path $w \in P(n, m)$ by discrete points in the plane. Namely, if if $w = w_0 w_1 \in P(n, m)$ with $w_0 = u_0 y$, $w_1 = x u_1$, and $|w_0| = \ell$ (so w has a North-East turn at position ℓ), then we can denote this North-East turn with the point $(|w_0|_x, |w_0|_y)$. In a similar way, we can describe East-North turns by discrete points in the plane.

Let us call a path an *upper zigzag* if every of its North-East turns is above the line $y = x + \frac{1}{2}$ while every of its East-North turns is below this line. Notice that a path is an upper zigzag if and only every North-East turn is of the form (x, y) with $x < y$ and every East-North turn is of the form (x, y) with $y \leq x$. This property is illustrated on the right of Fig. 3.

Theorem 17. *A path* $w \in P(n,n)$ *is idempotent if and only if it is an upper zigzag.*

The proof of the theorem is scattered into the next three lemmas.

Lemma 18. *An upper zigzag path is idempotent.*

Proof. Let w be an upper zigzag with $\{ (x_i, y_i) \mid i = 1, \ldots, k \}$ the set of its North-East turns. For $(i,j) \in \{0, \ldots, n-1\} \times \{1, \ldots, n\}$, let $e_{i,j} := x^i y^j x^{n-i} y^{n-j}$ be the path that has a unique North-East turn at (i,j). Notice that $w = \bigvee_{i=1,\ldots,k} e_{x_i, y_i}$. By Eq. (1),

$$w \otimes w = (\bigvee_{i=1,\ldots,k} e_{x_i, y_i}) \otimes (\bigvee_{j=1,\ldots,k} e_{x_j, y_j}) = \bigvee_{i,j=1,\ldots,k} e_{x_i, y_i} \otimes e_{x_j, y_j} . \qquad (3)$$

It is now enough to observe that $e_{a,b} \otimes e_{c,d} = e_{a,d}$ if $c < b$ and, otherwise, $e_{a,b} \otimes e_{c,d} = \bot$, where $\bot = x^n y^n$ is the least element of $P(n,n)$. Therefore, we have: (i) $e_{x_i, y_i} \otimes e_{x_i, y_i} = e_{x_i, y_i}$, since $x_i < y_i$, (ii) if $i < j$, then $e_{x_i, y_i} \otimes e_{x_j, y_j} = \bot$, since $y_i \leq x_j$, (iii) if $j < i$, then $e_{x_i, y_i} \otimes e_{x_j, y_j} = e_{x_i, y_j}$, since $x_j < y_j \leq y_i$; in the latter case, we also have $e_{x_i, y_j} \leq e_{x_i, y_i}$, since $y_j \leq y_i$. Consequently, the expression on the right of (3) evaluates to $\bigvee_{i=1,\ldots,k} e_{x_i, y_i} = w$.   □

Next, let us say that $i \in \mathbb{I}_n \setminus \{n\}$ is an *increase* of $f \in Q_\vee(\mathbb{I}_n, \mathbb{I}_m)$ if $f(i) < f(i+1)$. It is easy to see that the set of North-East turns of w is the set $\{ (i, \mathbf{blno}_{i+1}^{w,x}) \mid i \text{ is an increase of } \mathbf{blno}^{w,x} \}$.

Lemma 19. *Let* $f \in Q_\vee(\mathbb{I}_n, \mathbb{I}_n)$ *and let* g *be its right adjoint. Then* $i \in \mathbb{I}_n \setminus \{n\}$ *is an increase of* f *if and only if* $i \in Image(g) \setminus \{n\}$.

Proof. Suppose $i = g(j)$ for some $j \in \mathbb{I}_n$. If $f(i+1) \leq f(i)$, then $i+1 \leq g(f(i)) = g(f(g(j))) = g(j) = i$, a contradiction. Therefore $f(i) < f(i+1)$.

Conversely, if $f(i) < f(i+1)$, then $f(i+1) \not\leq f(i)$, $i+1 \not\leq g(f(i))$, and $g(f(i)) < i+1$. Since $i \leq g(f(i))$, then $g(f(i)) = i$, so $i \in Image(g)$.   □

Lemma 20. *The North-East turns of an idempotent path* $w \in P(n,n)$ *corresponding to the emmentaler* $\{0 = y_0 \leq x_0 < y_1, \ldots y_k \leq x_k = n\}$ *of* $\mathbb{I}_n$ *are of the form* $(x_\ell, y_{\ell+1})$, *for* $\ell = 0, \ldots, k-1$. *Its East-North turns are of the form* (x_ℓ, y_ℓ), *for* $\ell = 0, \ldots, k$. *Therefore* w *is an upper zigzag.*

Proof. For the first statement, since $Image(g_\mathcal{E}) = \{x_0, \ldots, x_{k-1}, n\}$ and using Lemma 19, we need to verify that $f_\mathcal{E}(x_\ell) = y_\ell$: this is Lemma 15, point 3. The last statement is a consequence of the fact that East-North turns are computable from North-East turns: if (x_i, y_i), $i = 1, \ldots, k$, are the North-East turns of w, with $x_i < x_j$ and $y_i < y_j$ for $i < j$, then East-North turns of w are of the form $(x_1, 0)$ (if $x_1 > 0$), (x_{i+1}, y_i), $i = 1, \ldots, k-1$, and (n, y_k) (if $y_k < n$).   □

5 Counting Idempotent Discrete Paths

The goal of this section is to exemplify how the characterizations of idempotent discrete paths given in Sect. 4 can be of use. It is immediate to establish a bijective correspondence between emmentalers of the chain $\mathbb{I}_n$ and words $w = w_0 \ldots w_n$ on the alphabet $\{\underline{1}, 0, 1\}$ that avoid the pattern $\underline{1}0^*\underline{1}$ and such that $w_0 = 1$ and $w_n \in \{1, \underline{1}\}$; this bijection can be exploited for the sake of counting. We prefer to count idempotents using the characterization given in Theorem 17. In the following, we provide a geometric/combinatorial proof of counting results [13,16] for the number of idempotent elements in the monoid $Q_\vee(\mathbb{I}_n)$ and, also, in the monoid $\mathcal{O}_n$ of order preserving maps from $\{1, \ldots, n\}$ to itself. Let us recall that the Fibonacci sequence is defined by $f_0 := 0$, $f_1 := 1$, and $f_{n+2} := f_{n+1} + f_n$. Howie [13] proved that $\phi_n = f_{2n}$ (for $n \geq 1$), where ϕ_n is the number of idempotents in the monoid $\mathcal{O}_n$. Laradji and Umar [16] proved that $\gamma_n = f_{2n-1}$ ($n \geq 1$), where now γ_n is the number of idempotent elements of $\mathcal{O}_n^n$, the submonoid in $\mathcal{O}_n$ of maps fixing n. Clearly, $\mathcal{O}_n^n$ is a monoid isomorphic (and anti-isomorphic as well) to $Q_\vee(\mathbb{I}_{n-1})$. We infer that the number ψ_n of idempotents in the monoid $Q_\vee(\mathbb{I}_n)$ equals f_{2n+1} (for $n \geq 0$).

Remark 21. It is argued in [13] that $\phi_n = \frac{1}{2^n \sqrt{5}}\{(3 + \sqrt{5})^n - (3 - \sqrt{5})^n\}$, which can easily be verified using the fact that $f_n = \frac{\theta_0^n - \theta_1^n}{\theta_0 - \theta_1}$ with $\theta_0 = \frac{1+\sqrt{5}}{2}$ and $\theta_1 = \frac{1-\sqrt{5}}{2}$, see [7]. In a similar way, we derive the following explicit formula:

$$\psi_n = \frac{1}{2^{n+1}\sqrt{5}}\{(3 + \sqrt{5})^n(1 + \sqrt{5}) - (3 - \sqrt{5})^n(1 - \sqrt{5})\}.$$

Let us observe that the monoid $\mathcal{O}_n$ can be identified with the submonoid of $Q_\vee(\mathbb{I}_n)$ of join-continuous maps f such that $1 \leq f(1)$. A path corresponds to such an f if and only if its first step is a North step. Having observed that $\psi_0 = \phi_1 = 1$, the following proposition suffices to assert that $\phi_n = f_{2n}$ and $\psi_n = f_{2n+1}$.

Proposition 22. *The following recursive relations hold:*

$$\phi_{n+1} = \psi_n + \phi_n, \qquad\qquad \psi_{n+1} = \phi_{n+1} + \psi_n.$$

Proof. Every discrete path from $(0,0)$ to $(n+1, n+1)$ ends with y—that is, it visits the point $(n+1, n)$—or ends with x—that is, it visits the point $(n, n+1)$. Consider now an upper zigzag path π from $(0,0)$ to $(n+1, n+1)$ that visits $(n+1, n)$, see Fig. 4. By clipping on the rectangle with left-bottom corner $(0,0)$ and right-up corner (n,n), we obtain an upper zigzag path π' from $(0,0)$ to (n, n). If π starts with y, then π' does as well. This proves the right part of the recurrences above, i.e. $\phi_{n+1} = \ldots + \phi_n$ and $\psi_{n+1} = \ldots + \psi_n$.

Consider now an upper zigzag path π ending with x, see Fig. 5. The reflection along the line $y = n - x$ sends (x, y) to $(n - y, n - x)$, so it preserves upper zigzag paths. Applying this reflection to π, we obtain an upper zigzag path from $(0,0)$ to $(n+1, n+1)$ whose first step is y. This proves the $\psi_{n+1} = \phi_{n+1} + \ldots$ part of the recurrences above.

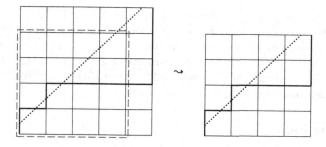

Fig. 4. An upper zigzag path to $(5,5)$ ending with y.

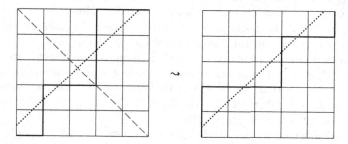

Fig. 5. An upper zigzag path to $(5,5)$ ending with x.

Consider now an upper zigzag path π ending with x and beginning with y, see Fig. 6. By clipping on the rectangle with left-bottom corner $(0,1)$ and right-up corner $(n, n+1)$ and then by applying the translation $x \mapsto x - 1$, we obtain a path whose all North-East turns are above the line $y = x - \frac{1}{2}$ and whose all East-North turns are below this line. By reflecting along diagonal, we obtain an upper zigzag path from $(0,0)$ to (n,n). This proves the $\phi_{n+1} = \psi_n + \ldots$ part of the recurrences above. $\qquad\square$

The geometric ideas used in the proof of Proposition 22 can be exploited further, so to show that the number of idempotent maps $f \in Q_\vee(\mathbb{I}_n)$ such that $f(n) = k$

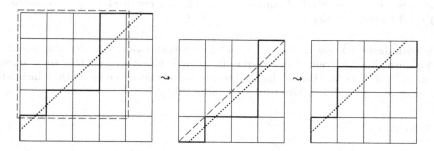

Fig. 6. An upper zigzag path to $(5,5)$ ending with x and beginning with y.

equals f_{2k}, see the analogous statement in [16, Corollary 4.5]. Indeed, if $f(n) = k$, then the path corresponding to f visits the points $(n-1, k)$ and (n, k); therefore, since it is an upper zigzag, also the points $(k-1, k)$ and (k, k). By clipping on the rectangle from $(0, 0)$ to (k, k), we obtain an upper zigzag path in $P(k, k)$ ending in x. As seen in the proof of Proposition 22, these paths bijectively correspond to upper zigzag paths in $P(k, k)$ beginning with y.

6   Conclusions

We have presented the monoid structure on the set $P(n, n)$ of discrete lattice paths (with North and East steps) that corresponds, under a well-known bijection, to the monoid $Q_\vee(\mathbb{I}_n)$ of join-continuous functions from the chain $\{0, 1, \ldots n\}$ to itself. In particular, we have studied the idempotents of this monoid, relying on a general characterization of idempotent join-continuous functions from a complete lattice to itself. This general characterization yields a bijection with a language of words on a three letter alphabet and a geometric description of idempotent paths. Using this characterization, we have given a geometric/combinatorial proof of counting results for idempotents in monoids of monotone endomaps of a chain [13, 16].

Our initial motivations for studying idempotents in $Q_\vee(\mathbb{I}_n)$ originates from the algebra of logic, see e.g. [14]. Willing to investigate congruences of $Q_\vee(\mathbb{I}_n)$ as a residuated lattice [10], it can be shown, using idempotents, that every subalgebra of a residuated lattice $Q_\vee(\mathbb{I}_n)$ is simple. This property does not generalize to infinite complete chains: if $\mathbb{I}$ is the interval $[0, 1] \subseteq \mathbb{R}$, then $Q_\vee(\mathbb{I})$ is simple but has subalgebras that are not simple [1]. Despite the results we presented are not related to our original motivations, we aimed at exemplifying how a combinatorial approach based on paths might be fruitful when investigating various kinds of monotone maps and the multiple algebraic structures these maps may carry.

We used the Online Encyclopedia of Integer Sequences [21] to trace related research. In particular, we discovered Howie's work [13] on the monoid $\mathcal{O}_n$ through the OEIS sequences A001906 and A088305. The sequence ψ_n is a shift of the sequence A001519. Related to this sequence is the doubly parametrized sequence A144224 collecting some counting results from [16] on idempotents. Relations with other kind of combinatorial objects counted by the sequence ψ_n still need to be understood.

Acknowledgment. The author is thankful to Srecko Brlek, Claudia Muresan, and André Joyal for the fruitful discussions he shared with them on this topic during winter 2018. The author is also thankful to the anonymous referees for their insightful comments and for pointing him to the reference [16].

References

1. Ball, R.N., Droste, M.: Normal subgroups of doubly transitive automorphism groups of chains. Trans. Am. Math. Soc. **290**(2), 647–664 (1985)

2. Bennett, M.K., Birkhoff, G.: Two families of Newman lattices. Algebra Univers. **32**(1), 115–144 (1994)
3. Birkhoff, G.: Lattice Theory, American Mathematical Society Colloquium Publications, vol. 25, 3rd edn. American Mathematical Society, Providence (1979)
4. Davey, B.A., Priestley, H.A.: Introduction to Lattices and Order, 2nd edn. Cambridge University Press, New York (2002)
5. Dzhumadil'daev, A.S.: Worpitzky identity for multipermutations. Math. Notes **90**(3), 448–450 (2011)
6. Engbers, J., Stocker, C.: Two combinatorial proofs of identities involving sums of powers of binomial coefficients. In: INTEGERS, vol. 16 (2016)
7. Fernandes, V.H., Gomes, G.M.S., Jesus, M.M.: The cardinal and the idempotent number of various monoids of transformations on a finite chain. Bull. Malays. Math. Sci. Soc. (2) **34**(1), 79–85 (2011)
8. Ferrari, L.: Dyck algebras, interval temporal logic, and posets of intervals. SIAM J. Discrete Math. **30**(4), 1918–1937 (2016)
9. Ferrari, L., Pinzani, R.: Lattices of lattice paths. J. Statist. Plann. Inference **135**(1), 77–92 (2005)
10. Galatos, N., Jipsen, P., Kowalski, T., Ono, H.: Residuated Lattices: An Algebraic Glimpse at Substructural Logics. Studies in Logic and the Foundations of Mathematics, vol. 151. Elsevier (2007)
11. Gessel, I., Stanley, R.P.: Stirling polynomials. J. Comb. Theory Ser. A **24**(1), 24–33 (1978)
12. Gouveia, M.J., Santocanale, L.: The continuous weak order, December 2018. Preprint https://hal.archives-ouvertes.fr/hal-01944759
13. Howie, J.M.: Products of idempotents in certain semigroups of transformations. Proc. Edinb. Math. Soc. **17**(2), 223–236 (1971)
14. Jipsen, P.: Relation algebras, idempotent semirings and generalized bunched implication algebras. In: Höfner, P., Pous, D., Struth, G. (eds.) RAMICS 2017. LNCS, vol. 10226, pp. 144–158. Springer, Cham (2017). https://doi.org/10.1007/978-3-319-57418-9_9
15. Krattenthaler, C.: The enumeration of lattice paths with respect to their number of turns. In: Balakrishnan, N. (ed.) Advances in Combinatorial Methods and Applications to Probability and Statistics. Statistics for Industry and Technology, pp. 29–58. Birkhäuser, Boston (1997). https://doi.org/10.1007/978-1-4612-4140-9_3
16. Laradji, A., Umar, A.: Combinatorial results for semigroups of order-preserving full transformations. Semigroup Forum **72**(1), 51–62 (2006)
17. Laradji, A., Umar, A.: Lattice paths and order-preserving partial transformations. Util. Math. **101**, 23–36 (2016)
18. Mühle, H.: A Heyting algebra on Dyck paths of type A and B. Order **34**(2), 327–348 (2017)
19. Rosenthal, K.: Quantales and Their Applications. Pitman Research Notes in Mathematics Series. Longman Scientific & Technical (1990)
20. Santocanale, L., Wehrung, F.: The equational theory of the weak Bruhat order on finite symmetric groups. J. Eur. Math. Soc. **20**(8), 1959–2003 (2018)
21. Sloane, N.: The On-Line Encyclopedia of Integer Sequences. https://oeis.org/

Author Index

Printed in the United States
By Bookmasters